Multiple Comparison Procedures

YOSEF HOCHBERG
Tel-Aviv University
Ramat-Aviv, Israel

AJIT C. TAMHANE
Northwestern University
Evanston, Illinois

JOHN WILEY & SONS

New York • Chichester • Brisbane • Toronto • Singapore

A NOTE TO THE READER
This book has been electronically reproduced from digital information stored at John Wiley & Sons, Inc. We are pleased that the use of this new technology will enable us to keep works of enduring scholarly value in print as long as there is a reasonable demand for them. The content of this book is identical to previous printings.

Library of Congress Cataloging in Publication Data:

Hochberg, Yosef.
Multiple comparison procedures.

(Wiley series in probability and mathematical statistics. Applied probability and statistics, ISSN 0271-6356)
Bibliography: p.
Includes index.
1. Muliple comparisons (Statistics) I. Tamhane, Ajit C. II. Title. III. Series.
QA278.4.H63 1987 519.5'35 87–6226

ISBN: 978-0-470-56833-0

10 9 8 7 6 5 4 3 2 1

To our parents

WILEY SERIES IN PROBABILITY AND MATHEMATICAL STATISTICS

ESTABLISHED BY WALTER A. SHEWHART AND SAMUEL S. WILKS
Editors
Vic Barnett, Ralph A. Bradley, J. Stuart Hunter, David G. Kendall, Rupert G. Miller, Jr., Adrian F. M. Smith, Stephen M. Stigler, Geoffrey S. Watson

Probability and Mathematical Statistics

ADLER • The Geometry of Random Fields
ANDERSON • The Statistical Analysis of Time Series
ANDERSON • An Introduction to Multivariate Statistical Analysis, *Second Edition*
ARNOLD • The Theory of Linear Models and Multivariate Analysis
BARNETT • Comparative Statistical Inference, *Second Edition*
BHATTACHARYYA and JOHNSON • Statistical Concepts and Methods
BILLINGSLEY • Probability and Measure, *Second Edition*
BOROVKOV • Asymptotic Methods in Queuing Theory
BOSE and MANVEL • Introduction to Combinatorial Theory
CASSEL, SARNDAL, and WRETMAN • Foundations of Inference in Survey Sampling
CHEN • Recursive Estimation and Control for Stochastic Systems
COCHRAN • Contributions to Statistics
COCHRAN • Planning and Analysis of Observational Studies
CONSTANTINE • Combinatorial Theory and Statistical Design
DOOB • Stochastic Processes
DUDEWICZ and MISHRA • Modern Mathematical Statistics
EATON • Multivariate Statistics: A Vector Space Approach
ETHIER and KURTZ • Markov Processes: Characterization and Convergence
FABIAN and HANNAN • Introduction to Probability and Mathematical Statistics
FELLER • An Introduction to Probability Theory and Its Applications, Volume I, *Third Edition*, Revised; Volume II, *Second Edition*
FULLER • Introduction to Statistical Time Series
FULLER • Measurement Error Models
GRENANDER • Abstract Inference
GUTTMAN • Linear Models: An Introduction
HAMPEL, RONCHETTI, ROUSSEEUW, and STAHEL • Robust Statistics: The Approach Based on Influence Functions
HANNAN • Multiple Time Series
HARRISON • Brownian Motion and Stochastic Flow Systems
HETTMANSPERGER • Statistical Inference Based on Ranks
HOEL • Introduction to Mathematical Statistics, *Fifth Edition*
HUBER • Robust Statistics
IMAN and CONOVER • A Modern Approach to Statistics
IOSIFESCU • Finite Markov Processes and Applications
JOHNSON and BHATTACHARYYA • Statistics: Principles and Methods, *Revised Printing*
LAHA and ROHATGI • Probability Theory
LARSON • Introduction to Probability Theory and Statistical Inference, *Third Edition*
LEHMANN • Testing Statistical Hypotheses, *Second Edition*
LEHMANN • Theory of Point Estimation
MATTHES, KERSTAN, and MECKE • Infinitely Divisible Point Processes
MUIRHEAD • Aspects of Multivariate Statistical Theory
PURI and SEN • Nonparametric Methods in General Linear Models
PURI and SEN • Nonparametric Methods in Multivariate Analysis

Probability and Mathematical Statistics (Continued)

PURI, VILAPLANA, and WERTZ • New Perspectives in Theoretical and Applied Statistics
RANDLES and WOLFE • Introduction to the Theory of Nonparametric Statistics
RAO • Linear Statistical Inference and Its Applications, *Second Edition*
RAO • Real and Stochastic Analysis
RAO and SEDRANSK • W.G. Cochran's Impact on Statistics
RAO • Asymptotic Theory of Statistical Inference
ROHATGI • An Introduction to Probability Theory and Mathematical Statistics
ROHATGI • Statistical Inference
ROSS • Stochastic Processes
RUBINSTEIN • Simulation and The Monte Carlo Method
SCHEFFE • The Analysis of Variance
SEBER • Linear Regression Analysis
SEBER • Multivariate Observations
SEN • Sequential Nonparametrics: Invariance Principles and Statistical Inference
SERFLING • Approximation Theorems of Mathematical Statistics
SHORACK and WELLNER • Empirical Processes with Applications to Statistics
TJUR • Probability Based on Radon Measures
ROGERS and WILLIAMS • Diffusions, Markov Processes, and Martingales, Volume II: Itô Calculus

Applied Probability and Statistics

ABRAHAM and LEDOLTER • Statistical Methods for Forecasting
AGRESTI • Analysis of Ordinal Categorical Data
AICKIN • Linear Statistical Analysis of Discrete Data
ANDERSON and LOYNES • The Teaching of Practical Statistics
ANDERSON, AUQUIER, HAUCK, OAKES, VANDAELE, and WEISBERG • Statistical Methods for Comparative Studies
ARTHANARI and DODGE • Mathematical Programming in Statistics
ASMUSSEN • Applied Probability and Queues
BAILEY • The Elements of Stochastic Processes with Applications to the Natural Sciences
BAILEY • Mathematics, Statistics and Systems for Health
BARNETT • Interpreting Multivariate Data
BARNETT and LEWIS • Outliers in Statistical Data, *Second Edition*
BARTHOLOMEW • Stochastic Models for Social Processes, *Third Edition*
BARTHOLOMEW and FORBES • Statistical Techniques for Manpower Planning
BECK and ARNOLD • Parameter Estimation in Engineering and Science
BELSLEY, KUH, and WELSCH • Regression Diagnostics: Identifying Influential Data and Sources of Collinearity
BHAT • Elements of Applied Stochastic Processes, *Second Edition*
BLOOMFIELD • Fourier Analysis of Time Series: An Introduction
BOX • R. A. Fisher, The Life of a Scientist
BOX and DRAPER • Empirical Model-Building and Response Surfaces
BOX and DRAPER • Evolutionary Operation: A Statistical Method for Process Improvement
BOX, HUNTER, and HUNTER • Statistics for Experimenters: An Introduction to Design, Data Analysis, and Model Building
BROWN and HOLLANDER • Statistics: A Biomedical Introduction
BUNKE and BUNKE • Statistical Inference in Linear Models, Volume I
CHAMBERS • Computational Methods for Data Analysis
CHATTERJEE and PRICE • Regression Analysis by Example
CHOW • Econometric Analysis by Control Methods
CLARKE and DISNEY • Probability and Random Processes: A First Course with Applications, *Second Edition*
COCHRAN • Sampling Techniques, *Third Edition*
COCHRAN and COX • Experimental Designs, *Second Edition*
CONOVER • Practical Nonparametric Statistics, *Second Edition*

Applied Probability and Statistics (Continued)

CONOVER and IMAN • Introduction to Modern Business Statistics
CORNELL • Experiments with Mixtures: Designs, Models and The Analysis of Mixture Data
COX • Planning of Experiments
COX • A Handbook of Introductory Statistical Methods
DANIEL • Biostatistics: A Foundation for Analysis in the Health Sciences, *Fourth Edition*
DANIEL • Applications of Statistics to Industrial Experimentation
DANIEL and WOOD • Fitting Equations to Data: Computer Analysis of Multifactor Data, *Second Edition*
DAVID • Order Statistics, *Second Edition*
DAVISON • Multidimensional Scaling
DEGROOT, FIENBERG and KADANE • Statistics and the Law
DEMING • Sample Design in Business Research
DILLON and GOLDSTEIN • Multivariate Analysis: Methods and Applications
DODGE • Analysis of Experiments with Missing Data
DODGE and ROMIG • Sampling Inspection Tables, *Second Edition*
DOWDY and WEARDEN • Statistics for Research
DRAPER and SMITH • Applied Regression Analysis, *Second Edition*
DUNN • Basic Statistics: A Primer for the Biomedical Sciences, *Second Edition*
DUNN and CLARK • Applied Statistics: Analysis of Variance and Regression, *Second Edition*
ELANDT-JOHNSON and JOHNSON • Survival Models and Data Analysis
FLEISS • Statistical Methods for Rates and Proportions, *Second Edition*
FLEISS • The Design and Analysis of Clinical Experiments
FOX • Linear Statistical Models and Related Methods
FRANKEN, KÖNIG, ARNDT, and SCHMIDT • Queues and Point Processes
GALLANT • Nonlinear Statistical Models
GIBBONS, OLKIN, and SOBEL • Selecting and Ordering Populations: A New Statistical Methodology
GNANADESIKAN • Methods for Statistical Data Analysis of Multivariate Observations
GREENBERG and WEBSTER • Advanced Econometrics: A Bridge to the Literature
GROSS and HARRIS • Fundamentals of Queueing Theory, *Second Edition*
GUPTA and PANCHAPAKESAN • Multiple Decision Procedures: Theory and Methodology of Selecting and Ranking Populations
GUTTMAN, WILKS, and HUNTER • Introductory Engineering Statistics, *Third Edition*
HAHN and SHAPIRO • Statistical Models in Engineering
HALD • Statistical Tables and Formulas
HALD • Statistical Theory with Engineering Applications
HAND • Discrimination and Classification
HOAGLIN, MOSTELLER and TUKEY • Exploring Data Tables, Trends and Shapes
HOAGLIN, MOSTELLER, and TUKEY • Understanding Robust and Exploratory Data Analysis
HOCHBERG and TAMHANE • Multiple Comparison Procedures
HOEL • Elementary Statistics, *Fourth Edition*
HOEL and JESSEN • Basic Statistics for Business and Economics, *Third Edition*
HOGG and KLUGMAN • Loss Distributions
HOLLANDER and WOLFE • Nonparametric Statistical Methods
IMAN and CONOVER • Modern Business Statistics
JESSEN • Statistical Survey Techniques
JOHNSON • Multivariate Statistical Simulation
JOHNSON and KOTZ • Distributions in Statistics
Discrete Distributions

(continued on back)

Multiple Comparison Procedures

Preface

The foundations of the subject of multiple comparisons were laid in the late 1940s and early 1950s, principally by David Duncan, S.N. Roy, Henry Scheffé, and John Tukey, although some of the ideas appeared much earlier in the works of Fisher, Student, and others; see Harter 1980 for a complete historical account. Tukey's (1953) mimeographed notes on the subject form a rich source of results (some of which are being rediscovered even today), but unfortunately they had only a limited circulation. Miller's (1966) book helped to popularize the use of multiple comparison procedures (MCPs) and provided an impetus to new research in the field. There have been a large number of review articles surveying the field, most published during the 1970s (e.g., Aitkin 1969, Chew 1976a, Dunnett 1970, Dunnett and Goldsmith 1981, Games 1971, Gill 1973, Miller 1977, 1985, O'Neill and Wetherill 1971, Ryan 1959, Shaffer 1986b, Spjøtvoll 1974, and Thomas 1973). Commonly used MCPs are discussed in many elementary and advanced texts. However, no unified comprehensive treatment of the subject that incorporates the developments in the last two decades and that plays a role similar to Miller's monograph is available. The aim of the present book is to fulfill this need.

Because MCPs have been around for quite some time and because their use is well accepted (and sometimes even required) in many disciplines, one might expect that there would be few, if any, unresolved questions and controversies. Such is not the case, however. Considerable confusion still exists in regard to what the different MCPs provide, which ones should be used, and when. There are many statisticians who even question the very need and appropriateness of any multiple comparison approach (see, e.g., the discussion following O'Neill and Wetherill 1971, and the papers by Carmer and Walker 1982, Dawkins 1983, Little 1978, O'Brien 1983, Perry 1986, and Petersen 1977).

Some of the controversial issues cut across the general subject of statistical inference and are not germane specifically to multiple comparisons. These include testing versus confidence estimation, and the choice of approach to inference (e.g., a classical Fisherian or a Neyman–Pearsonian approach, a decision-theoretic approach, a Bayesian approach, or an informal graphical approach). We prefer not to take rigid stands on these issues. Certainly applications can be found for each basic approach to inference, although some types of applications are more common than others. We have tried to present a variety of techniques based on different approaches, spelling out the pros and cons in each case. The emphasis, however, is on classical approaches and on confidence estimation.

Another line of criticism stems from the misuse of MCPs in practice. A common example of this is the application of an MCP (a popular choice being Duncan's 1955 stepwise procedure) for making pairwise comparisons among all treatments when the treatments have a certain structure, as is the case when they correspond to multifactorial combinations or to increasing levels of some quantitative factor (Chew 1976b). In such situations, comparisons other than pairwise comparisons may be of interest; for example, orthogonal contrasts, and a specially tailored MCP for such a family of comparisons may be required. Sometimes procedures other than MCPs are required, for example, when the researcher's goal is to select the "best" treatment or to explore the possible clustering patterns among the treatments. The key point is that statistical procedures that should be used in a given problem depend on the questions of interest and the nature of research. Problems of inference encountered in empirical research are of diverse nature and indiscriminate use of MCPs or, for that matter, any other statistical technique in all problems is clearly inappropriate. In Chapter 1 we have addressed the questions of when and why to use MCPs. This discussion should also help to answer other basic criticisms voiced against multiple comparisons.

The subject of multiple comparisons forms a part of the broader subject of simultaneous statistical inference. In this book we focus on problems involving multiplicity and selection ("data-snooping") of inferences when comparing treatments based on univariate responses. (Refer to Krishnaiah, Mudholkar, and Subbaiah 1980 and Krishnaiah and Reising 1985 for the corresponding multivariate techniques.) Roy's union–intersection method forms the unifying theme used to derive the various classical MCPs for these problems. We do not discuss the problems of simultaneous point estimation and simultaneous confidence bands in regression. We also do not discuss the related topic of ranking and selection procedures, on which there are full-length books by Gibbons, Olkin, and Sobel (1977) and Gupta and Panchapakesan (1979).

The following is a brief outline of the book. In Chapter 1 we elaborate on our philosophy and approach to multiple comparison problems, and discuss the basic notions of families, error rates, and control of error rates. The remainder of the book is divided into two parts. Part I, consisting of Chapters 2–6, deals with MCPs based on a classical error rate control approach for fixed-effects linear models under the usual normal theory assumptions. Part II, consisting of Chapters 7–11, deals with MCPs for other models and problems, and MCPs based on alternative approaches (e.g., decision-theoretic and Bayesian). There are three appendixes. Appendix 1 gives some general theory of MCPs that is not restricted to the setting of Part I (for which case the corresponding theory is given in Chapter 2). Appendix 2 reviews the probability inequalities that are used in deriving and computing conservative critical points for various MCPs. Finally, Appendix 3 discusses the probability distributions that arise in the context of some classical MCPs. Tables of the percentage points of these distributions are also included in this appendix.

Sections are numbered fresh starting in each chapter. Subsections, equations, examples, figures, and tables are numbered fresh starting in each section but with the corresponding section identification. In both cases there is no chapter identification provided. Thus, for example, within a given chapter, equation (3.10) is the tenth equation in Section 3 of that chapter. When the same equation is referenced in another chapter, it is referred to as equation (3.10) of such and such chapter.

This book is intended to serve the needs of researchers as well as practitioners. The comprehensive review of the published literature until 1986, as well as the many open problems noted throughout the book, should prove valuable to researchers. (It should be remarked, however, that the literature on multiple comparisons is vast and it is impossible to reference each and every publication. We have limited our references to those publications that are relevant to the discussions of the topics covered in the book.) Practitioners should find it helpful to have the steps involved in implementing the various MCPs clearly spelled out and illustrated by small numerical examples. Many of the examples are taken from the original papers where the corresponding MCPs first appeared. As mentioned before, the examples and discussion of Chapter 1 provide useful guidelines to the perennial problem of when to use MCPs; a need for such case studies was noted by Anscombe (1985).

We have assumed that the reader has had a course in mathematical statistics covering the basic concepts of inference and is familiar with matrix algebra, linear models, and experimental designs. For the benefit of those who may not have had training in the last two topics, we have included a brief review of the necessary ideas in Section 1 of Chapter 2. This book can be used as a text in a special topics course on multiple comparisons or for supplementary reading in a course on linear models and analysis of variance.

In conclusion, we hope that this book will meet its stated objectives. In addition, we hope that it will help to dispel some of the confusion and controversy that have surrounded the subject of MCPs, and encourage their correct use when appropriate in practice.

YOSEF HOCHBERG
AJIT C. TAMHANE

Tel Aviv, Israel
Evanston, Illinois

Acknowledgments

We would like to gratefully acknowledge the facilities provided principally by Northwestern University and also by Tel-Aviv and New York Universities, which made the work on this book possible. Early drafts of some of the chapters were written when the second author visited Cornell University in 1982–1983 on sabbatical leave with partial support from the U.S. Army Research Office, Durham Contract DAAG-29-81-K-0168.

We are also indebted to many individuals who contributed in different ways to this book. Bob Berk, Morton Brown, Tony Hayter, Jason Hsu, Tom Santner, Juliet Shaffer, Gary Simon, and Yung Liang Tong read sections of the manuscript and provided comments and suggestions. G.V.S. Gopal read the entire manuscript and noted several errors. Charlie Dunnett computed Tables 4–7 in Appendix 3 as well as some special percentage points of the multivariate *t*-distribution required in Examples 2.1 and 2.3 of Chapter 5. Bob Bechhofer and Ruben Gabriel gave timely advice and encouragement. Beatrice Shube, the senior editor of the statistics series at Wiley, was always patient and cooperative with us. The other staff at Wiley were also very helpful and efficient in carrying out a speedy and quality production of the book. Particular mention should be made of Margaret Comaskey, the editorial supervisor and Diana Cisek, the senior production supervisor. Finally, it is our pleasure to acknowledge the assistance of June Wayne, Carol Surma and Marianne Delaney of Northwestern University. June cheerfully typed and retyped numerous drafts of the manuscript (which required a considerable amount of patience and perseverance in spite of all the modern word processor features). Carol graciously pitched in whenever June was away. Marianne typed the author and subject indexes. We wish to express our most sincere thanks to them.

Contents

Multiple Comparison Procedures

CHAPTER 1

Introduction

Comparative studies are commonly employed in empirical research. Some examples of such studies are: clinical trials comparing different drug regimens in terms of their therapeutic values and side effects, agricultural field experiments comparing different crop varieties in terms of their yields, and sample surveys comparing different demographic groups in terms of their attitudes toward the issues of interest. Additional examples can be given from other fields of empirical research employing experimental or observational studies.

Traditionally, a common tool in analyzing data from such studies is a test of homogeneity of the groups (treatments) under investigation. By itself, however, such a test does not provide inferences on various detailed comparisons among the groups that are often of interest to the researcher, for example, pairwise comparisons among the groups. In practice, some of the comparisons may be *prespecified* (i.e., before looking at the data) while the others may be *selected* after looking at the data ("data-snooping" or "post-hoc selection").

One possible approach to making such *multiple comparisons* is to assess each comparison (prespecified or selected by data-snooping) separately by a suitable procedure (a hypothesis test or confidence estimate) at a level deemed appropriate for that single inference. This is referred to as a *per-comparison* or *separate inferences approach*. For instance, for detecting differences among the means of $k \geqq 3$ treatments, one could perform separate $\binom{k}{2}$ pairwise two-sided *t*-tests, each at level α appropriate for a single test. Such *multiple t-tests* (without a preliminary *F*-test of the overall homogeneity hypothesis H_0, say) are in fact used quite frequently in practice; see, for example, Godfrey's (1985) survey of selected medical research articles. But the difficulty with the per-comparison approach is that it does not account for the *multiplicity* or *selection effect* (Tukey 1977), which has been put as follows: "If enough statistics are computed, some of them will be sure to show structure" (Diaconis 1985). The same difficulty is present in many other statistical problems, for example, in selection of the "best" subset of variables in regression and in repeated significance testing. In the following we further clarify the notion of the multiplicity effect in the context of pairwise comparisons among group means. For simplicity, our discussion is given in terms of hypothesis testing; a similar discussion can be given in terms of confidence estimation.

Results of multiple t-tests are usually summarized by highlighting the significant pairwise differences. Any significant pairwise difference also implies *overall significance*, that is, rejection of H_0. It can readily be seen that with $\binom{k}{2}$ pairwise t-tests applied separately each at level α, the probability of concluding overall significance, when in fact H_0 is true, can be well in excess of α and will be close to 1 for sufficiently large k. The probability of concluding any false *pairwise significance* will equal α when exactly one pairwise null hypothesis is true, and will exceed α when two or more pairwise null hypotheses are true. (Under H_0 this probability is the same as the former error probability.) Thus with multiple t-tests, spurious overall and detailed (pairwise) significant results are obtained more frequently than is indicated by the per-comparison level α. This essentially is the import of the multiplicity effect.

Statistical procedures that are designed to take into account and properly control for the multiplicity effect through some combined or joint measure of erroneous inferences are called *multiple comparison procedures* (MCPs). Two early MCPs due to R. A. Fisher are discussed in Section 1. These MCPs are of different types. They help to illustrate the typology of MCPs followed in this book and the differences in the properties of the two types of MCPs. This section also informally introduces some terminology. Section 2 is devoted to the philosophy and basic notions of MCPs. Finally Section 3 gives a number of examples of multiple comparison problems, which are discussed in light of the basic notions developed in Section 2.

1 TWO EARLY MULTIPLE COMPARISON PROCEDURES

In order to discuss control of the multiplicity effect for the pairwise comparisons problem, we regard the $\binom{k}{2}$ tests of pairwise null hypotheses as a *family* and consider alternative Type I *error rates* (Tukey 1953) for the inferences in this family. (A more detailed discussion of these two concepts is given in Section 2.) The *per-comparison error rate* (PCE) is defined as the expected proportion of Type I errors (incorrect rejections of the true pairwise null hypotheses). It is obvious that multiple α-level t-tests control the PCE at level α but they do not control the probability of erroneous significance conclusions (overall or detailed) at level α. For this purpose, a more pertinent error rate is the *familywise error rate* (FWE), which is the probability of making *any* error in the given family of inferences. In the following we discuss an MCP proposed by Fisher (1935, Section 24) for making inferences on pairwise comparisons. This procedure has FWE $= \alpha$ under H_0. We also discuss another MCP proposed by Fisher (op. cit.) which has FWE $\leqq \alpha$ under *all* configurations of the true unknown means.

Fisher's first procedure consists of performing multiple t-tests each at level α only if the preliminary F-test is significant at level α. This can be viewed as a *two-step procedure* in which H_0 is tested at the first step by an α-level F-test. If the F-test is not significant, then the procedure terminates without making detailed inferences on pairwise differences; otherwise each pairwise difference is tested by an α-level t-test. This procedure is referred to in the literature as the *protected least significant difference* (LSD) *test*. We refer to it simply as the LSD or the LSD procedure.

The LSD controls the FWE at level α under H_0 because of the protection provided to this hypothesis by the preliminary F-test. However, at other configurations of the true means its FWE can be well in excess of α as the following argument shows. Suppose that one of the group means is far removed from the others, which are all equal. Then the preliminary F-test rejects almost surely and we essentially apply α-level multiple t-tests to $k-1$ homogeneous means. If $k-1>2$, then because of the multiplicity effect the FWE exceeds α. We therefore say that the LSD controls the FWE only in a *weak* sense (under H_0 but not under all configurations).

Fisher's second procedure is popularly known as the *Bonferroni procedure*, and it controls the FWE in the *strong* sense, that is, under all configurations. This is a *single-step procedure* that consists of performing multiple t-tests, each at level $\alpha^* = \alpha/\binom{k}{2}$ without the preliminary F-test. If we define the *per-family error rate* (PFE) as the expected number of errors in the family, then under H_0 the PFE of this procedure equals α. At any configuration other than H_0, since the number of true hypotheses is less than $\binom{k}{2}$, the PFE is readily seen to be less than α. By the Bonferroni inequality (which says that the probability of a union of events is less than the sum of the individual event probabilities; see Appendix 2) it follows that the probability of at least one error is always bounded above by the expected number of errors and hence FWE $\leqq$ PFE. Therefore the FWE of this procedure is strongly controlled at level α.

At this point we can briefly compare the LSD and the Bonferroni procedures. As noted above, the LSD controls the FWE weakly (under H_0 only), while the Bonferroni procedure controls the FWE strongly. However, the LSD is more powerful than the Bonferroni procedure. The key reason behind this is not that the LSD controls the FWE weakly nor that the Bonferroni procedure controls the PFE (which is an upper bound on the FWE) strongly, although these are contributing factors. In fact, as we see in later chapters, even after both procedures are modified so that they exercise the same α-level strong control of the FWE, the power superiority of the two-step procedure over the single-step procedure persists (although not to the same extent). The key reason is that the LSD uses the first step F-test to "learn" about the true configuration of means, and if they are indicated to be heterogeneous, it then uses liberal α-level multiple t-tests for pairwise comparisons. In this sense the two-step nature of the LSD makes it *adaptive* to configurations different from H_0. On the other hand, the Bonferroni procedure, being a single-step procedure, is not adaptive; although the PFE under configurations different from H_0 is less than α, the Bonferroni procedure does not use more liberal critical points for such configurations.

We also note that the Bonferroni test procedure can be readily inverted to obtain confidence intervals for all pairwise differences among the group means. All of these intervals cover the respective true pairwise differences with a joint coverage probability of at least $1 - \alpha$ (FWE $\leqq \alpha$) and are therefore referred to as $(1 - \alpha)$-level *simultaneous confidence intervals*. Such intervals cannot be obtained with the LSD procedure, which can be a serious shortcoming in many problems.

The Bonferroni and LSD procedures initiated the development of two distinct classes of MCPs, namely, *single-step** and *multistep*† (or *stepwise*) procedures, respectively. As just discussed, these two types of procedures are inherently different; as a result, generally the types of inferences (e.g., confidence estimates, tests, directional decisions on the signs of parameters under test) that can be made with them and their operating

* An important subclass of single-step procedures is referred to as *simultaneous procedures* and studied in detail by Gabriel (1969) and others. We reserve this terminology for that subclass rather than using it for any single-step procedure as is sometimes done in the literature. It should also be mentioned that, when referring to a set of inferences together, we use the terms "simultaneous," "joint," and "multiple" interchangeably.

† Many authors refer to multistep procedures as *multistage* or *sequential* procedures; we avoid this terminology because usually it is used to describe the nature of sampling and not the nature of testing.

characteristics are different. These distinctions will become clearer as we study these MCPs further for a variety of multiple comparison problems (different families of inferences, various experimental and observational study designs, different distributional models for the data, etc.).

2 BASIC NOTIONS AND PHILOSOPHY OF MULTIPLE COMPARISONS

2.1 Families

The classical statistical approach to inference is based on the premise of a separate experiment for each anticipated finding, which is assumed to be stated in advance. It is well recognized, however, that often a better scientific practice is to conduct one large experiment designed to answer multiple related questions. This practice is also statistically and economically more efficient. Perhaps the most cogent justification for large (factorial) experiments was that originally given by Fisher (1926):

> No aphorism is more frequently repeated . . . than that we must ask Nature . . . one question at a time. I am convinced that this view is wholly mistaken. Nature . . . will best respond to . . . a carefully thought out questionnaire; indeed if we ask her a single question, she will often refuse to answer until some other topic has been discussed.

In the previous section we saw that a separate inferences (per-comparison) approach to multiple related inferences can lead to too many false significances. On the other hand, it would be unwise to take the other extreme view, which attributes too many "apparently significant" results solely to chance and dismisses them as spurious. As a first step toward providing an objective approach to the problem of multiple inferences (e.g., for obtaining valid indicators of statistical significances of the inferences) we introduce the concept of a family: Any collection of inferences for which it is meaningful to take into account some combined measure of errors is called a *family*. For convenience, we also refer to the corresponding collection of inferential problems or even the corresponding collection of parameters on which the inferences are to be made as a family.

Just what constitutes a family depends to a great extent on the type of research—exploratory or confirmatory. In purely exploratory research the questions of interest (or lines of inquiry) are generated by data-snooping. In purely confirmatory research they are stated in advance. Most empirical studies combine aspects of both types of research, but for expositional purposes we treat them separately.

In exploratory research one must consider all inferences (including those that are actually made and those that potentially could have been made) together as a family in order to take into account the effect of selection. What are these "potential" inferences? Fisher (1935) has characterized them as those inferences that "Would have been from the start equally plausible." As Putter (1983) puts it, these are inferences, "Which would have made . . . (the researcher) sit up and take notice in a 'similar' way . . . ." Why is it necessary to consider the inferences that were not actually made? Because by controlling the FWE for the family of *all* potential inferences the probability of *any* error among the *selected* inferences is automatically controlled. (Putter 1983 has proposed an alternative appraoch based on controlling the error probability for *each* selected inference.)

There are different types of families. The collection of pairwise comparisons considered in the previous section is an example of a *finite* family. In some problems the researcher may wish to examine *any* contrast among the group means (i.e., a linear combination of the group means where the coefficients of the linear combination sum to zero and they are not identically zero) that might turn out to be "interesting" in light of the data. Then the family of potential inferences is *infinite* consisting of all contrasts. In some other types of exploratory research it may be impossible to specify in advance the family of all potential inferences that may be of interest. For example, a researcher may wish to find out the effects of a given treatment on a list of variables that cannot be fully specified in advance (e.g., side effects of a drug). New variables may be added to the list (within the framework of the same study) depending on what is learned from the previous tests. In this case, it is impossible to formally account for the multiplicity effect.

Let us now turn to confirmatory research. As noted earlier, in this type of research one usually has a finite number of inferences of interest specified prior to the study. If these inferences are unrelated in terms of their content or intended use (although they may be statistically dependent), then they should be treated separately and not jointly. If a decision (or conclusion) is to be based on these inferences and its accuracy depends on some joint measure of erroneous statements in the given set of inferences, then that collection of inferences should be considered jointly as a family.

To summarize, the following are the two key reasons for regarding a set of inferences as a family (Cox 1965):

(i) To take into account the selection effect due to data-snooping.
(ii) To ensure the simultaneous correctness of a set of inferences so as to guarantee a correct overall decision.

2.2 Error Rates

In thc previous section three different error rates were introduced. In the present section we define these error rates formally and discuss them further.

Let $\mathcal{F}$ denote a family of inferences and let $\mathcal{P}$ denote an MCP for this family. We assume that every inference that $\mathcal{P}$ makes (or can potentially make) is either right or wrong. Let $M(\mathcal{F}, \mathcal{P})$ be the random number of wrong inferences. We can now define the three error rates; the definition (2.3) of the PCE involves the cardinality $N(\mathcal{F})$ of $\mathcal{F}$ and is valid only in the case of finite families.

(i) Familywise Error Rate (FWE)*:

$$\mathrm{FWE}(\mathcal{F}, \mathcal{P}) = \Pr\{M(\mathcal{F}, \mathcal{P}) > 0\} . \tag{2.1}$$

(ii) Per-Family Error Rate (PFE)*:

$$\mathrm{PFE}(\mathcal{F}, \mathcal{P}) = E\{M(\mathcal{F}, \mathcal{P})\} . \tag{2.2}$$

(iii) Per-Comparison Error Rate (PCE)*:

$$\mathrm{PCE}(\mathcal{F}, \mathcal{P}) = \frac{E\{M(\mathcal{F}, \mathcal{P})\}}{N(\mathcal{F})} . \tag{2.3}$$

For the infinite family the FWE and PFE are well defined for any MCP (the PFE may equal infinity). For a given MCP if the PCE for every finite subfamily is the same, then we can consider that common value as the PCE of that MCP for the infinite family. For example, if each inference is erroneous with the same probability α, then the PCE $= \alpha$ for an infinite family.

* Many writers following Tukey (1953) refer to the familywise and per-family error rates as the *experimentwise* and *per-experiment error rates*, respectively. We prefer the former terminology (as in Miller 1981), since not all inferences made in a given experiment may constitute a single family. We also note that the per-comparison error rate is sometimes referred to as the *per-statement* or *comparisonwise error rate*.

TABLE 2.1. Error Rates as Functions of $N(\mathscr{F})$ when $\alpha = 0.05$ and (2.5) Holds

$N(\mathscr{F})$	PCE	FWE	PFE
1	0.05	0.05	0.05
5	0.05	0.23	0.25
10	0.05	0.40	0.50
20	0.05	0.64	1.00
50	0.05	0.92	2.50

In many problems only certain types of errors are counted and controlled. For example, in hypothesis testing problems it is common to count only Type I errors, say, $M'(\mathscr{F}, \mathscr{P})$ where $M'(\mathscr{F}, \mathscr{P}) \leqq M(\mathscr{F}, \mathscr{P})$. The corresponding error rates defined for $M'(\mathscr{F}, \mathscr{P})$ as in equations (2.1)–(2.3) are referred to as *Type I error rates*. In the sequel, unless otherwise specified, the error rates will be assumed to be of Type I.

From (2.1)–(2.3) the following relation among the error rates is easy to show:

$$\text{PCE} \leqq \text{FWE} \leqq \text{PFE} . \tag{2.4}$$

For finite families, the PFE is simply a multiple of the PCE, while the exact relationship between the PFE and FWE depends on $N(\mathscr{F})$ and on the nature and the extent of the dependencies induced by $\mathscr{P}$ among the inferences in $\mathscr{F}$. If the inferences are mutually independent and the error rate for an individual inference is α, then PCE $= \alpha$, PFE $= \alpha N(\mathscr{F})$, and

$$\text{FWE} = 1 - (1 - \alpha)^{N(\mathscr{F})} . \tag{2.5}$$

If α is close to zero and $N(\mathscr{F})$ is small, then the first order approximation gives FWE $\cong \alpha N(\mathscr{F}) =$ PFE. Table 2.1 shows the large differences that can occur between the FWE and PFE when $N(\mathscr{F})$ is large and (2.5) holds.

2.3 Control of Error Rates

The question of which error rate to control in a given multiple comparison problem has generated much discussion in the literature. We review below some prominent opinions that have been expressed.

Tukey (1953, Chapter 8) examined various error rates and reached the conclusion that control of the FWE "Should be standard, rarely will any other be appropriate." He rejected the relevance of the PCE in scientific work because of its underlying "Philosophy that errors are allowed to increase in proportion to the number of statements . . ." His main arguments for controlling the FWE rather than the PFE were as follows:

(i) Control of the FWE for the family of *all potential* inferences ensures that the probability of any error in the *selected* set of inferences is controlled.
(ii) For an infinite family the FWE can be controlled but not the PFE.
(iii) When the requirement of the simultaneous correctness of all inferences must be satisfied, the FWE is the only choice for control.

Miller (1981, p. 10) also recommended the FWE because

> The thought that all . . . statements are correct with high probability seems to afford . . . a greater serenity and tranquility of mind than a discourse on . . . expected number of mistakes.

Spjøtvoll (1972a) recommended the use of the PFE for finite families. He cited the following reasons in support of the PFE:

(i) The PFE is technically easier to work with than the FWE.
(ii) The PFE imposes a penalty in direct proportion to the number of errors, while the FWE corresponds to a zero–one loss function.
(iii) Because the PFE is an upper bound on the FWE, controlling the former also controls the latter.

Duncan (1955) advocated still another approach to error rate control in finite families. In this approach the PCE is controlled at traditional α-levels but the FWE is allowed to increase only as $1-(1-\alpha)^{N'(\mathcal{F})}$ where $N'(\mathcal{F})$ is the number of statistically independent comparisons in $\mathcal{F}$.

We now explain our approach to error rate control, which involves a careful consideration of the nature and purpose of the research. Again for simplicity, our discussion is phrased in terms of hypothesis testing. We first discuss the choice of an error rate to control in exploratory research. This choice is seen to depend on the level of statistical validity to be attached to exploratory findings and the type of family—infinite or finite. By "statistical validity" we refer to the levels of the Type I error probabilities. Here it is assumed that the conventional approach to inference is adopted, that is, a finding is said to be "established" if the negation of the finding is set up as a null hypothesis and is rejected using a suitable test. Type I errors are of concern in purely exploratory research when an effective screening of false positives is necessary. If this premise is not strictly true, then Type II errors (missing true leads) may be of greater concern. However, probabilistic measures of Type I errors are frequently chosen because they are easier to analyze and control. The probability of Type II errors is reduced by allowing a higher probability of Type I errors. (Alternatively, one can employ a decision theoretic approach that takes into account relative costs of Type I and Type II errors. See the discussion of Duncan's K-ratio t-tests in Chapter 11. Or one can design the experiment with the objective of controlling the probability of Type II errors; see Chapter 6.)

The results of exploratory studies are not always reported with the same level of detail. Some studies report all the hypotheses tested and how they happened to be selected (e.g., some hypotheses may have been of prior interest, some may have been selected because of certain patterns apparent in the data, and some others may have been selected by "trial and error," testing many candidate hypotheses until some significant ones are found). Some other studies only report the significant results and do not reveal the selection process that led to the reported results. In the former case the readers may perceive the research findings as less conclusive (needing additional confirmation) than in the latter case. Because in the latter case the findings convey a sense of greater conclusiveness, it becomes incumbent upon the researcher to apply stricter yardsticks of statistical validity.

No error rate can be controlled if the family of all potential inferences cannot be specified in advance. One practical remedy in this case (Diaconis 1985) is to publish the results without any P-values but with a careful and detailed description of how they were arrived at. If possible, the results should be verified by future confirmatory studies. From now on we assume that the family can be specified in advance.

In the case of an infinite family, the PFE is usually not amenable to control. (It cannot be bounded above when the same level α is used for individual inferences.) The PCE may be controlled at a desired level α (e.g., by making each separate inference at level α). However, this does not guarantee that the expected proportion of errors in the set of *selected* inferences is controlled at level α. In contrast to this, control of the FWE at level α provides an upper bound of α on the probability of any erroneous inference even in the selected set of inferences. Thus when high statistical validity must be attached to exploratory inferences, the FWE seems to be the more meaningful error rate to control. In other cases, the PCE may be controlled keeping in mind its limitations mentioned above.

In the case of a finite family all three error rates can be controlled. If high validity is to be attached to exploratory findings, then the FWE seems to be the natural candidate because its control provides an upper bound on the levels of any selected hypotheses. As a rule of thumb, one may adhere to controlling the FWE whenever the findings are presented without highlighting the selection process and/or without emphasizing the need for further confirmation.

Note that with any given error rate one has the freedom to choose the level α. For example, when high statistical validity is to be attached to exploratory findings, the FWE may be controlled at traditional α-levels for a single inference (e.g., $0.01 \leqq \alpha \leqq 0.10$). When low statistical validity can be tolerated, the FWE may be controlled at more liberal levels (e.g., $0.10 < \alpha \leqq 0.25$). Thus it may be prudent to uniformly adopt the FWE as the error rate to control with varying levels of α depending on the desired levels of statistical validity.

In the case of low desired statistical validity, other alternatives are also possible. If confirmation is anticipated, then (especially when the same organization or individual is to carry out both the exploratory and confirmatory studies) it might be desirable to control the PFE at a level that depends on the costs of confirmation. One can also control the PCE at traditional α-levels, which amounts to a separate inferences approach in which the researcher is ignoring the selection effect; the researcher is using the results of statistical inferences merely as indicators to suggest hypotheses of interest to be pursued in future confirmatory studies. It should be noted that control of the PCE at level α implies control of the PFE at level $\alpha N(\mathcal{F})$. Also, the level at which the PFE is controlled provides an upper bound on the FWE.

We next turn to confirmatory research. As discussed in Section 2.1, here we usually have a finite prespecified family. All three error rates are possible candidates for control, the choice being determined by the way the correctness of the final decisions or conclusions (which are based on multiple inferences) depends on the errors in the individual inferences. If, in order for a final decision to be correct, it is necessary that all inferences be simultaneously correct, then clearly the FWE must be controlled. If it is only necessary to have no more than a certain number or a certain proportion of errors, then the PFE or the PCE may be controlled.

To summarize, the particular error rate to be controlled should be determined by a detailed consideration of the nature of the given study and its place in the overall plan for research. In data analysis situations of varying complexity, it is hard to justify a fixed choice of one error rate in all situations. This point of view is demonstrated by the examples in Section 3.

Much of the discussion in the present book is oriented toward control of the FWE in the strong sense. This is done for the reasons cited above, and also because it enables us to uniformly calibrate different procedures to a common benchmark and thereby compare their operating characteristics in a fair manner.

3 EXAMPLES

The purpose of the following examples is to illustrate some of the ideas introduced in Section 2. Through these examples we also wish to bring out the potential pitfalls inherent in the "all or nothing" attitude taken by some with regard to multiple comparisons.

Example 3.1 **(*Comparing New Competing Treatments with a Control*).** One of the main activities in a pharmaceutical concern is the search for new drugs that are more effective ("better") than a standard drug or placebo. In the exploratory phase of research it is common to start with a large number of chemicals with structures related to a known active compound (Dunnett and Goldsmith 1981). Therefore a *screening* device is needed to weed out potential noncontenders. An MCP can be used for this purpose. (Alternatively, one may adopt the subset selection approach of Gupta and Sobel 1958.) The choice of error rate and associated α depends on the costs of confirmation, relative costs of Type I versus Type II errors, and prior experiences with similar problems. One choice that is reasonable in some situations is the Type I PFE whose control guarantees a bound on the expected number of false leads.

To obtain approval from the regulatory agency for marketing a new drug, the pharmaceutical company needs to demonstrate by a confirmatory study that the new drug is at least as good as the control (i.e., the standard drug or placebo). Typically several drugs that have passed the exploratory screening phase are tested against the control in the confirmatory phase. No Type I error must be made in any of these comparisons because otherwise a drug that is actually inferior to the control may be recommended. Thus in this case control of the Type I FWE is required. □

The next example is also concerned with treatment versus control comparisons but a careful analysis of how the final conclusion depends on the individual inferences shows that control of the PCE is required.

Example 3.2 **(*Comparing a New Treatment with Several Controls*).** Before a pharmaceutical company can market a combination drug, the regulatory agency requires that the manufacturer produce convincing evidence that the combination drug is better than every one of its m (say) subcombinations, which may be regarded as controls (Dunnett and Goldsmith 1981). Thus protection is needed against erroneously concluding that the combination drug is better than all of its subcombinations when in fact some of them are at least as good.

If separate one-sided tests are used for comparing the combination drug with each subcombination (and concluding that the combination drug is better when all m tests reject), then the probability of erroneously recommending the combination drug can be seen to achieve its maximum at a least favorable configuration where exactly one subcombination is equivalent to the combination drug and all the others are infinitely worse. This follows from Berger's (1982) general results on *intersection-union tests*. Thus to control the relevant Type I error probability it is only necessary to test each one of the m possible least favorable configurations at level α. This amounts to controlling the PCE by using separate α-level tests. □

Multiple comparison problems can also arise when only two treatments are compared. The next two examples illustrate such situations.

Example 3.3 **(*Comparing a New Treatment with a Control for Several Classes of Patients*).** The problem of comparing a new treatment with a control for different classes of patients is common in clinical research. Tukey (1977) noted that although such comparisons are presumably separate, still, "Special attention is given to the results for whichever class or classes of patients for whom the results appear most favorable for the intervention or therapy under test." Ethical considerations require that an inferior treatment should not be given in a follow-up confirmatory study or recommended for eventual use for *any* class of patients. Thus whether the study is exploratory or confirmatory, the FWE should be controlled.

A closely related problem is that of detecting interactions between the treatments and some covariates of interest, for example, age, sex, and prognosis. Here the patients may be divided into (say) two categories for each covariate and the difference in the effects of the treatment and the control for the two categories may be contrasted. This problem was studied by Shafer and Olkin (1983). □

Example 3.4 **(*Comparing a New Treatment with a Control Based on Multiple Measures*).** This example was used in Section 2.1 to illustrate exploratory situations where the family of all potential inferences cannot be specified in advance. The best approach then is to describe in detail how different measures (e.g., response variables, side effects) were selected in light of the data and/or prior knowledge so that the reader can appreciate the exploratory nature of the findings.

Now consider a confirmatory study and suppose that the list of measures of interest is finite and can be specified in advance. Which error rate to control would then depend on the criterion used for recommending the new treatment. For example, if the criterion is based on the number of measures on which the new treatment does better, then the PFE should be controlled. On the other hand, if the criterion requires the new treatment to do better on *all* the measures, then the FWE should be controlled. □

The following example deals with comparisons of *noncompeting* treatments with a common control and is in direct contrast with Example 3.1. Here control of the PCE is justified regardless of whether the study is exploratory or confirmatory.

Example 3.5 **(*Comparing Noncompeting Treatments with a Control*).** Consider a situation where a common control is included in the experiment for benchmark purposes and the various treatments that are combined in one experiment mainly for reasons of economy are not competitive. For example, different school boards may hire the same consulting organization to perform evaluations of their different educational programs. Clearly, each school board will only be concerned with the comparison of its own program with the control, and a board's interpretation of the data will be unaffected by the fact that other programs were included in the same experiment and were compared with the same control. Also the consulting organization would report separately to each school board—not together as a group—and therefore it would have no special interest in ensuring that the FWE for the collection of all comparisons is controlled. Here we have a simultaneous study but each inference is autonomous (although statistically dependent). Therefore the PCE should be controlled. □

The following example also involves one central organization testing data from several independent sources, but because of the nature of the problem the conclusions reached here are quite different.

Example 3.6 **(*Detailed Inferences Following an Overall Test*).** The Nuclear Regulatory Commission (NRC) of the United States is charged with the responsibility of maintaining nuclear materials safeguards at various nuclear facilities in this country. One of NRC's duties in this regard is to inspect balances of special nuclear materials at various locations in different periods and report if there are any losses (Goldman et al. 1982). Such inspections are usually followed by suitable actions, and thus the inferences should be regarded as confirmatory.

Suppose that there are $k \geqq 2$ locations where inspections are performed and material losses are measured. Let θ_i denote the true unknown material loss in the ith location in a given period $(1 \leqq i \leqq k)$. The k hypotheses to be tested are $\theta_i = 0$ against the alternative $\theta_i > 0$ $(1 \leqq i \leqq k)$. In this case, NRC may have a special interest in controlling the Type I FWE at some reasonably small level α so that not too many false alarms are raised. A way to proceed might be to first test the overall null hypothesis $\theta_1 = \theta_2 = \cdots = \theta_k = 0$ (which is quite plausible in practice) at level α, and if it is rejected, then follow it up with detailed inferences on individual θ_i's (in particular, their confidence estimates) aimed at identifying the potentially "culprit" locations that should be further examined in order to determine the reasons for diversion or losses of materials. □

The final example discusses two exploratory screening problems. They contrast two situations, one of which requires Type I error rate control, while the other requires Type II error rate control.

Example 3.7 ***(Screening Problems Requiring Type I or Type II Error Protection).*** In any pharmaceutical company drug screening is an exploratory ongoing program in which a large number of compounds are routinely tested for the presence of activity. In this case, many null configurations are *a priori* plausible because quite a few compounds lack activity. The resources available for confirmatory testing are generally scarce and so control of some Type I error rate is appropriate, especially in those situations where elimination of a few active compounds is not a serious error (e.g., when a large number of alternative compounds are developed and tested).

Let us now contrast this situation with one drawn from the area of carcinogenic testing. Consider an exploratory study for screening a large number of substances to detect those that are carcinogens. As in the previous example, here also many of the substances lack activity and therefore many null configurations are *a priori* plausible. However, here a Type II error is much more serious than a Type I error because if a true carcinogen goes undetected, then this could have serious consequences. One should employ an MCP for controlling a Type II error rate here. This could be achieved by a proper choice of the design of the study and of sample sizes. MCPs controlling Type I error rates are often used (inappropriately) because the data are already available. Usually liberal control of Type I error rates is employed in such situations (e.g., using a larger α, controlling the PCE rather than the FWE, using weak rather than strong control) so that a larger proportion of substances are carried into the confirmatory phase and thus chances of missing any true carcinogens are small. □

PART 1

Procedures Based on Classical Approaches for Fixed-Effects Linear Models with Normal Homoscedastic Independent Errors

CHAPTER 2

Some Theory of Multiple Comparison Procedures for Fixed-Effects Linear Models

Part I of the book covers multiple comparison procedures (MCPs) for fixed-effects linear models with independent homoscedastic normal errors. In this chapter we give some basic theory for single-step and stepwise (multistep) MCPs for such models. A more general theory of MCPs, which is applicable to models other than those considered in Part I (e.g., nonparametric models and mixed-effects linear models), is given in Appendix 1.

The theory of MCPs covered in Part I deserves a separate discussion for the following reasons:

(i) This theory is simpler in some aspects, and hence serves as an easy introduction to the subject. Much of it also readily extends to other models. The general theory discussed in Appendix 1, although applicable to problems not covered by the present chapter, involves other restrictive assumptions and sometimes more complicated arguments.

(ii) This theory and the associated MCPs are more fully developed than those for other models.

The outline of this chapter is as follows. In section 1 we give a brief review of some basic results for fixed-effects linear models. (See Scheffé 1959 for a fuller discussion.) Another purpose of this section is to set the stage for the discussion of the problem of multiple comparisons among the levels of a treatment factor of interest in designs involving other factors (e.g., other treatment factors or blocking or covariate factors). Some examples of such designs are given, which are used throughout Part I. In section 2 we discuss single-step MCPs for nonhierarchical families. In such families there are no implication relations among the inferences (in a sense made precise later). In Section 3 we discuss single-step MCPs for hierarchical families. These MCPs are required to make inferences satisfying certain implication relations. In Section 4 we discuss stepwise (actually step-down) MCPs for both these types of families.

1 A REVIEW OF FIXED-EFFECTS LINEAR MODELS

1.1 The Basic Model

Consider independent observations $Y_1, Y_2, \ldots, Y_N$ on a "response" variable Y. A linear model for Y_i postulates that

$$Y_i = X_{i1}\beta_1 + X_{i2}\beta_2 + \cdots + X_{ir}\beta_r + E_i \qquad (1 \leqq i \leqq N) \tag{1.1}$$

where the β_j's are unknown parameters, the X_{ij}'s are known constants (given values of "factors" X_j), and the E_i's are uncorrelated and identically distributed random variables (r.v.'s), each with mean zero and variance σ^2. Throughout Part I we assume that the E_i's are normally distributed. On letting $\mathbf{Y} = (Y_1, \ldots, Y_N)'$, $\mathbf{X} = \{X_{ij}\}$, $\boldsymbol{\beta} = (\beta_1, \ldots, \beta_r)'$, and $\mathbf{E} = (E_1, \ldots, E_N)'$, (1.1) can be written in matrix notation as

$$\mathbf{Y} = \mathbf{X}\boldsymbol{\beta} + \mathbf{E}. \tag{1.2}$$

The model (1.2) is called *linear* because it postulates that $E(\mathbf{Y})$ is a linear function of $\boldsymbol{\beta}$.

Each factor can be *qualitative*, that is, measured on a categorical scale, or *quantitative*, that is, measured on a numerical scale. The β_j's are often referred to as the *effects* of the corresponding factors X_j. Linear models are classified as *fixed-effects*, *random-effects*, or *mixed-effects* according to whether the β_j's are considered all fixed, all random, or some of each type. In the rest of this chapter and throughout Part I we assume that the β_j's are fixed unknown parameters.

1.2 Point Estimation

The method of least squares estimation is usually employed for estimating $\boldsymbol{\beta}$. A *least squares* (*LS*) *estimator* of $\boldsymbol{\beta}$ is defined as a vector $\hat{\boldsymbol{\beta}}$ that minimizes $\|\mathbf{Y} - \mathbf{X}\boldsymbol{\beta}\|^2$, the squared Euclidean norm of the vector $\mathbf{Y} - \mathbf{X}\boldsymbol{\beta}$. It can be shown that $\hat{\boldsymbol{\beta}}$ is an LS estimator if and only if it satisfies the set of normal equations: $\mathbf{X}'\mathbf{X}\hat{\boldsymbol{\beta}} = \mathbf{X}'\mathbf{Y}$. A solution $\hat{\boldsymbol{\beta}}$ is unique if and only if $\mathbf{X}$ is full column rank. (A convenient approach to obtaining a unique LS estimator $\hat{\boldsymbol{\beta}}$ when $\mathbf{X}$ is not full column rank is to introduce side-conditions on the β_j's. This approach is used in some of the examples in Section 1.4.) However, in general, there exists a class of scalar linear functions of $\boldsymbol{\beta}$, say $\mathbf{l}'\boldsymbol{\beta}$, such that for any solution $\hat{\boldsymbol{\beta}}$ of the normal equations, $\mathbf{l}'\hat{\boldsymbol{\beta}}$ is unique and is an unbiased estimator of $\mathbf{l}'\boldsymbol{\beta}$. Such a linear function is referred to as an *estimable parametric function* (or simply referred to as a *parametric function*). A vector parametric function is an ordered set of several scalar parametric functions.

The necessary and sufficient condition for $\mathbf{l}'\boldsymbol{\beta}$ to be an estimable function is that $\mathbf{l}$ be in the row space of $\mathbf{X}$. In particular, the *fitted vector* $\hat{\mathbf{Y}} = \mathbf{X}\hat{\boldsymbol{\beta}}$ gives a unique unbiased estimate of $E(\mathbf{Y}) = \mathbf{X}\boldsymbol{\beta}$. The vector $\hat{\mathbf{Y}}$ is the projection of $\mathbf{Y}$ on the column space of $\mathbf{X}$, and this projection is unique although $\hat{\boldsymbol{\beta}}$ may not be unique. The class of estimable parametric functions forms a linear subspace whose dimension equals rank $(\mathbf{X})$.

Every LS estimator $\hat{\boldsymbol{\beta}}$ can be written as $\hat{\boldsymbol{\beta}} = \mathbf{M}\mathbf{Y}$ for some matrix $\mathbf{M}$ satisfying $\mathbf{XMXM} = \mathbf{XM}$; if $\mathbf{X}$ is full column rank, then $\mathbf{M} = (\mathbf{X}'\mathbf{X})^{-1}\mathbf{X}'$. For any finite collection of parametric functions $(\mathbf{l}_1'\boldsymbol{\beta}, \mathbf{l}_2'\boldsymbol{\beta}, \ldots, \mathbf{l}_m'\boldsymbol{\beta})' = \mathbf{L}\boldsymbol{\beta}$ (say) (where $\mathbf{L} = (\mathbf{l}_1, \mathbf{l}_2, \ldots, \mathbf{l}_m)'$ is an $m \times r$ known matrix), $\mathbf{L}\hat{\boldsymbol{\beta}}$ has an m-variate normal distribution with mean vector $\mathbf{L}\boldsymbol{\beta}$ and covariance matrix $\sigma^2\mathbf{LMM}'\mathbf{L}'$ (denoted by $\mathbf{L}\hat{\boldsymbol{\beta}} \sim N(\mathbf{L}\boldsymbol{\beta}, \sigma^2\mathbf{LMM}'\mathbf{L}'))$. Thus for the purposes of making inferences on parametric functions of $\boldsymbol{\beta}$, we may assume that an estimator $\hat{\boldsymbol{\beta}}$ of $\boldsymbol{\beta}$ is available such that $\hat{\boldsymbol{\beta}} \sim N(\boldsymbol{\beta}, \sigma^2\mathbf{MM}')$.

In many problems multiple comparisons are sought among the parametric functions of the components of a subvector $\boldsymbol{\theta} : k \times 1$ of $\boldsymbol{\beta}$. Often $\boldsymbol{\theta}$ corresponds to the effects of a certain qualitative factor (or a combination of two or more qualitative factors), which we refer to as the *treatment factor*, and we refer to its $k \geqq 2$ levels as the *treatments* of main interest. Apart from $\boldsymbol{\theta}$, the parameter vector $\boldsymbol{\beta}$ possibly may contain the effects of other treatment factors. It may also contain the effects of factors such as blocks and covariates, included to account for the variability among the experimental (or observational) units and thus yield more precise comparisons among the treatment effects.

Let $\hat{\boldsymbol{\theta}}$ denote the corresponding subvector of $\hat{\boldsymbol{\beta}}$ and let $\mathbf{V} : k \times k$ be the submatrix of $\mathbf{MM}'$ corresponding to the $\hat{\boldsymbol{\theta}}$ part of the $\hat{\boldsymbol{\beta}}$ vector. Then it is true that for any finite collection of parametric functions $(\mathbf{l}_1'\boldsymbol{\theta}, \ldots, \mathbf{l}_m'\boldsymbol{\theta})' = \mathbf{L}\boldsymbol{\theta}$ (where $\mathbf{L}$ is an $m \times k$ known matrix), $\mathbf{L}\hat{\boldsymbol{\theta}}$ is distributed as $N(\mathbf{L}\boldsymbol{\theta}, \sigma^2\mathbf{LVL}')$. Thus for the purposes of making inferences on parametric functions of $\boldsymbol{\theta}$, we may assume that an estimator $\hat{\boldsymbol{\theta}}$ of $\boldsymbol{\theta}$ is available such that $\hat{\boldsymbol{\theta}} \sim N(\boldsymbol{\theta}, \sigma^2\mathbf{V})$.

Finally, we assume that an unbiased estimator S^2 of σ^2 is available that is distributed as $\sigma^2\chi^2_\nu/\nu$ independent of $\hat{\boldsymbol{\theta}}$. For S^2 we can always use the mean square error (MSE) given by

$$S^2 = \text{MSE} = \frac{\|\mathbf{Y} - \hat{\mathbf{Y}}\|^2}{\nu} \tag{1.3}$$

where $\nu = N - \text{rank}(\mathbf{X})$ is the error degrees of freedom (d.f.).

1.3 Confidence Estimation and Hypothesis Testing

Let $\gamma = \mathbf{l}'\boldsymbol{\theta} = \Sigma_{i=1}^k l_i\theta_i$ be an estimable scalar parametric function of interest where $\boldsymbol{\theta}$ is the subvector of $\boldsymbol{\beta}$ corresponding to the treatment effects. Let $\hat{\gamma} = \mathbf{l}'\hat{\boldsymbol{\theta}}$ be the LS estimator of γ; $\hat{\gamma}$ is distributed as $N(\gamma, \sigma^2_{\hat{\gamma}})$ where $\sigma^2_{\hat{\gamma}} = \sigma^2\mathbf{l}'\mathbf{V}\mathbf{l}$. Frequently γ is a *contrast* among the θ_i's, that is, $\Sigma_{i=1}^k l_i = 0$.

In this section we review standard statistical tests and confidence estimates for γ. These procedures are based on the likelihood ratio (LR) method and possess certain optimality properties. For example, if σ^2 is known, then the uniformly most powerful test of level α for testing $H_0^{(-)}: \gamma \leqq \gamma_0$ against the upper one-sided alternative $H_1^{(+)}: \gamma > \gamma_0$ (where γ_0 is a specified threshold that γ must exceed in order to constitute a research finding of practical importance) can be shown to be an LR test that rejects H_0 if

$$Z = \frac{\hat{\gamma} - \gamma_0}{\sigma_{\hat{\gamma}}} > Z^{(\alpha)}\ ; \tag{1.4}$$

here $Z^{(\alpha)}$ is the upper α point of the standard normal distribution. For the lower one-sided alternative ($H_0^{(+)}: \gamma \geqq \gamma_0$ vs. $H_1^{(-)}: \gamma < \gamma_0$) we have an analogous test. For testing $H_0: \gamma = \gamma_0$ versus the two-sided alternative $H_1: \gamma \neq \gamma_0$, the rejection region of the uniformly most powerful unbiased test of level α is given by

$$|Z| = \frac{|\hat{\gamma} - \gamma_0|}{\sigma_{\hat{\gamma}}} > Z^{(\alpha/2)}\ . \tag{1.5}$$

This can also be shown to be an LR test.

Corresponding to each of the above tests there is a $(1-\alpha)$-level confidence interval for γ consisting of the set of γ_0-values for which H_0 is not rejected by the given test. Thus corresponding to (1.4) we have a lower one-sided $(1-\alpha)$-level confidence interval for γ given by

$$\{\gamma : \gamma \geqq \hat{\gamma} - Z^{(\alpha)}\sigma_{\hat{\gamma}}\} \tag{1.6}$$

and corresponding to (1.5) we have a two-sided $(1-\alpha)$-level confidence interval for γ given by

$$\{\gamma : \gamma \in [\hat{\gamma} \pm Z^{(\alpha/2)}\sigma_{\hat{\gamma}}]\}\,. \tag{1.7}$$

By reversing the steps above we can obtain an α-level rejection region for any hypothesis on γ from any $(1-\alpha)$-level confidence region for γ. For example, the rejection region (1.4) for $H_0^{(-)}: \gamma \leqq \gamma_0$ corresponds to $\hat{\gamma}$-values for which the intersection of the set $(\gamma : \gamma \leqq \gamma_0)$ with the confidence interval (1.6) is empty. Similarly the rejection region (1.5) for $H_0 : \gamma = \gamma_0$ corresponds to $\hat{\gamma}$-values for which γ_0 does not fall in the confidence interval (1.7). In general, given an arbitrary $(1-\alpha)$-level confidence region for γ, a test of any hypothesis H_0 that restricts γ to a specified subset of the real line can be obtained by rejecting H_0 if the intersection of that subset with the confidence region is empty. The resulting test is known as the *confidence-region test* (Aitchison 1964) and has size $\leqq \alpha$ (see Theorem 3.3). This result enables us to use a given confidence region for γ to test any hypotheses (which may be specified after looking at the data) concerning γ.

If σ^2 is unknown but we have an independent estimator S^2 of σ^2 (as mentioned in the previous section), then the corresponding LR procedures use $S_{\hat{\gamma}} = S\sqrt{\mathbf{l}'\mathbf{V}\mathbf{l}}$ in place of $\sigma_{\hat{\gamma}}$, and critical points from Student's t-distribution in place of those from the standard normal distribution. Thus (1.4) is replaced by the test that rejects $H_0^{(-)}: \gamma \leqq \gamma_0$ if

$$T = \frac{\hat{\gamma} - \gamma_0}{S_{\hat{\gamma}}} > T_\nu^{(\alpha)}, \tag{1.8}$$

and (1.5) is replaced by the test that rejects $H_0 : \gamma = \gamma_0$ if

$$|T| = \frac{|\hat{\gamma} - \gamma_0|}{S_{\hat{\gamma}}} > T_\nu^{(\alpha/2)}, \tag{1.9}$$

where $T_\nu^{(\alpha)}$ is the upper α point of Student's t-distribution with ν d.f. These procedures also possess certain optimal properties; see, for example, Lehmann (1986).

Now consider a set of scalar parametric functions $\gamma_i = \mathbf{l}_i'\boldsymbol{\theta}$ and write

$\boldsymbol{\gamma} = (\mathbf{l}_1'\boldsymbol{\theta}, \mathbf{l}_2'\boldsymbol{\theta}, \ldots, \mathbf{l}_m'\boldsymbol{\theta})' = \mathbf{L}\boldsymbol{\theta}$ (say) where $\mathbf{L} = (\mathbf{l}_1, \mathbf{l}_2, \ldots, \mathbf{l}_m)'$ is an $m \times k$ known matrix. Let $\hat{\boldsymbol{\gamma}} = \mathbf{L}\hat{\boldsymbol{\theta}}$ be an LS estimator of $\boldsymbol{\gamma}$. As noted before, $\hat{\boldsymbol{\gamma}} \sim N(\boldsymbol{\gamma}, \sigma^2\mathbf{LVL}')$.

The LR test of $H_0: \boldsymbol{\gamma} = \boldsymbol{\gamma}_0$ versus $H_1: \boldsymbol{\gamma} \neq \boldsymbol{\gamma}_0$, where $\boldsymbol{\gamma}_0 = (\gamma_{01}, \gamma_{02}, \ldots, \gamma_{0m})'$ is a specified vector, is of the form

$$\text{reject } H_0 \text{ if} \quad \frac{(\hat{\boldsymbol{\gamma}} - \boldsymbol{\gamma}_0)'(\mathbf{LVL}')^{-1}(\hat{\boldsymbol{\gamma}} - \boldsymbol{\gamma}_0)}{mS^2} > \xi\,, \tag{1.10}$$

where we have assumed that $\mathbf{L}$ is a full row rank matrix. In (1.10) ξ is a critical constant to be chosen so as to make the level of the test equal to α. Standard distribution theory (see Scheffé 1959) shows that the test statistic on the left hand side of (1.10) has the F-distribution under H_0 with m and ν d.f. and hence $\xi = F_{m,\nu}^{(\alpha)}$, the upper α point of that distribution.

Associated with the test (1.10) there is a confidence region for $\boldsymbol{\gamma}$ with level $1 - \alpha$, which comprises of all vectors $\boldsymbol{\gamma}_0$ that are not rejected by (1.10). This confidence region is given by

$$\left\{\boldsymbol{\gamma}: \frac{(\hat{\boldsymbol{\gamma}} - \boldsymbol{\gamma})'(\mathbf{LVL}')^{-1}(\hat{\boldsymbol{\gamma}} - \boldsymbol{\gamma})}{mS^2} \leq F_{m,\nu}^{(\alpha)}\right\}. \tag{1.11}$$

1.4 Examples

Example 1.1 **(*One-Way Layout*).** Consider a single qualitative factor with $k \geq 2$ levels (treatments). Let Y_{ij} be the jth observation on the ith treatment and assume the linear model

$$Y_{ij} = \theta_i + E_{ij} \qquad (1 \leq i \leq k, 1 \leq j \leq n_i)\,. \tag{1.12}$$

In this case the unique LS estimates of the θ_i's are given by $\hat{\theta}_i = \bar{Y}_i = \Sigma_{j=1}^{n_i} Y_{ij}/n_i$ $(1 \leq i \leq k)$. The $\hat{\theta}_i$'s are independent normal with means θ_i and variances σ^2/n_i. Thus the matrix $\mathbf{V} = \text{diag}(1/n_1, \ldots, 1/n_k)$. An unbiased estimate S^2 of σ^2 is provided by $S^2 = \Sigma_{i=1}^k \Sigma_{j=1}^{n_i} (Y_{ij} - \bar{Y}_i)^2/(N - k)$ with $\nu = N - k$ d.f. where $N = \Sigma_{i=1}^k n_i$.

Throughout Part I we discuss MCPs for various families of parametric functions of the θ_i's for the one-way layout model. These procedures are applicable more widely to any design yielding LS estimates $\hat{\theta}_i$ that are independent of each other (i.e., the $\mathbf{V}$ matrix is diagonal) and of $S^2 \sim \sigma^2\chi_\nu^2/\nu$. We call such designs *balanced* if the $\hat{\theta}_i$'s have equal variances. For these designs we can use the procedures restricted to balanced one-way layouts (i.e., $n_1 = n_2 = \cdots = n_k$). For a more general notion of balance for pairwise comparisons of treatment means, see (2.9) of Chapter 3. □

Example 1.2 **(*One-Way Layout with a Fixed Covariate*).** Here the setup is the same as in the case of a one-way layout except that in addition to the treatment factor, we have a nonrandom quantitative factor X, called a covariate or concomitant. Usually the following linear model is postulated:

$$Y_{ij} = \theta_i + \beta(X_{ij} - \bar{X}_{..}) + E_{ij} \qquad (1 \leq i \leq k, 1 \leq j \leq n_i) \tag{1.13}$$

where (X_{ij}, Y_{ij}) is the observation on the jth experimental unit receiving the ith treatment and $\bar{X}_{..}$ is the grand mean of the X_{ij}'s.

The unique LS estimates of the θ_i's and β are given by

$$\hat{\theta}_i = \bar{Y}_{i.} - \hat{\beta}(\bar{X}_{i.} - \bar{X}_{..}) \ (1 \leq i \leq k), \qquad \hat{\beta} = \frac{S_{XY}}{S_{XX}} \tag{1.14}$$

where

$$S_{XY} = \sum_{i,j} (X_{ij} - \bar{X}_{i.})(Y_{ij} - \bar{Y}_{i.}), \qquad S_{XX} = \sum_{i,j} (X_{ij} - \bar{X}_{i.})^2,$$

$$S_{YY} = \sum_{i,j} (Y_{ij} - \bar{Y}_{i.})^2,$$

and $\bar{X}_{i.}$ and $\bar{Y}_{i.}$ are the sample means of the observations from the ith treatment. The vector $\hat{\boldsymbol{\theta}} \sim N(\boldsymbol{\theta}, \sigma^2\mathbf{V})$ where $\mathbf{V} = \{v_{ij}\}$ is given by

$$v_{ij} = \begin{cases} \dfrac{1}{n_i} + \dfrac{(\bar{X}_{i.} - \bar{X}_{..})^2}{S_{XX}} & \text{if } i = j \\ \dfrac{(\bar{X}_{i.} - \bar{X}_{..})(\bar{X}_{j.} - \bar{X}_{..})}{S_{XX}} & \text{if } i \neq j. \end{cases} \tag{1.15}$$

The MSE estimate of σ^2 is given by

$$S^2 = \{S_{YY} - S_{XY}^2/S_{XX}\}/(N - k - 1) \tag{1.16}$$

with $\nu = N - k - 1$ d.f. where $N = \Sigma_{i=1}^k n_i$. □

Example 1.3 **(*Balanced Incomplete Block (BIB) Design*).** In a BIB design k treatments are arranged in b blocks each of size $p < k$ such that each treatment is replicated r times (at most once in each block) and every pair of treatments occurs together in λ blocks. Here $r = bp/k$ and $\lambda = r(p-1)/(k-1)$.

The linear model for this design is given by

$$Y_{ij} = \mu + \theta_i + \beta_j + E_{ij} \qquad \forall (i,j) \in D \tag{1.17}$$

where Y_{ij} is the observation on the ith treatment in the jth block, and D is the set of all treatment × block combinations in the design. Note that this model assumes no treatment × block interaction.

The θ_i's and β_j's are not estimable but under the side conditions $\Sigma_{i=1}^{k} \theta_i = \Sigma_{j=1}^{b} \beta_j = 0$, the unique LS estimates are given by $\hat{\mu} = \bar{Y}_{..}$, $\hat{\beta}_j = \bar{Y}_{.j} - \bar{Y}_{..}$ $(1 \leqq j \leqq b)$, and

$$\hat{\theta}_i = \frac{pQ_i}{\lambda k} \qquad (1 \leqq i \leqq k)\,. \tag{1.18}$$

Here $\bar{Y}_{..}$ is the grand mean of the Y_{ij}'s, $\bar{Y}_{.j}$ is the mean of the Y_{ij}'s from the jth block, and Q_i is the ith "adjusted" treatment total given by the sum of all observations on the ith treatment $-(1/p) \times$ the sum of block totals for all blocks containing the ith treatment $(1 \leqq i \leqq k)$. It can be shown (see Scheffé 1959, page 167) that $\mathbf{V} = \{v_{ij}\}$ is given by

$$v_{ij} = \begin{cases} \dfrac{p(k-1)}{\lambda k^2} & \text{if } i = j \\[2ex] -\dfrac{p}{\lambda k^2} & \text{if } i \neq j\,. \end{cases} \tag{1.19}$$

The MSE estimator of σ^2 is given by

$$S^2 = \frac{\sum_{i,j \in D} (Y_{ij} - \hat{\mu} - \hat{\theta}_i - \hat{\beta}_j)^2}{(N - k - b + 1)} \tag{1.20}$$

with $\nu = N - k - b + 1$ d.f. where $N = bp$ is the total number of observations. □

Example 1.4 **(*Partially Balanced Incomplete Block* (*PBIB*) *Design*).** We consider a PBIB design with two associate classes. In this design there are k treatments arranged in b blocks each of size $p < k$ such that each treatment is replicated r times (at most once in each block) and two treatments that are ith associates of each other occur together in λ_i blocks $(i = 1, 2)$. If two treatments are ith associates, then let p^i_{jl} be the number of treatments that are jth associates of one treatment and lth associates of the other ($i, j, l = 1, 2$). The linear model for this design is the same as that for a BIB design given by (1.17).

Rao's solution vector (see John 1971, p. 257) for this design takes the form

$$\hat{\boldsymbol{\theta}} = \mathbf{MQ}\,. \tag{1.21}$$

Here $\mathbf{Q} = (Q_1, \ldots, Q_k)'$ is the vector of adjusted treatment totals Q_i as defined for a BIB design and the solution matrix $\mathbf{M} = \{m_{ij}\}$ is given by

$$m_{ij} = \begin{cases} \dfrac{p\delta_{22}}{\Delta} & \text{if } i = j \\ \dfrac{-p\delta_{12}}{\Delta} & \text{if } i \neq j \text{ are first associates} \\ 0 & \text{otherwise} \end{cases} \tag{1.22}$$

where

$$\Delta = \delta_{11}\delta_{22} - \delta_{12}\delta_{21}\,, \qquad \delta_{11} = r(p-1) + \lambda_2\,, \qquad \delta_{12} = \lambda_2 - \lambda_1\,,$$
$$\delta_{21} = (\lambda_2 - \lambda_1)p_{12}^2 \quad \text{and} \tag{1.23}$$
$$\delta_{22} = r(p-1) + \lambda_2 + (\lambda_2 - \lambda_1)(p_{11}^1 - p_{11}^2)\,.$$

For this design all contrasts $\mathbf{c}'\boldsymbol{\theta}$ are estimable, and

$$\mathrm{cov}(\mathbf{c}_1'\hat{\boldsymbol{\theta}}, \mathbf{c}_2'\hat{\boldsymbol{\theta}}) = \sigma^2 \mathbf{c}_1'\mathbf{M}\mathbf{c}_2 \qquad \forall\, \mathbf{c}_1, \mathbf{c}_2 \in \mathbb{C}^k$$

where $\mathbb{C}^k = \{\mathbf{c} \in \mathbb{R}^k : \Sigma_{i=1}^k c_i = 0\}$ is the k-dimensional contrast space. We can thus take $\mathbf{V} = \mathbf{M}$. We then have

$$\mathrm{var}(\hat{\theta}_i - \hat{\theta}_j) = \begin{cases} \dfrac{2p}{\Delta}(\delta_{22} + \delta_{12})\sigma^2 & \text{if } i \text{ and } j \text{ are first associates} \\ \dfrac{2p\delta_{22}}{\Delta}\sigma^2 & \text{if } i \text{ and } j \text{ are second associates}\,. \end{cases} \tag{1.24}$$

Therefore the design is unbalanced for pairwise comparisons (see (2.9) of Chapter 3) unless $\lambda_1 = \lambda_2$, in which case the design reduces to a BIB design. The MSE estimator of σ^2 for this design is given by (1.20) where $\hat{\mu} = \bar{Y}_{..}$, $\hat{\beta}_j = \bar{Y}_{.j} - \bar{Y}_{..}$ ($1 \leq j \leq b$), and the $\hat{\theta}_i$'s are given by (1.21). □

2 SINGLE-STEP PROCEDURES FOR NONHIERARCHICAL FAMILIES

In this section we restrict consideration to families of scalar parametric functions. Such families are nonhierarchical. Hierarchical families involving vector parametric functions are discussed in Section 3.1 and more generally in Section 1.1 of Appendix 1.

Before we proceed further it may be useful to illustrate the difference between a *single* inference on a vector parametric function and *multiple* inferences on its scalar components. For this purpose we use the following example from Lehmann (1952): Let θ_1 and θ_2 be the unknown proportions of items with major and minor defects in a lot. The lot is considered acceptable if $\theta_1 \leqq \theta_{01}$ *and* $\theta_2 \leqq \theta_{02}$, where θ_{01} and θ_{02} are preassigned specification limits. If rejected lots are completely screened, then the reason for rejection is immaterial, and a single test of the hypothesis $H_0: \theta_1 \leqq \theta_{01}$ *and* $\theta_2 \leqq \theta_{02}$ suffices. On the other hand, if a lot rejected for major defects is screened differently from one rejected for minor defects, then it is important to know the reason for rejection. Therefore separate tests on θ_1 and θ_2 are required. An MCP becomes necessary for performing these latter tests when they are regarded together as a family. The union-intersection method provides a general method for constructing such MCPs. We now discuss this method.

2.1 The Union-Intersection Method

Roy (1953) proposed a heuristic method of constructing a test of any hypothesis H_0 that can be expressed as an intersection of a family of hypotheses. Suppose that $H_0 = \bigcap_{i \in I} H_{0i}$ where I is an arbitrary index set. Further suppose that a suitable test of each H_{0i} is available. Then according to Roy's union-intersection (UI) method, the rejection region for H_0 is given by the union of rejection regions for the H_{0i}, $i \in I$, that is, H_0 is rejected if and only if at least one H_{0i} is rejected.

Roy and Bose (1953) showed that the single inference given by the UI test of H_0 implies multiple inferences (tests and confidence estimates) for the parameters on which the hypotheses H_{0i} are postulated. For example, suppose the hypotheses H_{0i} are postulated on scalar parameters γ_i, $i \in I$, and I is finite. Then by inverting the UI test of $H_0 = \bigcap_{i \in I} H_{0i}$ we obtain a confidence region for the vector parameter $\boldsymbol{\gamma}$. It is seen later that from this confidence region we can derive simultaneous tests and confidence intervals not only on the γ_i's but also on any functions of them. If the UI test of H_0 is of level α, then all such inferences derived from it have the familywise error rate (FWE) strongly controlled at level α. An MCP derived in this manner from a UI test is referred to as a *UI procedure*. Exactly how such UI procedures are derived is seen in Section 2.1.1 for finite families and in Section 2.1.2 for infinite families. In Section 2.3 we see that in addition to tests and confidence intervals we can also make directional decisions without exceeding the specified upper bound α on the FWE.

2.1.1 Finite Families

In this section we consider the problem of making multiple inferences on the components γ_i $(1 \leq i \leq m)$ of $\boldsymbol{\gamma} = \mathbf{L}\boldsymbol{\theta}$. We regard these inferences jointly as a family. If it is desired to control the per-family error rate (PFE) or the per-comparison error rate (PCE), then standard procedures for the individual γ_i's (as discussed in Section 1.3) can readily be constructed. The optimality of such procedures is discussed in Chapter 11. Here we focus on UI procedures for controlling the FWE. These latter procedures have been referred to as *finite intersection procedures* by Krishnaiah (1965, 1979).

Two-sided inferences are discussed in Section 2.1.1.1 and one-sided inferences are discussed briefly in Section 2.1.1.2.

2.1.1.1 Two-Sided Inferences. For testing $H_0 : \boldsymbol{\gamma} = \boldsymbol{\gamma}_0$ versus $H_1 : \boldsymbol{\gamma} \neq \boldsymbol{\gamma}_0$ an LR test was given in (1.10). Here we first give a UI test for the same problem based on a "natural" representation of H_0 as a finite intersection of hypotheses on the components γ_i of $\boldsymbol{\gamma}$. Consider a finite family of hypothesis testing problems:

$$H_{0i} : \gamma_i = \gamma_{0i} \text{ versus } H_{1i} : \gamma_i \neq \gamma_{0i} \quad (1 \leq i \leq m) . \tag{2.1}$$

Clearly, $H_0 = \bigcap_{i=1}^m H_{0i}$ and $H_1 = \bigcup_{i=1}^m H_{1i}$. For testing H_{0i} versus H_{1i} we can use the LR test (1.9) whose rejection region is of the form

$$|T_i| = \frac{|\hat{\gamma}_i - \gamma_{0i}|}{S\sqrt{\mathbf{l}_i'\mathbf{V}\mathbf{l}_i}} > \xi_i \quad (1 \leq i \leq m) . \tag{2.2}$$

The critical constants ξ_i can be determined as follows: The rejection region of the UI test of H_0 is the union of rejection regions (2.2). In order for this union to have size α, the ξ_i's must satisfy

$$\Pr_{H_0}\{|T_i| > \xi_i \text{ for some } i = 1, 2, \ldots, m\} = \alpha . \tag{2.3}$$

Usually we choose $\xi_i = \xi$ for all i for the following reasons:

(i) The m testing problems (2.1) are generally treated symmetrically with regard to the relative importance of Type I versus Type II errors. This implies that the marginal levels $\alpha_i = \Pr_{H_{0i}}\{|T_i| > \xi_i\}$ should be the same for $i = 1, 2, \ldots, m$. Since the T_i's have the same marginal distribution under the H_{0i}'s (namely, Student's t-distribution with ν d.f.), it follows that the ξ_i's should be equal.

(ii) For $m = 2$, Kunte and Rattihalli (1984) have shown that subject to (2.3) the choice $\xi_1 = \xi_2$ minimizes $\xi_1\xi_2$ among all choices of (ξ_1, ξ_2). This corresponds to choosing a $(1-\alpha)$-level rectangular confidence region for $\gamma = (\gamma_1, \gamma_2)'$ with the smallest area (from a certain class of rectangular confidence regions). One would conjecture that this result would also hold in higher dimensions.

(iii) Finally, of course, the task of computing the ξ_i's is greatly simplified if they are chosen to be equal.

Letting $\xi_i = \xi$ for all i, from (2.2) we see that the UI test of H_0 rejects if

$$\max_{1 \leq i \leq m} |T_i| > \xi \tag{2.4}$$

where (using (2.3)) ξ must be chosen so that

$$\Pr_{H_0}\{\max_{1 \leq i \leq m} |T_i| > \xi\} = \alpha . \tag{2.5}$$

From the standard distribution theory discussed in Appendix 3, it follows that under H_0, $T_1, T_2, \ldots, T_m$ have an m-variate t-distribution with ν d.f. and the associated correlation matrix $\{\rho_{ij}\}$ where

$$\rho_{ij} = \frac{\mathbf{l}_i'\mathbf{V}\mathbf{l}_j}{\{(\mathbf{l}_i'\mathbf{V}\mathbf{l}_i)(\mathbf{l}_j'\mathbf{V}\mathbf{l}_j)\}^{1/2}} \qquad (1 \leq i \neq j \leq m) . \tag{2.6}$$

The critical value ξ is the upper α point of $\max_{1 \leq i \leq m}|T_i|$, which is denoted by $|T|^{(\alpha)}_{m,\nu,\{\rho_{ij}\}}$.

Now (2.5) holds for any choice of γ_0. Therefore the set of all γ_0's that will not be rejected by the UI test (2.4) constitutes a $(1-\alpha)$-level confidence region for γ and is given by

$$\{\gamma : \gamma_i \in [\hat{\gamma}_i \pm |T|^{(\alpha)}_{m,\nu,\{\rho_{ij}\}} S\sqrt{\mathbf{l}_i'\mathbf{V}\mathbf{l}_i}] \ (1 \leq i \leq m)\} . \tag{2.7}$$

The critical points $|T|^{(\alpha)}_{m,\nu,\{\rho_{ij}\}}$ are extensively tabulated only in the equicorrelated case ($\rho_{ij} = \rho \ \forall\, i \neq j$); see Table 5 in Appendix 3. In other cases one must use some approximation. The simplest of such (conservative) approximations is the Bonferroni approximation given by $T_\nu^{(\alpha/2m)}$, which leads to the well-known *Bonferroni procedure*. A less conservative approximation based on the Dunn–Šidák inequality (see Appendix 2) uses $T_\nu^{(\alpha^*/2)}$ where $\alpha^* = 1-(1-\alpha)^{1/m}$. These approximations have been studied by Dunn (1958, 1959, 1961). (For a still better approximation, see Example 2.1 below.)

We now offer two examples, both of which involve special correlation structures $\{\rho_{ij}\}$, thus obviating the need for using an approximation.

Example 2.1. Suppose that the $\hat{\gamma}_i$'s are uncorrelated. In that case the critical constant $|T|_{m,\nu,\{0\}}^{(\alpha)}$ is denoted by $|M|_{m,\nu}^{(\alpha)}$ and is referred to as the upper α point of the Studentized maximum modulus distribution with parameter m and d.f. ν. The resulting UI procedure was independently proposed by Tukey (1953) and Roy and Bose (1953) and is referred to as the *Studentized maximum modulus procedure*. Various applications of this procedure are studied in Chapter 5.

We note that if the $\hat{\gamma}_i$'s are correlated with an arbitrary correlation matrix $\{\rho_{ij}\}$, then $|M|_{m,\nu}^{(\alpha)}$ provides an approximation to the desired two-sided critical point $|T|_{m,\nu,\{\rho_{ij}\}}^{(\alpha)}$ that is sharper than the Dunn–Šidák approximation $T_\nu^{(\alpha^*/2)}$ given above. □

Example 2.2. As another example of a finite family, consider the family of all $m = \binom{k}{2}$ pairwise differences between treatment means in the one-way layout model (1.12). The parametric functions of interest are $\theta_i - \theta_j$ and their LS estimates are $\bar{Y}_i - \bar{Y}_j$ with variances $\sigma^2(1/n_i + 1/n_j)$ $(1 \leq i < j \leq k)$. The exact critical point of the distribution of the maximum of the pairwise $|t|$-statistics $|T_{ij}| = |\bar{Y}_i - \bar{Y}_j|/S\sqrt{1/n_i + 1/n_j}$ $(1 \leq i < j \leq k)$ is difficult to determine in this case. However, for a balanced one-way layout $(n_1 = \cdots = n_k = n)$ it can be determined easily by noting that

$$\sqrt{2} \max_{1 \leq i < j \leq k} |T_{ij}| = \max_{1 \leq i < j \leq k} \frac{|(\bar{Y}_i - \theta_i) - (\bar{Y}_j - \theta_j)|}{S/\sqrt{n}}$$

is distributed as the range of k independent standard normal r.v.'s divided by an independent $\sqrt{\chi_\nu^2/\nu}$ r.v. This r.v. is denoted by $Q_{k,\nu}$ and is referred to as the Studentized range r.v. with parameter k and d.f. ν (see Appendix 3). Thus the desired critical point is $Q_{k,\nu}^{(\alpha)}/\sqrt{2}$, and the resulting $(1-\alpha)$-level simultaneous confidence intervals for all pairwise differences $\theta_i - \theta_j$ are

$$\theta_i - \theta_j \in [\bar{Y}_i - \bar{Y}_j \pm Q_{k,\nu}^{(\alpha)} S/\sqrt{n}] \qquad (1 \leq i < j \leq k)\,. \tag{2.8}$$

This is the well-known Tukey's (1953) *T-procedure* (also known as the *Studentized range procedure*), which is discussed in detail in Chapter 3. □

2.1.1.2 One-Sided Inferences. We consider the problem of constructing simultaneous one-sided confidence intervals on the γ_i's $(1 \leqq i \leqq m)$. Application of the UI method leads to the r.v.

$$\max_{1 \leqq i \leqq m} T_i = \max_{1 \leqq i \leqq m} \frac{\hat{\gamma}_i - \gamma_i}{S\sqrt{\mathbf{l}_i'\mathbf{V}\mathbf{l}_i}} \tag{2.9}$$

where $T, T_2, \ldots, T_m$ have an m-variate t-distribution with ν d.f. and the associated correlation matrix $\{\rho_{ij}\}$ given by (2.6). We denote the upper α point of (2.9) by $T^{(\alpha)}_{m,\nu,\{\rho_{ij}\}}$. The resulting $(1-\alpha)$-level simultaneous lower one-sided confidence intervals for the γ_i's are given by

$$\gamma_i \geqq \hat{\gamma}_i - T^{(\alpha)}_{m,\nu,\{\rho_{ij}\}} S\sqrt{\mathbf{l}_i'\mathbf{V}\mathbf{l}_i} \quad (1 \leqq i \leqq m) . \tag{2.10}$$

The intervals (2.10) can be used to test hypotheses $H^{(-)}_{0i}: \gamma_i \leqq \gamma_{0i}$ against upper one-sided alternatives $H^{(+)}_{1i}: \gamma_i > \gamma_{0i}$ where the γ_{0i}'s are specified constants $(1 \leqq i \leqq m)$. This test procedure rejects H_{0i} if

$$T_i = \frac{\hat{\gamma}_i - \gamma_{0i}}{S\sqrt{\mathbf{l}_i'\mathbf{V}\mathbf{l}_i}} > T^{(\alpha)}_{m,\nu,\{\rho_{ij}\}} \quad (1 \leqq i \leqq m) .$$

It is readily seen that the Type I FWE of this test procedure is controlled at level α.

The critical points $T^{(\alpha)}_{m,\nu,\{\rho_{ij}\}}$ are extensively tabulated only in the equicorrelated case ($\rho_{ij} = \rho \; \forall \, i \neq j$); see Table 4 in Appendix 3. In other cases one must use some approximation. As in the preceding section, one can use the Bonferroni approximation $T^{(\alpha/m)}_{\nu}$. When $\rho_{\min} > -1/(m-1)$, a less conservative approximation is given by $T^{(\alpha)}_{m,\nu,\rho_{\min}}$. This latter approximation is based on Slepian's (1962) inequality (see Appendix 2).

Example 2.3. One-sided inferences are often of interest when comparing treatments with a control. Again consider the one-way layout model (1.12) and suppose that the kth treatment is a control with which the first $k-1$ treatments are to be compared. The parametric functions of interest are $\gamma_i = \theta_i - \theta_k$ $(1 \leqq i \leqq k-1)$. The correlation coefficients ρ_{ij} of the LS estimators $\hat{\gamma}_i = \bar{Y}_i - \bar{Y}_k$ $(1 \leqq i \leqq k-1)$ are given by

$$\rho_{ij} = \left[\frac{n_i n_j}{(n_i + n_k)(n_j + n_k)} \right]^{1/2} \quad (1 \leqq i \neq j \leqq k-1) .$$

The resulting $(1-\alpha)$-level simultaneous lower one-sided confidence intervals for the treatment-control differences are given by

$$\theta_i - \theta_k \geqq \bar{Y}_i - \bar{Y}_k - T^{(\alpha)}_{k-1,\nu,\{\rho_{ij}\}} S\sqrt{\frac{1}{n_i} + \frac{1}{n_k}} \qquad (1 \leqq i \leqq k-1) . \tag{2.11}$$

This is the well-known Dunnett's (1955) procedure for one-sided comparisons with a control. The corresponding two-sided procedure uses the critical point $|T|^{(\alpha)}_{k-1,\nu,\{\rho_{ij}\}}$ in place of $T^{(\alpha)}_{k-1,\nu,\{\rho_{ij}\}}$. These procedures are discussed in detail in Chapter 5. □

We conclude this section by noting that to construct an MCP for a finite family one must prespecify not only the number of comparisons, m, but also the particular comparisons. Ury and Wiggins (1971) suggested that in a Bonferroni procedure some of the comparisons could be selected post-hoc as long as the total number of comparisons does not exceed the upper bound m. Rodger (1973) pointed out that the FWE is *not* controlled if the Bonferroni procedure is used in this manner. Ury and Wiggins (1974) offered a corrected version of their procedure in which the prespecified comparisons are made using the Bonferroni procedure at level α_1 (say), post-hoc comparisons are made using an appropriate procedure for the family from which these comparisons are selected at level α_2 (say) where $\alpha_1 + \alpha_2 \leqq \alpha$.

2.1.2 *Infinite Families*

Consider an infinite family of hypotheses testing problems:

$$H_{0\mathbf{a}} : \mathbf{a}'\boldsymbol{\gamma} = \mathbf{a}'\boldsymbol{\gamma}_0 \qquad \text{versus} \qquad H_{1\mathbf{a}} : \mathbf{a}'\boldsymbol{\gamma} \neq \mathbf{a}'\boldsymbol{\gamma}_0, \ \mathbf{a} \in \mathbb{R}^m . \tag{2.12}$$

Let $H_0 = \bigcap_{\mathbf{a}\in\mathbb{R}^m} H_{0\mathbf{a}}$ and $H_1 = \bigcup_{\mathbf{a}\in\mathbb{R}^m} H_{1\mathbf{a}}$. We now derive a UI test of H_0 using this representation, which will yield a UI procedure for the infinite family of parametric functions $\mathbf{a}'\boldsymbol{\gamma}$, $\mathbf{a} \in \mathbb{R}^m$.

For testing $H_{0\mathbf{a}}$ versus $H_{1\mathbf{a}}$, the LR test has the rejection region

$$|T_{\mathbf{a}}| = \frac{|\mathbf{a}'(\hat{\boldsymbol{\gamma}} - \boldsymbol{\gamma}_0)|}{S\sqrt{\mathbf{a}'\mathbf{LVL}'\mathbf{a}}} > \xi_{\mathbf{a}} \tag{2.13}$$

where $\xi_{\mathbf{a}}$ is some critical constant appropriate for testing $H_{0\mathbf{a}}$. If we choose $\xi_{\mathbf{a}} = \xi$ for all $\mathbf{a} \in \mathbb{R}^m$, then by the UI method the rejection region for H_0 is given by

$$\sup_{\mathbf{a}\in\mathbb{R}^m} |T_{\mathbf{a}}| > \xi . \tag{2.14}$$

Alternatively we can put

$$\mathbf{a}'\boldsymbol{\gamma} = \mathbf{a}'\mathbf{L}\boldsymbol{\theta} = \mathbf{l}'\boldsymbol{\theta} \tag{2.15}$$

where $\mathbf{l} = \mathbf{L}'\mathbf{a}$ is in $\mathscr{L}$, the row space of $\mathbf{L}$ with dimension m. We then have

$$|T_{\mathbf{a}}| = |T_{\mathbf{l}}| = \frac{|\mathbf{l}'(\hat{\boldsymbol{\theta}} - \boldsymbol{\theta}_0)|}{S\sqrt{\mathbf{l}'\mathbf{V}\mathbf{l}}} \tag{2.16}$$

where $\boldsymbol{\theta}_0$ is such that $\mathbf{a}'\boldsymbol{\gamma}_0 = \mathbf{l}'\boldsymbol{\theta}_0$. Thus (2.14) becomes

$$\sup_{\mathbf{l}\in\mathscr{L}} |T_{\mathbf{l}}| > \xi . \tag{2.17}$$

The following theorem shows that the UI test (2.14) (or equivalently (2.17)) of H_0 is the same as the LR test (1.10). A more general proof of this result when $\mathbf{LVL}'$ is singular is given by Altschul and Marcuson (1979).

Theorem 2.1 (Roy and Bose 1953). If $\mathbf{LVL}'$ is a nonsingular matrix, then

$$\sup_{\mathbf{a}\in\mathbb{R}^m} T_{\mathbf{a}}^2 = \sup_{\mathbf{l}\in\mathscr{L}} T_{\mathbf{l}}^2 = \frac{(\hat{\boldsymbol{\gamma}} - \boldsymbol{\gamma}_0)'(\mathbf{LVL}')^{-1}(\hat{\boldsymbol{\gamma}} - \boldsymbol{\gamma}_0)}{S^2} \tag{2.18}$$

Furthermore, if ξ in (2.14) (or equivalently (2.17)) is chosen equal to $\sqrt{mF_{m,\nu}^{(\alpha)}}$, then that test has level α.

Proof. Since $\mathbf{L}$ and $\mathbf{V}$ are full rank, it follows that $\mathbf{LVL}'$ is full rank and there exists a nonsingular $m \times m$ matrix $\mathbf{M}$ such that $\mathbf{MM}' = \mathbf{LVL}'$. Also there is a one-to-one correspondence between every $\mathbf{l} \in \mathscr{L}$ and every $\mathbf{a} \in \mathbb{R}^m$ such that $\mathbf{l} = \mathbf{L}'\mathbf{a}$.

For finding the supremum of $T_{\mathbf{a}}^2 = T_{\mathbf{l}}^2$, without loss of generality, we can restrict to $\mathbf{a}$ such that $\mathbf{a}'\mathbf{MM}'\mathbf{a} = \mathbf{a}'\mathbf{LVL}'\mathbf{a} = 1$. Now, from (2.16) we have

$$\begin{aligned} S^2T_{\mathbf{l}}^2 &= \frac{\{\mathbf{l}'(\hat{\boldsymbol{\theta}} - \boldsymbol{\theta}_0)\}^2}{\mathbf{l}'\mathbf{V}\mathbf{l}} = \frac{\{\mathbf{a}'\mathbf{MM}^{-1}\mathbf{L}(\hat{\boldsymbol{\theta}} - \boldsymbol{\theta}_0)\}^2}{\mathbf{a}'\mathbf{LVL}'\mathbf{a}} \\ &= \{(\mathbf{M}'\mathbf{a})'(\mathbf{M}^{-1}\mathbf{L}(\hat{\boldsymbol{\theta}} - \boldsymbol{\theta}_0))\}^2 . \end{aligned} \tag{2.19}$$

By the Cauchy–Schwarz inequality, the supremum of (2.19) is given by

$$\begin{aligned} &(\mathbf{a}'\mathbf{MM}'\mathbf{a})((\hat{\boldsymbol{\theta}} - \boldsymbol{\theta}_0)'\mathbf{L}'\mathbf{M}'^{-1}\mathbf{M}^{-1}\mathbf{L}(\hat{\boldsymbol{\theta}} - \boldsymbol{\theta}_0)) \\ &= (\hat{\boldsymbol{\theta}} - \boldsymbol{\theta}_0)'\mathbf{L}'(\mathbf{MM}')^{-1}\mathbf{L}(\hat{\boldsymbol{\theta}} - \boldsymbol{\theta}_0) = (\hat{\boldsymbol{\gamma}} - \boldsymbol{\gamma}_0)'(\mathbf{LVL}')^{-1}(\hat{\boldsymbol{\gamma}} - \boldsymbol{\gamma}_0) , \end{aligned}$$

which proves (2.18). Since under H_0, (2.18) is distributed as an $mF_{m,\nu}$ r.v., the second part of the theorem follows. □

From this theorem it is clear that the confidence region (1.11) for $\boldsymbol{\gamma}$ yields joint confidence intervals for the infinite family of all linear combinations $\mathbf{a}'\boldsymbol{\gamma}$, $\mathbf{a} \in \mathbb{R}^m$, or equivalently for all $\mathbf{l}'\boldsymbol{\theta}$, $\mathbf{l} \in \mathcal{L}$. In fact, we can write

$$\begin{aligned}\Pr\left\{\frac{[\mathbf{l}'(\hat{\boldsymbol{\theta}} - \boldsymbol{\theta})]^2}{S^2(\mathbf{l}'\mathbf{V}\mathbf{l})} \le mF_{m,\nu}^{(\alpha)} \ \forall \mathbf{l} \in \mathcal{L}\right\} \\ = \Pr\left\{\mathbf{l}'\boldsymbol{\theta} \in [\mathbf{l}'\hat{\boldsymbol{\theta}} \pm (mF_{m,\nu}^{(\alpha)})^{1/2} S(\mathbf{l}'\mathbf{V}\mathbf{l})^{1/2}] \ \forall \mathbf{l} \in \mathcal{L}\right\} \\ = 1 - \alpha . \qquad (2.20)\end{aligned}$$

Thus (2.20) provides simultaneous confidence intervals of level $1 - \alpha$ for all $\mathbf{l}'\boldsymbol{\theta}$, $\mathbf{l} \in \mathcal{L}$. This forms the basis of Scheffé's (1953) *S-procedure*, the varied applications of which are discussed in Chapter 3. Scheffé derived his procedure by using a geometric projection method that is equivalent to the algebraic result of Theorem 2.1. This method (frequently referred to as *Scheffé's projection method*) is discussed in the following section.

2.2 The Projection Method

Suppose that we have a $(1 - \alpha)$-level confidence region C for $\boldsymbol{\gamma} = \mathbf{L}\boldsymbol{\theta}$ such that C is convex and symmetric around $\hat{\boldsymbol{\gamma}}$, that is, $\hat{\boldsymbol{\gamma}} + \mathbf{b} \in C$ if and only if $\hat{\boldsymbol{\gamma}} - \mathbf{b} \in C$ for any $\mathbf{b} \in \mathbb{R}^m$. A set of $(1 - \alpha)$-level simultaneous confidence intervals for all $\mathbf{a}'\boldsymbol{\gamma}$, $\mathbf{a} \in \mathbb{R}^m$, or equivalently for all $\mathbf{l}'\boldsymbol{\theta}$, $\mathbf{l} \in \mathcal{L}$, can be derived from C by using the projection method based on the following simple identity:

$$|\mathbf{a}'(\hat{\boldsymbol{\gamma}} - \boldsymbol{\gamma})| = \|\mathbf{a}\| \cdot \|\text{projection of } \hat{\boldsymbol{\gamma}} - \boldsymbol{\gamma} \text{ on } \mathbf{a}\| .$$

Letting $C_0 = \{\hat{\boldsymbol{\gamma}} - \boldsymbol{\gamma} : \boldsymbol{\gamma} \in C\}$, we have

$$|\mathbf{a}'(\hat{\boldsymbol{\gamma}} - \boldsymbol{\gamma})| \le \|\mathbf{a}\| \cdot \max_{\hat{\boldsymbol{\gamma}} - \boldsymbol{\gamma} \in C_0} \|\text{projection of } \hat{\boldsymbol{\gamma}} - \boldsymbol{\gamma} \text{ on } \mathbf{a}\| . \qquad (2.21)$$

The maximum here is given by the line segment intercepted on the ray passing through $\mathbf{a}$ by two hyperplanes that are orthogonal to $\mathbf{a}$ and tangential to C_0. The inequality (2.21) holds for all $\mathbf{a} \in \mathbb{R}^m$ if and only if $\hat{\boldsymbol{\gamma}} - \boldsymbol{\gamma}$ lies between all such hyperplanes. Since C_0 is convex, by the supporting hyperplanes theorem it follows that the latter statement is true whenever $\hat{\boldsymbol{\gamma}} - \boldsymbol{\gamma} \in C_0$, an event with probability $1 - \alpha$. Thus (2.21) yields $(1 - \alpha)$-level simultaneous confidence intervals for all $\mathbf{a}'\boldsymbol{\gamma}$, $\mathbf{a} \in \mathbb{R}^m$.

Scheffé (1953) applied this method when C is the confidence ellipsoid (1.11). He showed that the value of the right hand side in (2.21) is given by

$$(mF^{(\alpha)}_{m,\nu})^{1/2}(S^2\mathbf{a}'\mathbf{LVL}'\mathbf{a})^{1/2} = (mF^{(\alpha)}_{m,\nu})^{1/2}(S^2\mathbf{l}'\mathbf{V}\mathbf{l})^{1/2} , \qquad (2.22)$$

which is the "allowance" (Tukey 1953) used in the intervals (2.20).

We next apply the projection method to the rectangular confidence region (2.7). In this case the maximal projection of $\hat{\boldsymbol{\gamma}} - \boldsymbol{\gamma}$ on any vector $\mathbf{a}$ is obtained when $\hat{\boldsymbol{\gamma}} - \boldsymbol{\gamma}$ is one of the "corner" vectors $(\pm\xi_1, \pm\xi_2, \ldots, \pm\xi_m)'$ where

$$\xi_i = |T|^{(\alpha)}_{m,\nu,\{\rho_{ij}\}} S\sqrt{\mathbf{l}_i'\mathbf{V}\mathbf{l}_i} \qquad (1 \leqq i \leqq m) .$$

Thus the right hand side of (2.21) can be written as $\Sigma_{i=1}^m |a_i| \cdot |\xi_i|$ and hence $(1 - \alpha)$-level simultaneous confidence intervals for all $\mathbf{a}'\boldsymbol{\gamma}$ are given by

$$\mathbf{a}'\boldsymbol{\gamma} \in \left[\mathbf{a}'\hat{\boldsymbol{\gamma}} \pm |T|^{(\alpha)}_{m,\nu,\{\rho_{ij}\}} S \sum_{i=1}^{k} |a_i|(\mathbf{l}_i'\mathbf{V}\mathbf{l}_i)^{1/2}\right] \qquad \forall\, \mathbf{a} \in \mathbb{R}^m . \qquad (2.23)$$

Richmond (1982) has given an algebraic proof of this result including the case where the $\mathbf{l}_i$'s are possibly linearly dependent.

Example 2.4. Consider two parameters γ_1 and γ_2 and assume for simplicity that their LS estimators $\hat{\gamma}_1$ and $\hat{\gamma}_2$ are uncorrelated with $\hat{\gamma}_i \sim N(\gamma_i, \sigma_i^2)$ where $\sigma_1^2 = 1$ and $\sigma_2^2 = 4$ are known. Thus in our notation, $\sigma^2 = 1$ and $\mathbf{LVL}' = \mathrm{diag}(1, 4)$.

From (1.11), a $(1 - \alpha)$-level confidence region for (γ_1, γ_2) is given by

$$\frac{(\hat{\gamma}_1 - \gamma_1)^2}{1} + \frac{(\hat{\gamma}_2 - \gamma_2)^2}{4} \leqq 2F^{(\alpha)}_{2,\infty} = \chi^2_2(\alpha) \qquad (2.24)$$

where $\chi^2_2(\alpha)$ is the upper α point of the χ^2 distribution with 2 d.f. This confidence region is shown in Figure 2.1 for $\alpha = 0.05$ $(\chi^2_2(\alpha) = 5.991)$.

A $(1 - \alpha)$-level confidence region for (γ_1, γ_2) based on (2.7) is given by

$$\{\gamma_1 \in [\hat{\gamma}_1 \pm Z^{(\alpha')}]\} \times \{\gamma_2 \in [\hat{\gamma}_2 \pm 2Z^{(\alpha')}]\} \qquad (2.25)$$

where $Z^{(\alpha')}$ is the upper $\alpha' = (1 - \sqrt{1 - \alpha})/2$ point of the standard

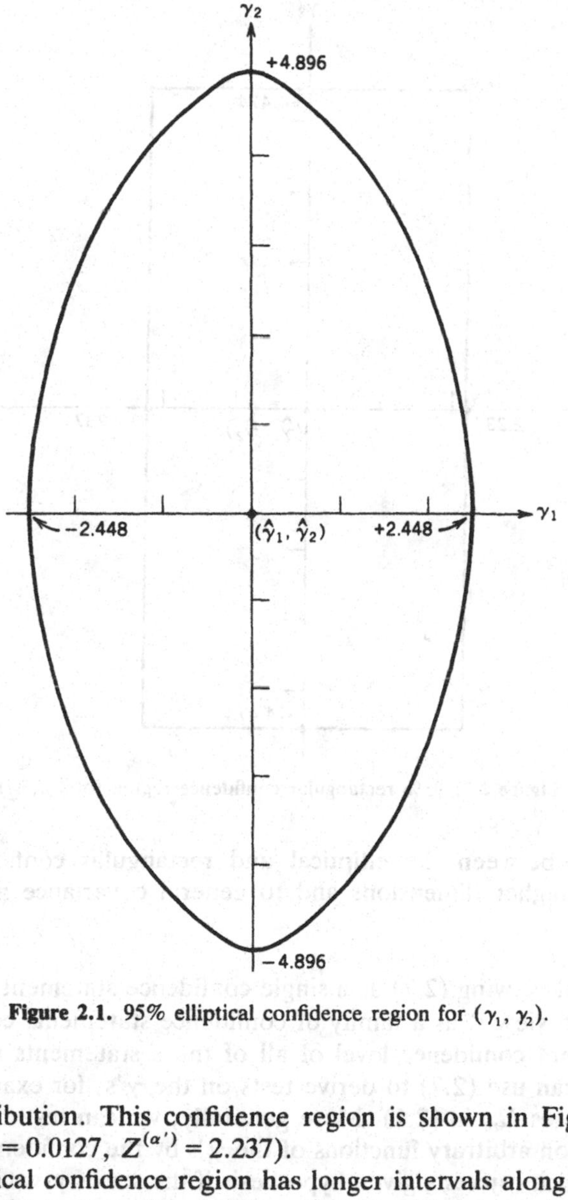

Figure 2.1. 95% elliptical confidence region for (γ_1, γ_2).

normal distribution. This confidence region is shown in Figure 2.2 for $\alpha = 0.05$ ($\alpha' = 0.0127$, $Z^{(\alpha')} = 2.237$).

The elliptical confidence region has longer intervals along both γ_1 and γ_2 axes than the rectangular confidence region. However, the area of the former is 37.56 while that of the latter is 40.03. Also, the maximum dimension of the former is along its major axis (γ_2-axis), which is 9.792, while that of the latter is along its diagonal, which is 10.004. This type of

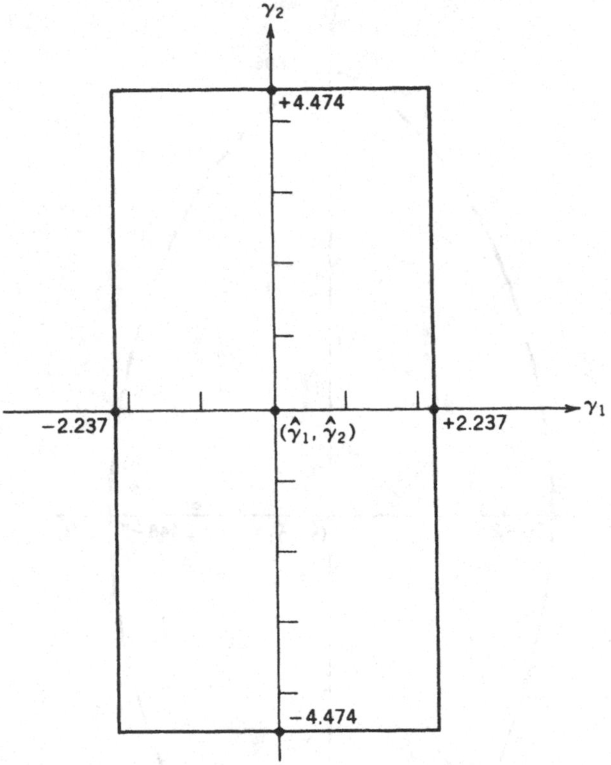

Figure 2.2. 95% rectangular confidence region for (γ_1, γ_2).

comparison between the elliptical and rectangular confidence regions extends to higher dimensions and to general covariance matrices of $\hat{\boldsymbol{\gamma}}$. □

Instead of viewing (2.7) as a single confidence statement concerning $\boldsymbol{\gamma}$ one can also view it as a family of confidence statements concerning the γ_i's. The joint confidence level of all of these statements is also clearly $1 - \alpha$. We can use (2.7) to derive tests on the γ_i's, for example, tests of the hypotheses H_{0i} in (2.1). More generally, we can test any number of hypotheses on arbitrary functions of the γ_i's by the confidence-region test method, which rejects a given hypothesis if the set of $\boldsymbol{\gamma}$-values postulated by it has an empty intersection with (2.7). In Theorem 3.3 it is shown that the FWE for all such inferences is controlled at α.

Example 2.5. We now consider a linear combination $\mathbf{a}'\boldsymbol{\gamma}$ where $\mathbf{a} = (1/\sqrt{5}, 2/\sqrt{5})'$. Projection of the elliptical confidence region (2.24) on a gives the interval $[(1/\sqrt{5})\hat{\gamma}_1 + (2/\sqrt{5})\hat{\gamma}_2 \pm 4.513]$ using (2.22). Projection of the rectangular confidence region (2.25) on **a** gives the interval $[(1/\sqrt{5})\hat{\gamma}_1 + (2/\sqrt{5})\hat{\gamma}_2 \pm ((1/\sqrt{5}) \times 2.237 + (2/\sqrt{5}) \times 4.474)] = [(1/\sqrt{5})\hat{\gamma}_1 + (2/\sqrt{5})\hat{\gamma}_2 \pm 5.002]$ using (2.23). Alternatively we may note that the vector **a** is along a diagonal of the rectangle (2.25) and therefore the length of the projection equals the length of the diagonal, which is 2×5.002. □

This example illustrates how different regions might be preferred for obtaining confidence intervals for different types of parametric functions.

2.3 Directional Decisions

In some practical applications it is desired to make decisions on the signs of parameters of interest. Such decisions are referred to as *directional decisions*. In Section 2.3.1 we consider this problem for the case of a single parameter and in Section 2.3.2 for the case of multiple parameters.

Directional decision problems differ from the inference problems considered thus far in this chapter in one important respect, namely, they require consideration of the so-called Type III errors (Kaiser 1960), which are errors of misclassifications of signs of nonnull parameter values. One approach in these problems is to control the Type III FWE, that is, the probability of any misclassification of signs. This approach is satisfactory if null parameter values are *a priori* ruled out or if protection against Type I errors is not required (as would be the case if there is very little loss associated with Type I errors, e.g., when comparing the mean effectiveness of two medical treatments that are about equal on other counts such as side effects and costs). In other cases it would be desirable to control the probability of making any Type I or Type III errors (Type I and Type III FWE).

If only two decisions are allowed for each parametric sign (i.e., the sign is positive or negative), then it is clear that as each parameter value approaches zero, the Type III FWE will be at least $1/2$ and thus cannot be controlled at $\alpha < 1/2$. Therefore we need a third decision of "inconclusive data" (Bohrer 1979) for each parameter. The MCPs discussed below have this three-decision feature.

2.3.1 A Single Parameter

Let us consider a single parameter γ whose sign is to be decided based on its estimate $\hat{\gamma}$ (the notation here is the same as in Section 1.3). The following two procedures may be used for this purpose.

(i) First test $H_0: \gamma = 0$ using the two-sided test (1.9) at level α. If H_0 is rejected, classify the sign of γ according to the sign of $\hat{\gamma}$. Otherwise decide that the data are inconclusive.

(ii) First observe the sign of $\hat{\gamma}$. If $\hat{\gamma} > 0$, test $H_0^{(-)}: \gamma \leqq 0$ (and vice versa) using the one-sided test (1.8) at level α. If $H_0^{(-)}$ is rejected, then decide that $\gamma > 0$. Otherwise decide that the data are inconclusive.

Both of these procedures have the same general form: Decide that $\gamma > 0$ or $\gamma < 0$, depending on whether $\hat{\gamma} > \xi S_{\hat{\gamma}}$ or $\hat{\gamma} < -\xi S_{\hat{\gamma}}$, otherwise (if $|\hat{\gamma}| \leqq \xi S_{\hat{\gamma}}$) decide that the data are inconclusive; here $\xi = T_\nu^{(\alpha/2)}$ for the first procedure and $\xi = T_\nu^{(\alpha)}$ for the second. They can be viewed as UI procedures for the family of two hypotheses testing problems: $H_0^{(-)}: \gamma \leqq 0$ versus $H_1^{(+)}: \gamma > 0$ and $H_0^{(+)}$: $\gamma \geqq 0$ versus $H_1^{(-)}: \gamma < 0$.

Using the UI nature of both the procedures it is easy to show that the maximum of

$$\Pr\{H_0^{(+)} \text{ or } H_0^{(-)} \text{ is falsely rejected}\} \tag{2.26}$$

is attained under $H_0 = H_0^{(+)} \cap H_0^{(-)}$: $\gamma = 0$. This maximum is the probability of a Type I error and equals α for the first procedure and 2α for the second procedure. If only $H_0^{(-)}$ (say) is true but $\gamma \neq 0$, then (2.26) is the probability of a Type III error and is given by $\Pr\{\hat{\gamma} > \xi S_{\hat{\gamma}} | \gamma < 0\}$. This probability is an increasing function of γ approaching the limit $\alpha/2$ for the first procedure and α for the second procedure as γ increases to zero. Thus the first procedure controls both the Type I and Type III error probabilities at level α while the second procedure controls only the latter.

For the problem of classifying the sign of a single parameter, Neyman (1935a,b) and Bahadur (1952–53) have shown the optimality of the procedures stated above, which are based on the Student t-test.

2.3.2 Multiple Parameters

Let $\gamma_i = \mathbf{l}_i'\boldsymbol{\theta}$, $i \in I$, be a collection of parametric functions on which directional decisions are desired; here I is an arbitrary index set. We first consider a generalization of the first procedure of the preceding section. This generalization involves augmenting a single-step test procedure for the family of two-sided testing problems:

$$H_{0i}: \gamma_i = 0 \qquad \text{versus} \qquad H_{1i}: \gamma_i \neq 0, \ i \in I \tag{2.27}$$

with directional decisions only on those γ_i's ($\gamma_i > 0$ or < 0 depending on whether $\hat{\gamma}_i > 0$ or <0) for which the corresponding H_{0i}'s are rejected. As in the preceding section we now show that such a multiple three-decision procedure controls the Type I and Type III FWE if the test procedure controls the Type I FWE.

Theorem 2.2. Consider a single-step test procedure that rejects H_{0i} of (2.27) if

$$|T_i| = \frac{|\hat{\gamma}_i|}{S_{\hat{\gamma}_i}} > \xi_i\,, \qquad i \in I \tag{2.28}$$

where $S_{\hat{\gamma}_i} = S\sqrt{\mathbf{l}_i'\mathbf{V}\mathbf{l}_i}$, and the critical constants $\xi_i > 0$ are chosen so that the maximum Type I FWE $= \alpha$. Then the multiple three-decision procedure that decides that $\gamma_i > 0$ or <0 according as $\hat{\gamma}_i > \xi_i S_{\hat{\gamma}_i}$ or $< -\xi_i S_{\hat{\gamma}_i}$ and makes no directional decision about γ_i if $|\hat{\gamma}_i| \leqq \xi_i S_{\hat{\gamma}_i}$, $i \in I$, has

$$\text{Type I and Type III FWE} = \Pr\{\text{any Type I or Type III error}\} \leqq \alpha \tag{2.29}$$

for all values of the γ_i's.

Proof. It is readily shown that the maximum Type I FWE of the procedure (2.28) is attained under $H_0 = \bigcap_{i \in I} H_{0i}$, and thus

$$\Pr_{H_0}\{|T_i| > \xi_i \text{ for some } i \in I\} = \alpha\,.$$

From this probability statement we obtain the following joint confidence statement concerning the γ_i's:

$$\Pr\{\gamma_i \in [\hat{\gamma}_i \pm \xi_i S\sqrt{\mathbf{l}_i'\mathbf{V}\mathbf{l}_i}]\ \forall\, i \in I\} = 1 - \alpha\,. \tag{2.30}$$

From Theorem 3.3 it follows that we can test any hypotheses on the γ_i's using this confidence region and the Type I FWE $\leqq \alpha$ for all such tests. In particular, we can test one-sided hypotheses $H_{0i}^{(-)}: \gamma_i \leqq 0$ and $H_{0i}^{(+)}: \gamma_i \geqq 0$. Using (2.30) we see that $H_{0i}^{(-)}$ (respectively, $H_{0i}^{(+)}$) is rejected when $\hat{\gamma}_i > \xi_i S_{\hat{\gamma}_i}$ (respectively, $< -\xi_i S_{\hat{\gamma}_i}$) and this rejection implies the decision $\gamma_i > 0$ (respectively, <0), $i \in I$. The Type I FWE for all such tests is the same as the Type III FWE for all directional decisions and hence the conclusion of the theorem follows. □

An alternative proof of the theorem can be given by noting that the directional decision procedure given above can be looked upon as a UI procedure derived by representing $H_0 = \bigcap_{i \in I} H_{0i} = \bigcap_{i \in I} (H_{0i}^{(+)} \cap H_{0i}^{(-)})$.

We next consider Bohrer's (1979) generalization of the second procedure of the preceding section. This generalization leads to a multiple three-decision procedure that controls only the Type III FWE; no protection is provided against false rejections of the hypotheses $H_{0i}: \gamma_i = 0$. Bohrer restricted attention to the special case of a finite family of parametric functions $\{\gamma_i = \mathbf{l}_i'\boldsymbol{\theta}\ (1 \leqq i \leqq m)\}$ and a common critical constant $\xi > 0$. The following theorem shows how ξ is determined to control the Type III FWE.

Theorem 2.3 (Bohrer 1979). Consider a multiple three-decision procedure that decides that $\gamma_i > 0$ or <0 according as $\hat{\gamma}_i > \xi S_{\hat{\gamma}_i}$ or $< -\xi S_{\hat{\gamma}_i}$ and makes no directional decision about γ_i if $|\hat{\gamma}_i| \leqq \xi S_{\hat{\gamma}_i}$ $(1 \leqq i \leqq m)$. This procedure has

$$\text{Type III FWE} = \Pr\{\text{at least one sign is misclassified}\} \leqq \alpha \quad (2.31)$$

for all values of the γ_i's if ξ is chosen so that

$$\min \Pr\{\pm T_i \leqq \xi\ (1 \leqq i \leqq m)\} = 1 - \alpha \quad (2.32)$$

where the minimum is taken over all 2^m combinations of the signs of the T_i's and where $T_1, T_2, \ldots, T_m$ have an m-variate t-distribution with ν d.f. and the associated correlation matrix $\{\rho_{ij}\}$, the ρ_{ij}'s being given by (2.6).

Proof. We have

$$\begin{aligned}
&\Pr\{\text{no sign is misclassified}\}\\
&\quad = \Pr\{(-\text{sgn}\ \gamma_i)\hat{\gamma}_i \leqq \xi S\sqrt{\mathbf{l}_i'\mathbf{V}\mathbf{l}_i}\ (1 \leqq i \leqq m)\}\\
&\quad = \Pr\{(-\text{sgn}\ \gamma_i)(\hat{\gamma}_i - \gamma_i) \leqq \xi S\sqrt{\mathbf{l}_i'\mathbf{V}\mathbf{l}_i} + |\gamma_i|\ (1 \leqq i \leqq m)\}\\
&\quad \geqq \Pr\{(-\text{sgn}\ \gamma_i)T_i \leqq \xi\ (1 \leqq i \leqq m)\} \qquad (2.33)
\end{aligned}$$

with equality attained if and only if $\gamma_i \to 0$ for $1 \leqq i \leqq m$. Here the T_i's are given by (2.9) and they have the joint distribution stated in the theorem.

To satisfy (2.31) ξ must be chosen so that (2.33) equals $1 - \alpha$. But the signs of the γ_i's are unknown. If the minimum of (2.33) over all 2^m possible combinations of the signs of the T_i's equals $1 - \alpha$, then it follows that for any true combination of the signs of the γ_i's, (2.31) will be satisfied. □

The solution ξ to (2.32) equals the maximum of the critical points $T^{(\alpha)}_{m,\nu,\{\pm\rho_{ij}\}}$ over all 2^m correlation matrices $\{\pm\rho_{ij}\}$ where $\{\rho_{ij}\}$ is a fixed correlation matrix given by (2.6). When $\rho_{ij}=0$ for all $i \neq j$, the desired ξ is simply $T^{(\alpha)}_{m,\nu,\{0\}}$, which is the upper α point of the Studentized (with ν d.f.) maximum of m independent standard normals. We refer to this r.v. as the Studentized maximum r.v. with parameter m and d.f. ν and denote it by $M_{m,\nu}$; its upper α point is denoted by $M^{(\alpha)}_{m,\nu} = T^{(\alpha)}_{m,\nu,\{0\}}$ (see Table 6, Appendix 3).

If the correlation matrix $\{\rho_{ij}\}$ is arbitrary, then an exact evaluation of ξ from (2.32) is difficult. In such cases by using the Bonferroni inequality one can obtain a conservative upper bound on ξ, which is $T^{(\alpha/m)}_{\nu}$. Bohrer et al. (1981) studied several conservative approximations to the critical constant ξ and found that this Bonferroni approximation is generally the best choice.

Recently Bofinger (1985) has proposed an MCP for ordering all pairs of treatment means (θ_i, θ_j) in a balanced one-way layout that controls the Type III FWE (the probability that any pair (θ_i, θ_j) is incorrectly ordered) at a designated level α. According to Theorem 2.2, the T-procedure can be used for ordering the treatment means (declare $\theta_i > \theta_j$ or $<\theta_j$ depending on whether $\bar{Y}_i - \bar{Y}_j > Q^{(\alpha)}_{k,\nu}S/\sqrt{n}$ or $< Q^{(\alpha)}_{k,\nu}S/\sqrt{n}$, and do not order (θ_i, θ_j) if $|\bar{Y}_i - \bar{Y}_j| \leq Q^{(\alpha)}_{k,\nu}S/\sqrt{n})$ with simultaneous control of the Type I and Type III FWE. Bofinger's procedure has the same multiple three-decision structure, but it uses a smaller critical constant since it is required to control only the Type III FWE; as a result, it is more powerful. The exact critical constant for Bofinger's procedure is difficult to evaluate for $k>3$. She has provided selected exact values for $k=2$ and 3 and upper bounds for $k>3$.

Optimality and admissibility questions concerning directional decision procedures have been studied by Bohrer and Schervish (1980), Bohrer (1982), Hochberg (1987) and Hochberg and Posner (1986).

3 SINGLE-STEP PROCEDURES FOR HIERARCHICAL FAMILIES

When comparing a given set of treatments, it is sometimes of interest to see whether some subsets can be regarded as homogeneous. Thus it is of interest to test hypotheses of the type $H_P: \theta_{i_1} = \theta_{i_2} = \cdots = \theta_{i_p}$ where $P = \{i_1, i_2, \ldots, i_p\} \subseteq \{1, 2, \ldots, k\}$ is of cardinality p $(2 \leq p \leq k)$. The hypothesis H_P can be expressed as $\boldsymbol{\gamma}_P = \mathbf{0}$ where $\boldsymbol{\gamma}_P: (p-1) \times 1$ is a vector of any linearly independent contrasts among the θ_i's, $i \in P$ and $\mathbf{0}$

denotes a $(p-1) \times 1$ null vector. The family of all hypotheses H_P is referred to as the *family of subset* (*homogeneity*) *hypotheses*. In this family, a hypothesis H_Q implies another hypothesis H_P if $P \subset Q$. Families with such *implication relations* are referred to as *hierarchical families*. Note that if H_Q implies H_P, then $\boldsymbol{\gamma}_P = \mathbf{A}_{PQ}\boldsymbol{\gamma}_Q$ for some $(p-1) \times (q-1)$ matrix $\mathbf{A}_{PQ}$. This motivates the following setting for hierarchical families of inferences on vector parametric functions; a more general (and simpler) description of hierarchical families is given in Appendix 1.

3.1 Hierarchical Families

Consider a family of vector parametric functions $\boldsymbol{\gamma}_i = \mathbf{L}_i\boldsymbol{\theta}$, $i \in I$, where $\mathbf{L}_i$ is an $m_i \times k$ full row rank matrix and I is an arbitrary index set. Some of the $\boldsymbol{\gamma}_i$'s may be scalar (i.e., $m_i = 1$) but not all. To avoid trivialities further assume that for any two matrices $\mathbf{L}_i$ and $\mathbf{L}_j$ with $m_i = m_j$ we do not have $\mathbf{L}_i = \mathbf{A}_{ij}\mathbf{L}_j$ for some nonsingular matrix $\mathbf{A}_{ij}$; that is, $\boldsymbol{\gamma}_i$ and $\boldsymbol{\gamma}_j$ are not nonsingular linear transformations of each other. However, if $m_i < m_j$, then there may exist a matrix $\mathbf{A}_{ij} : m_i \times m_j$ such that $\mathbf{L}_i = \mathbf{A}_{ij}\mathbf{L}_j$, that is, a $\boldsymbol{\gamma}_i$ of lower dimension may be obtained by linearly transforming a $\boldsymbol{\gamma}_j$ of higher dimension. In this case we say that $\boldsymbol{\gamma}_i$ is a *linear function* of $\boldsymbol{\gamma}_j$.

Consider a family of hypotheses $H_i : \boldsymbol{\gamma}_i \in \Gamma_i$, $i \in I$, where the Γ_i's are specified subsets of $\mathbb{R}^{m_i}$. Suppose that for some pair $(\boldsymbol{\gamma}_i, \boldsymbol{\gamma}_j)$, $\boldsymbol{\gamma}_i$ is a linear function of $\boldsymbol{\gamma}_j$ with $\boldsymbol{\gamma}_i = \mathbf{A}_{ij}\boldsymbol{\gamma}_j$. If

$$\mathbf{A}_{ij}(\Gamma_j) = \{\mathbf{A}_{ij}\boldsymbol{\gamma} : \boldsymbol{\gamma} \in \Gamma_j\} \subseteq \Gamma_i , \tag{3.1}$$

then we say that H_j implies H_i. In other words, the parameter values postulated by H_j form a subset of the parameter values postulated by H_i. A family of hypotheses is said to be *hierarchical* if an implication relation holds between at least two hypotheses.

If a hypothesis H_j implies H_i, then H_i is said to be a *component* (Gabriel 1969) of H_j. A hypothesis with no components is called *minimal*; all other hypotheses are called *nonminimal* (Gabriel 1969). For example, in the family of subset hypotheses, the pairwise hypotheses $H_{ij} : \theta_i = \theta_j$ $(1 \leqq i < j \leqq k)$ are minimal while all other subset hypotheses are nonminimal. All subset hypotheses are implied by the overall null hypothesis $H_0 : \theta_1 = \theta_2 = \cdots = \theta_k$ and are thus components of it.

3.2 Coherence and Consonance

Any MCP for a hierarchical family of hypotheses $\{H_i, i \in I\}$ is generally required to possess the following logical consistency property: For any pair of hypotheses (H_i, H_j) such that H_j implies H_i, if H_j is not rejected then H_i is also not rejected. This requirement is called *coherence* and was introduced by Gabriel (1969). The same requirement under the name *compatibility* was introduced earlier by Lehmann (1957b). An MCP that satisfies this requirement is called *coherent*. A coherent MCP avoids the inconsistency of rejecting a hypothesis without also rejecting all hypotheses implying it.

The coherence requirement for simultaneous confidence procedures can be stated in an analogous manner as follows: An MCP producing confidence regions $\Gamma_i(\mathbf{Y})$ for $\boldsymbol{\gamma}_i$, $i \in I$, is said to be *coherent* if for any pair $(\boldsymbol{\gamma}_i, \boldsymbol{\gamma}_j)$ such that $\boldsymbol{\gamma}_i = \mathbf{A}_{ij}\boldsymbol{\gamma}_j$, we have

$$\{\mathbf{A}_{ij}\boldsymbol{\gamma}_j : \boldsymbol{\gamma}_j \in \Gamma_j(\mathbf{Y})\} \subseteq \Gamma_i(\mathbf{Y}) . \tag{3.2}$$

In other words, any $\boldsymbol{\gamma}_j$ in the confidence set $\Gamma_j(\mathbf{Y})$ when linearly transformed using the transformation matrix $\mathbf{A}_{ij}$ gives a $\boldsymbol{\gamma}_i$ that is admitted by the confidence set $\Gamma_i(\mathbf{Y})$.

Example 3.1. Consider a family of three parametric functions γ_1, γ_2, and $\boldsymbol{\gamma} = (\gamma_1, \gamma_2)'$. Suppose that the $\hat{\gamma}_i$'s are independently distributed as $N(\gamma_i, 1)$. As in Example 2.4, a $(1-\alpha)$-level confidence region for (γ_1, γ_2) is given by

$$|\hat{\gamma}_i - \gamma_i| \leqq Z^{(\alpha')} , \qquad i = 1, 2 \tag{3.3}$$

where $\alpha' = (1 - \sqrt{1-\alpha})/2$. A procedure that uses the projection of (3.3) on the corresponding axis as the confidence interval for each γ_i is clearly coherent according to (3.2) since the values of the γ_i's $(i = 1, 2)$ admitted by their individual confidence intervals are exactly the ones admitted by the confidence region (3.3).

A related simultaneous hypotheses testing problem involves testing $H_1: \gamma_1 = 0$, $H_2: \gamma_2 = 0$, and their intersection $H_0: \gamma_1 = 0,\ \gamma_2 = 0$ against two-sided alternatives. A single-step test procedure can be based on (3.3). This test procedure, which rejects H_0 if $\max\{|\hat{\gamma}_1|, |\hat{\gamma}_2|\} > Z^{(\alpha')}$ and which rejects H_i if $|\hat{\gamma}_i| > Z^{(\alpha')}$ $(i = 1, 2)$, is also coherent.

Now consider a simultaneous confidence procedure for the same problem, which gives

$$(\hat{\gamma}_1 - \gamma_1)^2 + (\hat{\gamma}_2 - \gamma_2)^2 \leqq \chi_2^2(\alpha) , \tag{3.4}$$

as a $(1-\alpha)$-level confidence region for (γ_1, γ_2) and

$$|\hat{\gamma}_i - \gamma_i| \leqq Z^{(\alpha/2)} \tag{3.5}$$

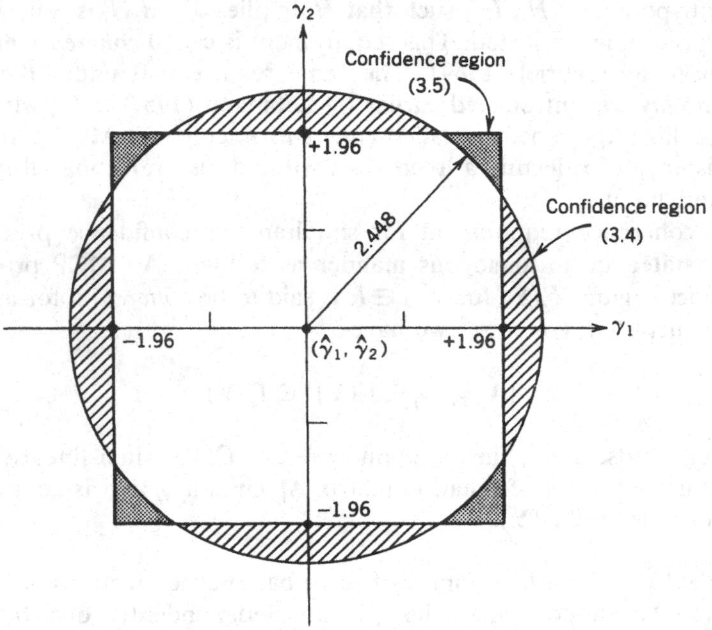

Figure 3.1. Confidence regions (3.4) and (3.5) for $1-\alpha=0.95$.

as a $(1-\alpha)$-level confidence interval for γ_i $(i=1,2)$. Note that $\sqrt{\chi_2^2(\alpha)} > Z^{(\alpha/2)}$ and thus condition (3.2) is violated. Therefore this procedure is not coherent. The confidence regions (3.4) and (3.5) are displayed in Figure 3.1 for $\alpha=0.05$ ($\sqrt{\chi_2^2(\alpha)}=2.448$, $Z^{(\alpha/2)}=1.96$).

If we base tests of $\boldsymbol{\gamma}$ and γ_i $(i=1,2)$ on the respective confidence regions (3.4) and (3.5), then it is easy to see that the resulting test procedure is also not coherent. For example, consider any $\boldsymbol{\gamma}_0=(\gamma_{01},\gamma_{02})'$ in the hatched region. Clearly, $H_0:\boldsymbol{\gamma}=\boldsymbol{\gamma}_0$ will be accepted but its component hypotheses $H_1:\gamma_1=\gamma_{01}$ and $H_2:\gamma_2=\gamma_{02}$ will be rejected. □

We now turn to another desirable property for MCPs. For a hierarchical family of hypotheses $\{H_i, i\in I\}$, *consonance* (Gabriel 1969) refers to the property that whenever any nonminimal H_i is rejected, at least one of its components is also rejected. An MCP that has this property is called *consonant*.

It should be noted that nonconsonance does not imply logical contradictions as noncoherence does. This is because the failure to reject a hypothesis is not usually interpreted as its acceptance. Also sometimes the failure to reject a component of H_i may be due to noninclusion of enough components of H_i in the family. Thus whereas coherence is an essential requirement, consonance is only a desirable property.

Example 3.1 ***(Continued).*** We show that the noncoherent MCP depicted in Figure 3.1 is also nonconsonant. Consider any $\boldsymbol{\gamma}_0 = (\gamma_{01}, \gamma_{02})'$ in the shaded region. Then clearly, $H_0 : \boldsymbol{\gamma} = \boldsymbol{\gamma}_0$ will be rejected since $\boldsymbol{\gamma}_0$ falls outside the confidence region (3.4), but its components $H_i : \gamma_i = \gamma_{0i}$ will not be rejected because γ_{0i} falls in the confidence interval (3.5) for γ_i $(i = 1, 2)$. □

3.3 Simultaneous Test and Confidence Procedures

Gabriel (1969) gave a theory of a special class of single-step test procedures for hierarchical families of hypotheses, which he referred to as *simultaneous test procedures*.* We first discuss this theory in Section 3.3.1. (The extensions and limitations of the theory for more general models and testing problems than those considered in Part I are discussed in Section 1 of Appendix 1.) In Section 3.3.2 we discuss the confidence analogs of simultaneous test procedures. We also discuss the confidence-region test method of Aitchison (1964), which can be used to simultaneously test any number of additional hypotheses.

3.1.1 Simultaneous Test Procedures

Let $\{H_i : \gamma_i \in \Gamma_i, i \in I\}$ be a hierarchical family of hypotheses. A simultaneous test procedure for this family is characterized by a collection of test statistics Z_i, $i \in I$, and a common critical constant ξ such that the procedure rejects H_i if $Z_i > \xi$, $i \in I$. Usually we require the test statistics Z_i to be *monotone*, that is, $Z_j \geqq Z_i$ with probability 1 whenever H_j implies H_i. The following theorem makes clear why this is necessary.

Theorem 3.1 **(Gabriel 1969).** The simultaneous test procedure stated above is coherent for any choice of the critical constant ξ if and only if the test statistics Z_i are monotone.

Proof. If the test statistics are monotone, then it is clear that if H_j implies H_i and if $Z_i > \xi$, then with probability 1, $Z_j > \xi$ and hence H_j is rejected. If the test statistics are not monotone, then for some value of ξ we can have $Z_j < \xi < Z_i$ with positive probability. Hence the implied

* The abbreviation STP is often used in the literature for a simultaneous test procedure. We do not use this abbreviation because it can be confused with stepwise test procedures discussed later in this chapter and in other parts of the book.

hypothesis H_i is rejected but not the implying hypothesis H_j, and thus the procedure is not coherent. □

From now on we assume that the test statistics satisfy the monotonicity property. Note that because of this property a simultaneous test procedure can test all hypotheses in a given hierarchical family in a single step; no stepwise ordering of the tests is required to ensure coherence. The problem of determining the common critical constant ξ is addressed in Section 3.3.2.

We now discuss two standard methods of obtaining monotone test statistics.

3.3.1.1 Likelihood Ratio Statistics. An LR test statistic Z_i for H_i is a monotone decreasing function of

$$\lambda_i = \frac{\sup_{H_i} L}{\sup L}$$

where L is the likelihood function of the data, the numerator supremum is over the part of the parameter space postulated by H_i, while the denominator supremum is unrestricted, that is, over the entire parameter space. It is readily seen (Gabriel 1969, Lemma A) that the LR test statistics $\{Z_i, i \in I\}$ are monotone. Hence by Theorem 3.1, any simultaneous test procedure based on them (i.e., a procedure that rejects H_i if $Z_i > \xi$ for some common critical constant ξ) is coherent. However, such a procedure is not necessarily consonant. We see in Section 3.3.1.2 that a simultaneous test procedure must be based on UI statistics in order to be both coherent and consonant.

Example 3.2. Consider the hierarchical family of subset hypotheses in the context of the one-way layout design of Example 1.1. For any subset $P = \{i_1, i_2, \ldots, i_p\} \subseteq \{1, 2, \ldots, k\}$ of size $p \geqq 2$, the simultaneous test procedure based on LR statistics rejects the hypothesis $H_P: \theta_{i_1} = \cdots = \theta_{i_p}$ if

$$Z_P = \frac{\sum_{i \in P} n_i(\bar{Y}_i - \bar{\bar{Y}}_P)^2}{S^2} > \xi \tag{3.6}$$

where $\bar{\bar{Y}}_P = \Sigma_{i \in P} n_i \bar{Y}_i / \Sigma_{i \in P} n_i$. We see in the sequel that the common critical constant ξ must be chosen equal to $(k-1)F^{(\alpha)}_{k-1,\nu}$ in order to control the Type I FWE at a designated level α. Furthermore, this procedure is implied by Scheffé's S-procedure (2.20) where $\mathscr{L}$ is taken to be the space $\mathbb{C}^k$ of all k-dimensional contrasts.

This procedure is coherent but not consonant. For example, the test of the overall null hypothesis $H_0: \theta_1 = \cdots = \theta_k$ may be significant according to (3.6), but this does not imply that there is at least some subset of the treatments for which the homogeneity hypothesis will be rejected using (3.6). □

3.3.1.2 Union-Intersection Statistics. A general presentation of the UI method for constructing a single-step test procedure can be given as follows: Let $I_{\min}$ be the index set of all minimal hypotheses in $\{H_i, i \in I\}$ and let $I_{\min}^{(j)}$ be the index set of the minimal components of a nonminimal H_j. We assume that every nonminimal H_j can be expressed as

$$H_j = \bigcap_{i \in I_{\min}^{(j)}} H_i, \qquad j \in I - I_{\min}.$$

Let $\phi_i(\mathbf{Y})$ be the indicator test function of H_i when data vector $\mathbf{Y}$ is observed, that is, $\phi_i(\mathbf{Y}) = 1$ if H_i is rejected and 0 if H_i is not rejected. Given the indicator test functions $\phi_i(\mathbf{Y})$ for all $i \in I_{\min}$, the UI test of any nonminimal H_j is given by

$$\phi_j(\mathbf{Y}) = \max_{i \in I_{\min}^{(j)}} \{\phi_i(\mathbf{Y})\} = 1 - \prod_{i \in I_{\min}^{(j)}} \{1 - \phi_i(\mathbf{Y})\}. \tag{3.7}$$

The collection of test functions $\{\phi_i(\mathbf{Y}), i \in I\}$ obtained using (3.7) is referred to as a *UI procedure*. Note that this is a single-step test procedure but not necessarily a simultaneous test procedure.

A UI procedure is derived by first constructing tests for H_i, $i \in I_{\min}$, and then obtaining tests of nonminimal H_j by using (3.7). If the critical region of each H_i for $i \in I_{\min}$ is of the form $(Z_i > \xi)$, then from (3.7) we see that the test statistic for any nonminimal H_j will be $Z_j = \max_{i \in I_{\min}^{(j)}} Z_i$ and its critical region will be $(Z_j > \xi)$. The Z_j's obtained in this manner are referred to as *UI statistics*. They are monotone because if H_j implies H_i, then $I_{\min}^{(i)} \subseteq I_{\min}^{(j)}$ and therefore $Z_j = \max_{l \in I_{\min}^{(j)}} Z_l \geq Z_i = \max_{l \in I_{\min}^{(i)}} Z_l$. This implies coherence. The following theorem shows that a coherent and consonant single-step test procedure is equivalent to a UI procedure.

Theorem 3.2 (Gabriel 1969). For a hierarchical family $\{H_i, i \in I\}$, a single-step test procedure is coherent and consonant if and only if it is a UI procedure.

Proof. We have coherence and consonance if and only if for every nonminimal H_j,

$$\phi_j(\mathbf{Y}) = 1 \Leftrightarrow \phi_i(\mathbf{Y}) = 1 \text{ for some } i \in I_{\min}^{(j)} .$$

But this condition is satisfied if and only if (3.7) holds, which is the defining property of a UI procedure. □

Example 3.3. Consider the setting of Example 3.2 but for the balanced case ($n_1 = \cdots = n_k = n$). The minimal hypotheses in the family of subset hypotheses are pairwise null hypotheses $H_{ij}: \theta_i = \theta_j$ $(1 \leqq i < j \leqq k)$. Suppose for testing each H_{ij} we use the t-test, which rejects H_{ij} if

$$\sqrt{2}|T_{ij}| = \frac{|\bar{Y}_i - \bar{Y}_j|\sqrt{n}}{S} > \xi .$$

The UI procedure then rejects any nonminimal hypothesis $H_P: \theta_{i_1} = \cdots = \theta_{i_p}$ if

$$Z_P = \max_{i,j \in P} \sqrt{2}|T_{ij}| = \frac{\max_{i,j \in P} |\bar{Y}_i - \bar{Y}_j|\sqrt{n}}{S} > \xi .$$

By Theorem 3.2 this procedure is coherent and consonant. We see in the sequel that the common critical constant ξ must be chosen equal to $Q_{k,\nu}^{(\alpha)}$ in order to control the Type I FWE at a designated level α. Furthermore, this procedure is implied by Tukey's T-procedure (2.8). □

3.3.2 *Simultaneous Confidence Procedures and Confidence-Region Tests*

Let $\{Z_i(\gamma_i), i \in I\}$ be a collection of scalar pivotal r.v.'s for the family of parametric functions $\{\gamma_i = \mathbf{L}_i\boldsymbol{\theta}, i \in I\}$, that is, the joint distribution of the $Z_i(\gamma_i)$'s, $i \in I$, is free of unknown parameters $\boldsymbol{\theta}$ (and hence of the γ_i's) and σ^2. In analogy with simultaneous test procedures, a simultaneous confidence procedure for the γ_i's, $i \in I$, can be based on a common critical constant ξ. The associated $(1-\alpha)$-level simultaneous confidence regions for the γ_i's are given by

$$\{\gamma_i : Z_i(\gamma_i) \leqq \xi\} , \qquad i \in I , \tag{3.8}$$

if ξ is chosen to be the upper α point of the distribution of $\max_{i \in I} Z_i(\gamma_i)$, that is, if

$$\Pr\{Z_i(\gamma_i) \leqq \xi \ \forall\, i \in I\} = 1 - \alpha . \tag{3.9}$$

In analogy with Theorem 3.1, this simultaneous confidence procedure will be coherent if and only if the pivotal r.v.'s $Z_i(\gamma_i)$ are monotone, that is, if $Z_j(\gamma_j) \geqq Z_i(\gamma_i)$ with probability 1 whenever γ_i is a linear function of γ_j.

The intersection of regions (3.8) gives a $(1-\alpha)$-level confidence region $\Theta(\mathbf{Y})$ (say) for $\boldsymbol{\theta}$. We can test arbitrary hypotheses $H_i : \gamma_i \in \Gamma_i$ by the *confidence-region test method* (Aitchison 1964), which

$$\text{rejects } H_i \text{ if } \Gamma_i \cap \Theta(\mathbf{Y}) = \phi , \qquad i \in I ; \tag{3.10}$$

here ϕ denotes an empty set. It is easy to see that this is equivalent to a simultaneous test procedure that

$$\text{rejects } H_i \text{ if } Z_i = \inf_{\gamma_i \in \Gamma_i} Z_i(\gamma_i) > \xi , \qquad i \in I. \tag{3.11}$$

Theorem 3.3. The simultaneous test procedure (3.10) (or equivalently (3.11)) controls the Type I FWE at level α if ξ is chosen to be the upper α point of the distribution of $\max_{i \in I} Z_i(\gamma_i)$.

Proof. Let $H_i, i \in \tilde{I}$, be the set of true null hypotheses where $\tilde{I} \subseteq I$. Without loss of generality we can take $\tilde{I}$ to be nonempty since there is no Type I error if $\tilde{I}$ is empty. Let γ_i^0 denote the true value of γ_i for $i \in I$; clearly $\gamma_i^0 \in \Gamma_i$ for $i \in \tilde{I}$. The Type I FWE of the simultaneous test procedure (3.11) is given by

$$\begin{aligned}
\Pr\{\Gamma_i \cap \Theta(\mathbf{Y}) = \phi \text{ for some } i \in \tilde{I}\} &= \Pr\{Z_i(\gamma_i) > \xi \ \ \forall \gamma_i \in \Gamma_i \text{ for some } i \in \tilde{I}\} \\
&\leqq \Pr\{Z_i(\gamma_i^0) > \xi \text{ for some } i \in \tilde{I}\} \\
&\leqq \Pr\{Z_i(\gamma_i^0) > \xi \text{ for some } i \in I\} \\
&= \alpha \text{ (from (3.9))} .
\end{aligned}$$

Since this inequality holds for all $\tilde{I} \subseteq I$, the theorem follows. □

Suppose that we are given the statistics Z_i (derived using the methods in Section 3.3.1, for instance) associated with the simultaneous test procedure that rejects $H_i : \gamma_i \in \Gamma_i$ if $Z_i > \xi, i \in I$. To determine the common critical constant ξ so as to control the FWE at a designated level α, we can apply the preceding theorem by identifying the pivotal r.v.'s $Z_i(\gamma_i)$ underlying the test statistics Z_i and then choosing ξ to be the upper α point of $\max_{i \in I} Z_i(\gamma_i)$. A more direct and equivalent approach is to choose ξ to be the upper α point of the distribution of $\max_{i \in I} Z_i$ under the overall hypothesis $H_0 = \bigcap_{i \in I} H_i$. Using this approach it can be readily verified that the appropriate critical constant ξ equals $(k-1)F_{k-1,\nu}^{(\alpha)}$ in Example 3.2 and $Q_{k,\nu}^{(\alpha)}$ in Example 3.3.

We now give two examples of simultaneous confidence procedures and confidence-region tests.

Example 3.4. Consider the collection of pivotal r.v.'s

$$Z_i(\gamma_i) = \frac{(\hat{\gamma}_i - \gamma_i)'(\mathbf{L}_i\mathbf{V}\mathbf{L}_i')^{-1}(\hat{\gamma}_i - \gamma_i)}{m_i S^2}, \qquad i \in I, \qquad (3.12)$$

which are correlated $m_i F_{m_i,\nu}$ r.v.'s (the so-called augmented F r.v.'s with m_i and ν d.f.). From Theorem 2.1 we see that $Z_i(\gamma_i)$ can be expressed as

$$Z_i(\gamma_i) = \sup_{\mathbf{l} \in \mathscr{L}_i} \left\{ \frac{[\mathbf{l}'(\hat{\boldsymbol{\theta}} - \boldsymbol{\theta})]^2}{S^2(\mathbf{l}'\mathbf{V}\mathbf{l})} \right\}$$

where $\mathscr{L}_i$ is the vector space of dimension m_i spanned by the rows of $\mathbf{L}_i$ (recall that $\gamma_i = \mathbf{L}_i\boldsymbol{\theta}$). Now if γ_i is a linear function of γ_j, then $\mathbf{L}_i = \mathbf{A}_{ij}\mathbf{L}_j$ and thus $\mathscr{L}_i$ is a subspace of $\mathscr{L}_j$. Therefore it follows that $Z_j(\gamma_j) \geqq Z_i(\gamma_i)$ with probability 1 and the pivotal r.v.'s (3.12) satisfy the monotonicity property. For testing the hypotheses $H_i: \gamma_i = \gamma_{0i}$ where the γ_{0i}'s are specified vectors, it is clear that the test statistics $Z_i = Z_i(\gamma_{0i})$ also satisfy the monotonicity property. □

Example 3.5. The previous example dealt with two-sided inferences. In the present example we consider a hierarchical family of one-sided inferences on $\gamma_1 = \mathbf{l}_1'\boldsymbol{\theta}$, $\gamma_2 = \mathbf{l}_2'\boldsymbol{\theta}$, and $\boldsymbol{\gamma} = (\gamma_1, \gamma_2)'$.

For γ_1 and γ_2 we can choose the pivotal r.v.'s

$$Z_i(\gamma_i) = \frac{\hat{\gamma}_i - \gamma_i}{S\sqrt{\mathbf{l}_i'\mathbf{V}\mathbf{l}_i}} \qquad (i = 1, 2).$$

If we choose $Z_0(\boldsymbol{\gamma}) = \max\{Z_1(\gamma_1), Z_2(\gamma_2)\}$ as the pivotal r.v. for $\boldsymbol{\gamma}$, then clearly the monotonicity condition is satisfied. The upper α point of the distribution of $\max\{Z_0(\boldsymbol{\gamma}), Z_1(\gamma_1), Z_2(\gamma_2)\} = Z_0(\boldsymbol{\gamma})$ is given by $T_{2,\nu,\rho}^{(\alpha)}$ where

$$\rho = \frac{\mathbf{l}_1'\mathbf{V}\mathbf{l}_2}{\sqrt{(\mathbf{l}_1'\mathbf{V}\mathbf{l}_1)(\mathbf{l}_2'\mathbf{V}\mathbf{l}_2)}}.$$

This gives the following simultaneous $(1-\alpha)$-level lower one-sided confidence regions for γ_1, γ_2, and $\boldsymbol{\gamma}$:

$$\gamma_i \geqq \hat{\gamma}_i - T_{2,\nu,\rho}^{(\alpha)} S\sqrt{\mathbf{l}_i'\mathbf{V}\mathbf{l}_i} \qquad (i = 1, 2). \qquad (3.13)$$

Suppose we want to test hypotheses $H_1: \gamma_1 \leqq 0$, $H_2: \gamma_2 \leqq 0$, and $H_0 = H_1 \cap H_2$. Then the confidence-region test based on (3.13) rejects H_i if

$$Z_i = \inf_{\gamma_i \leqq 0} Z_i(\gamma_i) = \frac{\hat{\gamma}_i - 0}{S\sqrt{\mathbf{l}_i'\mathbf{V}\mathbf{l}_i}} > T_{2,\nu,\rho}^{(\alpha)} \qquad (i = 1, 2),$$

and rejects H_0 if

$$Z_0 = \inf_{\gamma_1 \leqq 0, \gamma_2 \leqq 0} Z_0(\boldsymbol{\gamma}) = \max(Z_1, Z_2) > T_{2,\nu,\rho}^{(\alpha)}.$$

This simultaneous test procedure controls the Type I FWE at level α and is coherent. □

More generally, if we have a $(1-\alpha)$-level confidence region $\Theta(\mathbf{Y})$, then we can test any number of arbitrary hypotheses on $\boldsymbol{\theta}$ by the confidence-region test method. Moreover, using the projection method of Section 2.2 we can obtain confidence regions for any functions of $\boldsymbol{\theta}$. Since all such inferences are deduced from the same probability statement (namely, $\Pr\{\boldsymbol{\theta} \in \Theta(\mathbf{Y})\} = 1 - \alpha$), the Type I FWE for them is controlled at α. In particular, we can classify the signs of any scalar functions $f(\boldsymbol{\theta})$ by testing the hypotheses $f(\boldsymbol{\theta}) \leqq 0$ and $f(\boldsymbol{\theta}) \geqq 0$ by the confidence-region test method. The Type I and Type III FWE of such a directional decision procedure is also controlled at α (see Theorem 2.2).

4 STEP-DOWN PROCEDURES

Stepwise procedures can be divided into two types—*step-down* and *step-up*. A step-down procedure begins by testing the overall intersection hypothesis and then steps down through the hierarchy of implied hypotheses. If any hypothesis is not rejected, then all of its implied hypotheses are retained without further tests; thus a hypothesis is tested if and only if all of its implying hypotheses are rejected. On the other hand, a step-up procedure begins by testing all minimal hypotheses and then steps up through the hierarchy of implying hypotheses. If any hypothesis is rejected, then all of its implying hypotheses are rejected without further tests; thus a hypothesis is tested if and only if all of its implied hypotheses are retained. Typically a step-down procedure uses a nonincreasing sequence of critical constants for successive steps of testing while a step-up procedure uses a nondecreasing sequence.

In this section we study some theory of step-down procedures. An analogous theory is not available for step-up procedures; in fact, there has been only one such procedure proposed in the literature (see Section 4 of Chapter 4).

As noted in Chapter 1, step-down procedures are generally more powerful than the corresponding single-step procedures. There are some drawbacks associated with the former, however. First, their application is limited, for the most part, to hypothesis testing problems; only in a few cases it is known how to invert them to obtain simultaneous confidence intervals (see Section 4.2.4). Second, only for a few step-down test procedures has it been shown that they can be augmented with directional decisions in the usual manner with control of both the Type I and Type III FWE (see Section 4.2.3). Third, they can only be used for *finite* families of hypotheses and the inferences cannot be extended to larger families than those initially specified. As seen in Sections 2 and 3 these limitations do not apply to single-step procedures.

A general method for constructing step-down procedures is given in Section 4.1. Step-down procedures for nonhierarchical families are studied in Section 4.2 and for hierarchical families in Section 4.3.

4.1 The Closure Method

A general method for constructing step-down test procedures was proposed by Marcus, Peritz, and Gabriel (1976). This method is referred to as the *closure method* and the resulting procedures are referred to as *closed testing procedures*. Peritz (1970) originally applied this method to the problem of testing all subset homogeneity hypotheses in a one-way layout (see Section 4.3.1).

Let $\{H_i \ (1 \leqq i \leqq m)\}$ be a finite family of hypotheses (on scalar or vector parametric functions). Form the closure of this family by taking all nonempty intersections $H_P = \bigcap_{i \in P} H_i$ for $P \subseteq \{1, 2, \ldots, m\}$. If an α-level test of each hypothesis H_P is available, then the closed testing procedure rejects any hypothesis H_P if and only if every H_Q is rejected by its associated α-level test for all $Q \supseteq P$. The following theorem shows that this procedure controls the Type I FWE.

Theorem 4.1 (Marcus et al. 1976). The closed testing procedure stated above strongly controls the Type I FWE at level α.

Proof. Let $\{H_i, i \in P\}$ be any collection of true null hypotheses and let $H_P = \bigcap_{i \in P} H_i$ where P is some unknown subset of $\{1, 2, \ldots, m\}$. If P is empty, then there can be no Type I error, so let P be nonempty. Let A be the event that at least one true H_i is rejected and B be the event that H_P is rejected. The closed testing procedure rejects a true H_i if and only if all hypotheses implying H_i, in particular H_P, are tested at level α and are rejected, and the test of H_i is also significant at level α. So $A = A \cap B$ and thus under H_P,

$$\mathrm{FWE} = \Pr\{A\} = \Pr\{A \cap B\} = \Pr\{B\}\Pr\{A|B\} \leqq \alpha . \quad (4.1)$$

The last inequality follows since $\Pr\{B\} = \alpha$ when H_P is true. Because (4.1) holds under any H_P, the theorem follows. □

As is evident from the proof above, this theorem is applicable more generally and is not restricted to the normal theory linear models setup of the present chapter.

4.2 Nonhierarchical Families

We assume the setting of Section 2.1.1 and consider the problem of testing a nonhierarchical family of hypotheses $H_i: \gamma_i \in \Gamma_i$ $(1 \leqq i \leqq m)$ where $\gamma_i = \mathbf{l}_i'\boldsymbol{\theta}$ and the Γ_i's are specified subsets of the real line. The number of tests in a closed testing procedure increases exponentially with m. Therefore it is of interest to develop a shortcut version of the closed testing procedure. This topic is discussed next.

4.2.1 A Shortcut Version of the Closed Testing Procedure

Suppose that for any $P \subseteq \{1, 2, \ldots, m\}$ the set of parameter points for which all H_i's, $i \in P$, are true and all H_j's, $j \not\subseteq P$, are false is nonempty for any choice of P. This means that every partition of the m hypotheses into two subsets such that the hypotheses in one subset are true and those in the other subset are false is possible for at least some point in the parameter space. This is referred to as the *free combinations condition* by Holm (1979a); this condition is assumed to hold (except when noted otherwise) in the present section. Note that this condition is not satisfied for the family of pairwise null hypotheses because, for example, the set $(\theta_1 - \theta_2 = 0) \cap (\theta_2 - \theta_3 = 0) \cap (\theta_1 - \theta_3 \neq 0)$ is empty.

Consider a closed testing procedure that uses UI statistics for testing all intersection hypotheses $H_P = \bigcap_{i \in P} H_i$ (i.e., the test statistic for every intersection hypothesis H_P is derived from those for the H_i's by the UI method). Such a closed testing procedure can be applied in a shortcut manner because the UI tests have the property that whenever any intersection hypothesis H_P is rejected, at least one of the H_i's implied by H_P is rejected. Therefore, in order to make a rejection decision on any H_i, it is not necessary to test all the intersections H_P containing that H_i; one simply needs to test only the latter. But this must be done in a step-down manner by ordering the H_i's to ensure the coherence condition that a hypothesis is automatically retained if any intersection hypothesis implying that hypothesis is retained.

If the UI related test statistics for the hypotheses H_P are of the form $Z_P = \max_{i \in P} Z_i$ (which is the case if the rejection regions for the individual H_i's are of the form $Z_i > \xi$), then the requirement mentioned above can be ensured if the H_i's are tested in the order of the magnitudes of the corresponding test statistics Z_i's from the largest to the smallest. Thus the hypothesis corresponding to the largest Z_i is tested first. (Note that testing this hypothesis is equivalent to testing the overall intersection hypothesis.) If it is rejected, then any intersection hypothesis containing that hypothesis will clearly be rejected and therefore that hypothesis can be set aside as being rejected without further tests. Next the hypothesis corresponding to the second largest Z_i is tested. This procedure is continued until some Z_i is found to be not significant. At that point all the hypotheses whose test statistic values are less than or equal to the current Z_i are automatically retained.

According to Theorem 4.1, this procedure will control the Type I FWE at level α if the individual tests at different steps are of level α. Suppose that at some step the hypotheses H_i, $i \in P$, are still to be tested. Then an α-level test is obtained by comparing the test statistic $Z_P = \max_{i \in P} Z_i$ against the upper α point of its distribution under $H_P = \bigcap_{i \in P} H_i$.

This shortcut version of the closed testing procedure was proposed by Holm (1979a) (also independently proposed earlier in special contexts by Hartley 1955, Williams 1971, and Naik 1975) who labeled it a *sequentially rejective procedure*. We now give two examples of this procedure.

Example 4.1. Suppose that the estimates $\hat{\gamma}_i$ $(1 \leqq i \leqq m)$ are independent and we wish to test $H_i: \gamma_i = \gamma_{0i}$ $(1 \leqq i \leqq m)$ against two-sided alternatives. In that case, as seen in Example 2.1, the distribution of $\max_{1 \leqq i \leqq m} |T_i|$ under $\bigcap_{i=1}^m H_i$ is the Studentized maximum modulus distribution with parameter m and d.f. ν. The upper α point of this distribution is denoted by $|M|_{m,\nu}^{(\alpha)}$. More generally, the upper α point of the distribution of $\max_{i \in P} |T_i|$ under H_P is given by $|M|_{p,\nu}^{(\alpha)}$ where $p = \text{card}(P)$.

The closed testing procedure can be applied in a shortcut form as follows:

(i) First order the $|T_i|$-statistics as $|T|_{(1)} \leqq |T|_{(2)} \leqq \cdots \leqq |T|_{(m)}$. Let $H_{(1)}, H_{(2)}, \ldots, H_{(m)}$ be the corresponding ordered hypotheses H_i.

(ii) Reject $H_{(m)}$ if $|T|_{(m)} > |M|_{m,\nu}^{(\alpha)}$; otherwise retain all hypotheses without further tests.

(iii) In general, reject $H_{(j)}$ if $|T|_{(i)} > |M|_{i,\nu}^{(\alpha)}$ for $i = m, m-1, \ldots, j$. If $H_{(j)}$ is not rejected, then also retain $H_{(j-1)}, \ldots, H_{(1)}$ without further tests.

Note that the corresponding single-step test procedure mentioned in Example 2.1 uses a common critical point $|M|_{m,\nu}^{(\alpha)}$, which is strictly greater than $|M|_{i,\nu}^{(\alpha)}$ for $i = 1, 2, \ldots, m-1$. Therefore that single-step test procedure is less powerful than the step-down test procedure stated above. □

Example 4.2. We next describe an application of the closed testing procedure to the problem of making one-sided comparisons with a control (Naik 1975, Marcus et al. 1976). Consider the finite family of hypotheses $\{H_i : \theta_i - \theta_k \leqq 0 \ (1 \leqq i \leqq k-1)\}$ against upper one sided alternatives. Assume that $n_1 = \cdots = n_{k-1} = n$ and let $\rho = n/(n + n_k)$. The closed testing procedure can be applied in a shortcut form as follows:

(i) Calculate the statistics

$$T_i = \frac{\bar{Y}_i - \bar{Y}_k}{S\sqrt{1/n + 1/n_k}} \qquad (1 \leqq i \leqq k-1),$$

and order them $T_{(1)} \leqq T_{(2)} \leqq \cdots \leqq T_{(k-1)}$. Let $H_{(1)}, H_{(2)}, \ldots, H_{(k-1)}$ be the corresponding hypotheses.

(ii) Reject $H_{(k-1)}$ if $T_{(k-1)} > T_{k-1,\nu,\rho}^{(\alpha)}$; otherwise retain all hypotheses without further tests.

(iii) In general, reject $H_{(j)}$ if $T_{(i)} > T_{i,\nu,\rho}^{(\alpha)}$ for $i = k-1, k-2, \ldots, j$. If $H_{(j)}$ is not rejected, then also retain $H_{(j-1)}, \ldots, H_{(1)}$ without further tests.

It is readily seen that this step-down procedure is more powerful than the single-step test procedure based on (2.11), which uses a common critical constant $T_{k-1,\nu,\rho}^{(\alpha)} > T_{i,\nu,\rho}^{(\alpha)}$ for $i = 1, \ldots, k-2$. □

4.2.2 A Sequentially Rejective Bonferroni Procedure

In the preceding section we gave two examples of step-down test procedures obtained by the closure method. In both cases the correlation structure $\{\rho_{ij}\}$ was particularly simple, which enabled us to easily obtain the exact critical point for an α-level test at each step. Furthermore, the critical point used at each step depended on the $\hat{\gamma}_i$'s still under test only through their number. When the correlation structure of the $\hat{\gamma}_i$'s is arbitrary, this is no longer the case. In general, suppose that at a given step a set of hypotheses $H_i : \gamma_i = \gamma_{0i}$ for $i \in P$ is still to be tested (against, say, two-sided alternatives). Then for comparing with the $\max_{i \in P} |T_i|$-statistic, one must use the critical constant $|T|^{(\alpha)}_{p,\nu,\mathbf{R}_P}$ where $\mathbf{R}_P$ is the correlation matrix of the $\hat{\gamma}_i$'s for $i \in P$ and $p = \text{card}(P)$. Computation of these critical constants poses a formidable problem. Furthermore, for each different subset P of hypotheses, the critical point must be calculated anew.

A simple, conservative solution to this problem can be based on the Bonferroni inequality. One can thus use $T_\nu^{(\alpha/2p)}$ in place of $|T|^{(\alpha)}_{p,\nu,\mathbf{R}_P}$ in the procedure stated above. This is the basic idea behind Holm's (1979a) so-called *sequentially rejective Bonferroni procedure*. This procedure can be implemented as follows once the P-values for the test statistics for the H_i $(1 \leqq i \leqq m)$ have been calculated. Order the P-values $P_{(1)} \geqq P_{(2)} \geqq \cdots \geqq P_{(m)}$ and let $H_{(1)}, H_{(2)}, \ldots, H_{(m)}$ be the corresponding hypotheses. First check if $P_{(m)} \geqq \alpha/m$, in which case retain all the hypotheses without further tests; otherwise (if $P_{(m)} < \alpha/m$) reject $H_{(m)}$ and proceed to test $H_{(m-1)}$. Next check if $P_{(m-1)} \geqq \alpha/(m-1)$, in which case retain $H_{(m-1)}, \ldots, H_{(1)}$ without further tests; otherwise (if $P_{(m-1)} < \alpha/(m-1)$) reject $H_{(m-1)}$ and proceed to test $H_{(m-2)}$, and so on. This step-down Bonferroni procedure based on the P-values may be contrasted with the corresponding single-step Bonferroni procedure in which all the P-values are compared to the common value α/m. Clearly, the step-down procedure is more powerful.

Shaffer (1986a) proposed a modification of the sequentially rejective Bonferroni procedure when the free combinations condition is not assumed to hold and thus the truth or falsity of certain hypotheses implies the truth or falsity of some others. For example, if we have three hypotheses: $H_1: \theta_1 = \theta_2$, $H_2: \theta_1 = \theta_3$, and $H_3: \theta_2 = \theta_3$, and if, say, H_1 is known to be false, then it follows that at least one of H_2 and H_3 must also be false. Such information can be utilized in sharpening the sequentially rejective Bonferroni procedure as follows: In the usual procedure, after the j hypotheses corresponding to the j smallest P-values (viz., $P_{(m)} \leqq P_{(m-1)} \leqq \cdots \leqq P_{(m-j+1)}$) have been rejected, the $(j+1)$th hypothesis is tested by comparing $P_{(m-j)}$ with $\alpha/(m-j)$. Shaffer suggested that instead of $\alpha/(m-j)$ one should use α/j^* where j^* equals the maximum possible number of true hypotheses given that at least j hypotheses are false. She also noted that a further sharpening can be obtained (at the expense of greater complexity) by determining j^* given that at least j *specific* hypotheses are false. Conforti and Hochberg (1986) showed that this latter modified procedure is equivalent to Peritz's (1970) closed testing procedure. They also gave an efficient computational algorithm based on graph theory for calculating j^* for the family of pairwise comparisons. However, this algorithm is derived under quite a restrictive assumption, which often is not satisfied. Therefore Shaffer's simpler modification is recommended in practice.

4.2.3 Directional Decisions

Shaffer (1980) considered the problem of supplementing a closed testing procedure having Type I FWE $\leqq \alpha$ for the family $\{H_i : \gamma_i = 0\ (1 \leqq i \leqq m)\}$ with directional decisions for any null hypotheses that are rejected in favor of two-sided alternatives. (Thus, e.g., if $H_i : \gamma_i = 0$ is rejected, then decide the sign of γ_i depending on the observed sign of $\hat{\gamma}_i$.) She showed that if the test statistics for the individual hypotheses are independently distributed, then such a directional decision procedure controls both the Type I and Type III FWE at level α; for a further discussion of this result, see Appendix 1, Section 3.3. For single-step procedures this result was shown in Theorem 2.2 without assuming independence of the test statistics.

Shaffer's condition of independence of test statistics is rather restrictive. It does not hold in most problems that we have discussed thus far. For example, even if the $\hat{\gamma}_i$'s are independent ($\rho_{ij} = 0\ \forall\, i \neq j$), the statistics T_i are correlated because they all share the common estimate S^2 of σ^2. In this case, for the step-down procedure based on the Studentized maximum modulus distribution given in Example 4.1, Holm (1979b) has given a proof to show that Shaffer's result holds even though the independence condition is violated.

Very little work has been done on step-down procedures for controlling only the Type III FWE when making directional decisions. See, however, Bohrer and Schervish (1980) and Hochberg and Posner (1986).

4.2.4 Simultaneous Confidence Estimation

Kim, Stefánsson, and Hsu (1987) proposed a method* for obtaining simultaneous confidence intervals associated with some step-down testing procedures. Their method applies to general location parameter families of distributions. But at present it is only known how to apply it to test procedures for one-sided hypothesis testing problems satisfying the free combinations condition. Thus, for example, it is not known whether one can use this method to derive two-sided simultaneous confidence intervals associated with, say, Fisher's least significant difference (LSD) procedure (after adjusting it so that it controls the Type I FWE; see Example 4.6) for all pairwise differences of treatment means in a one-way layout.

Let $\hat{\boldsymbol{\theta}} \sim N(\boldsymbol{\theta}, \sigma^2\mathbf{V})$ and let $\{H_i, i \in I\}$ be a family of linear hypotheses

* Our discussion of this method is based on Stefánsson and Hsu (1985), which is an earlier version of Kim, Stefánsson, and Hsu (1987).

on $\boldsymbol{\theta}$ where I is an arbitrary index set. Let $\phi_i(\hat{\boldsymbol{\theta}})$ be the indicator test function for H_i corresponding to the given test procedure where $\phi_i(\hat{\boldsymbol{\theta}}) = 1$ (respectively, $=0$) means that H_i is rejected (respectively, not rejected) when $\hat{\boldsymbol{\theta}}$ is observed, $i \in I$. The method is based on the following straightforward result.

Theorem 4.2 (Kim, Stefánsson, and Hsu 1987). Define

$$C(\hat{\boldsymbol{\theta}}) = \{\boldsymbol{\theta} : \phi_{g(\boldsymbol{\theta})}(\hat{\boldsymbol{\theta}} - \boldsymbol{\theta}) = 0\} \tag{4.2}$$

where $g(\cdot)$ is an arbitrary function mapping $\mathbb{R}^k$ into I. If

$$\Pr\{\phi_i(\hat{\boldsymbol{\theta}}) = 1 | \boldsymbol{\theta} = \mathbf{0}\} = \alpha \qquad \forall\, i \in I\,, \tag{4.3}$$

then

$$\Pr\{\boldsymbol{\theta} \in C(\hat{\boldsymbol{\theta}})\} = 1 - \alpha \qquad \forall\, \boldsymbol{\theta} \in \mathbb{R}^k\,.$$

Thus $C(\hat{\boldsymbol{\theta}})$ provides a $(1 - \alpha)$-level confidence region for $\boldsymbol{\theta}$. □

Note that the sole probability assumption (4.3) states that the size of each test is α when $\boldsymbol{\theta} = \mathbf{0}$ although each H_i can be a composite hypothesis (including the parameter point $\boldsymbol{\theta} = \mathbf{0}$). By making a judicious choice of $g(\boldsymbol{\theta})$ and the test functions $\phi_i(\hat{\boldsymbol{\theta}})$, one can obtain simultaneous confidence regions for any desired parametric functions of $\boldsymbol{\theta}$. We now illustrate the use of this theorem to obtain lower one-sided simultaneous confidence intervals for $\theta_i - \theta_k$ $(1 \leqq i \leqq k-1)$ by inverting the one-sided step-down test procedure of Example 4.2.

Example 4.3. Recall that the hypotheses under test are $H_i : \theta_i - \theta_k \leqq 0$ $(1 \leqq i \leqq k-1)$ and the ordered hypotheses $H_{(1)}, H_{(2)}, \ldots, H_{(k-1)}$ correspond to the ordered statistics: $T_{(1)} \leqq T_{(2)} \leqq \cdots \leqq T_{(k-1)}$ or equivalently the ordered sample means $\bar{Y}_{(1)} \leqq \bar{Y}_{(2)} \leqq \cdots \leqq \bar{Y}_{(k-1)}$. Let $P = \{(1), (2), \ldots, (p)\}$ and $H_P = \bigcap_{i=1}^{p} H_{(i)}$. Note that I consists of all nonempty subsets P. The indicator test function of H_P for the given step-down procedure is given by

$$\phi_P(\bar{\mathbf{Y}}) = \begin{cases} 0 & \text{if } \bar{Y}_{(i)} \leqq \bar{Y}_k + \xi_p \quad i = 1, \ldots, p \\ 1 & \text{otherwise} \end{cases} \tag{4.4}$$

where

$$\xi_p = T_{p,\nu,\rho}^{(\alpha)} S \sqrt{\frac{1}{n} + \frac{1}{n_k}} \qquad (1 \leqq p \leqq k-1)\,. \tag{4.5}$$

We know that $\phi_P(\bar{\mathbf{Y}})$ is a test of level α of the hypothesis H_P. In particular, $\Pr\{\phi_P(\bar{\mathbf{Y}})=1\}=\alpha$ for all subsets P when $\theta_1=\cdots=\theta_{k-1}=\theta_k$. Thus (4.3) is satisfied and we can apply Theorem 4.2 with

$$g(\boldsymbol{\theta})=\{i:\theta_i\leqq\theta_k\ (1\leqq i\leqq k-1)\}.$$

We obtain from (4.2)

$$\begin{aligned}C(\bar{\mathbf{Y}})&=\{\boldsymbol{\theta}:\phi_{g(\boldsymbol{\theta})}(\bar{\mathbf{Y}}-\boldsymbol{\theta})=0\}\\&=\bigcup_P\{\boldsymbol{\theta}:g(\boldsymbol{\theta})=P,\ \phi_P(\bar{\mathbf{Y}}-\boldsymbol{\theta})=0\}\\&=\bigcup_P\{\boldsymbol{\theta}:\theta_i-\theta_k\leqq 0\ \forall i\in P,\ \theta_i-\theta_k>0\ \forall i\not\in P;\\&\qquad\theta_i-\theta_k\geqq\bar{Y}_i-\bar{Y}_k-\xi_p\ \forall i\in P\}\\&=\bigcup_P\{\boldsymbol{\theta}:\bar{Y}_i-\bar{Y}_k-\xi_p\leqq\theta_i-\theta_k\leqq 0\ \forall i\in P,\ \theta_i-\theta_k>0\ \forall i\not\in P\}\end{aligned}\tag{4.6}$$

where the unions are taken over all nonempty subsets P of $\{1,2,\ldots,k-1\}$.

The confidence set (4.6) is rather cumbersome to use. It can be restated in a more convenient form as follows: Let p be the index such that

$$\bar{Y}_{(i)}>\bar{Y}_k+\xi_i\ (i=k-1,\ldots,p+1)\qquad\text{and}\qquad\bar{Y}_{(p)}\leqq\bar{Y}_k+\xi_p,$$

that is, the step-down test procedure rejects $H_{(k-1)},\ldots,H_{(p+1)}$ and retains $H_{(p)},\ldots,H_{(1)}$. Then

$$\theta_i-\theta_k\geqq\left[\bar{Y}_i-\bar{Y}_k-T^{(\alpha)}_{p,\nu,\rho}S\sqrt{\frac{1}{n}+\frac{1}{n_k}}\right]^-\qquad(1\leqq i\leqq k-1)\tag{4.7}$$

provide $(1-\alpha)$-level simultaneous lower one-sided confidence intervals; here $x^-=\min(x,0)$.

Comparing (4.7) with the intervals (2.11) (for $n_1=\cdots=n_{k-1}=n$ and $\rho_{ij}=\rho=n/(n+n_k)$ for all $i\neq j$) associated with the corresponding single-step test procedure, we see that the confidence bounds in the former are not always as sharp as those in the latter for all $i=1,2,\ldots,k-1$. In fact, for those treatments for which H_i is rejected (thus implying that $\theta_i-\theta_k>0$) the confidence bound given by (4.7) is zero while that given by (2.11) is positive. The lack of positive lower bound for $\theta_i-\theta_k$ in this case is a drawback of (4.7). For those treatments for which H_i is not rejected, the confidence bounds given by (4.7) are at least as sharp as those given by (2.11). □

4.3 Hierarchical Families

In Chapter 1 we introduced Fisher's LSD, which is a step-down procedure for the hierarchical family consisting of the overall null hypothesis and all pairwise null hypotheses. In this section we study the theory of step-down procedures for the larger family of all subset hypotheses. After introducing some basic preliminaries in Section 4.3.1 we discuss the application of the closure method to this family in Section 4.3.2. The resulting closed testing procedure is not easy to use because it involves too many tests. Simpler step-down procedures that are not of closed type are studied in Section 4.3.3.

4.3.1 Preliminaries

Let $\hat{\boldsymbol{\theta}} \sim N(\boldsymbol{\theta}, \sigma^2\mathbf{V})$ where $\mathbf{V}$ is a known matrix and let $S^2 \sim \sigma^2\chi^2_\nu/\nu$ independent of $\hat{\boldsymbol{\theta}}$. Assume that, for testing each subset homogeneity hypothesis $H_P: \theta_i = \theta_j$ for all $i, j \in P$, we have a test statistic Z_P whose distribution under H_P is free of unknown parameters $\boldsymbol{\theta}$ and σ^2. If large values of Z_P indicate heterogeneity of the subset P, then we need an appropriately chosen upper critical point of the null (under H_P) distribution of Z_P for testing H_P. This critical point may, in general, depend on P and we denote it by ξ_P.

Later we need to consider hypotheses of the type $H_{\mathbf{P}} = \bigcap_{i=1}^r H_{P_i}$ where $\mathbf{P} = (P_1, P_2, \ldots, P_r)$ is a "partition" of $K = \{1, 2, \ldots, k\}$, that is, the P_i's are disjoint subsets of K with cardinalities $p_i \geqq 2$ such that $\Sigma_{i=1}^r p_i \leqq k$; here r, the number of homogeneous subsets, is between 1 and $k/2$. These hypotheses correspond to parameter configurations $\boldsymbol{\theta}$ with at least two of the θ_i's equal and are referred to as *multiple subset hypotheses*. Consideration of such parameter configurations becomes necessary for studying whether a given MCP controls the FWE strongly. Note that the family $\{H_{\mathbf{P}}, \forall \mathbf{P}\}$ forms the closure of the family $\{H_P, \forall P\}$ and is the family addressed by the closed testing procedure of Section 4.3.2.

Theorems 4.3 and 4.4 give the choice of the so-called nominal significance levels to be used for each test in the procedures of Sections 4.3.2 and 4.3.3, respectively. Both of these theorems require the following condition to hold: For any $H_{\mathbf{P}} = \bigcap_{i=1}^r H_{P_i}$,

$$\Pr_{H_{\mathbf{P}}}\{Z_{P_i} \leqq \xi_{P_i}\ (1 \leqq i \leqq r)\} \geqq \prod_{i=1}^{r} \Pr_{H_{P_i}}\{Z_{P_i} \leqq \xi_{P_i}\} \tag{4.8}$$

for the given set of test statistics $\{Z_P\}$ and critical constants $\{\xi_P\}$. From Kimball (1951) it follows that for a one-way layout design, (4.8) is satisfied when the Z_P's are Studentized range or F-statistics. Moreover, there is equality in (4.8) if σ^2 is known. More general conditions for (4.8) to hold can be obtained from Theorem 2.3 in Appendix 2 and are as follows:

(i) For every given S, the set $\{Z_{P_i} \leqq \xi_{P_i}\}$ is convex and symmetric about the origin in the space of the $\hat{\theta}_j$, $j \in P_i$ $(1 \leqq i \leqq r)$.

(ii) The matrix $\mathbf{V}$ has the product structure

$$v_{ij} = \lambda_i \lambda_j \qquad (1 \leqq i \neq j \leqq k) \tag{4.9}$$

for some real constants λ_i.

It is conjectured (see Conjecture 2.3.1, Tong 1980, p. 27) that (4.8) holds without the condition (4.9). However, Khatri's (1970) proof of this conjecture has been shown to be in error by Šidák (1975).

4.3.2 A Closed Testing Procedure

The basic principle behind the closure method was explained in Section 4.1. There it was shown (see Theorem 4.1) that given a finite family of hypotheses, if one tests each intersection hypothesis at level α subject to the coherence requirement, then the Type I FWE is strongly controlled at level α. Peritz (1970) applied this method to the family of all subset hypotheses. The resulting closed testing procedure tests every multiple subset hypothesis $H_{\mathbf{P}}$ at nominal level α subject to the coherence requirement which is satisfied if, whenever a hypothesis $H_{\mathbf{P}}$ is not rejected, then all $H_{\mathbf{Q}}$'s that are implied by $H_{\mathbf{P}}$ are retained automatically without further tests. Note that if $\mathbf{P} = (P_1, P_2, \ldots, P_r)$ and $\mathbf{Q} = (Q_1, Q_2, \ldots, Q_s)$, then $H_{\mathbf{P}}$ implies $H_{\mathbf{Q}}$ if and only if every $P_i \supseteq Q_j$ for some j $(1 \leqq i \leqq r, 1 \leqq j \leqq s)$.

Let $\alpha_p = \Pr_{H_P}\{Z_P > \xi_P\}$ be the nominal level of the test of H_P; note that α_P is assumed to depend on P only through its cardinality p. The following theorem shows that Peritz's choice (4.10) for the α_p's results in nominal level $\leqq \alpha$ for each multiple subset hypothesis $H_{\mathbf{P}}$. Before stating the theorem we introduce the following terminology (due to Begun and Gabriel 1981): A hypothesis $H_{\mathbf{P}} = \bigcap_{i=1}^{r} H_{P_i}$ is said to be significant at nominal level $\alpha_{\mathbf{p}}$ if and only if $Z_{P_i} > \xi_{P_i}$ for some i $(1 \leqq i \leqq r)$ where $\mathbf{p} = (p_1, p_2, \ldots, p_r)'$ denotes the vector of cardinalities of $P_1, P_2, \ldots, P_r$.

Theorem 4.3 (Begun and Gabriel 1981). Suppose that we use the following nominal levels for tests of individual hypotheses $H_{\mathbf{P}}$: If $\mathbf{P}$ is a singleton, say $\mathbf{P} = (P)$, then test $H_\mathbf{P}$ at level α. If $\mathbf{P}$ is not a singleton, say $\mathbf{P} = (P_1, P_2, \ldots, P_r)$ for $r \geqq 2$, then test each H_{P_i} at level

$$\alpha_{P_i} = 1 - (1-\alpha)^{p_i/k} \qquad (1 \leqq i \leqq r)\,. \tag{4.10}$$

Then $\alpha_\mathbf{P} \leqq \alpha$ for all $H_\mathbf{P}$ if (4.8) is satisfied. Hence the resulting closed testing procedure controls the Type I FWE strongly at level α.

Proof. If $\mathbf{P}$ is a singleton, then we clearly have $\alpha_\mathbf{P} = \alpha$. If $\mathbf{P}$ is not a singleton, then

$$\begin{aligned}
\alpha_\mathbf{P} &= \Pr_{H_\mathbf{P}}\left\{\bigcup_{i=1}^{r} (Z_{P_i} > \xi_{P_i})\right\} \\
&\leqq 1 - \prod_{i=1}^{r} (1 - \alpha_{P_i}) && \text{(from (4.8))} \\
&= 1 - \prod_{i=1}^{r} \{1 - [1 - (1-\alpha)^{p_i/k}]\} && \text{(from (4.10))} \\
&= 1 - (1-\alpha)^{\Sigma p_i/k} \\
&\leqq \alpha \qquad \left(\text{since } \sum_{i=1}^{r} p_i \leqq k\right).
\end{aligned}$$

Hence from Theorem 4.1 it follows that the Type I FWE is strongly controlled at level α. □

As noted above, the closed testing procedure involves testing every multiple subset hypothesis $H_\mathbf{P}$ unless it is retained by implication. A systematic and computationally feasible algorithm for implementing this procedure was given by Begun and Gabriel (1981), and is discussed in Section 3 of Chapter 4.

4.3.3 A Class of Other Step-Down Procedures

Many step-down procedures proposed in the literature for the family of subset hypotheses share a common testing scheme that is much simpler than that of the closed testing procedure. Einot and Gabriel's (1975) description of this testing scheme is given below. By making different choices for the test statistics Z_P and the critical points ξ_P used to test subset hypotheses H_P, different step-down procedures in this class can be obtained.

4.3.3.1 A General Testing Scheme. Coherence requires that a subset hypothesis H_P be rejected if and only if all implying subset hypotheses H_Q

are rejected for $Q \supseteq P$. This is ensured by carrying out the tests in a step-down fashion as follows:

Step 1. Test the overall null hypothesis $H_K: \theta_1 = \theta_2 = \cdots = \theta_k$.
(a) If $Z_K \leqq \xi_K$, then retain all subset hypotheses H_P, $P \subseteq K$, and stop testing.
(b) If $Z_K > \xi_K$, then reject H_K and proceed to Step 2.

Step 2. Test the subset hypotheses H_P for all subsets P of size $k-1$.
(a) If $Z_P \leqq \xi_P$, then retain all H_Q's, $Q \subseteq P$. If $Z_P \leqq \xi_P$ for all P of size $k-1$, then stop testing.
(b) If $Z_P > \xi_P$, then reject H_P. If H_P is rejected for some P, then proceed to Step 3.

Step 3. In general, if H_P is rejected for P of size $p \geqq 3$, then test all of its subsets of size $p-1$ (except those that were retained by implication at an earlier step). If H_P is retained, then retain H_Q for all $Q \subseteq P$. Continue in this manner until no subsets remain to be tested.

A simultaneous test procedure uses a common critical value $\xi_P = \xi$ for all $P \subseteq K$ and statistics Z_P having the monotonicity property that $Z_P \geqq Z_Q$ whenever $Q \subseteq P$. As we saw in Theorem 3.1, this ensures coherence without a step-down testing scheme as above.

Now consider a balanced one-way layout. Suppose we use the Studentized range statistic

$$R_P = \frac{\sqrt{n}(\max_{i \in P} \bar{Y}_i - \min_{i \in P} \bar{Y}_i)}{S} \qquad (4.11)$$

for testing the hypothesis H_P. Because of symmetry considerations, here each critical point ξ_P depends on P only through its cardinality p; let us denote it by ξ_p ($2 \leqq p \leqq k$). Lehmann and Shaffer (1977) noted that in this case whenever a hypothesis H_P is rejected using the step-down procedure given above, every H_Q implied by H_P such that Q contains the two treatments yielding $\max_{i \in P} \bar{Y}_i$ and $\min_{i \in P} \bar{Y}_i$ will also be rejected (and hence need not be separately tested) if the ξ_p's satisfy the following necessary and sufficient condition:

$$\xi_2 \leqq \xi_3 \leqq \cdots \leqq \xi_k . \qquad (4.12)$$

In particular, the two treatments corresponding to $\max_{i \in P} \bar{Y}_i$ and $\min_{i \in P} \bar{Y}_i$ can be declared significantly different from each other when H_P is rejected. This enables us to use a shortcut step-down procedure as illustrated in Example 4.4 below. Note that, as in Section 4.2.1, this shortcut version is possible because the statistics (4.11) are UI statistics. See Section 3.1 of Appendix 1 for a further discussion of this problem in the setting of general location parameter families of distributions.

Example 4.4. For a balanced one-way layout design the statistic R_P given by (4.11) is distributed as a Studentized range r.v. $Q_{p,\nu}$ under H_P. Suppose that we test each H_P at the same (nominal) level α by using the critical point $\xi_p = Q^{(\alpha)}_{p,\nu}$. These critical points satisfy (4.12) because $Q^{(\alpha)}_{p,\nu}$ is increasing in p for fixed α and ν. Therefore a shortcut step-down procedure can be used as follows: Let $\bar{Y}_{(1)} \leqq \bar{Y}_{(2)} \leqq \cdots \leqq \bar{Y}_{(k)}$ be the ordered sample means.

Step 1. (a) Test H_K first. If $\sqrt{n}(\bar{Y}_{(k)} - \bar{Y}_{(1)})/S \leqq Q^{(\alpha)}_{k,\nu}$, then retain all subset hypotheses H_P, $P \subseteq K$, and stop testing.
(b) If $\sqrt{n}(\bar{Y}_{(k)} - \bar{Y}_{(1)})/S > Q^{(\alpha)}_{k,\nu}$, then reject H_K (and all H_Q's for $Q \subseteq K$ containing both (1) and (k)) and proceed to Step 2.

Step 2. (a) Test H_{P_1} and H_{P_2} where $P_1 = \{(1), \ldots, (k-1)\}$ and $P_2 = \{(2), \ldots, (k)\}$. If $\sqrt{n}(\bar{Y}_{(k-1)} - \bar{Y}_{(1)})/S \leqq Q^{(\alpha)}_{k-1,\nu}$, then retain all subset hypotheses H_P, $P \subseteq P_1$. Similarly if $\sqrt{n}(\bar{Y}_{(k)} - \bar{Y}_{(2)})/S \leqq Q^{(\alpha)}_{k-1,\nu}$, then retain all subset hypotheses H_P, $P \subseteq P_2$. If both H_{P_1} and H_{P_2} are not rejected, then stop testing.
(b) If at least one of H_{P_1} or H_{P_2} is rejected, then proceed to Step 3.

Step 3. In general, if H_P for subset $P = \{(i), \ldots, (j)\}$ with cardinality $p = j - i + 1 \geqq 3$ is rejected, then also reject all subset hypotheses H_Q for $Q \subseteq P$ containing both (i) and (j). Test H_{P_1} and H_{P_2} where $P_1 = \{(i+1), \ldots, (j)\}$ and $P_2 = \{(i), \ldots, (j-1)\}$ unless they are retained as homogeneous by implication at a previous step. For testing H_{P_1} use the statistic $\sqrt{n}(\bar{Y}_{(j)} - \bar{Y}_{(i+1)})/S$ and for testing H_{P_2} use the statistic $\sqrt{n}(\bar{Y}_{(j-1)} - \bar{Y}_{(i)})/S$. Compare both the statistics against the same critical point $Q^{(\alpha)}_{j-i,\nu}$. If H_{P_i} is not rejected, then retain all subset hypotheses H_P for $P \subseteq P_i$ ($i = 1, 2$). Stop testing when no further subsets remain to be tested.

This procedure was proposed by Newman (1939) and later reinvented by Keuls (1952) and is hence known as the *Newman–Keuls (NK) procedure*. We study it in more detail in Chapter 4. □

4.3.3.2 *Nominal and True Levels.* The *nominal levels* α_P of a step-down procedure are given by

$$\alpha_P = \Pr_{H_P}(Z_P > \xi_P), \qquad P \subseteq K, \tag{4.13}$$

and the corresponding *true levels* α'_P are given by

$$\alpha'_P = \sup_{H_P}\left\{\Pr_{H_P}\left[\bigcap_{Q \supseteq P}(Z_Q > \xi_Q)\right]\right\}, \qquad P \subseteq K. \tag{4.14}$$

(Note that the supremum in (4.14) is needed because the given probability depends on the θ_i's, $i \not\in P$, and hence is not the same under all $\boldsymbol{\theta}$'s satisfying H_P. In contrast, no supremum is needed in (4.13).) A true level α'_P takes into account the fact that the hypothesis H_P can be rejected if and only if the hypotheses H_Q are rejected for all $Q \supseteq P$; thus α'_P gives the supremum of the true probability of rejecting H_P when H_P is true. Clearly, we have $\alpha'_K = \alpha_K$ and $\alpha'_P \leqq \alpha_P$ for all $P \subseteq K$. Under the so-called *separability condition* (stated in Appendix 1), one can show that $\alpha'_P \rightarrow \alpha_P$ when $|\theta_i - \theta_j| \rightarrow \infty$ for all $i \in P$, $j \not\in P$. This condition is satisfied by the standard procedures based on Studentized range and F-statistics (see Example 3.1 of Appendix 1) that we study in Part I.

4.3.3.3 Control of the Type I Familywise Error Rate. Define, for the given MCP,

$$\alpha^*(\mathbf{P}) = \sup_{H_\mathbf{P}}[\Pr_{H_\mathbf{P}}\{\text{any } H_{P_i}\ (1 \leqq i \leqq r) \text{ is rejected}\}]. \tag{4.15}$$

Then the given MCP controls the Type I FWE strongly at level α if

$$\alpha^* = \max_{\mathbf{P}}\{\alpha^*(\mathbf{P})\} \leqq \alpha. \tag{4.16}$$

To simplify the presentation, from now on we make the assumption that each α_P depends on P only through p and denote it by α_p. Theorem 4.4 below gives a simple upper bound on $\alpha^*(\mathbf{P})$.

Theorem 4.4 (Tukey 1953). Under the separability condition referred to above and condition (4.8), for any partition $\mathbf{P} = (P_1, P_2, \ldots, P_r)$ we have

$$\alpha^*(\mathbf{P}) \leqq 1 - \prod_{i=1}^{r}(1 - \alpha_{p_i}), \tag{4.17}$$

and hence

$$\alpha^* \leqq \max_{(p_1, \ldots, p_r)}\left[1 - \prod_{i=1}^{r}(1 - \alpha_{p_i})\right] \tag{4.18}$$

where the maximum is taken over all sets of integers $p_1, \ldots, p_r$ satisfying $p_i \geqq 2$, $\Sigma_{i=1}^r p_i \leqq k$, and $1 \leqq r \leqq k/2$.

Proof. First note that the probability of at least one false rejection under $H_{\mathbf{P}} = \bigcap_{i=1}^r H_{P_i}$ is maximized under a (limiting) least favorable configuration of the θ_i's when the homogeneous subsets $P_1, \ldots, P_r$ are spread infinitely apart from each other. This is so because in that case, with probability one, none of the sets P_i will be retained by implication. Under this least favorable configuration we can write

$$\begin{aligned}
\alpha^*(\mathbf{P}) &= \sup_{H_{\mathbf{P}}} \Pr_{H_{\mathbf{P}}}\left\{\bigcup_{i=1}^{r}(Z_{P_i} > \xi_{P_i})\right\} \\
&= \sup_{H_{\mathbf{P}}}\left[1 - \Pr_{H_{\mathbf{P}}}\left\{\bigcap_{i=1}^{r}(Z_{P_i} \leqq \xi_{P_i})\right\}\right] \\
&\leqq \sup_{H_{\mathbf{P}}}\left[1 - \prod_{i=1}^{r} \Pr_{H_{P_i}}\{Z_{P_i} \leqq \xi_{P_i}\}\right] \qquad \text{(from (4.8))} \\
&\leqq 1 - \prod_{i=1}^{r} \inf_{H_{\mathbf{P}}}[\Pr_{H_{P_i}}\{Z_{P_i} \leqq \xi_{P_i}\}] \\
&= 1 - \prod_{i=1}^{r} \inf_{H_{P_i}}[\Pr_{H_{P_i}}\{Z_{P_i} \leqq \xi_{P_i}\}] \\
&= 1 - \prod_{i=1}^{r}(1 - \alpha_{p_i}) .
\end{aligned}$$

Hence (4.17) is proved. The upper bound (4.18) requires no proof. □

It is worth noting that the upper bound obtained depends only on the nominal significance levels α_p $(2 \leqq p \leqq k)$ and not on the choice of the test statistics Z_P.

If there is any doubt about the validity of assumption (4.8), then one can instead use the first order Bonferroni inequality to obtain the following upper bound on $\alpha^*(\mathbf{P})$:

$$\alpha^*(\mathbf{P}) \leqq \sum_{i=1}^{r} \alpha_{p_i} . \tag{4.19}$$

This upper bound is more conservative than (4.17) but it is always valid. It is easily seen that if the nominal levels α_p are chosen as

$$\alpha_p = \frac{\alpha p}{k} \qquad (2 \leqq p \leqq k) , \tag{4.20}$$

then (4.16) is satisfied. This specification of the α_p's was suggested by Ryan (1960). Tukey (1953) and Welsch (1972) independently proposed a slightly improved (still satisfying (4.16)) specification

$$\alpha_p = \frac{\alpha p}{k} \ (2 \leqq p \leqq k-2) \,, \qquad \alpha_{k-1} = \alpha_k = \alpha \,. \tag{4.21}$$

If assumption (4.8) holds, then (4.21) can be further improved upon to obtain

$$\alpha_p = 1-(1-\alpha)^{p/k} \ (2 \leqq p \leqq k-2) \,, \qquad \alpha_{k-1} = \alpha_k = \alpha \,. \tag{4.22}$$

We refer to (4.22) as the *Tukey–Welsch (TW) specification*.† We see in Appendix 1 that this specification is very close to the "optimum" specification of the α_p's derived by Lehmann and Shaffer (1979). Because of its near optimality and ease of use, we adopt this specification in the sequel.

Example 4.5. As seen in Example 4.4, the NK-procedure uses the nominal levels $\alpha_2 = \cdots = \alpha_k = \alpha$. From (4.18) we obtain the following upper bound on the maximum Type I FWE of the NK-procedure:

$$\begin{aligned} \alpha^* &\leqq \max_{(p_1, \ldots, p_r)} [1-(1-\alpha)^r] \\ &= \max_{1 \leqq r \leqq k/2} [1-(1-\alpha)^r] \\ &= 1-(1-\alpha)^{\lfloor k/2 \rfloor} \end{aligned} \tag{4.23}$$

where $\lfloor x \rfloor$ is the greatest integer $\leqq x$. If σ^2 is known, then we have equality in (4.23). Since (4.23) exceeds α for $k>3$, it follows that the NK-procedure does not strongly control the FWE at the designated level α. This fact was noted by Tukey (1953) and Hartley (1955).

From (4.23) we see that if we use the nominal level $\alpha' = 1-(1-\alpha)^{1/\lfloor k/2 \rfloor}$ for each test in the NK-procedure, then the resulting modified procedure will control the Type I FWE in the strong sense at level α. □

Example 4.6. We now obtain an upper bound on α^* for Fisher's LSD introduced in Chapter 1. For simplicity we assume a balanced one-way layout setting. As in the proof of Theorem 4.4, it can be readily shown

† Einot and Gabriel (1975) attribute (4.22) (with α_{k-1} given by the first part of the formula) to Ryan (1960).

that under $H_{\mathbf{P}} = \bigcap_{i=1}^{r} H_{P_i}$,

$$\alpha^*(\mathbf{P}) \leqq 1 - \prod_{i=1}^{r} \Pr\{Q_{p_i,\nu} \leqq \sqrt{2}T_\nu^{(\alpha/2)}\} \tag{4.24}$$

where $Q_{p_i,\nu}$ is a Studentized range r.v. with parameter p_i and d.f. ν $(1 \leqq i \leqq r)$. Spjøtvoll (1971) conjectured and Hayter (1986) proved that the maximum of the right hand side of (4.24) over all $\mathbf{P}$ is achieved when $k-1$ of the θ's are equal and the remaining one is removed at a distance tending to infinity; in fact, we get equality in that case and

$$\alpha^* = \Pr\{Q_{k-1,\nu} > \sqrt{2}T_\nu^{(\alpha/2)}\}\,. \tag{4.25}$$

It is clear that α^* exceeds α for $k > 3$. For example, for $\nu = \infty$ and $\alpha = 0.05$, Hayter's (1986) calculations show that this upper bound increases from 0.1222 for $k = 4$ to 0.9044 for $k = 20$. Thus the LSD does not strongly control the FWE at the designated level α for $k > 3$. From (4.25) we see that if at the second step we compare each pairwise $|t|$-statistic with the critical point $Q_{k-1,\nu}^{(\alpha)}/\sqrt{2}$ instead of $T_\nu^{(\alpha/2)}$, then the FWE will be strongly controlled at level α. □

Example 4.7. Duncan (1955) proposed the following nominal levels:

$$\alpha_p = 1 - (1-\alpha)^{p-1} \qquad (2 \leqq p \leqq k)\,. \tag{4.26}$$

Duncan's $\alpha_p \geqq \alpha$ for $p \geqq 2$ with equality holding if and only if $p = 2$, and thus they are more liberal than those of Newman and Keuls when both use the same α appropriate for a single comparison. Since the NK-procedure does not control the FWE at level α, it is clear that neither does the *Duncan (D) procedure* based on (4.26). In fact, from (4.18) it follows that for the D-procedure,

$$\alpha^* \leqq 1 - (1-\alpha)^{k-1}\,. \tag{4.27}$$

As in Example 4.5, we have equality in (4.27) if σ^2 is known.

Duncan's nominal levels are intended to provide α-level protection separately for each pairwise comparison (PCE $= \alpha$). Duncan's reasons for his liberal choice of nominal levels are stated in his 1955 paper (pp. 14–18) and are convincingly critiqued by Miller (1981, p. 89). If it is desired to control the FWE at level α using α_p's that increase with p according to (4.26), then from (4.27) we see that the following modified nominal levels must be used:

$$\alpha_p = 1 - (1 - \alpha)^{(p-1)/(k-1)} \qquad (2 \leqq p \leqq k) . \qquad (4.28)$$

This modification was suggested by Einot and Gabriel (1975). □

Before concluding this section we remark that if one makes a directional decision in the usual manner whenever a pair of treatments is found significant using a step-down procedure, then it is not known (no mathematical proof is available) whether the Type I *and* Type III FWE is still controlled.

CHAPTER 3

Single-Step Procedures for Pairwise and More General Comparisons among All Treatments

Consider the general linear model setting of Section 1.1 of Chapter 2 and, as assumed there, let $\boldsymbol{\theta} = (\theta_1, \theta_2, \ldots, \theta_k)'$ be the vector of parameters of interest corresponding to the k treatment effects. Let $\hat{\boldsymbol{\theta}}$ be the least squares (LS) estimator of $\boldsymbol{\theta}$ such that $\hat{\boldsymbol{\theta}} \sim N(\boldsymbol{\theta}, \sigma^2\mathbf{V})$ where $\mathbf{V}$ is a known positive definite symmetric matrix. We also have an estimator S^2 of σ^2 that is distributed independently of $\hat{\boldsymbol{\theta}}$ as a $\sigma^2\chi^2_\nu/\nu$ random variable (r.v.).

The families of comparisons considered here include (i) pairwise comparisons, (ii) all contrasts, (iii) subset homogeneity hypotheses, and (iv) multiple subset homogeneity hypotheses. These families were introduced in Chapter 2. Note that (i) is a subfamily of (ii) and (iii), while (iv) forms a closure of (iii). Thus the latter three families can be regarded as generalizations of the family of pairwise comparisons. For families (i) and (ii) either simultaneous confidence estimates or hypotheses tests (or both) may be of interest. We primarily concentrate on simultaneous confidence estimates. As noted in Chapter 2, the associated simultaneous tests are readily obtained by the confidence-region test method. Families (iii) and (iv) involve only hypotheses tests. Stepwise procedures for these two families are discussed in Chapter 4.

In Section 1 we discuss Scheffé's procedure and illustrate its use in different families. Tukey's procedure for balanced designs is discussed in Section 2. Various extensions of the Tukey procedure for unbalanced designs are discussed in Section 3. The procedures described in the first three sections are compared in Section 4. Some extensions of these procedures and further comments on them are given in Section 5.

1 SCHEFFÉ'S S-PROCEDURE

1.1 General Balanced or Unbalanced Designs

Scheffé's (1953) S-procedure in its general form was given as a simultaneous confidence statement (2.20) in Chapter 2. For convenience, we reproduce it here: Exact $(1-\alpha)$-level simultaneous confidence intervals for all $\mathbf{l}'\boldsymbol{\theta}$, $\mathbf{l} \in \mathscr{L}$ (where $\mathscr{L}$ is a specified subspace of dimension $m \leqq k$) are given by

$$\mathbf{l}'\boldsymbol{\theta} \in [\mathbf{l}'\hat{\boldsymbol{\theta}} \pm (mF_{m,\nu}^{(\alpha)})^{1/2}S(\mathbf{l}'\mathbf{V}\mathbf{l})^{1/2}] \qquad \forall\, \mathbf{l} \in \mathscr{L} \tag{1.1}$$

where $F_{m,\nu}^{(\alpha)}$ denotes the upper α point of the F-distribution with m and ν degrees of freedom (d.f.). A derivation of (1.1) was given in Theorem 2.1 of Chapter 2.

Any hypothesis $H_1: \mathbf{l}'\boldsymbol{\theta} = 0$ (or equal to any other specified constant) for $\mathbf{l} \in \mathscr{L}$, whether specified *a priori* or selected *a posteriori* by data-snooping, can be tested by the confidence-region test method using (1.1). Similarly one-sided inferences and directional decisions can also be made. The familywise error rate (FWE) for all such inferences is controlled at α; see Theorem 3.3 of Chapter 2.

An optimality property of the intervals (1.1) was shown by Bohrer (1973). He showed that under certain conditions (which are commonly satisfied), the intervals (1.1) have the smallest average (with respect to a uniform measure over an ellipsoidal region in the space $\mathscr{L}$) width among all simultaneous confidence procedures with the same confidence level. See Naiman (1984) for a generalization of Bohrer's result.

In the present section we discuss some applications of the basic S-procedure (1.1). The versatility of the S-procedure stems from the flexibility available in the choice of space $\mathscr{L}$ and its applicability to any design following the general linear model of Section 1.1 of Chapter 2. It forms the basis of techniques for constructing simultaneous confidence bands in regression problems, a topic that is not covered in the present book. In fact, Working and Hotelling's (1929) method of constructing a simultaneous confidence band in simple linear regression was a precursor of Scheffé's procedure.

As a first example, we consider the one-way layout design discussed in Example 1.1 of Chapter 2. As noted there, the formulas in this case apply more generally whenever a design yields LS estimates $\hat{\theta}_i$ that are independently distributed $N(\theta_i, v_{ii}\sigma^2)$ r.v.'s (where the v_{ii}'s are known constants) and an unbiased estimate S^2 of σ^2 is distributed independently of the $\hat{\theta}_i$'s as a $\sigma^2\chi^2_\nu/\nu$ r.v.

By choosing $\mathscr{L}$ to be the contrast space $\mathbb{C}^k = \{\mathbf{c} \in \mathbb{R}^k : \Sigma_{i=1}^k c_i = 0\}$ and applying (1.1) we obtain the following $(1-\alpha)$-level simultaneous confidence intervals for all contrasts among the treatment means in a one-way layout (for notation see Example 1.1 of Chapter 2):

$$\sum_{i=1}^k c_i\theta_i \in \left[\sum_{i=1}^k c_i\bar{Y}_i \pm \{(k-1)F_{k-1,\nu}^{(\alpha)}\}^{1/2} S\left(\sum_{i=1}^k \frac{c_i^2}{n_i}\right)^{1/2}\right] \quad \forall\, \mathbf{c} \in \mathbb{C}^k \tag{1.2}$$

where $\bar{Y}_i$ is the sample mean for the ith treatment $(1 \leqq i \leqq k)$ and $S^2 = MS_{\text{error}}$ with $\nu = \Sigma_{i=1}^k n_i - k$ d.f. For pairwise comparisons, the intervals given by (1.2) simplify to the following conservative $(1-\alpha)$-level simultaneous confidence intervals:

$$\theta_i - \theta_j \in \left[\bar{Y}_i - \bar{Y}_j \pm \{(k-1)F_{k-1,\nu}^{(\alpha)}\}^{1/2} S\left(\frac{1}{n_i} + \frac{1}{n_j}\right)^{1/2}\right] \quad (1 \leqq i < j \leqq k)\,. \tag{1.3}$$

The test procedure based on (1.3) is often referred to in the literature as the *fully significant difference* or *globally significant difference* (the FSD or GSD) test. These names reflect the larger critical value, $\{(k-1)F_{k-1,\nu}^{(\alpha)}\}^{1/2}$, which the pairwise $|t|$-statistic, $|T_{ij}| = |\bar{Y}_i - \bar{Y}_j|/S\sqrt{1/n_i + 1/n_j}$, must exceed in order to be significant, in contrast to Fisher's *least significant difference* (LSD) test (see Chapter 1), which uses the smaller critical value $T_\nu^{(\alpha/2)}$.

As another example, consider the one-way layout design with a fixed covariate discussed in Example 1.2 of Chapter 2. In the following we use the notation defined there, in particular, the $\mathbf{V}$ matrix for this design given by (1.15) of Chapter 2. The exact $(1-\alpha)$-level simultaneous confidence intervals for all contrasts $\Sigma_{i=1}^k c_i\theta_i$ using the S-procedure work out to be (see Halperin and Greenhouse 1958):

$$\sum_{i=1}^k c_i\theta_i \in \left[\sum_{i=1}^k c_i\hat{\theta}_i \pm \{(k-1)F_{k-1,\nu}^{(\alpha)}\}^{1/2} S \right.$$
$$\left. \times \left\{\sum_{i=1}^k \frac{c_i^2}{n_i} + \frac{1}{S_{XX}}\left[\sum_{i=1}^k c_i(\bar{X}_{i\cdot} - \bar{X}_{\cdot\cdot})\right]^2\right\}^{1/2}\right] \quad \forall\, \mathbf{c} \in \mathbb{C}^k\,. \tag{1.4}$$

For pairwise differences $\theta_i - \theta_j$, the intervals of (1.4) simplify to the following conservative $(1-\alpha)$-level simultaneous confidence intervals:

$$\theta_i - \theta_j \in \left[\hat{\theta}_i - \hat{\theta}_j \pm \{(k-1)F^{(\alpha)}_{k-1,\nu}\}^{1/2} S \right.$$
$$\left. \times \left\{\frac{1}{n_i} + \frac{1}{n_j} + \frac{(\bar{X}_{i.} - \bar{X}_{j.})^2}{S_{XX}}\right\}^{1/2}\right] \quad (1 \leq i < j \leq k). \tag{1.5}$$

Similar formulas apply for estimating the contrasts (general and pairwise) among the treatment effects in a randomized block design with a fixed covariate and fixed block effects.

We now discuss the relation between the S-procedure and the analysis of variance (ANOVA) F-test. As seen in Theorem 2.1 of Chapter 2, the simultaneous confidence intervals (1.1) are based on showing that

$$\sup_{\mathbf{l} \in \mathcal{L}} \left\{\frac{[\mathbf{l}'(\hat{\boldsymbol{\theta}} - \boldsymbol{\theta})]^2}{S^2(\mathbf{l}'\mathbf{V}\mathbf{l})}\right\} = \frac{(\hat{\boldsymbol{\theta}} - \boldsymbol{\theta})'\mathbf{L}'(\mathbf{L}\mathbf{V}\mathbf{L}')^{-1}\mathbf{L}(\hat{\boldsymbol{\theta}} - \boldsymbol{\theta})}{S^2} \sim mF_{m,\nu} \tag{1.6}$$

where $\mathcal{L}$ is a specified subspace of dimension $m \leq k$ and $\mathbf{L}: m \times k$ is any matrix whose rows form a basis for $\mathcal{L}$. In particular, if $\mathcal{L}$ is the contrast subspace $\mathbb{C}^k$ and if the overall null hypothesis $H_0: \theta_1 = \theta_2 = \cdots = \theta_k$ holds, then (1.6) becomes

$$\sup_{\mathbf{c} \in \mathbb{C}^k} \left\{\frac{(\mathbf{c}'\hat{\boldsymbol{\theta}})^2}{S^2(\mathbf{c}'\mathbf{V}\mathbf{c})}\right\} = \frac{\hat{\boldsymbol{\theta}}'\mathbf{C}'(\mathbf{C}\mathbf{V}\mathbf{C}')^{-1}\mathbf{C}\hat{\boldsymbol{\theta}}}{S^2} \sim (k-1)F_{k-1,\nu}; \tag{1.7}$$

here $\mathbf{C}: (k-1) \times k$ is any matrix whose rows are linearly independent contrasts. For a one-way layout, (1.7) becomes

$$\sup_{\mathbf{c} \in \mathbb{C}^k} \frac{\left(\sum_{i=1}^{k} c_i \bar{Y}_i\right)^2}{S^2 \sum_{i=1}^{k} (c_i^2/n_i)} = \frac{\sum_{i=1}^{k} n_i(\bar{Y}_i - \bar{\bar{Y}})^2}{S^2} \tag{1.8}$$

where $\bar{\bar{Y}} = \Sigma_{i=1}^k n_i \bar{Y}_i / \Sigma_{i=1}^k n_i$ is the grand mean of the Y_{ij}'s. Note that the statistic in (1.7) is simply $(k-1)$ times the ANOVA F-statistic for testing H_0 and therefore the ANOVA F-test rejects H_0 at level α if and only if

$$\frac{(\mathbf{c}'\hat{\boldsymbol{\theta}})^2}{S^2(\mathbf{c}'\mathbf{V}\mathbf{c})} > (k-1)F^{(\alpha)}_{k-1,\nu} \qquad \text{for some } \mathbf{c} \in \mathbb{C}^k. \tag{1.9}$$

But (1.9) is precisely the test of significance of any contrast $\mathbf{c}'\hat{\boldsymbol{\theta}}$ based on the S-procedure (1.1) with $\mathcal{L} = \mathbb{C}^k$. This means that the ANOVA F-test for H_0 is significant at level α if and only if at least some contrast $\mathbf{c}'\hat{\boldsymbol{\theta}}$ (in particular, the one giving the supremum in (1.7)) is significant at FWE = α using the S-procedure. For a one-way layout, the contrast $\mathbf{c}$ that gives the supremum in (1.8) can be shown to be given by

$$c_i = an_i(\bar{Y}_i - \bar{\bar{Y}}) \qquad (1 \leqq i \leqq k)$$

where a is an arbitrary nonzero constant. However, this contrast may not be of any practical interest and the S-procedure does not provide a method for hunting other significant contrasts. Gabriel (1964) proposed a different way of applying the S-procedure to make detailed inferences implicit in a significant ANOVA F-test. This is discussed in the following section.

1.2 A Simultaneous Test Procedure for the Family of Subset Hypotheses

For convenience, we restrict our attention to the one-way layout setting in this section. To test a subset hypothesis $H_P: \theta_i = \theta_j \ \forall\, i, j \in P$, the test associated with the S-procedure (1.2) rejects if the supremum of the test statistic

$$\frac{\left(\sum_{i=1}^{k} c_i \bar{Y}_i\right)^2}{S^2 \sum_{i=1}^{k} (c_i^2/n_i)}$$

over the subspace of $\mathbb{C}^k$ such that $\Sigma_{i \in P}\, c_i = 0$, $c_j = 0$ for $j \not\in P$ exceeds the critical constant $(k-1)F_{k-1,\nu}^{(\alpha)}$. In analogy with (1.8), this supremum is given by the left hand side of (1.10) below and the resulting simultaneous test procedure rejects H_P if

$$(p-1)F_P = \frac{\sum_{i \in P} n_i(\bar{Y}_i - \bar{\bar{Y}}_P)^2}{S^2} > (k-1)F_{k-1,\nu}^{(\alpha)} \qquad (1.10)$$

where $\bar{\bar{Y}}_P = (\Sigma_{i \in p}\, n_i \bar{Y}_i)/(\Sigma_{i \in P}\, n_i)$. Since all of these tests for the subset hypotheses H_P are derived from a common $(1-\alpha)$-level confidence statement (1.2), it is clear that the FWE of this simultaneous test procedure is strongly controlled at level α. In Example 3.2 of Chapter 2 we noted that this simultaneous test procedure is coherent, being based on likelihood ratio (LR) test statistics. Observe that the ANOVA F-test for the overall null hypothesis $H_0 = \bigcap_P H_P: \theta_1 = \theta_2 = \cdots = \theta_k$ is included in (1.10). This procedure readily extends to more general designs.

Gabriel (1964) pointed out that because of coherence, results of the tests (1.10) on all sets P can be concisely summarized either in the form of a list of *minimal significant sets*, that is, significant sets that have no significant proper subsets, or in the form of a list of *maximal nonsignificant sets*, that is, nonsignificant sets that are not proper subsets of other nonsignificant sets. Significance or nonsignificance of any set P can be readily ascertained from such a list.

Peritz (1965) noted that every minimal significant set P determined using (1.10) can be divided into two disjoint subsets $Q = \{i \in P: \bar{Y}_i > \bar{\bar{Y}}_P\}$ and $R = \{j \in P: \bar{Y}_j \leqq \bar{\bar{Y}}_P\}$ such that one can make the statement

$$\max_{i \in Q} \{\theta_i\} > \min_{j \in R} \{\theta_j\},$$

and the probability that any such statement is erroneous is controlled at α. This permits a partial ordering of the θ_i's.

Now consider the family of multiple subset hypotheses $\{H_{\mathbf{P}} = \bigcap_{i=1}^{r} H_{P_i}\}$ where $\mathbf{P} = (P_1, P_2, \ldots, P_r)$ is a "partition" of $\{1, 2, \ldots, k\}$, that is, the P_i's are disjoint subsets of $\{1, 2, \ldots, k\}$ with cardinalities $p_i \geqq 2$ and $\sum_{i=1}^{r} p_i \leqq k$. Note that this is a hierarchical family since if $\mathbf{Q} = (Q_1, Q_2, \ldots, Q_s)$ is a subpartition of $\mathbf{P} = (P_1, P_2, \ldots, P_r)$ (i.e., if every Q_j is a subset of some P_i), then $H_{\mathbf{P}}$ implies $H_{\mathbf{Q}}$.

A simultaneous test procedure for this family can be constructed by applying the union-intersection (UI) method to tests of component hypotheses H_{P_i} given by (1.10). The resulting procedure rejects $H_{\mathbf{P}}$ if

$$\max_{1 \leqq i \leqq r} (p_i - 1)F_{P_i} > (k-1)F_{k-1,\nu}^{(\alpha)} . \tag{1.11}$$

A more powerful procedure in this case rejects $H_{\mathbf{P}}$ if

$$\sum_{i=1}^{r} (p_i - 1)F_{P_i} > (k-1)F_{k-1,\nu}^{(\alpha)} .$$

This procedure is coherent and controls the FWE at level α, but is not consonant.

1.3 Examples

Example 1.1 ***(Randomized Block Design).*** Duncan (1955) gave the data shown in Table 1.1 on the mean yields (in bushels per acre) of seven varieties of barley, labeled A through G, which were each replicated six times in a randomized block design. Duncan also gave an analysis of variance, which shows that $MS_{\text{varieties}} = 366.97$ with 6 d.f. and $MS_{\text{error}} = 79.64$ with 30 d.f. Thus $S = \sqrt{79.64} = 8.924$. Also $F = 4.61$, which is highly significant ($P < 0.005$). We now apply the S-procedure to make detailed comparisons among the varietal means to find where the significant differences lie.

TABLE 1.1. Varietal Means (in Bushels per Acre)

Variety	A	B	C	D	E	F	G
$\bar{Y}$	49.6	71.2	67.6	61.5	71.3	58.1	61.0

Source: Duncan (1955).

From (1.3) we obtain the critical value to be used for testing pairwise differences with (conservative) FWE $= \alpha = 0.05$ as

$$\sqrt{(k-1)F^{(\alpha)}_{k-1,\nu}} \times S\sqrt{\frac{2}{n}} = \sqrt{6 \times 2.42} \times 8.924\sqrt{\frac{2}{6}} = 19.63\,.$$

Using this critical value we find two pairwise differences significant: $E - A = 71.3 - 49.6 = 21.7$ and $B - A = 71.2 - 49.6 = 21.6$. Since this is a simultaneous test procedure we can also make directional decisions (see Theorem 2.2 of Chapter 2) and claim that A is significantly "worse" than both B and E.

After looking at the data the researcher may wish to compare the average yield of the varieties A, F, and G, which produce the three lowest sample mean yields, with the average yield of the varieties C, B, and E, which produce the three highest sample mean yields. Since this contrast is suggested by the data, it is appropriate to use formula (1.2). For $1 - \alpha = 0.95$ we obtain the desired interval to be

$$\frac{67.6 + 71.2 + 71.3}{3} - \frac{49.6 + 58.1 + 61.0}{3} \pm \sqrt{6 \times 2.42} \times 8.924 \times \sqrt{\frac{6}{9 \times 6}}$$
$$= 13.80 \pm 11.34 = (2.46, 25.14)\,.$$

We next illustrate use of the S-procedure for the family of all subset homogeneity hypotheses. In carrying out this simultaneous test procedure we need not test any subsets containing varieties A and E or A and B, since such subsets will be automatically significant because the pairs (A, E) and (A, B) have been found significant. The critical value to be used in this procedure for $\alpha = 0.05$ is $(k-1)F^{(\alpha)}_{k-1,\nu} = 6 \times 2.42 = 14.52$. We begin with the largest subset not containing these two pairs, which is $P = (F, G, D, C, B, E)$. For this subset $(p-1)F_P = 12.10 < 14.52$, and therefore the subset is nonsignificant (i.e., the null hypothesis H_P is not rejected). The next largest subset that remains to be tested is $P = (A, F, G, D, C)$ for which $(p-1)F_P = 12.94 < 14.52$, and therefore this subset is also nonsignificant. In fact, these two subsets are maximal nonsignificant sets, any subsets of them being also nonsignificant. Decisions on individual subsets can be combined to make a decision on any multiple subset hypothesis as indicated in Section 1.2. □

Example 1.2 **(*Randomized Block Design with a Fixed Covariate*).** Steel and Torrie (1980, p. 412) give data on the ascorbic acid content (measured in mg per 100 g of dry weight) for 11 varieties of lima beans planted in $n = 5$ randomized blocks. To keep the example short we consider only the first $k = 6$ varieties. From past experience it was known that the ascorbic acid content was negatively correlated with maturity of the plants at harvest. Percentage of dry matter (from 100 g of freshly harvested beans) was measured as an index of maturity and used as a covariate. This covariate is clearly not fixed but we regard it to be so for illustration purposes in this example and also in Example 3.2 of this chapter. (See Example 2.3 of Chapter 8 where the same data are used, but the covariate is regarded as random.) The data are reproduced in Table 1.2. Table 1.3 gives the sums of squares and cross products of X and Y for the given data. The formulas for the analysis of covariance for a randomized block design are slightly different from the ones given in Example 1.2 of Chapter 2 for a one-way layout, and are given in (2.37)–(2.41) of Chapter 8. Using those formulas we get $SS_{\text{error}} = S_{YY} - S_{XY}^2/S_{XX} = 959.3$ with $\nu = (k-1)(n-1) - 1 = 19$ d.f., and hence $MS_{\text{error}} = S^2 =$

TABLE 1.2. Ascorbic Acid Content[a] (Y) and Percentage of Dry Matter[b] (X) for Lima Beans

	Block									
	1		2		3		4		5	
Variety	X	Y	X	Y	X	Y	X	Y	X	Y
1	34.0	93.0	33.4	94.8	34.7	91.7	38.9	80.8	36.1	80.2
2	39.6	47.3	39.8	51.5	51.2	33.3	52.0	27.2	56.2	20.6
3	31.7	81.4	30.1	109.0	33.8	71.6	39.6	57.5	47.8	30.1
4	37.7	66.9	38.2	74.1	40.3	64.7	39.4	69.3	41.3	63.2
5	24.9	119.5	24.0	128.5	24.9	125.6	23.5	129.0	25.1	126.2
6	30.3	106.6	29.1	111.4	31.7	99.0	28.3	126.1	34.2	95.6

[a]In mg per 100 g of dry weight.
[b]Based on 100 g of freshly harvested beans.
Source: Steel and Torrie (1980, p. 412).

TABLE 1.3. Sums of Squares and Cross Products for Data in Table 1.2

	XX	XY	YY
Varieties	$A_{XX} = 1552.8$	$A_{XY} = -6198.4$	$A_{YY} = 25409.6$
Blocks	$B_{XX} = 234.0$	$B_{XY} = -654.6$	$B_{YY} = 2045.7$
Error	$S_{XX} = 257.1$	$S_{XY} = -714.8$	$S_{YY} = 2946.6$
Total	$T_{XX} = 2043.9$	$T_{XY} = -7567.8$	$T_{YY} = 30401.9$

TABLE 1.4. Adjusted Varietal Means for Data in Table 1.2

Variety	1	2	3	4	5	6	
$\bar{X}_{i\cdot}$	35.42	47.76	36.60	39.38	24.48	30.72	$\bar{X}_{\cdot\cdot} = 35.73$
$\bar{Y}_{i\cdot}$	88.10	35.98	69.92	67.64	125.76	107.74	
$\hat{\theta}_i$	87.24	69.43	72.34	77.80	94.49	93.82	

TABLE 1.5. Simultaneous 90% Scheffé Intervals for All Pairwise Differences for Data in Table 1.2

(i, j)	$\hat{\theta}_i - \hat{\theta}_j \pm$ Allowance	(i, j)	$\hat{\theta}_i - \hat{\theta}_j \pm$ Allowance
(1, 2)	17.81 ± 23.38	(2, 6)	−24.39 ± 28.96
(1, 3)	14.90 ± 14.93	(3, 4)	−5.46 ± 15.43
(1, 4)	9.44 ± 15.95	(3, 5)	−22.15 ± 23.02
(1, 5)	−7.25 ± 21.79	(3, 6)	−21.48 ± 17.07
(1, 6)	−6.58 ± 16.32	(4, 5)	−16.69 ± 26.39
(2, 3)	−2.91 ± 22.13	(4, 6)	−16.02 ± 19.51
(2, 4)	−8.37 ± 19.21	(5, 6)	0.67 ± 17.43
(2, 5)	−25.06 ± 37.11		

$959.3/19 = 50.49$ ($S = 7.106$). The estimate of the common slope is $\hat{\beta} = S_{XY}/S_{XX} = -2.78$.

Table 1.4 shows the calculation of the adjusted varietal means $\hat{\theta}_i = \bar{Y}_{i\cdot} - \hat{\beta}(\bar{X}_{i\cdot} - \bar{X}_{\cdot\cdot})$. In (1.5) we use $\alpha = 0.10$, $F_{5,19}^{(.10)} = 2.18$ and thus obtain 90% Scheffé intervals for all pairwise comparisons between the varieties, which are given in Table 1.5. Note that these intervals are rather wide. Based on these intervals we find that only varieties 3 and 6 are significantly different (variety 6 is significantly better than variety 3) and pair (1, 3) is on the borderline of significance. One can use (1.4) to construct a confidence interval or to test the significance for any contrast $\Sigma_{i=1}^{k} c_i\theta_i$. □

2 TUKEY'S T-PROCEDURE FOR BALANCED DESIGNS

2.1 Balanced One-Way Layouts

2.1.1 Pairwise Comparisons

Tukey's T-procedure was given as a simultaneous confidence statement (2.8) in Chapter 2 for the family of pairwise comparisons in a balanced one-way layout. For convenience, we reproduce it here: Exact $(1-\alpha)$-level simultaneous confidence intervals for all pairwise differences $\theta_i - \theta_j$ are given by

$$\theta_i - \theta_j \in [\bar{Y}_i - \bar{Y}_j \pm Q_{k,\nu}^{(\alpha)} S/\sqrt{n}] \qquad (1 \leqq i < j \leqq k) \tag{2.1}$$

where $Q_{k,\nu}^{(\alpha)}$ is the upper α point of the Studentized range distribution with parameter k and ν d.f.

One can test the significance of any pairwise difference using the intervals (2.1) with the FWE controlled at α. The resulting procedure is sometimes referred to in the literature as the *honestly significant difference* or the *wholly significant difference* (the HSD or the WSD) test.

We now mention some optimality properties of the T-procedure. Gabriel (1970) showed that in a balanced one-way layout, among all procedures that give equal width intervals for all pairwise differences with joint confidence level $\geqq 1 - \alpha$, the T-procedure (2.1) gives the shortest intervals. This follows from the union-intersection (UI) nature of the T-procedure noted in Example 2.2 of Chapter 2. Genizi and Hochberg (1978) showed that in the class of all transformation procedures (see Section 3.2.4) having joint confidence level $\geqq 1 - \alpha$, the T-procedure gives the shortest interval for *every* pairwise difference. Based on the result of Kunte and Rattihalli (1984) for the two-dimensional case, one may conjecture another optimality property for the T-procedure for $k > 2$, namely, that among all $(1 - \alpha)$-level simultaneous confidence procedures with confidence intervals of the form $\hat{\theta}_i - \hat{\theta}_j \pm d_{ij} S$, the T-procedure (2.1) minimizes the quantity $\Pi_{1 \leqq i < j \leqq k} d_{ij}$, which is proportional to the volume of the joint confidence parallelepiped for the pairwise differences $\theta_i - \theta_j$ $(1 \leqq i < j \leqq k)$.

2.1.2 All Contrasts

The intervals (2.1) can be extended to the family of all contrasts $\Sigma_{i=1}^{k} c_i \theta_i$, yielding the following $(1 - \alpha)$-level simultaneous confidence intervals:

$$\sum_{i=1}^{k} c_i \theta_i \in \left[\sum_{i=1}^{k} c_i \bar{Y}_i \pm Q_{k,\nu}^{(\alpha)} \frac{S}{\sqrt{n}} \sum_{i=1}^{k} \frac{|c_i|}{2} \right] \qquad \forall\, \mathbf{c} \in \mathbb{C}^k . \tag{2.2}$$

This extension is based on the following lemma.

Lemma 2.1 (Tukey 1953). Let $\mathbf{x} = (x_1, \ldots, x_k)'$ be any real vector and let ξ_{ij} $(1 \leqq i < j \leqq k)$ be nonnegative numbers. Then

$$|x_i - x_j| \leqq \xi_{ij} \qquad \forall\, i < j \Leftrightarrow |\mathbf{c}'\mathbf{x}| \leqq \frac{\sum_{i=1}^{k} \sum_{j=1}^{k} c_i^+ c_j^- \xi_{ij}}{\frac{1}{2} \sum_{i=1}^{k} |c_i|} \qquad \forall\, \mathbf{c} \in \mathbb{C}^k \tag{2.3}$$

where $c_i^+ = \max(c_i, 0)$ and $c_j^- = -\min(c_j, 0)$. When $\xi_{ij} = \xi \geqq 0$, (2.3) simplifies to

$$|x_i - x_j| \leqq \xi \qquad \forall\, i<j \Leftrightarrow |\mathbf{c}'\mathbf{x}| \leqq \xi \sum_{i=1}^{k} \frac{|c_i|}{2} \qquad \forall\, \mathbf{c} \in \mathbb{C}^k . \tag{2.4}$$

□

We obtain (2.2) by letting $x_i = \bar{Y}_i - \theta_i$ $(1 \leqq i \leqq k)$ and $\xi = Q_{k,\nu}^{(\alpha)} S/\sqrt{n}$ in (2.4), and by making use of the fact that (2.1) holds with probability $1 - \alpha$ and hence (2.2) does too.

2.1.3 All Linear Combinations

We now consider an extension of the T-procedure to the family of all linear combinations of the θ_i's; this extension is also due to Tukey (1953). Arbitrary linear combinations become of interest when, for example, the experimenter wants simultaneous inferences on the θ_i's in addition to contrasts $\mathbf{c}'\boldsymbol{\theta}$, $\mathbf{c} \in \mathbb{C}^k$. The following algebraic result enables us to extend the T-procedure to all linear combinations of the θ_i's.

Lemma 2.2 (Tukey 1953). Let $\mathbf{x} = (x_1, \ldots, x_k)'$ be an arbitrary real vector and let $x_0 \equiv 0$. Define a norm for vector $\mathbf{l} \in \mathbb{R}^k$ as

$$M(\mathbf{l}) = \max\left\{ \sum_{i=1}^{k} l_i^+ , \sum_{i=1}^{k} l_i^- \right\} . \tag{2.5}$$

Then

$$\sup_{\mathbf{l} \in \mathbb{R}^k} \frac{|\mathbf{l}'\mathbf{x}|}{M(\mathbf{l})} = \max_{0 \leqq i < j \leqq k} |x_i - x_j| \tag{2.6}$$

$$= \max\{ \max_{1 \leqq i<j \leqq k} |x_i - x_j|, \max_{1 \leqq i \leqq k} |x_i| \} .$$

□

If we let $\mathbf{x} = (\bar{\mathbf{Y}} - \boldsymbol{\theta})\sqrt{n}/S$ where $\bar{\mathbf{Y}} = (\bar{Y}_1, \ldots, \bar{Y}_k)'$, then it follows from (2.6) that

$$\Pr\left\{ \sup_{\mathbf{l} \in \mathbb{R}^k} \frac{\sqrt{n}|\mathbf{l}'(\bar{\mathbf{Y}} - \boldsymbol{\theta})|}{S M(\mathbf{l})} \leqq \xi \right\} = \Pr\{ Q'_{k,\nu} \leqq \xi \}$$

where

$$Q'_{k,\nu} = \max\left\{ \max_{1 \leqq i<j \leqq k} \frac{|(\bar{Y}_i - \theta_i) - (\bar{Y}_j - \theta_j)|}{S/\sqrt{n}} , \max_{1 \leqq i \leqq k} \frac{|\bar{Y}_i - \theta_i|}{S/\sqrt{n}} \right\} .$$

The first term inside the braces is a Studentized range r.v. $Q_{k,\nu}$ and the second term is a Studentized maximum modulus r.v. $|M|_{k,\nu}$. The r.v. $Q'_{k,\nu}$

has the *Studentized augmented range* distribution (see Appendix 3). If $Q_{k,\nu}'^{(\alpha)}$ denotes the upper α point of $Q_{k,\nu}'$, then exact $(1-\alpha)$-level simultaneous confidence intervals for all linear combinations $\Sigma_{i=1}^{k} l_i\theta_i$ are given by

$$\sum_{i=1}^{k} l_i\theta_i \in \left[\sum_{i=1}^{k} l_i\bar{Y}_i \pm \frac{S}{\sqrt{n}} Q_{k,\nu}'^{(\alpha)} M(\mathbf{l})\right] \quad \forall \mathbf{l} \in \mathbb{R}^k . \tag{2.7}$$

We have not included the tables of $Q_{k,\nu}'^{(\alpha)}$ in this book since their use is extremely limited; the interested reader is referred to Stoline (1978). Tukey (1953) noted that $Q_{k,\nu}^{(\alpha)}$ is a good approximation for $Q_{k,\nu}'^{(\alpha)}$ provided $k \geqq 3$ and $\alpha \leqq 0.05$.

2.2 General Balanced Designs

Thus far we have restricted the discussion of the T-procedure to the balanced one-way layout setting for which it was originally proposed by Tukey (1953). However, Tukey also extended it to designs where the $\hat{\theta}_i$'s have a constant variance and are equicorrelated, that is, the covariance matrix $\sigma^2\mathbf{V}$ of the $\hat{\theta}_i$'s is a "uniform" matrix with $v_{ii} = v$ (say) for $1 \leqq i \leqq k$ and $v_{ij} = u$ (say) for $1 \leqq i \neq j \leqq k$. Thus the common correlation coefficient ρ equals u/v with $-1/(k-1) \leqq \rho \leqq 1$. The resulting exact $(1-\alpha)$-level simultaneous confidence intervals for all contrasts $\Sigma_{i=1}^{k} c_i\theta_i$ are given by

$$\sum_{i=1}^{k} c_i\theta_i \in \left[\sum_{i=1}^{k} c_i\hat{\theta}_i \pm Q_{k,\nu}^{(\alpha)} S\sqrt{v(1-\rho)} \sum_{i=1}^{k} \frac{|c_i|}{2}\right] \quad \forall \mathbf{c} \in \mathbb{C}^k \tag{2.8}$$

A proof of (2.8) may be found in Miller (1981, p. 46) or Scheffé (1959, p. 75). An application of (2.8) to multiple comparisons in a balanced incomplete block design is given in Section 2.4.

Hochberg (1974c) proved a more general result that if all pairwise differences $(\hat{\theta}_i - \hat{\theta}_j)$ have the same variance,

$$\operatorname{var}(\hat{\theta}_i - \hat{\theta}_j) = \sigma^2(v_{ii} + v_{jj} - 2v_{ij}) = 2\sigma^2 v_0 \text{ (say)} , \tag{2.9}$$

then exact $(1-\alpha)$-level simultaneous confidence intervals for all contrasts are given by

$$\sum_{i=1}^{k} c_i\theta_i \in \left[\sum_{i=1}^{k} c_i\hat{\theta}_i \pm Q_{k,\nu}^{(\alpha)} S\sqrt{v_0} \sum_{i=1}^{k} \frac{|c_i|}{2}\right] \quad \forall \mathbf{c} \in \mathbb{C}^k . \tag{2.10}$$

Any design satisfying (2.9) is referred to as *pairwise balanced*.

Scheffé (1959, p. 29) and Miller (1981, p. 42) stated that the simultaneous confidence intervals (2.7) for all linear combinations can be similarly extended to the case where the $\hat{\theta}_i$'s have equal variances and correlations, but Hochberg (1976b) showed that this assertion is incorrect.

2.3 A Simultaneous Test Procedure for the Family of Subset Hypotheses

The T-procedure (2.1) implies the following simultaneous test procedure for the family of subset hypotheses in a balanced one-way layout design: Reject $H_P: \theta_i = \theta_j \ \forall \ i, j \in P$ if

$$R_P = \frac{\max_{i \in P} \bar{Y}_i - \min_{i \in P} \bar{Y}_i}{S/\sqrt{n}} > Q_{k,\nu}^{(\alpha)} . \tag{2.11}$$

In Example 3.3 of Chapter 2 we see that this procedure is coherent, consonant, and controls the FWE at level α. For a general pairwise balanced design (satisfying (2.9)), (2.11) takes the form: Reject H_P if

$$R_P = \frac{\max_{i \in P} \hat{\theta}_i - \min_{i \in P} \hat{\theta}_i}{S\sqrt{v_0}} > Q_{k,\nu}^{(\alpha)} . \tag{2.12}$$

This simultaneous test procedure can be extended to the family of multiple subset hypotheses by the UI method. The resulting procedure rejects any multiple subset hypothesis $H_{\mathbf{P}} = \bigcap_{i=1}^{r} H_{P_i}$ if

$$\max_{1 \leq i \leq r} R_{P_i} > Q_{k,\nu}^{(\alpha)} . \tag{2.13}$$

This procedure is also coherent, consonant, and controls the FWE at level α.

2.4 Examples

***Example 2.1** (**Randomized Block Design**; **Continuation of Example 1.1**).* We now apply the T-procedure to the barley data of Example 1.1. For making pairwise comparisons at $\alpha = 0.05$ the appropriate critical value is $Q_{k,\nu}^{(\alpha)} S/\sqrt{n} = 4.46 \times 8.924/\sqrt{6} = 16.26$. (Compare this critical value with 19.63 for the S-procedure.) Using this critical value we find one additional significant difference as compared to the S-procedure, namely $C - A = 67.6 - 49.6 = 18$.

We next construct a 95% confidence interval for the contrast $\{(C +$

$B + E) - (A + F + G)\}/3$ (selected *a posteriori*), which was also analyzed by the S-procedure. Using (2.2) we obtain the desired confidence interval as

$$13.80 \pm 16.26 = (-2.46, 30.06) .$$

Note that although the T-procedure gives narrower confidence intervals than the S-procedure for pairwise comparisons, for the higher order contrast considered here it gives a substantially wider confidence interval.

By using the range procedure (2.11) at $\alpha = 0.05$ we find the maximal nonsignificant sets to be (A, F, G, D) and (F, G, D, C, B, E). Recall that in Example 1.1 we obtained (A, F, G, D, C) and (F, G, D, C, B, E) as the maximal nonsignificant sets by using the procedure (1.10). Thus in this example (2.11) declares some additional subsets significant in comparison to (1.10). This is, however, not always the case. For a further discussion of the powers of the two procedures see Section 4.1. □

Example 2.2 **(*Balanced Incomplete Block Design*).** From (1.19) of Chapter 2 we see that a balanced incomplete block (BIB) design has a **V** matrix with $v_{ii} = v = p(k-1)/\lambda k^2$, $v_{ij} = u = -p/\lambda k^2$ for $i \neq j$, and thus $\rho = -1/(k-1)$. Because of the common correlation we can use (2.8) to obtain the following simultaneous $100(1-\alpha)\%$ confidence intervals for all pairwise comparisons among the θ_i's:

$$\theta_i - \theta_j \in \left[\hat{\theta}_i - \hat{\theta}_j \pm Q_{k,\nu}^{(\alpha)} S \sqrt{\frac{p}{\lambda k}}\right] \qquad (1 \leq i < j \leq k) \qquad (2.14)$$

where the $\hat{\theta}_i$'s and $S^2 = MS_{\text{error}}$ with $\nu = bp - b - k + 1$ d.f. are given, respectively, by (1.18) and (1.20) of Chapter 2. □

3 MODIFICATIONS OF THE T-PROCEDURE FOR UNBALANCED DESIGNS

The family of pairwise comparisons is of common interest in practice. We have seen that the T-procedure is optimal for this family according to various criteria for balanced designs. Thus it is natural to seek modifications of the T-procedure for making pairwise comparisons in unbalanced designs. Although the S-procedure can be used for this purpose (see (1.3)), it is known to be overly conservative in this case.

Many of the proposed modifications of the T-procedure are designed to address the family of pairwise comparisons directly and give simultaneous confidence intervals of the form

$$\theta_i - \theta_j \in [\hat{\theta}_i - \hat{\theta}_j \pm \xi S\sqrt{d_{ij}}] \qquad (1 \leqq i < j \leqq k) \tag{3.1}$$

where

$$d_{ij} = v_{ii} + v_{jj} - 2v_{ij} \qquad (1 \leqq i < j \leqq k)\,. \tag{3.2}$$

By using (2.3) these intervals can be extended to the family of all contrasts as follows:

$$\sum_{i=1}^{k} c_i\theta_i \in \left[\sum_{i=1}^{k} c_i\hat{\theta}_i \pm \xi S \frac{\sum_{i=1}^{k}\sum_{j=1}^{k} c_i^+ c_j^- \sqrt{d_{ij}}}{\frac{1}{2}\sum_{i=1}^{k} |c_i|} \right] \quad \forall \mathbf{c} \in \mathbb{C}^k\,. \tag{3.3}$$

Depending upon whether the critical constant ξ is chosen to guarantee that the coverage probability of (3.1) is exactly or approximately equal to the nominal confidence level $1-\alpha$, we have an exact or approximate modification of the T-procedure. Most of the approximate modifications are in fact conservative, that is, they use upper bounds on the exact value of ξ and thus guarantee that the associated coverage probability is at least $1-\alpha$. Exact procedures are discussed in Section 3.1 and approximate/conservative procedures are discussed in Section 3.2. Illustrative examples are given in Section 3.3.

3.1 Exact Procedures

3.1.1 Pairwise Comparisons

In order for the simultaneous confidence intervals (3.1) to have a joint confidence level exactly $1-\alpha$, the critical constant ξ must be the upper α point of $\max_{1\leqq i<j\leqq k} |T_{ij}|$ where

$$T_{ij} = \frac{\hat{\theta}_i - \hat{\theta}_j - (\theta_i - \theta_j)}{S\sqrt{d_{ij}}} \qquad (1 \leqq i < j \leqq k)\,. \tag{3.4}$$

The T_{ij}'s have a joint $\binom{k}{2}$-variate singular t-distribution with ν d.f. and the associated correlation matrix, which can be readily computed. Thus in principle, it is possible to numerically evaluate the required critical constant $\xi = \xi_{k,\nu}^{(\alpha)}$* by solving the equation

* Here we have suppressed the dependence of ξ on $\mathbf{V}$; it can be shown that the correlation matrix associated with the T_{ij}'s depends only on the d_{ij}'s, and thus ξ depends on $\mathbf{V}$ only through the d_{ij}'s.

$$\Pr\{\max_{1\leq i<j\leq k} |T_{ij}| \leq \xi\} = 1 - \alpha . \tag{3.5}$$

However, this is generally a difficult computational task. Spurrier and Isham (1985) have carried out this task for unbalanced one-way layout designs (in which case $d_{ij} = 1/n_i + 1/n_j$ for $i \neq j$) for $k = 3$. Their method as extended by Hayter (1985) to general unbalanced designs is outlined in the following paragraph.

Denote $T_1 = T_{12}$, $T_2 = T_{13}$, and $T_3 = T_{23}$. Then

$$T_3 = \frac{\sqrt{d_{13}}T_2 - \sqrt{d_{12}}T_1}{\sqrt{d_{23}}} \tag{3.6}$$

and (T_1, T_2) have a bivariate t-distribution with ν d.f. and the associated correlation coefficient

$$\rho = \frac{v_{11} - v_{12} - v_{13} + v_{23}}{\sqrt{d_{12}d_{13}}}$$

$$= \frac{d_{12} + d_{13} - d_{23}}{2\sqrt{d_{12}d_{13}}} .$$

Using (3.6) we can write (3.5) as

$$\int_{-\xi}^{\xi} \int_{l(t_1,\xi)}^{u(t_1,\xi)} f_\nu(t_1, t_2; \rho)\, dt_2\, dt_1 = 1 - \alpha \tag{3.7}$$

where $f_\nu(t_1, t_2; \rho)$ is the bivariate t density function of (T_1, T_2),

$$l(t_1, \xi) = \max\left\{-\xi, \frac{\sqrt{d_{12}}t_1 - \sqrt{d_{23}}\xi}{\sqrt{d_{13}}}\right\}$$

and

$$u(t_1, \xi) = \min\left\{\xi, \frac{\sqrt{d_{12}}t_1 + \sqrt{d_{23}}\xi}{\sqrt{d_{13}}}\right\} .$$

Note that $|\sqrt{d_{12}} - \sqrt{d_{23}}| \leq \sqrt{d_{13}}$ always and this ensures that $u(t_1, \xi) \geq l(t_1, \xi)$ for $-\xi \leq t_1 \leq \xi$. Spurrier and Isham (1985) tabulated the solutions $\xi_{k,\nu}^{(\alpha)}$ to equation (3.7) for $k = 3$, $\alpha = 0.01$, 0.05, and 0.10, and n_1, n_2, n_3 satisfying $n_i \geq 3$, $10 \leq N = \Sigma_{i=1}^{3} n_i \leq 29$, and $\nu = N - 3$. For $N \geq 30$ they suggested the following approximation:

$$\xi_{3,\nu}^{(\alpha)} \cong \xi_{3,\infty}^{(\alpha)}\left\{\frac{Q_{3,\nu}^{(\alpha)}}{Q_{3,\infty}^{(\alpha)}}\right\} . \tag{3.8}$$

The constant $\xi_{3,\infty}^{(\alpha)}$ depends on the proportions n_i/N $(1 \leqq i \leqq 3)$ and is tabulated in their paper for $\alpha = 0.01, 0.05$, and 0.10. Thus for unbalanced one-way layouts with three treatments we now have the tables available to construct exact simultaneous pairwise confidence intervals (which can be extended to all contrasts using (3.3)) for practical values of n_1, n_2, n_3, and α. For a general unbalanced design with given values of the d_{ij}'s (defined in (3.2)) we can find (by rounding off to the nearest integers) the "equivalent" one-way layout sample sizes

$$n_1 = \frac{2}{d_{12} + d_{13} - d_{23}}, \; n_2 = \frac{2}{d_{12} + d_{23} - d_{13}}, \; n_3 = \frac{2}{d_{13} + d_{23} - d_{12}}. \tag{3.9}$$

However, the error d.f. ν for the given design will not generally equal $\Sigma_{i=1}^{3} n_i - 3$ for which case Spurrier and Isham (1985) have prepared their tables. If ν is large ($\geqq 27$), then the approximation given by (3.8) can be employed with the proportions $n_i/\Sigma_{i=1}^{3} n_i$ (which are required to find $\xi_{3,\infty}^{(\alpha)}$) computed from (3.9).

The preceding method can be extended to higher values of k but the computations become more formidable and tedious. Hayter (1985) has given the necessary formulas for $k = 4$.

3.1.2 *General Contrasts*

Let $C \subseteq \mathbb{C}^k$ be a specified set of contrasts. For pairwise comparisons $C = \{\mathbf{e}_i - \mathbf{e}_j \; (1 \leqq i < j \leqq k)\}$ where $\mathbf{e}_i : k \times 1$ is the ith unit vector. Uusipaikka (1985) considered the problem of constructing exact $(1-\alpha)$-level simultaneous confidence intervals for contrasts $\mathbf{c}'\boldsymbol{\theta}$, $\mathbf{c} \in C$. This requires the evaluation of the distribution of

$$\sup_{\mathbf{c} \in C} \frac{|\mathbf{c}'(\hat{\boldsymbol{\theta}} - \boldsymbol{\theta})|}{S\sqrt{\mathbf{c}'\mathbf{V}\mathbf{c}}}.$$

Uusipaikka showed that the distribution function of this r.v. is given by

$$H(x) = \int_{\zeta}^{1} F\left[\frac{x^2}{z^2(k-1)}\right] g(z)\, dz \tag{3.10}$$

where $F(\cdot)$ is the distribution function of a $F_{k-1,\nu}$ r.v. and $g(\cdot)$ is the density function of the r.v.

$$Z = \sup_{\mathbf{c} \in C^*} |\mathbf{c}'\mathbf{U}|.$$

Here $\mathbf{U}$ has the uniform distribution on the unit hypersphere in $k-1$ dimensions,

$$\mathcal{S}^{k-1} = \{\mathbf{x} \in \mathbb{R}^{k-1} : \mathbf{x}'\mathbf{x} = 1\} ,$$

and C^* is any subset of $\mathcal{S}^{k-1}$ that is in one-to-one correspondence with C via the relation

$$\mathbf{c}_1^{*\prime}\mathbf{c}_2^* = \frac{\mathbf{c}_1'\mathbf{V}\mathbf{c}_2}{\{(\mathbf{c}_1'\mathbf{V}\mathbf{c}_1)(\mathbf{c}_2'\mathbf{V}\mathbf{c}_2)\}^{1/2}}$$

for every two pairs of $\mathbf{c}_1, \mathbf{c}_2 \in C$ and $\mathbf{c}_1^*, \mathbf{c}_2^* \in C^*$. The real number ζ in (3.10) is equal to

$$\inf_{\mathbf{x} \in \mathcal{S}^{k-1}} \sup_{\mathbf{c} \in C^*} |\mathbf{c}'\mathbf{x}|$$

so that $\Pr\{\zeta \leqq Z \leqq 1\} = 1$.

To obtain exact $(1-\alpha)$-level simultaneous confidence intervals

$$\mathbf{c}'\boldsymbol{\theta} \in [\mathbf{c}'\hat{\boldsymbol{\theta}} \pm \xi S\sqrt{\mathbf{c}'\mathbf{V}\mathbf{c}}] \quad \forall \mathbf{c} \in C ,$$

we need to find the critical constant $\xi = \xi_{k,\nu}^{(\alpha)}$ (where the dependence on $\mathbf{V}$ is again suppressed), which is the solution in x to the equation obtained by setting (3.10) equal to $1-\alpha$. The main obstacle in implementing this method is that the density function $g(\cdot)$ of the r.v. Z is extremely complicated and hence it is difficult to evaluate $H(\cdot)$ except in some special cases.

One such special case is $C = \mathbb{C}^k$, in which case $C^* = \mathcal{S}^{k-1}$ and hence $Z = 1$ with probability one. Then from (3.10) we have $H(x) = F[x^2/(k-1)]$ so that

$$\xi_{k,\nu}^{(\alpha)} = \{(k-1)F_{k-1,\nu}^{(\alpha)}\}^{1/2} ,$$

which gives Scheffé's S-procedure for simultaneous confidence intervals for all contrasts $\mathbf{c}'\boldsymbol{\theta}$, $\mathbf{c} \in \mathbb{C}^k$. More commonly we are interested in a proper (usually finite) subset of $\mathbb{C}^k$. Uusipaikka (1985) has evaluated the exact distribution $g(\cdot)$ for $k = 3$ and 4 when C consists of $k-1$ pairwise linearly independent contrasts. This gives us a method for constructing exact simultaneous confidence intervals for pairwise differences in any unbalanced design with three or four treatments.

3.1.3 All Linear Combinations

Spjøtvoll and Stoline (1973) proposed an exact extension to unbalanced one-way layouts of the T-procedure for the family of all linear combinations $\mathbf{l}'\boldsymbol{\theta}$ (see (2.7)). Their procedure is based on writing

$$\sum_{i=1}^{k} l_i(\bar{Y}_i - \theta_i) = \sum_{i=1}^{k} l_i n_i^{1/2}\{n_i^{-1/2}(\bar{Y}_i - \theta_i)\} \tag{3.11}$$

so that the $(\bar{Y}_i - \theta_i)$'s are transformed to $n_i^{-1/2}(\bar{Y}_i - \theta_i)$'s, which are independent $N(0, \sigma^2)$ r.v.'s. As the vector $\mathbf{l}$ ranges over $\mathbb{R}^k$, so does the vector $(l_1 n_1^{1/2}, \ldots, l_k n_k^{1/2})$. This gives the following exact $(1-\alpha)$-level simultaneous confidence intervals for all linear combinations $\mathbf{l}'\boldsymbol{\theta}$:

$$\mathbf{l}'\boldsymbol{\theta} \in \left[\mathbf{l}'\bar{\mathbf{Y}} \pm Q_{k,\nu}'^{(\alpha)} S \max\left(\sum_{i=1}^{k} \frac{l_i^+}{\sqrt{n_i}}, \sum_{i=1}^{k} \frac{l_i^-}{\sqrt{n_i}}\right)\right] \quad \forall \mathbf{l} \in \mathbb{R}^k \tag{3.12}$$

where $Q_{k,\nu}'^{(\alpha)}$ is the upper α critical point of the Studentized augmented range distribution with parameter k and d.f. ν. Note that if we put $\mathbf{U} = \mathrm{diag}(1/\sqrt{n_1}, \ldots, 1/\sqrt{n_k})$, then

$$\max\left(\sum_{i=1}^{k} \frac{l_i^+}{\sqrt{n_i}}, \sum_{i=1}^{k} \frac{l_i^-}{\sqrt{n_i}}\right) = M(\mathbf{U}'\mathbf{l})$$

where $M(\mathbf{U}'\mathbf{l})$ is the norm (defined in (2.5)) of the vector $\mathbf{U}'\mathbf{l}$. When the n_i's are equal to n, (3.12) reduces to (2.7). Spjøtvoll and Stoline (1973) proposed using the intervals (3.12) for pairwise comparisons, in which case they take the form

$$\theta_i - \theta_j \in [\bar{Y}_i - \bar{Y}_j \pm Q_{k,\nu}'^{(\alpha)} S \max(1/\sqrt{n_i}, 1/\sqrt{n_j})] \qquad (1 \leqq i < j \leqq k)\,. \tag{3.13}$$

This procedure based on the transformation (3.11) was mentioned by Tukey (1953, p. 322) (and studied by his student Kurtz 1956, pp. 15–16). Therefore we refer to it as the *Tukey–Spjøtvoll–Stoline* (*TSS*) *procedure*. Tukey considered this procedure not so satisfactory because, "It transforms pairwise comparisons into linear combinations which are not even contrasts."

The TSS-procedure can be extended to general unbalanced designs by writing (in analogy with (3.11))

$$\mathbf{l}'(\hat{\boldsymbol{\theta}} - \boldsymbol{\theta}) = \mathbf{l}'\mathbf{U}\mathbf{U}^{-1}(\hat{\boldsymbol{\theta}} - \boldsymbol{\theta}) \tag{3.14}$$

where $\mathbf{U}\mathbf{U}' = \mathbf{V}$ and thus the transformed vector $\mathbf{U}^{-1}(\hat{\boldsymbol{\theta}} - \boldsymbol{\theta})$ consists of independent $N(0, \sigma^2)$ r.v.'s. This was noted by Hochberg (1975a), who gave the following generalization of the simultaneous confidence intervals (3.12):

$$\mathbf{l}'\boldsymbol{\theta} \in [\mathbf{l}'\hat{\boldsymbol{\theta}} \pm Q'^{(\alpha)}_{k,\nu} SM(\mathbf{U}'\mathbf{l})] \qquad \forall \mathbf{l} \in \mathbb{R}^k \tag{3.15}$$

The matrix $\mathbf{U}$ that satisfies $\mathbf{UU}' = \mathbf{V}$ is not unique. This raises the question of choosing $\mathbf{U}$ in some optimal manner, for example, to minimize the lengths of the confidence intervals (3.15) in some sense. Another consideration in the choice of $\mathbf{U}$ is the following: Any contrast $\mathbf{c} \in \mathbb{C}^k$ in $\hat{\boldsymbol{\theta}}$ is transformed by (3.14) to a linear combination $\mathbf{U}'\mathbf{c}$ in $\mathbf{U}^{-1}\hat{\boldsymbol{\theta}}$. To allay Tukey's concern (mentioned above) that the latter may not be a contrast, can one choose $\mathbf{U}$ so that $\mathbf{U}'\mathbf{c} \in \mathbb{C}^k$? Hochberg (1975a) answered the latter question in the affirmative by noting that this is achieved by choosing $\mathbf{U}$ with equal row sums. Genizi and Hochberg (1978) called such a matrix a contrast set preserving (CSP) matrix. Hochberg (1975a) showed that if $\mathbf{U}$ is a CSP matrix, then exact $(1-\alpha)$-level simultaneous confidence intervals for all contrasts are obtained by using the critical constant $Q^{(\alpha)}_{k,\nu}$ in (3.15) in place of the larger constant $Q'^{(\alpha)}_{k,\nu}$.

The problem of choosing an "optimal" CSP matrix was considered by Felzenbaum et al. (1983). Given any non-CSP matrix $\mathbf{U}$ satisfying $\mathbf{UU}' = \mathbf{V}$, they proposed a specific transformation of $\mathbf{U}$ into a CSP matrix and gave a sufficient condition for a uniform improvement (shorter confidence intervals for all contrasts) of the transformed procedure over the original. We omit the details of this work since the Tukey–Kramer (TK) procedure (discussed in Section 3.2.1), which is now known to guarantee the designated confidence level conservatively, gives uniformly shorter confidence intervals than those given by this optimal procedure for all pairwise contrasts; it is also much easier to implement.

3.2 Approximate/Conservative Procedures

As noted in the previous section, the critical constant ξ needed for an exact simultaneous confidence procedure for a finite family of contrasts is very difficult to obtain numerically except when the number of treatments is small. For this reason many approximate/conservative modifications of the T-procedure that are easier to implement have been proposed in the literature. In this section we review these modified procedures.

3.2.1 The Tukey–Kramer Procedure

Historically this was the first modification, originally proposed by Tukey (1953) and later proposed independently by Kramer* (1956, 1957) in the

* Actually Duncan's (1957) modification of the step-down range procedure for unbalanced designs corresponds more accurately to Tukey's modification of the single-step range procedure; see Section 2.2 of Chapter 4.

context of step-down procedures. This modification (referred to as the *Tukey–Kramer procedure* or the *TK-procedure*) uses $Q_{k,\nu}^{(\alpha)}/\sqrt{2}$ as an approximation to the upper α point of the distribution of $\max_{1 \leqq i < j \leqq k} |T_{ij}|$ where the T_{ij}'s are given by (3.4) (if $\mathbf{V}$ is pairwise balanced, then of course no approximation is involved). The resulting $(1-\alpha)$-level simultaneous confidence intervals for all pairwise differences are

$$\theta_i - \theta_j \in \left[\hat{\theta}_i - \hat{\theta}_j \pm \frac{Q_{k,\nu}^{(\alpha)}}{\sqrt{2}} S\sqrt{d_{ij}} \right] \quad (1 \leqq i < j \leqq k) \qquad (3.16)$$

where the d_{ij}'s are defined by (3.2). For an unbalanced one-way layout, (3.16) becomes

$$\theta_i - \theta_j \in \left[\bar{Y}_i - \bar{Y}_j \pm Q_{k,\nu}^{(\alpha)} S \sqrt{\tfrac{1}{2}\left(\frac{1}{n_i} + \frac{1}{n_j} \right)} \right] \quad (1 \leqq i < j \leqq k). \qquad (3.17)$$

Tukey (1953, p. 39) conjectured that the approximation in the use of (3.17) "... is apparently in the conservative direction," that is,

$$\Pr\left\{ \theta_i - \theta_j \in \left[\bar{Y}_i - \bar{Y}_j \pm Q_{k,\nu}^{(\alpha)} S \sqrt{\tfrac{1}{2}\left(\frac{1}{n_i} + \frac{1}{n_j} \right)} \right] \quad (1 \leqq i < j \leqq k) \right\} \geqq 1 - \alpha \qquad (3.18)$$

for all values of the n_i's (or more generally for all diagonal covariance matrices $\sigma^2\mathbf{V}$). This is referred to as the *Tukey conjecture*. For the case of general unbalanced designs with nondiagonal $\mathbf{V}$ matrices, Tukey (1953, p. 333) stated that the properties of the approximate procedure (3.16) are "even less clear."

In a doctoral dissertation under Tukey's supervision, Kurtz (1956) proved the inequality (3.18) for all n_i's when $k = 3$ and for nearly equal n_i's when $k = 4$. Kurtz (1956) examined two additional cases for arbitrary k and found the Tukey conjecture to be true in both. These cases are:

(i) The ratio between any two n_i's either tends to 0 or ∞.

(ii) All but one of the n_i's are equal and that one is extremely large.

In case (i) when k is large, Kurtz found that, "The true error rate can be much smaller than the nominal." For example, when $k = 50$, the true FWE is 0.0032 with a nominal error rate of 0.05.

Dunnet (1980a) carried out an extensive simulation study that provided quite conclusive evidence in support of the Tukey conjecture for moderate to large imbalances among the n_i's. For small to moderate imbalances his results were less conclusive, but in most cases the estimated error rates were either close to or lower than the nominal. Previously several workers including Keselman, Murray, and Rogan (1976), Keselman, Toothaker, and Shooter (1975), and Smith (1971) had verified by simulation the conservative nature of the TK-procedure. Guided by Dunnett's simulations, Brown (1979) gave an analytical proof of the Tukey conjecture for $k = 3$, 4, and 5. The complete resolution of the validity of the Tukey conjecture for all k was achieved by Hayter (1984). Hayter proved that the simultaneous coverage probability (3.18) is strictly minimized when the n_i's are equal. Thus if the n_i's are not all equal, then we have strict inequality in (3.18).

A generalized Tukey conjecture would extend inequality (3.18) to nondiagonal covariance matrices **V**, which may result from other unbalanced designs, for example, a one-way layout with a fixed covariate. Brown (1984) has shown that this generalized conjecture is true when $k = 3$. Brown's proof based on Kurtz's (1956) work is geometrical in nature. Hayter (1985) has shown that the conjecture holds for any $k \geqq 3$ if the d_{ij}'s given by (3.2) satisfy

$$d_{ij} = a_i + a_j \qquad (1 \leqq i < j \leqq k) \tag{3.19}$$

for some positive numbers a_i.

To summarize, the TK-procedure offers a very simple and readily implementable solution to the problem of making pairwise comparisons in unbalanced designs. This solution is conservative (but not overly) for designs with diagonal **V**. For designs with nondiagonal **V** the solution is known to be conservative only for some special cases. Because of its simplicity and nearly accurate control of the FWE, in Section 4 we recommend the use of the TK-procedure over other alternative procedures proposed for pairwise comparisons. We now discuss these other procedures briefly.

3.2.2 The Miller–Winer Procedure

For a one-way layout with unequal sample sizes Miller (1981, p. 43) suggested using some "average" of the n_i's in place of n in the T-procedure (2.1). Winer (1971) recommended the use of the harmonic mean of the n_i's.

The simulation results of Keselman, Toothaker, and Shooter (1975) indicate that this procedure can be liberal (i.e., its FWE is greater than nominal α or its joint confidence level is less than nominal $1 - \alpha$) in moderate to highly imbalanced designs. In fact, the liberal nature of the Miller–Winer (MW) procedure can be analytically shown as follows.

The simultaneous coverage probability of all pairwise intervals for the MW-procedure with nominal joint confidence level $1-\alpha$ is

$$\Pr\left\{|T_{ij}| \leqq Q_{k,\nu}^{(\alpha)}\left[\tilde{n}\left(\frac{1}{n_i}+\frac{1}{n_j}\right)\right]^{-1/2} \quad (1 \leqq i < j \leqq k)\right\} \qquad (3.20)$$

where each T_{ij} is t-distributed with ν d.f. and $\tilde{n}$ is the harmonic mean of the n_i's. Consider one of the $\binom{k}{2}$ statements in (3.20), say for $i=1, j=2$, and let $1/n_1 + 1/n_2 = \delta$, $\Sigma_{i=3}^{k}(1/n_i) = \Delta$. Then the coverage probability for that single statement is

$$\Pr\left\{|T_{12}| \leqq Q_{k,\nu}^{(\alpha)}\sqrt{\frac{1+\Delta/\delta}{k}}\right\}. \qquad (3.21)$$

It is clear that if the n_i's are sufficiently unequal, then Δ/δ becomes small enough to make

$$Q_{k,\nu}^{(\alpha)}\sqrt{\frac{1+\Delta/\delta}{k}} < T_{\nu}^{(\alpha/2)}. \qquad (3.22)$$

For example, when $k=3$, $\alpha=0.05$, and $\nu=\infty$, we have $Q_{k,\nu}^{(\alpha)}=3.31$, $T_{\nu}^{(\alpha/2)}=1.96$, and it is easy to verify that (3.22) is satisfied when $n_1 = n_2 < 0.1038n_3$. However, (3.22) implies that (3.21), and hence (3.20), is less than $1-\alpha$, thus proving that the MW-procedure is liberal. Because of this drawback of the MW-procedure we drop it from further consideration.

3.2.3 *Procedures Based on the Bonferroni and Related Inequalities*

These procedures involve finding progressively less conservative upper bounds on the upper α point of the distribution of $\max_{1\leqq i<j\leqq k}|T_{ij}|$ where the T_{ij}'s are given by (3.4). The Bonferroni-type inequalities on which these bounds are based are discussed in Appendix 2.

The *Bonferroni procedure* uses the first order Bonferroni inequality:

$$\Pr\{\max_{1\leqq i<j\leqq k}|T_{ij}| \leqq \xi\} \geqq 1 - \sum_{1\leqq i<j\leqq k}\Pr\{|T_{ij}| > \xi\}. \qquad (3.23)$$

By equating the right hand side of (3.23) to $1-\alpha$, the Bonferroni upper bound on ξ is found to be $T_{\nu}^{(\alpha/2k^*)}$ where $k^* = \binom{k}{2}$. Thus conservative $(1-\alpha)$-level simultaneous confidence intervals for all pairwise differences are given by

$$\theta_i - \theta_j \in [\hat{\theta}_i - \hat{\theta}_j \pm T_{\nu}^{(\alpha/2k^*)} S\sqrt{d_{ij}}] \quad (1 \leqq i < j \leqq k). \qquad (3.24)$$

Hunter (1976) proposed using an optimal second order Bonferroni upper bound (see Appendix 2, Theorem 1.1) on ξ, the calculation of which requires a specialized computer program (Stoline and Mitchell 1981). This makes the use of Hunter's method not worthwhile, particularly because the upper bound must be calculated separately for every imbalanced design and because, as we see later, the TK-procedure usually provides a sharper upper bound that can be read directly from standard tables.

The *Dunn–Šidák procedure* proposed by Dunn (1958) is based on the sharper (than (3.23)) inequality

$$\Pr\{\max_{1 \leq i<j \leq k} |T_{ij}| \leq \xi\} \geq \prod_{1 \leq i<j \leq k} \Pr\{|T_{ij}| \leq \xi\}, \qquad (3.25)$$

which was proved by Šidák (1967). This inequality yields a less conservative upper bound on ξ, namely $T_\nu^{(\alpha'/2)}$, where $\alpha' = 1-(1-\alpha)^{1/k^*}$. The resulting intervals are given by

$$\theta_i - \theta_j \in [\hat{\theta}_i - \hat{\theta}_j \pm T_\nu^{(\alpha'/2)} S\sqrt{d_{ij}}] \quad (1 \leq i<j \leq k). \qquad (3.26)$$

Hochberg (1974b) exploited the following more refined version of the inequality (3.25) also due to Šidák (1967):

$$\Pr\{\max_{1 \leq i<j \leq k} |T_{ij}| \leq \xi\} \geq \Pr\{|M|_{k^*,\nu} \leq \xi\}, \qquad (3.27)$$

which yields a further improved upper bound on ξ, namely $|M|_{k^*,\nu}^{(\alpha)}$, the upper α critical point of the Studentized maximum modulus distribution with parameter k^*, and d.f. ν. Whereas (3.25) is obtained by regarding the T_{ij}'s as independently distributed t r.v.'s, (3.27) is obtained by regarding only their jointly normally distributed numerators as independent (all sharing the common denominator S). We thus get the following conservative $(1-\alpha)$-level simultaneous confidence intervals for all pairwise differences:

$$\theta_i - \theta_j \in [\hat{\theta}_i - \hat{\theta}_j \pm |M|_{k^*,\nu}^{(\alpha)} S\sqrt{d_{ij}}] \qquad (1 \leq i<j \leq k). \qquad (3.28)$$

Hochberg (1974b) referred to this procedure as the *GT2-procedure*.

We note that $|M|_{k^*,\nu}^{(\alpha)} \geq Q_{k,\nu}^{(\alpha)}/\sqrt{2}$ with equality holding if and only if $k=2$, and thus for $k \geq 3$ the GT2-procedure is more conservative than the TK-procedure*. Since the TK-procedure has now (Hayter 1984) been

* Spurrier (1981) proposed an improvement of the GT2-procedure for $k=3$ that gives shorter confidence intervals than the TK-procedure for some cases of large imbalances among the n_i's in a one-way layout. But because this improvement is too specialized and needs extra tables, we do not consider it here.

analytically shown to control the FWE for all k when $\mathbf{V}$ is diagonal and for $k = 3$ when $\mathbf{V}$ is nondiagonal, it should be the preferred choice for making pairwise comparisons in those cases. In other cases, the GT2-procedure may be used if guaranteed control of the FWE is desired. For further comparisons and recommendations, particularly for applications involving general contrasts, see Section 4.

3.2.4 Procedures Designed for Graphical Display

In recent years several graphical procedures have been proposed that attempt to control the FWE (at least approximately) for the family of all pairwise comparisons among the means. Every such procedure consists of plotting the sample means along with the so-called "uncertainty" intervals around them, so that any two sample means can be declared significantly different from each other (at a designated level α for the FWE) if and only if their uncertainty intervals do not overlap. For example, Tukey's T-procedure for a balanced one-way layout can be implemented graphically as follows: Construct uncertainty intervals with half-width equal to $Q_{k,\nu}^{(\alpha)}(S/2\sqrt{n})$ around each sample mean $\bar{Y}_i$. It is easy to verify that $\bar{Y}_i - \bar{Y}_j$ are significantly different at level α using the T-procedure, that is, $|\bar{Y}_i - \bar{Y}_j| > Q_{k,\nu}^{(\alpha)}S/\sqrt{n}$, if and only if the uncertainty intervals $\bar{Y}_i \pm Q_{k,\nu}^{(\alpha)}S/2\sqrt{n}$ and $\bar{Y}_j \pm Q_{k,\nu}^{(\alpha)}S/2\sqrt{n}$ do not overlap. Note that this graphical procedure is exact. These uncertainty intervals plotted around the ordered sample means for the barley data of Duncan (1955) are shown in Fig. 3.1. From this figure we see that varieties A and B, A and C, and A and E are significantly different from each other at $\alpha = 0.05$.

For an unbalanced one-way layout, graphical procedures can be devised corresponding to each simultaneous confidence procedure for the family of pairwise comparisons described in the preceding sections. Since for this family the TK-procedure is now known to be generally preferred to all of its contenders, we present our discussion with reference to it.

According to the TK-procedure, $\bar{Y}_i - \bar{Y}_j$ are declared significantly different at FWE $\leqq \alpha$ if

$$|\bar{Y}_i - \bar{Y}_j| > Q_{k,\nu}^{(\alpha)} S \sqrt{\tfrac{1}{2}\left(\frac{1}{n_i} + \frac{1}{n_j}\right)} \qquad (1 \leqq i < j \leqq k) . \tag{3.29}$$

If in the graphical display we use uncertainty intervals of half-width $Q_{k,\nu}^{(\alpha)} S W_i/\sqrt{2}$ around $\bar{Y}_i$ $(1 \leqq i \leqq k)$, then using the criterion of nonoverlapping uncertainty intervals, we would declare $\bar{Y}_i - \bar{Y}_j$ significantly different if

$$|\bar{Y}_i - \bar{Y}_j| > (Q_{k,\nu}^{(\alpha)}/\sqrt{2}) S (W_i + W_j) \qquad (1 \leqq i < j \leqq k) . \tag{3.30}$$

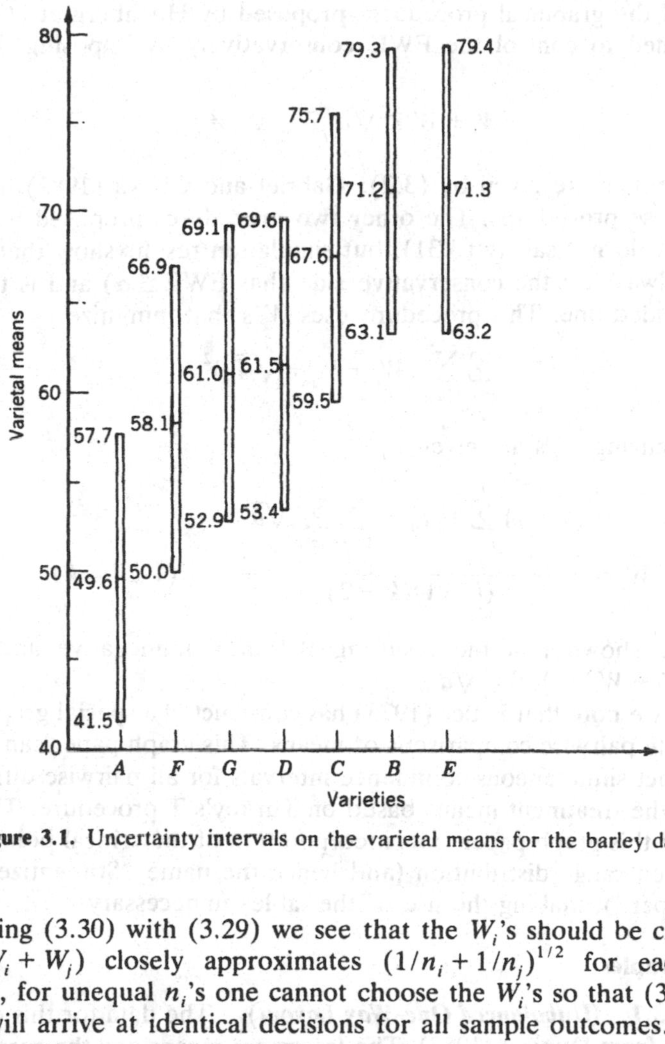

Figure 3.1. Uncertainty intervals on the varietal means for the barley data.

Comparing (3.30) with (3.29) we see that the W_i's should be chosen so that $(W_i + W_j)$ closely approximates $(1/n_i + 1/n_j)^{1/2}$ for each $i \neq j$. (Clearly, for unequal n_i's one cannot choose the W_i's so that (3.29) and (3.30) will arrive at identical decisions for all sample outcomes. In fact, Hochberg, Weiss, and Hart (1982) noted that for certain configurations of the $\bar{Y}_i$'s and n_i's, no W_i's exist for which the decisions according to (3.29) and (3.30) are identical.)

Gabriel (1978a) offered a simple choice for the W_i's, namely, $W_i = 1/\sqrt{2n_i}$ $(1 \leq i \leq k)$. This choice can be liberal since $1/\sqrt{2n_i} + 1/\sqrt{2n_j}$ can be less than $(1/n_i + 1/n_j)^{1/2}$ for some (n_i, n_j)-values. Hochberg et al. (1982) and Gabriel and Gheva (1982) proposed some approximate graphical procedures that improve upon the earlier ones of Gabriel (1978a) and Andrews, Snee, and Sarner (1980). We now discuss these improved procedures.

Two of the graphical procedures proposed by Hochberg et al. (1982) are designed to control the FWE conservatively by imposing the constraints

$$W_i + W_j \geqq \sqrt{d_{ij}} \qquad \forall\, i \neq j \tag{3.31}$$

where the d_{ij}'s are given by (3.2). Gabriel and Gheva (1982) modified one of these procedures. The other two procedures proposed by Hochberg et al. do not satisfy (3.31), but simulation results show that one of them is always on the conservative side (has FWE $\leqq \alpha$) and is thus the recommended one. This procedure uses W_i's that minimize

$$\sum_{1 \leqq i < j \leqq k}\sum (W_i + W_j - \sqrt{d_{ij}})^2 .$$

The minimizing W_i's are given by

$$W_i = \frac{(k-1)\sum_{\substack{j=1 \\ j \neq i}}^{k} \sqrt{d_{ij}} - \sum_{1 \leqq j < l \leqq k}\sum \sqrt{d_{jl}}}{(k-1)(k-2)} \qquad (1 \leqq i \leqq k) . \tag{3.32}$$

It can be shown that the resulting W_i's are nonnegative and satisfy $\Sigma\,\Sigma_{i \neq j}\,(W_i + W_j) = \Sigma\,\Sigma_{i \neq j}\,\sqrt{d_{ij}}$.

Finally we note that Feder (1975) has constructed a special graph paper to facilitate pairwise comparisons of means. This graph paper can be used to construct simultaneous confidence intervals for all pairwise differences between the treatment means based on Turkey's T-procedure. The main feature of this graph paper is its incorporation of the critical points of the Studentized range distribution (and hence the name "Studentized range graph paper"), making the use of the tables unnecessary.

3.3 Examples

Example 3.1 ***(Unbalanced One-Way Layout).*** The data for this example are taken from Duncan (1957). The treatment means and the corresponding sample sizes are given in Table 3.1. For these data we have $MS_{\text{error}} = S^2 = 5{,}395$ ($S = 73.45$) with $\nu = 16$ d.f.

TABLE 3.1. Data for Example 3.1

Treatment	A	B	C	D	E	F	G
$\bar{Y}_i$	743	851	873	680	902	734	945
n_i	5	5	3	3	2	2	3

Source: Duncan (1957).

TABLE 3.2. Simultaneous 95% TK-Intervals for All Pairwise Differences for Data in Table 3.1

(i, j)	$\bar{Y}_i - \bar{Y}_j \pm$ Allowance	(i, j)	$\bar{Y}_i - \bar{Y}_j \pm$ Allowance
(A, B)	-108 ± 155.7	(C, D)	193 ± 201.1
(A, C)	-130 ± 179.8	(C, E)	-29 ± 224.8
(A, D)	63 ± 179.8	(C, F)	139 ± 224.8
(A, E)	-159 ± 206.0	(C, G)	-72 ± 201.1
(A, F)	9 ± 206.0	(D, E)	-222 ± 224.8
(A, G)	-202 ± 179.8	(D, F)	-54 ± 224.8
(B, C)	-22 ± 179.8	(D, G)	-265 ± 201.1
(B, D)	171 ± 179.8	(E, F)	168 ± 246.2
(B, E)	-51 ± 206.0	(E, G)	43 ± 224.8
(B, F)	117 ± 206.0	(F, G)	-211 ± 224.8
(B, G)	-94 ± 179.8		

We first illustrate the application of the TK-procedure for constructing simultaneous confidence intervals for all pairwise comparisons. Using (3.17) with $Q_{7,16}^{(.05)} = 4.74$ we obtain simultaneous 95% TK-intervals, which are given in Table 3.2.

From these intervals only comparisons (A, G) and (D, G) are found to yield significant differences. This is not surprising in view of the fact that the sample sizes are so small.

The S-intervals given by (1.3) will be uniformly (for all pairwise differences) longer by a factor of

$$\frac{\{(k-1)F_{6,16}^{(.05)}\}^{1/2}}{Q_{7,16}^{(.05)}/\sqrt{2}} = \frac{\{6 \times 2.74\}^{1/2}}{4.74/\sqrt{2}} = 1.210\,,$$

the corresponding factor for the GT2-intervals (see (3.28)) being

$$\frac{|M|_{21,16}^{(.05)}}{Q_{7,16}^{(.05)}/\sqrt{2}} = \frac{3.52}{4.74/\sqrt{2}} = 1.050\,.$$

However, the S-intervals would become competitive for higher order contrasts. For example, if the contrast $(A+B+C+D)/4-(E+F+G)/3$ is selected post hoc, then the 95% TK-interval for this contrast using (3.3) and (3.17) is $[-73.58 \pm 207.1]$, while the corresponding S-interval using (1.2) is $[-73.58 \pm 138.0]$.

We next illustrate the use of the graphical procedure (3.30) with the choice of the W_i's given by (3.32). (Recall that some of the resulting pairwise intervals may be shorter than the TK-intervals (3.17) since condition (3.31) is not imposed.) Calculations yield the following values: $W_1 = 0.3231$, $W_2 = 0.3231$, $W_3 = 0.4076$, $W_4 = 0.4076$, $W_5 = 0.5061$, $W_6 =$

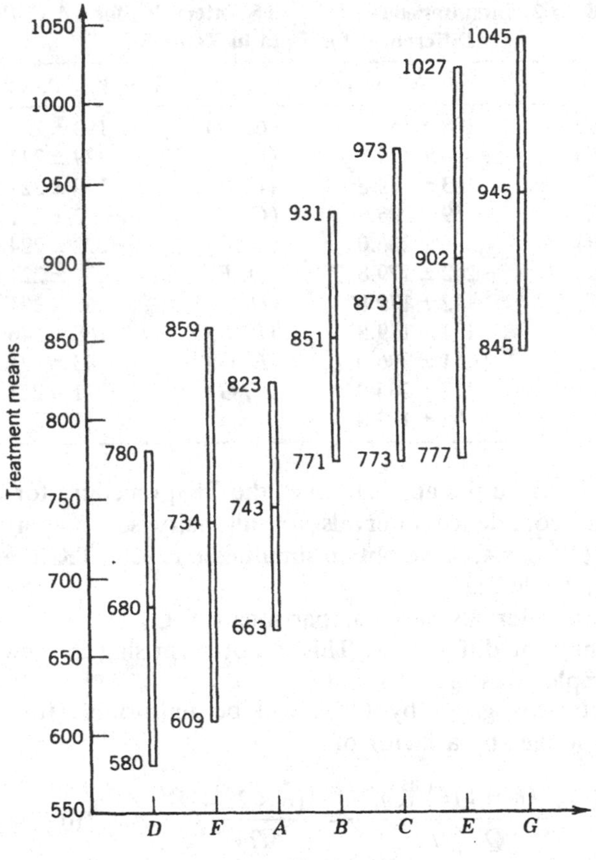

Figure 3.2. Uncertainty intervals on the treatment means for data in Table 3.1.

0.5061, and $W_7 = 0.4076$. The uncertainty intervals $\bar{Y}_i \pm (Q_{k,\nu}^{(\alpha)}/\sqrt{2})\, SW_i$ for the seven treatment means are displayed in Figure 3.2. Notice that these uncertainty intervals are of unequal lengths.

From this graphical display we see that comparisons (A, G) and (D, G) are found to yield significant differences. Also note that comparisons (B, D), (C, D), and (D, E) barely fail to be significant. □

Example 3.2 (Randomized Block Design with a Fixed Covariate; Continuation of Example 1.2). We calculate simultaneous 90% confidence intervals for all pairwise differences using the TK-procedure. The TK-allowance for the (i, j)th pairwise comparison is given by

$$\frac{Q_{k,\nu}^{(\alpha)} S\sqrt{d_{ij}}}{2} = Q_{k,\nu}^{(\alpha)} S\sqrt{\tfrac{1}{2}\left[\frac{2}{n} + \frac{(\bar{X}_{i\cdot} - \bar{X}_{j\cdot})^2}{S_{XX}}\right]} \quad (1 \leq i < j \leq k) .$$

For $\alpha = 0.10$, $k = 6$, and $\nu = 19$, $Q_{k,\nu}^{(\alpha)} = 3.97$; also $S = 7.106$, $n = 5$, $S_{XX} = 257.1$, and the $\bar{X}_{i\cdot}$'s are as given in Table 1.4. For any pairwise comparison the ratio of the width of the S-interval to that of the TK-interval is

$$\frac{\{(k-1)F_{k-1,\nu}^{(\alpha)}\}^{1/2}}{Q_{k,\nu}^{(\alpha)}/\sqrt{2}} = \frac{\{5 \times 2.18\}^{1/2}}{3.97/\sqrt{2}} = 1.176 .$$

Using this factor we find that comparisons (1, 3) and (3, 5), which are barely nonsignificant according to the S-procedure (see Table 1.5), turn out to be significant according to the TK-procedure, the corresponding confidence intervals being $[14.90 \pm 12.68]$ for the (1, 3) comparison and $[-22.15 \pm 19.55]$ for the (3, 5) comparison. In addition, the comparisons (2, 6) and (4, 6) are on the borderline of significance, the corresponding confidence intervals being $[-24.39 \pm 24.63]$ and $[-16.02 \pm 16.59]$, respectively.

The graphical procedure (3.30) can be applied by calculating the W_i's from (3.32) with $d_{ij} = 2/n + (\bar{X}_{i\cdot} - \bar{X}_{j\cdot})^2/S_{XX}$. We omit the details. □

Example 3.3 ***(Partially Balanced Incomplete Block Design).*** This example is taken from Hochberg (1975a). Consider a group divisible partially balanced incomplete block (PBIB) design with two associate classes for six treatments in six blocks: (1, 2, 3), (3, 4, 5), (2, 5, 6), (1, 2, 4), (3, 4, 6), and (1, 5, 6). For this design we have $k = 6$, $b = 6$, $p = 3$, $r = 3$, $\nu = 7$, $\lambda_1 = 2$, $\lambda_2 = 1$, and

$$\{p_{ij}^1\} = \begin{bmatrix} 0 & 0 \\ 0 & 4 \end{bmatrix}, \qquad \{p_{ij}^2\} = \begin{bmatrix} 0 & 1 \\ 1 & 2 \end{bmatrix}.$$

Pairs (1, 2), (3, 4), and (5, 6) are first associates and all other pairs are second associates.

Using (1.24) of Chapter 2 we readily obtain

$$\begin{aligned} \mathrm{var}(\hat{\theta}_i - \hat{\theta}_j) &= d_{ij}\sigma^2 \\ &= \begin{cases} \frac{3}{4}\sigma^2 & \text{if } i \text{ and } j \text{ are first associates ,} \\ \frac{7}{8}\sigma^2 & \text{if } i \text{ and } j \text{ are second associates .} \end{cases} \end{aligned}$$

To construct 95% TK-intervals for all pairwise differences we use

(3.16) with $Q_{6,7}^{(.05)} = 5.36$ and the d_{ij}'s given by the expression above, and obtain

$$\theta_i - \theta_j \in [\hat{\theta}_i - \hat{\theta}_j \pm 3.282S] \quad \text{if } i \text{ and } j \text{ are first associates,}$$

$$\theta_i - \theta_j \in [\hat{\theta}_i - \hat{\theta}_j \pm 3.545S] \quad \text{if } i \text{ and } j \text{ are second associates.} \quad \square$$

4 COMPARISONS AMONG SINGLE-STEP PROCEDURES

4.1 Balanced Designs

For balanced designs essentially there are only two contenders: the S-procedure and the T-procedure. The ratio of the length of the S-interval to that of the T-interval for a contrast $\mathbf{c}'\boldsymbol{\theta}$ is given by

$$R = \frac{\{(k-1)F_{k-1,\nu}^{(\alpha)}\}^{1/2}}{Q_{k,\nu}^{(\alpha)}} \cdot \frac{\left\{\sum_{i=1}^{k} c_i^2\right\}^{1/2}}{\left\{\sum_{i=1}^{k} |c_i|/2\right\}}$$
$$= R_1(k, \nu, \alpha) \cdot R_2(\mathbf{c}) \quad \text{(say)}. \tag{4.1}$$

Note that R greater than unity favors the T-procedure and vice versa. Scheffé (1953) investigated this ratio. His calculations show that $R_1(k, \nu, \alpha)$ remains nearly constant as α varies between 0.01 and 0.10, and decreases very slightly as ν increases. So over their practical ranges, α and ν have little effect on R, and we need to consider its dependence only on k and c.

The most common comparisons among the treatment means are what we call "$p:q$ comparisons," which involve contrasts of the type

$$\frac{1}{p}\sum_{i\in P}\theta_i - \frac{1}{q}\sum_{j\in Q}\theta_j$$

where P and Q are disjoint subsets of $\{1, 2, \ldots, k\}$ with cardinalities p and q, respectively ($p + q \leq k$). For such a $p:q$ contrast $R_2(\mathbf{c})$ equals $(1/p + 1/q)^{1/2}$. It is readily apparent that $R_2(\mathbf{c})$ will be maximum (the T-procedure will have the maximum advantage over the S-procedure) when $p = q = 1$, that is, for pairwise comparisons (in which case the T-procedure is known to provide the shortest intervals). As p and q increase, which results in "higher order" contrasts, $R_2(\mathbf{c})$ decreases and the T-procedure loses its advantage over the S-procedure. Scheffé (1959, pp. 75–77) gives a numerical comparison for $k = 6$, $\alpha = 0.05$, and $\nu = \infty$ between the two procedures for a balanced one-way layout that shows that the T-procedure is preferred for low order contrasts such as 1 : 1 and 1:2 comparisons, while the S-procedure is preferred for higher order contrasts. Scheffé's calculations show that the ratio R decreases from 1.17 for 1 : 1 comparisons to 0.67 for 3 : 3 comparisons.

If an experimenter's primary interest lies in pairwise contrasts, although he or she may wish to make inferences on few additional ones, then the T-procedure may offer a better choice. However, an experimenter who does not have a special interest in specific contrasts, but rather wishes to select *any* contrast that may be suggested by data, is probably better off using the S-procedure. Thus the S-procedure is suitable for "data-snooping." When intermediate contrasts are of primary interest, the procedure described in Section 5.1 may be used.

We next compare, for the case of a balanced one-way layout, the powers of the simultaneous test procedures based on the S-procedure (see Section 1.2) and the T-procedure (see Section 2.3) for the family of subset hypotheses. Fix a subset $P \subseteq K = \{1, 2, \ldots, k\}$ of size p $(2 \leq p \leq k)$ and let $\boldsymbol{\tau}_P$ denote the p-vector of deviations $(\theta_i - \bar{\theta}_P)/\sigma$, $i \in P$, where $\bar{\theta}_P = (1/p) \sum_{i \in P} \theta_i$. Under H_P, $\boldsymbol{\tau}_P$ is a null vector. Define the *P-subset power* (Einot and Gabriel 1975), denoted by β_P, as the probability of rejecting H_P when it is false. The powers β_P of both of these procedures depend only on θ_i, $i \in P$, and not on θ_j, $j \notin P$. Furthermore, since these procedures are location and scale invariant, β_P for each depends on θ_i, $i \in P$, and σ only through $\boldsymbol{\tau}_P$. The P-subset power for the S-procedure is given by

$$\beta_P^S(\boldsymbol{\tau}_P) = \Pr\{(p-1)F_{p-1,\nu}(\delta_P^2) > (k-1)F_{k-1,\nu}^{(\alpha)}\} \tag{4.2}$$

where $F_{\nu_1,\nu_2}(\delta^2)$ is a noncentral F r.v. with d.f. ν_1, ν_2 and the noncentrality parameter δ^2, and $\delta_P^2 = \boldsymbol{\tau}_P' \boldsymbol{\tau}_P$. The P-subset power for the T-procedure is given by

$$\beta_P^T(\boldsymbol{\tau}_P) = \Pr\{Q_{p,\nu}(\boldsymbol{\tau}_P) > Q_{k,\nu}^{(\alpha)}\} \tag{4.3}$$

where $Q_{p,\nu}(\boldsymbol{\tau}_P)$ is the range of p independent $N(\tau_i, 1)$ r.v.'s $(i \in P)$ divided by an independent $\sqrt{\chi_\nu^2/\nu}$ r.v. Note that β_P^S depends on $\boldsymbol{\tau}_P$ only through the scalar quantity $\delta_P^2 = \boldsymbol{\tau}_P' \boldsymbol{\tau}_P$, while such a simplification is not possible for β_P^T.

For the overall null hypothesis $H_0 = H_K : \theta_1 = \cdots = \theta_k$, the S-procedure gives the usual F-test of the analysis of variance. The power function of this test, $\beta_K^S(\tau_K)$, has been extensively studied. David, Lachenbruch, and Brandis (1972) made a numerical investigation of the power function $\beta_K^T(\tau_K)$ for testing $H_0 = H_K$. Their results show that $\beta_K^T(\tau_K)$ is approximately maximized and exceeds $\beta_K^S(\tau_K)$ when the τ_i's are so situated on the sphere $\Sigma_{i=1}^k \tau_i^2 = k\phi^2$ that their range is maximum. The range of the τ_i's is maximum in the configuration

$$\tau_{(1)} = -\frac{k}{2}\phi,\ \tau_{(k)} = \frac{k}{2}\phi,\ \tau_{(i)} = 0 \qquad (2 \leqq i \leqq k-1) \tag{4.4}$$

where $\tau_{(1)} \leqq \cdots \leqq \tau_{(k)}$. David, Lachenbruch, and Brandis (1972) have shown that $\beta_K^T(\tau_K)$ has a *stationary point* at (4.4), but it has not yet been analytically proven that the *maximum* of $\beta_K^T(\tau_K)$ is attained at (4.4).

The P-subset powers (4.2) and (4.3) have not been studied for other subsets (i.e., for $2 \leqq p \leqq k-1$). Gabriel (1964) has studied (4.2) and (4.3) under H_P, in which case they are called *p-mean significance levels* (Duncan 1955) denoted by α_p^S and α_p^T, respectively $(2 \leqq p \leqq k)$. For $k = 8$, $\nu = 40$, and $\alpha = 0.10$, Gabriel's calculations are summarized in Table 4.1. This table shows that when the two procedures are designed to have FWE $= \alpha$ (i.e., $\alpha_k^S = \alpha_k^T = \alpha$), then $\alpha_p^T > \alpha_p^S$ for $2 \leqq p \leqq k-1$, and the relative difference between α_p^T and α_p^S increases as p decreases. This implies that for *local* alternatives to H_P, the T-procedure will be more powerful than the S-procedure and this dominance of the T-procedure over the S-procedure would persist over larger regions of local alternatives when the subset P being tested is a relatively small subset of the whole set K (which corresponds to lower order contrasts).

Ramachandran and Khatri (1957) considered the problem of testing a "pair slippage" hypothesis

$$\theta_i = \theta - \Delta,\ \theta_j = \theta + \Delta,\ \theta_l = \theta \qquad (l \neq i \neq j,\ 1 \leqq l \leqq k) \tag{4.5}$$

for some pair (θ_i, θ_j) against the null hypothesis $H_K: \theta_1 = \cdots \theta_k$ in a

TABLE 4.1. *p*-Mean Significance Levels for the S- and T-Procedures

	p						
	8	7	6	5	4	3	2
α_p^S	0.10	0.079	0.048	0.021	0.009	0.003	0.001
α_p^T	0.10	0.081	0.062	0.045	0.030	0.016	0.006

Source: Gabriel (1964). Reproduced with the kind permission from the Biometrics Society.

balanced one-way layout. In (4.5), θ is unspecified and $\Delta > 0$ is given. They showed that in the class of symmetric, and location and scale invariant procedures with given error probability α under H_K, the following test is optimal: Reject H_K if

$$T = \frac{n(\bar{Y}_{\max} - \bar{Y}_{\min})}{\left[\sum_{i=1}^{k}\sum_{j=1}^{n}(Y_{ij} - \bar{\bar{Y}})^2\right]^{1/2}} > \xi \tag{4.6}$$

where ξ is the upper α point of the test statistic T under H_K. If H_K is rejected, then conclude that pair (θ_i, θ_j) has slipped as in (4.5) where $\bar{Y}_i = \bar{Y}_{\min}$ and $\bar{Y}_j = \bar{Y}_{\max}$. Ramachandran and Khatri mistakenly regard the test statistic T in (4.5) as the Studentized range statistic. It is not the Studentized range statistic since it uses the total sum of squares rather than the within groups sum of squares in the denominator. In fact, this is an instance where the T-procedure is *not* optimal.

The T-procedure is reasonably robust but perhaps less so than the S-procedure against departures from normality (particularly in cases of extreme nonnormality) as shown by the simulation studies of Ramseyer and Tcheng (1973) and Brown (1974). If there are serious doubts about the normality assumption, then one recourse is to employ an appropriate distribution-free or robust procedure; see Chapter 9 for a discussion of these procedures. The T- and S-procedures are much less robust in the presence of heterogeneous error variances. Special procedures to deal with this situation are discussed in Chapter 7. Departures from the assumption of independence have not been studied in as much detail as the other two. Box (1954) found that the effect of correlated errors on the performance of the F-test can be serious in terms of increased probability of Type I errors. An analogous study of the test based on the Studentized range statistic is lacking.

4.2 Unbalanced Designs

We first consider the problem of confidence estimation of pairwise contrasts in an unbalanced one-way layout. For this problem the TK-procedure is unarguably the best alternative available at this time (apart from the few cases for which exact procedures are available as discussed in Section 3.1). It is now analytically proven to be conservative. Yet the extent of conservatism in the TK-procedure is quite small even for cases of rather severe imbalance as shown by the numerical results of Uusipaikka (1985) and Spurrier and Isham (1985), and simulation results of

Dunnett (1980a). It provides uniformly shorter intervals than most of its competitors. This is readily clear for the GT2-procedure (and hence for the Bonferroni and Dunn-Šidák procedures) because of the inequality $Q_{k,\nu}^{(\alpha)}/\sqrt{2} < |M|_{k^*,\nu}^{(\alpha)}$, and for the S-procedure because of the inequality $Q_{k,\nu}^{(\alpha)}/\sqrt{2} < \{(k-1)F_{k-1,\nu}^{(\alpha)}\}^{1/2}$ for $k > 2$. With regard to the transformation procedures, Genizi and Hochberg (1978) proved that for any pairwise contrast the TK-interval is not longer than that provided by any transformation procedure. The only procedures that may beat the TK-procedure for certain configurations of the n_i's are the Hunter procedure based on the optimal second order Bonferroni inequality and the graphical procedure (3.30) of Hochberg et al. (1982) based on the choice (3.32) for the W_i's. However, as noted before, for the Hunter procedure the critical constant must be evaluated separately for each combination of the n_i's by running a special computer program. On the other hand, the TK-procedure can be applied very easily since its critical constant $Q_{k,\nu}^{(\alpha)}$ does not depend on the particular combination of the n_i's and is extensively tabulated (Harter 1960, 1969). Regarding the graphical procedure (3.30), it is not known whether it controls the FWE conservatively.

We next consider the problem of confidence estimation of general contrasts in an unbalanced one-way layout. For this problem the S-procedure and the transformation procedures are viable alternatives to the TK-procedure. The comparison between the TK- and S-procedures is very similar to that between the T- and S-procedures for a balanced one-way layout. In this case also we can write an expression analogous to (4.1) for the ratio of the lengths of the S- and TK-intervals by using (1.2) and (3.3), respectively. The dependence of this ratio on the n_i's (in addition to the c_i's) makes the comparison more difficult. However, the general conclusion that the TK-intervals are shorter than the S-intervals for low order contrasts, and the S-intervals are shorter for moderate to high order contrasts still holds. The transformation procedures can also be shown to yield shorter intervals than the TK-intervals for high order contrasts, but they involve much more computation than the S-procedure and hence the latter is preferred for high order contrasts. Tse (1983) gave some specific numerical criteria for choosing one procedure over the other for different contrasts. (Earlier Ury 1976 had given similar criteria for pairwise contrasts for choosing between the S-, TSS-, GT2-, and the Dunn–Šidák procedures. In view of the established superiority of the TK-procedure in this case, these criteria are no longer of interest.) It cannot be overemphasized, however, that in a given application one must use only one procedure for all contrasts. That procedure may be chosen depending on which contrasts are of primary interest. If different procedures are used to construct intervals for different contrasts (e.g., if for each contrast of interest the procedure that gives the shortest interval for that contrast is used), then clearly the designated joint confidence level is not guaranteed.

For general unbalanced designs the TK-procedure would again be the most preferred choice for making pairwise comparisons if it can be shown to control the FWE conservatively. At the present time this is known to be true only for $k = 3$ and for $k \geqq 4$ when the variances of the pairwise differences have the additive structure (3.19). In other cases, if guaranteed control of the FWE is desired, then we recommend the use of the GT2-procedure. The ratio of the length of the GT2-interval to that of the TK-interval is proportional to $\sqrt{2}|M|_{k^*,\nu}^{(\alpha)}/Q_{k,\nu}^{(\alpha)}$. Stoline (1981) computed this ratio and found that it lies in the range 1.01–1.04 for $\alpha = 0.01$, 1.02–1.05 for $\alpha = 0.05$, 1.03–1.06 for $\alpha = 0.10$, and 1.05–1.09 for $\alpha =$ 0.20. Thus the GT2-intervals are only fractionally longer than the TK-intervals. For higher order contrasts we recommend the use of the S-procedure for the same reasons that were mentioned earlier.

The robustness performance of the TK-procedure is similar to that of the T-procedure. In the case of variance heterogeneity the TK-procedure tends to be liberal and the extent of liberalism increases if larger variances are paired with smaller sample sizes and vice versa (Keselman, Games, and Rogan 1978). Violation of the normality assumption does not seriously affect the Type I FWE of the TK-procedure but violation of the independence assumption does.

5 ADDITIONAL TOPICS

5.1 An Intermediate Procedure for $p:q$ Comparisons

In Section 4.1 we saw that for a balanced one-way layout, the T-procedure is preferred for all 1:1 and few other low order (e.g., all 1:2) contrasts, while the S-procedure is preferred for high order contrasts. Can a procedure be constructed that would give shorter intervals for intermediate contrasts than those given by both of these procedures? Hochberg and Rodriguez (1977) addressed this question. They noted that both the T- and S-procedures are UI procedures (see Sections 2.1.1 and 2.1.2 of Chapter 2); the T-procedure is obtained by writing $H_0: \theta_1 = \theta_2 = \cdots = \theta_k$ as a finite intersection $\bigcap_{1 \leq i < j \leq k}(\theta_i - \theta_j = 0)$, while the S-procedure is obtained by writing H_0 as an infinite intersection $\bigcap_{\mathbf{c} \in \mathbb{C}^k}(\mathbf{c}'\boldsymbol{\theta} = 0)$. More generally, one can choose a finite subset $C \subseteq \mathbb{C}^k$ of interest such that

$H_0 = \bigcap_{\mathbf{c} \in C} (\mathbf{c}'\boldsymbol{\theta} = 0)$ and base the UI procedure on the distribution of the r.v.

$$\sup_{\mathbf{c} \in C} \frac{|\mathbf{c}'(\bar{\mathbf{Y}} - \boldsymbol{\theta})|\sqrt{n}}{S(\mathbf{c}'\mathbf{c})^{1/2}} \tag{5.1}$$

where $\bar{\mathbf{Y}} = (\bar{Y}_1, \bar{Y}_2, \ldots, \bar{Y}_k)'$.

When C is chosen to be some subset of all $p:q$ comparisons (including all pairwise comparisons), Hochberg and Rodriguez showed how the resolution of the resulting procedure can be "optimally" extended to the family of all $p:q$ comparisons, and further to the family of all contrasts. The upper α point of the r.v. (5.1) (denoted by $\xi_{k,\nu}^{(\alpha)}(C)$) falls between $Q_{k,\nu}^{(\alpha)}/\sqrt{2}$ (obtained when $C =$ the set of all pairwise contrasts) and $\{(k-1)F_{k-1,\nu}^{(\alpha)}\}^{1/2}$ (obtained when $C = \mathbb{C}^k$). The resulting intervals are shorter than both the S- and T-intervals for intermediate contrasts but longer than the T-intervals for pairwise contrasts and longer than the S-intervals for high order contrasts.

As in Uusipaikka's (1985) work, the critical constant $\xi_{k,\nu}^{(\alpha)}(C)$ is difficult to evaluate for arbitrary C. Hochberg and Rodriguez employed Siotani's (1964) second order Bonferroni approximation and tabulated the critical constants $\xi_{k,\nu}^{(\alpha)}(C)$ for $C =$ {all 1:1 and 1:2 comparisons} and {all 1:1 and 2:2 comparisons} for selected values of k, α, and ν.

5.2 Multiple Comparisons Following a Significant *F*-Test

Often in practice, multiple comparisons between treatment effects (e.g., confidence estimates or tests for selected contrasts) are pursued only when the F-test for the equality of the treatment effects is significant. In fact, Scheffé (1953, 1959) prescribed the use of his procedure in this manner. Olshen (1973) argued that when the S-intervals (1.1) are used in this manner, their conditional confidence level (conditioned on the event that the augmented F-statistic (1.6) for the hypothesis that $\mathbf{l}'\boldsymbol{\theta} = 0 \ \forall \mathbf{l} \in \mathcal{L}$ is significant at level α) is more pertinent than their unconditional confidence level; the latter equals the nominal value $1-\alpha$. He proved that the conditional confidence level is less than $1-\alpha$ (and can be substantially less than $1-\alpha$ for some values of the unknown parameters $\boldsymbol{\theta}$ and σ^2) if $\nu \geqq 2$ and $mF_{m,\nu}^{(\alpha)} \geqq 3\nu$. Thus the nominal confidence level $1-\alpha$ is not guaranteed if the intervals (1.1) are used conditionally.

This problem is, of course, not unique to the S-procedure. Bernhardson (1975) studied by Monte Carlo methods the Type I error rates of five multiple comparison procedures (MCPs) (including the S-procedure and some step-down procedures) conditional on the significant outcome of a preliminary F-test. He found the same phenomenon, namely, that the conditional Type I error rates exceeded the nominal levels for all five MCPs.

In response to Olshen's (1973) result, Scheffé (1977) suggested that in genuine multiple comparison problems, any inferences of interest based on (1.1) should be pursued regardless of the outcome of the preliminary F-test. Essentially, he suggested unconditional use of the S-procedure when it is applicable. The F-test of the hypothesis $\bigcap_{\mathbf{l} \in \mathscr{L}} (\mathbf{l}'\boldsymbol{\theta} = 0)$ (if $\mathscr{L} = \mathbb{C}^k$, then this is the overall null hypothesis $\theta_1 = \theta_2 = \cdots = \theta_k$) belongs to the family of inferences based on the simultaneous intervals (1.1) and all such inferences (including the F-test, confidence estimates for selected $\mathbf{l}'\boldsymbol{\theta}$, directional decisions on $\mathbf{l}'\boldsymbol{\theta}$, etc.) are correct unconditionally with probability $1 - \alpha$. Clearly, if the preliminary F-test is not significant at level α, then the confidence interval for every $\mathbf{l}'\boldsymbol{\theta}$ ($\mathbf{l} \in \mathscr{L}$) would cover zero. For a further exchange of views on the conditional use of the S-procedure in practice, see the comment on Scheffé's (1977) article by Olshen and the rejoinder by Scheffé.

CHAPTER 4

Stepwise Procedures for Pairwise and More General Comparisons among All Treatments

The previous chapter dealt with single-step procedures for families of pairwise and more general comparisons among all treatments. The present chapter is devoted to stepwise procedures for those same families except the family of all contrasts. This latter family is infinite and stepwise procedures for infinite families are not available.

Two types of stepwise procedures are discussed in this chapter—step-down and step-up. The distinction between the two is explained in Section 4 of Chapter 2. These procedures are generally more powerful than their single-step counterparts in Chapter 3. However, they suffer from two shortcomings: (i) they can be used only for testing and not for confidence estimation, and (ii) it is not known whether they control the Type I and Type III familywise error rate (FWE) if directional decisions are made in the usual manner on the signs of any pairwise differences that are found significant. As a result their applicability is somewhat limited.

We now give a brief summary of the present chapter. The first three sections are devoted to step-down procedures. Section 1 discusses step-down procedures based on F-statistics for the family of subset hypotheses, while Section 2 discusses the corresponding procedures based on Studentized range statistics. A practical algorithm for implementing the closed step-down testing procedure of Section 4.3.2 of Chapter 2 is given in Section 3. This procedure addresses the closed family of multiple subset hypotheses. In Section 4 a step-up procedure due to Welsch (1977) is presented. Finally Section 5 gives a comparison between various single-step and stepwise procedures.

1 STEP-DOWN PROCEDURES BASED ON F-STATISTICS

In this section we consider step-down procedures that have the same general testing scheme as in Section 4.3.3.1 of Chapter 2 and that use F-statistics for testing the subset homogeneity hypotheses H_P. As noted there, these procedures can be shown to satisfy the separability condition. Furthermore, assumption (4.8) of Chapter 2 can be shown to hold for these procedures, as seen in the sequel. Therefore we can use Theorem 4.4 of Chapter 2 to find a choice of the nominal levels α_p's to control the Type I FWE strongly at a designated level α. Among such choices we adopt the Tukey–Welsch (TW) choice (see (4.22) of Chapter 2):

$$\alpha_p = 1 - (1-\alpha)^{p/k} \; (2 \leqq p \leqq k-2), \qquad \alpha_{k-1} = \alpha_k = \alpha \tag{1.1}$$

for the reasons cited there. We refer to these procedures as *step-down F procedures* (also referred to in the literature as *multiple F procedures*).

Note that a shortcut testing scheme, as in the case of the Newman–Keuls (NK) procedure (see Example 4.4 of Chapter 2) cannot be used in conjunction with a step-down F procedure. Thus if a subset P with cardinality $p \geqq 3$ is found significant, then generally no statement can be made about the significance of its subsets without going through the full hierarchy of tests on those subsets. This is true regardless of whether the monotonicity condition (4.12) of Chapter 2 is satisfied or not.

We discuss step-down F procedures for one-way layouts in Section 1.1 and then indicate their extensions to general balanced or unbalanced designs in Section 1.2.

1.1 One-Way Layouts

We follow the notation of Example 1.1 of Chapter 2. In a step-down F-procedure, for testing any subset homogeneity hypothesis $H_P: \theta_i = \theta_j$ $\forall i, j \in P$ we use the F-statistic:

$$F_P = \frac{\left\{ \sum_{i \in P} n_i \bar{Y}_i^2 - \left(\sum_{i \in P} n_i \bar{Y}_i \right)^2 \Big/ \sum_{i \in P} n_i \right\}}{(p-1)S^2} \tag{1.2}$$

where p is the cardinality of set P. The statistic F_P is distributed as an $F_{p-1,\nu}$ random variable (r.v.) under H_P. So the nominal level is exactly α_p when the statistic F_P is compared against the critical point $\xi_p = F_{p-1,\nu}^{(\alpha_p)}$. Assumption (4.8) of Chapter 2 can be shown to hold in the present case by using Kimball's (1951) inequality. Therefore the Type I FWE is strongly controlled at level α, as noted at the beginning of Section 1, if the α_p's are chosen according to (1.1).

We now give a small example to illustrate this procedure.

Example 1.1 **(*Balanced One-Way Layout*).** Consider a balanced one-way layout with $k = 5$ treatments and $n = 9$ observations on each treat-

TABLE 1.1. Results of the Step-Down F Procedures Using the TW and NK Nominal Levels

		TW α_p's		NK α_p's	
Set P	F_P	$\xi_p = F^{(\alpha_p)}_{p-1,\nu}$	Decision[a]	$\xi_p = F^{(\alpha)}_{p-1,\nu}$	Decision
ABCDE	11.16	2.61	*S*	2.61	*S*
ABCD	7.06	2.84	*S*	2.84	*S*
ABCE	12.92		*S*		*S*
ABDE	14.56		*S*		*S*
ACDE	14.17		*S*		*S*
BCDE	7.06		*S*		*S*
ABC	3.57	3.69	*NS*	3.23	*S*
ABD	10.57		*S*		*S*
ACD	10.57		*S*		*S*
ABE	19.32		*S*		*S*
ACE	19.00		*S*		*S*
ADE	20.32		*S*		*S*
BCD	3.57		*NS*		*S*
BCE	9.81		*S*		*S*
BDE	8.07		*S*		*S*
CDE	6.84		*S*		*S*
AB	4.59	5.85	*NSI*	4.08	*S*
AC	6.00		*NSI*		*S*
AD	21.09		*S*		*S*
AE	37.50		*S*		*S*
BC	0.09		*NSI*		*NS*
BD	6.00		*NSI*		*S*
BE	15.84		*S*		*S*
CD	4.59		*NSI*		*S*
CE	13.50		*S*		*S*
DE	2.34		*NS*		*NS*

[a] *S* = significant, *NS* = not significant, *NSI* = not significant by implication.

ment. The ordered treatment means are

A	B	C	D	E
10.0	13.5	14.0	17.5	20.0

and $S^2 = 12$ with $\nu = 40$ degrees of freedom (d.f.).

We use two different choices of nominal levels—the TW-specification (1.1) and the NK-specification $\alpha_p \equiv \alpha$ for all p with $\alpha = 0.05$. The latter choice is, of course, known to be liberal but it is used here for two reasons: First, we wish to demonstrate the additional significant results obtained using a liberal procedure. Second, the results of both the procedures using the TW and the NK specifications of the α_p's are needed to apply the Peritz procedure in Example 3.1 where the same data are used. These results are summarized in Table 1.1. The maximal nonsignificant sets obtained by the step-down F procedures employing the two different specifications of nominal levels are then as follows:

TW-specification: $(A, B, C), (B, C, D), (D, E)$
NK-specification: $(B, C), (D, E)$.

Note the additional significant decisions obtained using the NK nominal levels. □

1.2 General Balanced or Unbalanced Designs

In this case we have $\hat{\boldsymbol{\theta}} \sim N(\boldsymbol{\theta}, \sigma^2\mathbf{V})$ and $S^2 \sim \sigma^2\chi^2_\nu/\nu$ independently of each other where $\mathbf{V}$ is an arbitrary positive definite symmetric matrix. The likelihood ratio (LR) test of $H_P: \theta_i = \theta_j \ \forall i, j \in P$ results in an F-statistic that can be written as

$$F_P = \frac{\hat{\boldsymbol{\theta}}_P'\mathbf{C}_P'(\mathbf{C}_P\mathbf{V}_P\mathbf{C}_P')^{-1}\mathbf{C}_P\hat{\boldsymbol{\theta}}_P}{(p-1)S^2} \tag{1.3}$$

where $\hat{\boldsymbol{\theta}}_P$ is a p-vector consisting of the $\hat{\theta}_i$'s for $i \in P$, $\mathbf{V}_P$ is a $p \times p$ submatrix of $\mathbf{V}$ corresponding to the subset P, and $\mathbf{C}_P$ is a $(p-1) \times p$ matrix whose rows form a basis for the contrast space $\mathbb{C}^P$.

Note that F_P is distributed as an $F_{p-1,\nu}$ r.v. under H_P and thus the nominal level is exactly α_p when (1.3) is referred to the critical point $F^{(\alpha_p)}_{p-1,\nu}$. Condition (4.8) of Chapter 2 can be shown to hold by using the following argument: For any $P \subseteq (1, 2, \ldots, k\}$ with cardinality $p \geqq 2$, the set $\{F_P \leqq \text{constant}\}$ is an ellipsoid centered at the origin in the space of the $\hat{\theta}_i$'s for $i \in P$, and hence the condition of convexity and symmetry about the origin is satisfied. The condition that $\mathbf{V}$ has the product structure (see (4.9) of Chapter 2) may not be satisfied for some designs. In such cases the validity of (4.8) is conditional on the truth of the conjecture referred to there.

2 STEP-DOWN PROCEDURES BASED ON STUDENTIZED RANGE STATISTICS

2.1 Balanced Designs

We consider the general setting of Section 2.1 of Chapter 3 where we assumed that $\hat{\boldsymbol{\theta}} \sim N(\boldsymbol{\theta}, \sigma^2\mathbf{V})$, $S^2 \sim \sigma^2\chi^2_\nu/\nu$ independently of $\hat{\boldsymbol{\theta}}$, and the matrix $\mathbf{V}$ satisfies the pairwise balance condition (see (2.9) of Chapter 3):

$$\mathrm{var}(\hat{\theta}_i - \hat{\theta}_j) = \sigma^2(v_{ii} + v_{jj} - 2v_{ij}) = 2\sigma^2 v_0 \qquad \text{(say)} \qquad \forall i \neq j\,. \tag{2.1}$$

We now study step-down procedures that are based on Studentized range statistics

$$R_P = \frac{(\max_{i\in P} \hat{\theta}_i - \min_{i\in P} \hat{\theta}_i)}{S\sqrt{v_0}}\,, \qquad P \subseteq K \tag{2.2}$$

for testing the subset hypotheses H_P, and that follow the general testing scheme of Section 4.3.3.1 of Chapter 2. We refer to them as *step-down range procedures* (also referred to in the literature as *multiple range procedures*). The NK-procedure given in Example 4.4 of Chapter 2 is such a procedure. That procedure uses a constant nominal level $\alpha_p = \alpha$ for testing each H_P. We use the TW-specification (1.1).

As noted in Chapter 2, step-down range procedures satisfy the separability condition. When the $\hat{\theta}_i$'s are independent (e.g., in a balanced one-way layout), condition (4.8) of Chapter 2 is satisfied because of Kimball's (1951) inequality. Hence Theorem 4.4 of Chapter 2 applies and the Type I FWE is strongly controlled using the TW-specification of α_p's. In the general nonindependence case it is not known whether the latter condition is satisfied or not.

In order to implement the shortcut version of the step-down range procedure (which permits one to declare any subset containing the two treatments yielding $\max_{i\in P} \hat{\theta}_i$ and $\min_{i\in P} \hat{\theta}_i$ as significant whenever a subset P is declared significant), the critical points $\xi_p = Q^{(\alpha_p)}_{p,\nu}$ must satisfy the monotonicity condition (4.12) of Chapter 2 as in the case of the NK-procedure. If the α_p's are chosen according to (1.1), the ξ_p's are not always monotonically nondecreasing in p, but we enforce monotonicity by setting $\xi_p = Q_{p-1,\nu}^{(\alpha_{p-1})}$ whenever $Q_{p,\nu}^{(\alpha_p)} < Q_{p-1,\nu}^{(\alpha_{p-1})}$ $(3 \leqq p \leqq k)$. Notice that this makes the procedure more conservative. In the sequel we do not reemphasize this point but such adjustments of critical points are assumed.

We do not describe the step-down range procedure in detail because structurally it is the same as the NK-procedure given in Example 4.4 of Chapter 2. This procedure can be implemented by using a method due to Duncan (1955) that makes it easier to interpret the results. In this method the $\hat{\theta}_i$'s are first ordered: $\hat{\theta}_{(1)} \leqq \hat{\theta}_{(2)} \leqq \cdots \leqq \hat{\theta}_{(k)}$, and the treatment labels are written in the order of the $\hat{\theta}_{(i)}$'s. Whenever any p-range, $\hat{\theta}_{(i+p-1)} - \hat{\theta}_{(i)}$, is tested and found nonsignificant, the set of treatments $\{(i), (i+1), \ldots, (i+p-1)\}$ is underscored by drawing a line connecting (i) to $(i+p-1)$ $(1 \leqq i \leqq k-p+1, 2 \leqq p \leqq k)$. Any subset of treatments sharing a common line is retained as homogeneous by implication without actually testing it. Any subset of treatments that is not connected by a common line is declared as heterogeneous. We now illustrate this method by an example.

Example 2.1 ***(Randomized Block Design; Continuation of Example 1.1 of Chapter 3).*** For these data we have $k = 7$, $n = 6$, $S^2 = 79.64$, and $\nu = 30$. Calculations of the critical values $Q_{p,\nu}^{(\alpha_p)} S/\sqrt{n}$ where the α_p's are given by (1.1) is shown in Table 2.1 for $\alpha = 0.05$; the upper α_p points needed in these calculations were obtained by linearly interpolating in $\log_e \alpha$ in Table 8 in Appendix 3. The ordered sample means are as follows.

A	*F*	*G*	*D*	*C*	*B*	*E*	
49.6	58.1	61.0	61.5	67.8	71.2	71.3	(2.3)

We now apply the step-down range procedure.

Step 1. A and E are significantly different since $71.3 - 49.6 = 21.7 > 16.26$.

TABLE 2.1. Calculation of the Critical Values

p	2	3	4	5	6	7
α_p	0.015	0.022	0.029	0.036	0.05	0.05
$Q_{p,\nu}^{(\alpha_p)}$	3.66	4.00	4.19	4.30	4.30	4.46
$Q_{p,\nu}^{(\alpha_p)} S/\sqrt{n}$	13.35	14.57	15.25	15.67	15.67	16.26

Step 2. (a) A and B are significantly different since $71.2 - 49.6 = 21.6 > 15.67$.
(b) F and E are not significantly different since $71.3 - 58.1 = 13.2 < 15.67$. Underscore the treatments F, G, D, C, B, E.

Step 3. A and C are significantly different since $67.6 - 49.6 = 18.0 > 15.67$.

Step 4. A and D are not significantly different since $61.5 - 49.6 = 11.9 < 15.25$. Underscore the treatments A, F, G, D.

This concludes the procedure since no sets remain to be tested that are not contained in underscored (homogeneous) sets. The final outcome of these tests is summarized by the underscorings shown in (2.3). Note that we obtained the same result earlier by using the T-procedure (see Example 2.1 of Chapter 3). □

2.2 Unbalanced Designs

We now consider general unbalanced designs for which condition (2.1) is not satisfied. For this setup, Kramer (1957) and Duncan (1957) proposed two different step-down range procedures. We describe the two procedures below and then indicate the reasons for the superiority of Duncan's procedure. (Both of these procedures were originally proposed in conjunction with Duncan's specification of α_p's (see (4.26) of Chapter 2); however, we use the TW-specification (1.1).)

2.2.1 *The Kramer Procedure*

Kramer's procedure can be applied in a shortcut manner using Duncan's method of underscoring indicated homogeneous subsets as described in Section 2.1. For testing the significance of a p-range, $\hat{\theta}_{(i+p-1)} - \hat{\theta}_{(i)}$, the procedure uses the critical value

$$Q_{p,\nu}^{(\alpha_p)} S\left(\frac{d_{(i),(i+p-1)}}{2}\right)^{1/2} \qquad (1 \leq i \leq k-p+1, 2 \leq p \leq k)$$

where, as in (3.2) of Chapter 3, $d_{ij} = v_{ii} + v_{jj} - 2v_{ij}$ $(1 \leq i \neq j \leq k)$. Note that this is equivalent to defining the Studentized range statistic for any set $P = \{i_1, i_2, \ldots, i_p\}$ such that $\hat{\theta}_{i_1} \leq \hat{\theta}_{i_2} \leq \cdots \leq \hat{\theta}_{i_p}$ by

$$R_P^{(K)} = \frac{\hat{\theta}_{i_p} - \hat{\theta}_{i_1}}{S\sqrt{d_{i_1 i_p}/2}},$$

which is compared with the critical point $Q_{p,\nu}^{(\alpha_p)}$; here the superscript K on

R_P stands for Kramer. In the special case of an unbalanced one-way layout, $R_P^{(K)}$ is given by

$$R_P^{(K)} = \frac{\bar{Y}_{i_p} - \bar{Y}_{i_1}}{S\sqrt{\frac{1}{2}\left(\frac{1}{n_{i_p}} + \frac{1}{n_{i_1}}\right)}} .$$

This statistic does not have the Studentized range distribution under H_P. The nature of approximation that is involved is discussed in Section 2.2.2.

2.2.2 *The Duncan Procedure*

Duncan (1957) defined the Studentized range statistic for testing the significance of any subset P of size $p \geqq 2$ by

$$R_P^{(D)} = \max_{i,j \in P} \left\{ \frac{|\hat{\theta}_i - \hat{\theta}_j|}{S\sqrt{d_{ij}/2}} \right\} ,$$

which is compared with the critical point $Q_{p,\nu}^{(\alpha_p)}$; here the superscript D on R_P stands for Duncan. Note that $R_P^{(D)}$ also does not have the Studentized range distribution under H_P.

Duncan's statistic $R_P^{(D)}$ is a union-intersection (UI) test statistic (for the family $H_{ij}: \theta_i = \theta_j \; \forall i, j \in P$) while Kramer's statistic $R_P^{(K)}$ is not. Also, for any subset P, $R_P^{(D)}$ is never smaller than $R_P^{(K)}$, and hence if the two statistics are referred to the same critical point $Q_{p,\nu}^{(\alpha_p)}$, then Duncan's procedure will be at least as powerful as Kramer's procedure. Therefore the former will be preferred if it is known to control the FWE. This point is discussed later in this section.

Since for any subset P we always have $R_P^{(D)} \geqq R_P^{(K)}$, it follows that if P is significant by Kramer's procedure, then it will also be significant by Duncan's procedure. In addition, some sets which are Kramer-nonsignificant may be Duncan-significant. Therefore, to apply Duncan's procedure one can first apply Kramer's procedure and note all maximal nonsignificant sets. Each one of these sets must be further scrutinized to check whether there is a pair of $\hat{\theta}$-values in that set such that their Studentized difference exceeds the critical value appropriate for that set. Thus suppose $P = \{1, \ldots, p\}$ is the set under question and further suppose that $\hat{\theta}_1 \leqq \hat{\theta}_2 \leqq \cdots \leqq \hat{\theta}_p$. Then one must check whether there is a pair (i, j) such that

$$\frac{\hat{\theta}_j - \hat{\theta}_i}{S\sqrt{d_{ij}/2}} > Q_{p,\nu}^{(\alpha_p)} \qquad (1 \leqq i < j \leqq p) . \tag{2.4}$$

If no such pair is found, then set P and all of its subsets are retained. If at least one such pair is found, then set P is declared significant and treatments i and j, as significantly different. If set P is declared significant, then its subsets must again be tested for significance. If pair (i, j) satisfies (2.4), then the two subsets of P to be tested at the next step are $P - \{i\}$ and $P - \{j\}$. Both of these sets have cardinality $p - 1$ and therefore the appropriate critical point to be used is $\xi_{p-1} = Q_{p-1,\nu}^{(\alpha_{p-1})}$. This procedure is applied recursively until no sets remain to be tested. (Duncan proposed a different algorithm for carrying out the tests but the final outcome is the same. The purpose of Duncan's algorithm is to simplify the task of computing the statistics $R_P^{(D)}$. But if the number of treatments is not too large, then the statistics $R_P^{(D)}$ can be obtained from a table of all pairwise $|t|$-statistics; in this case it is not even necessary to apply the Kramer procedure first. This is illustrated in Examples 2.2 and 2.3.)

We note that the underscoring method used to display homogeneous subsets cannot be used with this procedure. This is because when the treatments are ordered according to their $\hat{\theta}$-values, it does not necessarily follow that if two treatments are in the same homogeneous set then all treatments internal to those two treatments are also in the same set; in fact, the θ's of the internal treatments can be significantly different from each other.

We next address the question of control of the FWE using this procedure. (The Kramer procedure is more conservative than the Duncan procedure, and hence if the latter controls the FWE, then the former does, too.) From Theorem 4.4 of Chapter 2 we see that Duncan's procedure with the TW-specification of the α_p's will control the FWE if (i) assumption (4.8) of Chapter 2 is satisfied, and (ii)

$$\Pr_{H_P}\{R_P^{(D)} > Q_{p,\nu}^{(\alpha_p)}\} \leqq \alpha_p \qquad \forall P \subseteq K . \tag{2.5}$$

The conditions for (i) to hold have been discussed in Section 4.3.1 of Chapter 2. The convexity and symmetry condition on the sets $\{R_P^{(D)} \leqq \text{constant}\}$ is satisfied for any design, but the product structure condition (4.9) of Chapter 2 may be satisfied only for some designs, for example, trivially for unbalanced one-way layouts. If the conjecture referred to there is true, then (i) holds in all cases. Next inequality (2.5) corresponds to the Tukey conjecture discussed in Section 3.2.1 of Chapter 3. We saw there that this conjecture has been proved for unbalanced one-way layouts, but for general unbalanced designs it has not been proved except for $k = 3$. Thus at least for unbalanced one-way layouts (i) and (ii) are both satisfied and hence Duncan's procedure with the TW-specification of α_p's controls the FWE at level α.

We now give two examples to illustrate the Duncan procedure.

Example 2.2 **(*Unbalanced One-Way Layout; Continuation of Example 3.1 of Chapter 3*).** In this example we have $k=7$, $S^2=5{,}395$, and $\nu=16$. The sample means and sample sizes are given in Table 3.1 of Chapter 3.

We wish to apply Duncan's procedure to these data at $\alpha=0.05$ using the TW-specification of α_p's. The corresponding values of $Q_{p,\nu}^{(\alpha_p)}$ obtained by linearly interpolating with respect to $\log_e \alpha_p$ in Table 8 in Appendix 3 are shown in Table 2.2.

In this example the number of treatments is modest and we can easily compute all the pairwise statistics $\sqrt{2}|T_{ij}|$ $(1 \leq i < j \leq k)$ from which the range statistics $R_P^{(D)}$ can be readily obtained for any set P. These pairwise statistics are given in Table 2.3.

In the following sequence of tests made by the Duncan procedure, if some set (and a particular pair of treatments in that set) is found significant at any step, then at the following step two subsets of that set are tested for significance. As explained earlier, the two subsets are obtained by deleting one treatment of the significant pair at a time from the set.

Step 1. For $P=(D,F,A,B,C,E,G)$, $R_P^{(D)}=R_{(G,D)}=6.249>4.75$. Therefore G and D are significantly different.

Step 2. (a) For $P=(F,A,B,C,E,G)$, $R_P^{(D)}=R_{(G,A)}=5.326>4.57$. Therefore G and A are significantly different.

(b) For $P=(D,F,A,B,C,E)$, $R_P^{(D)}=R_{(E,D)}=4.682>4.57$. Therefore E and D are significantly different.

TABLE 2.2. Calculation of the Critical Values

p	2	3	4	5	6	7
α_p	0.015	0.022	0.029	0.036	0.05	0.05
$Q_{p,\nu}^{(\alpha_p)}$	3.87	4.25	4.44	4.57	4.57[a]	4.75

[a]For $p=6$ the correct value of $Q_{p,\nu}^{(\alpha_p)}$ is 4.56. We use the value for $p=5$ so that the critical points are monotone in p.

TABLE 2.3. Values of $\sqrt{2}|T_{ij}|$ Statistics

	D	F	A	B	C	E
F	1.139	—	—	—	—	—
A	1.661	0.207	—	—	—	—
B	4.508	2.693	3.288	—	—	—
C	4.551	2.932	3.427	0.580	—	—
E	4.682	3.234	3.659	1.174	0.612	—
G	6.249	4.450	5.326	2.478	1.698	0.907

Step 3. (a) For $P = (F, B, C, E, G)$, $R_P^{(D)} = R_{(G,F)} = 4.450 < 4.57$. Therefore retain the set (F, B, C, E, G) as homogeneous.
(b) For $P = (F, A, B, C, E)$, $R_P^{(D)} = R_{(E,A)} = 3.659 < 4.57$. Therefore retain the set (F, A, B, C, E) as homogeneous.
(c) For $P = (D, F, A, B, C)$, $R_P^{(D)} = R_{(C,D)} = 4.551 < 4.57$. Therefore retain the set (D, F, A, B, C) as homogeneous.

This completes the application of the Duncan procedure. The maximal nonsignificant sets retained by the Duncan procedure at $\alpha = 0.05$ are thus (F, B, C, E, G), (F, A, B, C, E), and (D, F, A, B, C). Two treatments not appearing together in any of the retained sets are significantly different; two treatments appearing together in any retained set are not significantly different.

Duncan (1957) obtained smaller maximal nonsignificant sets, namely (D, F, A), (F, B, C), and (B, C, E, G) (thereby obtaining additional significances between some treatments, e.g., A and B) because he used his more liberal choice for α_p's given by (4.26) of Chapter 2. Also note that the pair (D, E) was found nonsignificant by the Tukey–Kramer (TK) single-step procedure (see Example 3.1 of Chapter 3) while it is found significant by the present step-down procedure. □

***Example 2.3* (*Randomized Block Design with a Fixed Covariate; Continuation of Example 1.2 of Chapter 3*).** In this example we have $k = 6$ treatments randomized in each of $n = 5$ blocks. The adjusted treatment means $\hat{\theta}_i$ are given in Table 1.4 of Chapter 3. We also have $S^2 = 50.49$ with $\nu = 19$ d.f.

The values of $Q_{p,\nu}^{(\alpha_p)}$ needed to apply the Duncan procedure at $\alpha = 0.10$ are given in Table 2.4. The pairwise statistics $\sqrt{2}|T_{ij}| = |\hat{\theta}_i - \hat{\theta}_j| / S\sqrt{d_{ij}/2}$ where $d_{ij} = 2/n + (\bar{X}_{i\cdot} - \bar{X}_{j\cdot})^2 / S_{XX}$ are given in Table 2.5 (note that here the varieties $1, 2, \ldots, 6$ are labeled $A, B, \ldots, F$, respectively).

The steps in the Duncan procedure are as follows:

Step 1. For $P = (B, C, D, A, F, E)$, $R_P^{(D)} = R_{(C,F)} = 5.847 > 3.97$. Therefore C and F are significantly different.

TABLE 2.4. Calculation of the Critical Values

p	2	3	4	5	6
α_p	0.035	0.051	0.068	0.10	0.10
$Q_{p,\nu}^{(\alpha_p)}$	3.22	3.57	3.76	3.76[a]	3.97

[a]For $p = 5$ the correct value of $Q_{p,\nu}^{(\alpha_p)}$ is 3.75. We use the value for $p = 4$ so that the critical points are monotone in p.

TABLE 2.5. Values of $\sqrt{2}|T_{ij}|$ Statistics

	B	C	D	A	F
C	0.616	—	—	—	—
D	2.030	1.657	—	—	—
A	3.558	4.657	2.767	—	—
F	3.925	5.847	3.833	1.879	—
E	3.149	4.473	2.955	1.551	0.180

Step 2. (a) For $P = (B, D, A, F, E)$, $R_P^{(D)} = R_{(B,F)} = 3.925 > 3.76$. Therefore B and F are significantly different.
(b) For $P = (B, C, D, A, E)$, $R_P^{(D)} = R_{(A,C)} = 4.657 > 3.76$. Therefore A and C are significantly different.

Step 3. (a) For $P = (D, A, F, E)$, $R_P^{(D)} = R_{(D,F)} = 3.833 > 3.76$. Therefore D and F are significantly different.
(b) For $P = (B, D, A, E)$, $R_P^{(D)} = R_{(A,B)} = 3.558 < 3.76$. Therefore retain the set (B, D, A, E) as homogeneous.
(c) For $P = (B, C, D, E)$, $R_P^{(D)} = R_{(C,E)} = 4.473 > 3.76$. Therefore C and E are significantly different.

Step 4. (a) For $P = (A, F, E)$, $R_P^{(D)} = R_{(A,F)} = 1.879 < 3.57$. Therefore retain the set (A, F, E) as homogeneous.
(b) For $P = (B, C, D)$, $R_P^{(D)} = R_{(B,D)} = 2.030 < 3.57$. Therefore retain the set (B, C, D) as homogeneous.
(c) Retain the sets (D, A, E) and (B, D, E) as homogeneous because they are contained in the set (B, D, A, E), which was retained at Step 3b.

This completes the application of the Duncan procedure. The maximal nonsignificant sets retained at $\alpha = 0.10$ are (B, D, A, E), (A, F, E), and (B, C, D). □

3 PERITZ'S CLOSED STEP-DOWN PROCEDURE

In this section we describe a practical algorithm due to Begun and Gabriel (1981) for implementing Peritz's (1970) closed step-down procedure for the family of multiple subset hypotheses. The theory underlying this procedure was given in Section 4.3.2 of Chapter 2. There we saw that this procedure controls the Type I FWE strongly at level α by testing each multiple subset hypothesis $H_{\mathbf{P}} = \bigcap_{i=1} H_{P_i}$ at nominal level α (subject to the coherence requirement). If any $H_{\mathbf{P}}$ is regarded as significant when at least some H_{P_i} is significant at level α_{p_i} $(1 \leqq i \leqq r)$ where

$$\alpha_{p_i} = \begin{cases} \alpha & \text{if } \mathbf{P} = (P_i) \text{ is a singleton} \\ 1-(1-\alpha)^{p_i/k} & \text{if } \mathbf{P} \text{ is not a singleton}, \end{cases} \tag{3.1}$$

then from Theorem 4.3 of Chapter 2 we know that the nominal level of this test of $H_{\mathbf{P}}$ is controlled at α. Although this basic procedure is simple to state, it is difficult to implement because of the complicated implication relations among the hypotheses $H_{\mathbf{P}}$.

It may be observed that the first part of (3.1) corresponds to the NK-choice of nominal levels for testing individual subset hypotheses H_P, while the second part of (3.1) corresponds to the TW-choice (1.1). If we refer to the corresponding step-down procedures as the NK- and TW-procedures, respectively, then the following statements can be made (note that Begun and Gabriel refer to the TW-procedure as the Ryan procedure):

(i) Any H_P that is retained by the NK-procedure will also be retained by the closed procedure but the latter may retain some more H_P's by implication. This follows because the nominal levels used by the closed procedure for testing individual H_P's are always less than or equal to those used by the NK-procedure. For example, for $k = 4$, $H_{(1,2)}$ may be significant at level α, but $H_{(1,2)}$ and $H_{(3,4)}$ may each be nonsignificant at level $1-(1-\alpha)^{1/2}$, thus causing $H_{(1,2)}$ and $H_{(3,4)}$ to be retained.

(ii) Any H_P that is rejected by the TW-procedure will also be rejected by the closed procedure, but the latter may reject some more H_P's. This follows because the nominal levels used by the closed procedure for testing individual H_P's are always greater than or equal to those used by the TW-procedure.

Begun and Gabriel's (1981) algorithm is based on these two observations and it proceeds as follows:

Step 1. Retain all H_P's that are retained by the NK-procedure.

Step 2. Reject all H_P's that are rejected by the TW-procedure.

Step 3. Classify as contentious those H_P's that are retained by the TW-procedure but rejected by the NK-procedure. Make decisions on contentious H_P's starting from the largest P in the following manner: Retain a contentious H_P if

(a) a contentious H_Q is retained and $P \subseteq Q$, or if

(b) H_P is nonsignificant at level $\alpha_p = 1-(1-\alpha)^{p/k}$ and for some Q in the complement of P with cardinality $q \geqq 2$, H_Q is nonsignificant at level $\alpha_q = 1-(1-\alpha)^{q/k}$.

Reject a contentious H_P if (a) and (b) are both violated.

Notice that although the closed procedure addresses the family of multiple subset hypotheses, this algorithm makes decisions on subset hypotheses (which are of primary or sometimes of sole interest). Decisions on multiple subset hypotheses can be obtained in the usual manner by rejecting $H_{\mathbf{P}} = \cap_{i=1}^{r} H_{P_i}$ if and only if at least some H_{P_i} is rejected. These decisions on the H_P's are identical to the ones obtained by stepping down through the hierarchy of the $H_{\mathbf{P}}$'s and using the nominal levels given in (3.1) for testing each $H_{\mathbf{P}}$.

Another noteworthy feature of the closed procedure is that a decision on any set P is dependent on sets Q in the complement of P. Thus if the θ's outside a set P are spread out, it can help in the rejection of H_P.

We now give two examples to illustrate Peritz's closed procedure. The first example uses F-statistics while the second example uses Studentized range statistics.

Example 3.1 ***(Balanced One-Way Layout; Continuation of Example 1.1).*** We use the data of Example 1.1 where the TW and NK step-down F procedures were applied at $\alpha = 0.05$. From the results given in Table 1.1 we find that the largest contentious sets are (A, B, C) and (B, C, D). Using the algorithm for Peritz's closed procedure we make decisions on these and smaller contentious sets as follows:

Retain (A, B, C) as homogeneous since (A, B, C) is nonsignificant at level $\alpha_3 = 1 - (1 - \alpha)^{3/5}$ and since (D, E), which is in the complement of (A, B, C), is also nonsignificant at level $\alpha_2 = 1 - (1 - \alpha)^{2/5}$.

Declare (B, C, D) as heterogeneous since it is not contained in any retained contentious set and since (A, E), which is in the complement of (B, C, D), is significant at level α_2.

We now make decisions on the smaller contentious sets (A, B), (A, C), (B, D), and (C, D). The sets (A, B) and (A, C) are contained in the retained contentious set (A, B, C) and therefore are retained. The sets (B, D) and (C, D) are contained in rejected contentious sets (B, C, D) and therefore they must be further examined.

Declare (B, D) as heterogeneous since (B, D) is significant at level α_2 and since (A, C), (A, E), (C, E), and (A, C, E), which are all in the complement of (B, D), are also significant at levels α_q where $q = 2$ for the first three sets and $q = 3$ for the last set.

Retain (C, D) as homogeneous since (C, D) is nonsignificant at level α_2 and since (A, B), which is in the complement of (C, D), is also nonsignificant at level α_2.

In addition to the above decisions, the Peritz procedure also provides decisions on all multiple subset hypotheses. For example, $H_{\{(A,B);(C,D,E)\}}$ is rejected because (C, D, E) is significant at level α_3. On the other hand,

$H_{\{(A, B, C);(D,E)\}}$ is retained because (A, B, C) is nonsignificant at level α_3 and (D, E) is nonsignificant at level α_2. □

Example 3.2 **(*Randomized Block Design*; *Continuation of Example 2.1*).** The TW range procedure was applied in Example 2.1 at level $\alpha = 0.05$. The maximal nonsignificant sets identified by that procedure were (A, F, G, D) and (F, G, D, C, B, E).

We now apply the NK range procedure to the same data. The critical values needed for this procedure are given in Table 3.1 for $\alpha = 0.05$. Applying the tests in the same manner as in Example 2.1 we obtain the same maximal nonsignificant sets as obtained by the TW range procedure, namely (A, F, G, D) and (F, G, D, C, B, E). Thus there are no contentious sets in this example and therefore the Peritz range procedure makes the same decisions as the TW and NK range procedures.

TABLE 3.1. Critical Values for the NK Range Procedure

p	2	3	4	5	6	7
$Q^{(\alpha)}_{p,\nu}$	2.89	3.49	3.85	4.11	4.31	4.47
$Q^{(\alpha)}_{p,\nu} S/\sqrt{n}$	10.52	12.70	14.00	14.94	15.67	16.26

□

4 STEP-UP PROCEDURES

4.1 Balanced Designs

There is just one example of a step-up procedure in the multiple comparison literature, which was proposed by Welsch (1977) for the family of subset hypotheses in balanced one-way layouts. His procedure is based on Studentized range statistics. Determination of the critical points ξ_p $(2 \leqq p \leqq k)$ required for controlling the Type I FWE at designated level α using this procedure is discussed later in this section. We first describe the steps in the procedure:

Step 1. Order the treatment means $\bar{Y}_{(1)} \leqq \bar{Y}_{(2)} \leqq \cdots \leqq \bar{Y}_{(k)}$. Begin by testing all "gaps" or "2-ranges," $(\bar{Y}_{(i+1)} - \bar{Y}_{(i)})$ $(1 \leqq i \leqq k-1)$, by comparing them with the critical value $\xi_2 S/\sqrt{n}$. If $\bar{Y}_{(i+1)} - \bar{Y}_{(i)} > \xi_2 S/\sqrt{n}$, then declare that gap as significant and the corresponding pair of treatments as different. Also declare all sets of treatments containing that pair as heterogeneous and the corresponding p-ranges containing that gap as significant by implication without further tests. Proceed to Step 2 if at least one gap is not declared as significant.

Step 2. In general, test a p-range $(\bar{Y}_{(p+i-1)} - \bar{Y}_{(i)})$ $(1 \leqq i \leqq k - p + 1, 2 \leqq p \leqq k)$, using the critical value $\xi_p S/\sqrt{n}$ if that p-range is not declared significant by implication at an earlier step. If $\bar{Y}_{(p+i-1)} - \bar{Y}_{(i)} > \xi_p S/\sqrt{n}$, then declare that p-range as significant and the pair of treatments corresponding to $\bar{Y}_{(p+i-1)}$ and $\bar{Y}_{(i)}$ as different. Also declare all sets of treatments containing that subset of p treatments as heterogeneous and the corresponding q-ranges containing that p-range for all $q > p$ as significant by implication without further tests. Continue in this manner until no ranges remain to be tested that are not already declared as significant.

Clearly, this step-up procedure is both coherent and consonant. Also it would be expected to possess good power properties (at least for some parameter configurations) since it concludes *significance* by implication in contrast to step-down procedures, which conclude *nonsignificance* by implication. A more detailed comparison between the two types of procedures is given in Section 5.

We next indicate how Welsch (1977) determined the critical points ξ_p so as to achieve strong control of the Type I FWE. Welsch's result is stated in the following theorem.

Theorem 4.1 (Welsch 1977). Let $Z_{(1)} \leqq Z_{(2)} \leqq \cdots \leqq Z_{(k)}$ be an ordered random sample from a standard normal distribution and let U be distributed independently as a $\sqrt{\chi^2_\nu/\nu}$ r.v. where ν is the error d.f. on which the estimate S^2 is based. Define

$$R(p, q) = \max_{1 \leqq i \leqq p-q+1} (Z_{(i+q-1)} - Z_{(i)})$$

and

$$\alpha_p = \Pr\left\{\bigcup_{q=2}^{p} [R(p, q) > \xi_q U]\right\} \qquad (2 \leqq p \leqq k)$$

where $\xi_2 \leqq \cdots \leqq \xi_k$ are critical constants used by the step-up procedure. If these constants are determined so that

$$\sum_{i=1}^{r} \alpha_{p_i} \leqq \alpha \tag{4.1}$$

for all r-tuples $(1 \leqq r \leqq k/2)$ of integers $(p_1, p_2, \ldots, p_r)$ with $p_i \geqq 2$ and $\Sigma_{i=1}^{r} p_i \leqq k$, then the step-up procedure controls the Type I FWE strongly at level α.

Proof. Using the Bonferroni inequality it follows that the Type I FWE under any $H_{\mathbf{P}} = \bigcap_{i=1}^{r} H_{P_i}$ is bounded above by

$$\sum_{i=1}^{r} \Pr_{H_{P_i}}\{H_{P_i} \text{ is rejected}\} . \tag{4.2}$$

We next obtain an upper bound on a general summand in (4.2). To simplify the notation denote H_{P_i} for a general summand by H_P where $P = \{1, \ldots, p\}$ with $2 \leqq p \leqq k$. Let $\bar{Y}_{(1)} \leqq \cdots \leqq \bar{Y}_{(p)}$ be the ordered means in set P. The hypothesis H_P will be rejected at the first step (when testing 2-ranges) if

$$\max_{1 \leqq i \leqq p-1} (\bar{Y}_{(i+1)} - \bar{Y}_{(i)}) > \frac{\xi_2 S}{\sqrt{n}} .$$

Next, for $p \geqq 3$ consider testing a j-range $(3 \leqq j \leqq p)$. In that case rejection of H_P is possible if for some q $(2 \leqq q \leqq j)$,

$$\max_{1 \leqq i \leqq p-q+1} (\bar{Y}_{(i+q-1)} - \bar{Y}_{(i)}) > \frac{\xi_j S}{\sqrt{n}} . \tag{4.3}$$

Note that rejection of H_P is possible using (4.3) when a q-range $(\bar{Y}_{(i+q-1)} - \bar{Y}_{(i)})$ from set P exceeds $\xi_j S/\sqrt{n}$ (which is the appropriate critical value for a j-range) for $j > q$ because $(j-q)$ sample means $\bar{Y}_l$ for $l \not\in P$ could lie between $\bar{Y}_{(i+q-1)}$ and $\bar{Y}_{(i)}$ (and thus make it a j-range). Now since $\xi_j \geqq \xi_q$ for $j \geqq q$, it follows that event (4.3) is contained in the event

$$\max_{1 \leqq i \leqq p-q+1} (\bar{Y}_{(i+q-1)} - \bar{Y}_{(i)}) > \frac{\xi_q S}{\sqrt{n}} .$$

We thus obtain

$$\begin{aligned}
&\Pr_{H_P}\{H_P \text{ is rejected}\} \\
&\leqq \Pr_{H_P}\left\{\bigcup_{q=2}^{p}\left[\max_{1 \leqq i \leqq p-q+1} (\bar{Y}_{(i+q-1)} - \bar{Y}_{(i)}) > \frac{\xi_q S}{\sqrt{n}}\right]\right\} \\
&= \Pr_{H_P}\left\{\bigcup_{q=2}^{p}\left[\max_{1 \leqq i \leqq p-q+1} \left(\frac{\sqrt{n}\bar{Y}_{(i+q-1)}}{\sigma} - \frac{\sqrt{n}\bar{Y}_{(i)}}{\sigma}\right) > \xi_q \frac{S}{\sigma}\right]\right\} \\
&= \Pr\left\{\bigcup_{q=2}^{p} [R(p, q) > \xi_q U]\right\} = \alpha_p .
\end{aligned}$$

Substituting this in (4.2) we see that (4.1) must hold for all multiple subset hypotheses $H_{\mathbf{P}}$, and hence the theorem follows. □

Welsch (1977) proposed the choice

$$\alpha_p = \frac{\alpha p}{k} \qquad (2 \leqq p \leqq k-2), \qquad \alpha_{k-1} = \alpha_k = \alpha ,$$

which clearly guarantees (4.1). For this choice of α_p's and subject to the monotonicity constraint $\xi_2 \leqq \xi_3 \leqq \cdots \leqq \xi_k$, Welsch (1977) tabulated the critical constants ξ_p $(2 \leqq p \leqq k)$ for $k = 2(1)10$, $\nu = 5$, 20, 40, ∞, and $\alpha = 0.05$ which he obtained using Monte Carlo simulations.

For a general balanced design satisfying the pairwise balance condition (2.1), the same procedure as described above and the same tables of the critical constants ξ_p can be used. But, of course, for testing a p-range $(\hat{\theta}_{(p+i-1)} - \hat{\theta}_{(i)})$, the critical value must be modified to $\xi_p S\sqrt{v_0}$.

We now demonstrate the use of this step-up procedure by an example.

Example 4.1 **(*Randomized Block Design*; *Continuation of Example 2.1*).** For these data we have $S^2 = 79.64$ with $\nu = 30$ d.f. The critical values for $\alpha = 0.05$, $\nu = 30$ excerpted from Welsch's tables are given in Table 4.1.

The ordered sample means are

A	*F*	*G*	*D*	*C*	*B*	*E*
49.6	58.1	61.0	61.5	67.6	71.2	71.3 .

The sequence of tests in the Welsch procedure is as follows:

Step 1. The largest gap (2-range) is $F - A = 58.1 - 49.6 = 8.5$. Thus none of the 2-ranges exceeds the critical value 13.52.

Step 2. The largest 3-range is $G - A = 61.0 - 49.6 = 11.4$. Thus none of the 3-ranges exceeds the critical value 14.86.

Step 3. The largest 4-range is $D - A = 61.5 - 49.6 = 11.9$. Thus none of the 4-ranges exceeds the critical value 15.45.

Step 4. The largest 5-range is $C - A = 67.6 - 49.6 = 18.0 > 15.85$. Thus the set (A, F, G, D, C) is significant and, in particular, A and C

TABLE 4.1. Critical Values for the Step-Up Procedure

p	2	3	4	5	6	7
ξ_p	3.71	4.08	4.24	4.35	4.35	4.52
$\xi_p S/\sqrt{n}$	13.52	14.86	15.45	15.85	15.85	16.47

are different. From this it follows that sets (A, F, G, D, C, B) and (A, F, G, D, C, B, E) are also significant and need not be tested. Also, A and B, and A and E are different.

Step 5. The only 6-range that remains to be tested is $E - F = 71.3 - 58.1 = 13.2 < 15.85$, which is thus nonsignificant.

We thus obtain the following maximal nonsignificant sets: (A, F, G, D) and (F, G, D, C, B, E). Note that in Example 2.1, the TW step-down range procedure yielded the same result. □

4.2 Unbalanced Designs

Welsch's step-up procedure has not yet been extended to general unbalanced designs, that is, designs that do not satisfy the pairwise balance condition (2.1). One possible way of developing an approximate extension is as follows: Order the parameter estimates: $\hat{\theta}_{(1)} \leqq \hat{\theta}_{(2)} \leqq \cdots \leqq \hat{\theta}_{(k)}$. Use Welsch's step-up procedure with these ordered estimates but with the critical value $\xi_p S(d_{(i)(i+p-1)}/2)^{1/2}$ when testing the significance of a p-range $\hat{\theta}_{(i+p-1)} - \hat{\theta}_{(i)}$ $(1 \leqq i \leqq k - p + 1, 2 \leqq p \leqq k)$ where $d_{ij} = v_{ii} + v_{jj} - 2v_{ij}$ and the ξ_p's $(2 \leqq p \leqq k)$ depend, of course, on the error d.f. ν, the number of treatments k, and level α. It seems worthwhile to study the nature of the approximation (whether it is conservative or liberal) involved in this extension. Another way to extend Welsch's step-up procedure to unbalanced designs is by using F-statistics instead of range statistics.

5 A COMPARISON OF SINGLE-STEP AND STEPWISE PROCEDURES

5.1 Criteria for Comparison

We first discuss the criteria on which the comparisons between different multiple comparison procedures (MCPs) are based. Since confidence estimation analogs of the stepwise procedures of this chapter are not available, we restrict our comparisons to tests. The principal criterion is the power for testing subset homogeneity hypotheses. Other criteria include convenience of use and ease of interpreting the results. Strictly speaking, all of the comparisons apply only to balanced one-way layouts because the various simulation studies reported in the literature have been carried out only for this case.

One can define the power of an MCP in a variety of ways. Einot and Gabriel (1975) introduced the notion of the *P-subset power*, which, for a given subset $P \subseteq \{1, 2, \ldots, k\}$, is the probability of rejecting the subset homogeneity hypothesis $H_P: \theta_i = \theta_j \ \forall i, j \in P$ when it is false. In general, for arbitrary $\boldsymbol{\theta}$, an MCP has a collection of $2^k - k - 1$ P-subset powers, one for each subset P of cardinality $p \geqq 2$. Einot and Gabriel (1975) based their comparisons on this notion of power.

Ramsey (1978) introduced two different notions of power. The first one, called the *any-pair power*, is the probability of detecting at least one true difference among all pairs. The second one, called the *all-pairs power*, is the probability of detecting all true differences among all pairs. One can extend these notions to any-triplet power and all-triplets power, any-quadruplet power and all-quadruplets power, and so forth. Ramsey based his comparisons primarily on the all-pairs power.

The P-subset power focuses attention on detecting the heterogeneity of a specific subset. Thus if P denotes a given pair, then the corresponding subset power, called the *per-pair power*, refers to detecting the significance of that particular pair. There is some debate (Gabriel 1978b) about which one of these three notions of power, namely the per-pair power, any-pair power, and all-pairs power (and their corresponding extensions to larger subsets) is the most appropriate. In our view, the all-pairs power would generally be the right choice in those situations where control of the Type I FWE is appropriate. The per-pair power would generally be the right choice in those situations where control of the Type I per-family error rate (PFE) or the Type I per-comparison error rate (PCE) is appropriate.

Finally, there is the notion of the *overall power*, which is the probability of making no Type II error, that is, the probabiity of rejecting all false subset hypotheses. This criterion was used by Carmer and Swanson (1973) and Welsch (1977).

For assessing the interpretability of the results of an MCP, Shaffer (1981) introduced a criterion called *complexity*. Essentially this criterion considers the ability of an MCP to produce (via reject/retain decisions) a simple partition of means into distinct nonoverlapping homogeneous subsets. Not much work has been done to compare different MCPs based on this criterion. We refer the reader to Shaffer's article for additional details.

5.2 Procedures to Be Compared

Comparisons are restricted to those single-step and stepwise procedures that strongly control the Type I FWE. Among the different stepwise procedures for balanced one-way layouts, only the step-down procedures of Tukey–Welsch and Peritz and the step-up procedure of Welsch satisfy this requirement, and hence are included in the comparison. Step-down procedures based on Studentized range statistics as well as F-statistics are considered. For the Welsch step-up procedure only Studentized range statistics are considered, since a corresponding procedure based on F-statistics has not been developed yet.

From the description of Peritz's closed procedure it is clear that it will be more powerful (using any definition of power) than the TW-procedure, which in turn will be more powerful than the corresponding single-step procedure; the appropriate single-step procedure to compare is the T-procedure (respectively, the S-procedure) if the step-down procedures are based on Studentized range statistics (respectively, F-statistics). However, we still retain all of these MCPs in our comparison because they are not quite equal on other counts. For example, single-step procedures are easier to use, yield confidence intervals, and allow directional decisions. These advantages are not available with stepwise procedures. Also, the Peritz procedure is less convenient to apply (requires more computations and cross-checking) than the TW-procedure. Therefore, we would like to assess the power differences between these MCPs in detail, and ascertain whether they are large enough to justify the use of the more complicated procedures. We study the performances of these MCPs in conjunction with Studentized range and F-statistics to determine the relative advantages associated with each type of statistic under different parameter configurations.

5.3 Discussion

Prior to Einot and Gabriel's (1975) work, several authors (Balaam 1963, Petrinovich and Hardyck 1969, Boardman and Moffitt 1971, Carmer and Swanson 1973, Thomas 1974) carried out simulation studies to compare the performances of different MCPs. They found that single-step procedures (the S-procedure and the T-procedure) have the least power, while the Duncan (D) procedure (see Example 4.7 of Chapter 2) has the highest power. However, we have seen that the D-procedure is the most liberal of all MCPs and does not strongly control the Type I FWE. Thus the reported differences in powers were mainly due to different choices of nominal significance levels made by different MCPs and had nothing to do with the intrinsic properties of the MCPs. Einot and Gabriel (1975) were the first to highlight this simple fact and to postulate that comparisons be restricted to only those MCPs that strongly control the Type I FWE at the same level.

As mentioned in Section 5.1, Einot and Gabriel studied the P-subset powers of various MCPs by means of simulation. Their study was confined to balanced one-way layouts with $k = 3, 4, 5, n = 9$, and selected configurations $\boldsymbol{\theta}$. The essence of their findings is that the S- and T-procedures do not lose much in terms of the P-subset power in comparison to the TW and the Peritz step-down procedures. More specifically, the gain in power due to the use of the step-down procedures in comparison to the single-step procedures is not substantial enough to offset some of the disadvantages associated with the former, such as more complicated implementation and unavailability of confidence intervals.

Einot and Gabriel found very small differences in the powers of the MCPs based on Studentized range statistics. The power differences between the T-procedure and the TW-procedure were roughly of the order of 0.01 to 0.08 and those between the T-procedure and the Peritz procedure were of the order of 0.04 to 0.09. Smaller differences in power were observed when the P-subset power of the T-procedure was fixed at 0.50 and when P was a larger subset of the whole set of treatments while larger differences in power were observed when the P-subset power of the T-procedure was fixed at 0.75 and when P was a smaller subset of the whole set. The power differences were somewhat larger for the MCPs based on F-statistics. The power differences between the S-procedure and the TW-procedure were of the order of 0.05 to 0.14 and those between the S-procedure and the Peritz procedure were of the order of 0.07 to 0.15 with the extreme differences occurring under the same conditions as in the case of the procedures based on Studentized range statistics.

On the basis of these results Einot and Gabriel recommended the use of the T-procedure in most practical problems. For situations where confidence intervals are not of interest, they recommended the TW step-down procedure based on F-statistics over the Peritz procedure because the latter involves "impractically complicated" tests. However, the implementation of the Peritz procedure has now become feasible through the computer program developed by Begun and Gabriel (1981).

Ramsey (1978) studied the all-pairs powers of various procedures via simulation and came up with conclusions opposite to those of Einot and Gabriel (1975). The essence of Ramsey's findings is that the Peritz procedure (particularly the one based on F-statistics) can be considerably more powerful than the T-procedure under different configurations, the advantage being greatest when the means are widely separated. In a large number of cases Ramsey found that the all-pairs power of the Peritz procedure exceeds that of the T-procedure by more than 0.20 and in extreme cases by more than 0.50. The TW Studentized range procedure also performed quite well except under the configuration of equispaced means when the advantage of the Peritz procedure (based on Studentized range or F-statistics) was quite large. The performance of the Peritz procedure based on F-statistics was uniformly superior to the performance of the same procedure based on Studentized range statistics except when the means were equispaced, in which case the powers of the two procedures were roughly equal.

Einot and Gabriel (1975) and Ramsey (1978), using two different definitions of power, reached quite different conclusions: Einot and Gabriel found that the extra power associated with step-down procedures is not worth the trouble and it is better to use single-step procedures, while Ramsey found that this extra power can be quite substantial. Gabriel (1978b) offered the following explanation of why step-down procedures dominate single-step procedures substantially in terms of the all-pairs power but not in terms of the P-subset powers, in particular, the per-pair power: In a step-down procedure (particularly the Peritz procedure) the rejection of any subset hypothesis becomes more likely if any of the other subsets are rejected, while in a single-step procedure the decision on a given subset is independent of the decisions on other subsets. Therefore, step-down procedures enjoy a significant advantage over single-step procedures for joint rejection decisions on several nonhomogeneous subsets, but not for a single rejection decision on a particular nonhomogeneous subset.

Welsch (1977) compared his step-up procedure with selected competitors and found that it generally has the best performance in terms of the overall power, followed closely by the TW step-down procedure based on Studentized range statistics. The Peritz procedure was not included in Welsch's comparison. The gain in power over the TW-procedure is not substantial enough to justify the use of the step-up procedure when one takes into account the drawbacks associated with it, namely, the lack of extensive tables of critical constants needed for its implementation and its presently limited applicability to only balanced designs.

To summarize, if confidence estimates of pairwise differences or directional decisions about them are desired, then the only option is to use one of the single-step procedures. A comparison between different single-step procedures is given in Section 4 of Chapter 3. If only the tests of subset homogeneity hypotheses are of interest, then stepwise procedures offer the potential of significant gains in power. Among the step-down procedures, the Peritz procedure based on F-statistics is generally the most powerful choice, but it is rather cumbersome to apply by hand. The TW procedure is somewhat less powerful but it is much easier to apply, particularly when Studentized range statistics are used in a balanced one-way layout. The step-up procedure of Welsch dominates the TW step-down procedure (both based on Studentized range statistics) by a small amount, and it may be used if the required critical constants are available in Welsch (1977).

CHAPTER 5

Procedures for Some Nonhierarchical Finite Families of Comparisons

In Chapters 3 and 4 we discussed procedures for some families of pairwise and more general comparisons among all treatments. In this chapter we discuss procedures for some other families of comparisons. A few of these families also involve comparisons among all treatments (e.g., the family of orthogonal comparisons); however, they are treated here separately because they are not generalizations of the family of pairwise comparisons. The families considered are for the most part finite and nonhierarchical. The theory underlying the procedures for such families is covered in Sections 2.1.1 and 4.2 of Chapter 2 for single-step and step-down procedures, respectively.

In Section 1 we consider the family of orthogonal comparisons. The problem of comparing the treatment means in a one-way layout with *known* standard values or benchmarks provides a simple example of this family. In Section 2 we consider the family of comparisons of treatment means with a single benchmark corresponding to the *unknown* mean of a control treatment. In Section 3 we consider the family of comparisons of treatment means with the mean of the *unknown* "best" treatment, that is, the treatment having the largest mean. Finally in Section 4 we discuss procedures for two miscellaneous families: the family of deviations from the average of all treatment means and the family of successive differences in treatment means when the means have an *a priori* ordering.

1 ORTHOGONAL COMPARISONS

In this section we are concerned with making simultaneous inferences on a finite family of parametric functions $\mathbf{l}_i'\boldsymbol{\theta}$, $i = 1, 2, \ldots, m$, such that their least squares (LS) estimators $\mathbf{l}_i'\hat{\boldsymbol{\theta}}$ are uncorrelated, that is,

$$\text{cov}(\mathbf{l}_i'\hat{\boldsymbol{\theta}}, \mathbf{l}_j'\hat{\boldsymbol{\theta}}) = \sigma^2 \mathbf{l}_i'\mathbf{V}\mathbf{l}_j = 0 \qquad (1 \leq i \neq j \leq m). \tag{1.1}$$

Because of the joint normality of the $\mathbf{l}_i'\hat{\boldsymbol{\theta}}$'s, (1.1) implies their mutual independence.

The simplest example of orthogonal comparisons is what Tukey (1953) refers to as "batch" comparisons in a one-way layout. These are the comparisons of batch or treatment means θ_i with specified values θ_{0i} or more generally, simultaneous confidence intervals for the θ_i's $(1 \leq i \leq k)$. Here the parametric functions of interest are $\theta_i = \mathbf{l}_i'\boldsymbol{\theta}$ where $\mathbf{l}_i$ is a k-vector with unity in the ith place and zero everywhere else, $\mathbf{V} = \text{diag}(1/n_1, \ldots, 1/n_k)$, and thus condition (1.1) is clearly satisfied. More generally, condition (1.1) is satisfied for a one-way layout if the vectors $\mathbf{l}_i$ satisfy

$$\sum_{h=1}^{k} \frac{l_{ih} l_{jh}}{n_h} = 0 \qquad (1 \leq i \neq j \leq m). \tag{1.2}$$

For a balanced one-way layout, condition (1.2) becomes $\mathbf{l}_i'\mathbf{l}_j = 0$ for all $i \neq j$. In that case, if the $\mathbf{l}_i$'s are contrast vectors $\mathbf{c}_i \in \mathbb{C}^k$, then the $\mathbf{c}_i'\boldsymbol{\theta}$'s $(1 \leq i \leq m)$ are called orthogonal contrasts and m can be at most $k - 1$. For example, the coefficients of orthogonal polynomials constitute such a set of orthogonal contrasts among the treatment means when the treatments consist of equispaced levels of a quantitative factor.

Another application of orthogonal comparisons arises in 2^d factorial experiments. Assume that the 2^d factorial combinations are randomized within each of n complete blocks. The vector $\boldsymbol{\theta}$ consists of the true means of the $k = 2^d$ factorial combinations. The vector $\hat{\boldsymbol{\theta}}$, which is the corresponding vector of sample means (across blocks), is normally distributed with mean $\boldsymbol{\theta}$ and variance-covariance matrix $(\sigma^2/n)\mathbf{I}$. The $2^d - 1$ comparisons consisting of d main effects, $\binom{d}{2}$ first order interactions, and so on, can be expressed as contrasts $\mathbf{c}_i'\boldsymbol{\theta}$ $(1 \leq i \leq 2^d - 1)$ where $\mathbf{c}_i'\mathbf{c}_j = 0$ for all $i \neq j$ and each component of $\mathbf{c}_i$ equals $\pm(1/2)^{d-1}$ so that $\mathbf{c}_i'\mathbf{c}_i = (1/2)^{d-2}$. These are orthogonal comparisons because $\text{cov}(\mathbf{c}_i'\hat{\boldsymbol{\theta}}, \mathbf{c}_j'\hat{\boldsymbol{\theta}}) = (\sigma^2/n)\mathbf{c}_i'\mathbf{c}_j = 0$ for all $i \neq j$. Each $\mathbf{c}_i'\hat{\boldsymbol{\theta}}$ has the same variance, which equals $(\sigma^2/n)\mathbf{c}_i'\mathbf{c}_i = \sigma^2/(2^{d-2}n)$. The finite set that an experimenter might wish to specify in advance could be the whole set of $2^d - 1$ comparisons or any subset of special interest (e.g., main effects only or higher order interactions only).

A two-sided single-step procedure for the family of orthogonal comparisons is given in Example 2.2 of Chapter 2 and the corresponding step-down test procedure is given in Example 4.1 of Chapter 2. Both of these procedures are based on the Studentized maximum modulus distribution. This is the distribution of $\max|T_i|$ where the pivotal random variables (r.v.'s)

$$T_i = \frac{\mathbf{l}_i'(\hat{\boldsymbol{\theta}} - \boldsymbol{\theta})}{S\sqrt{\mathbf{l}_i'\mathbf{V}\mathbf{l}_i}} \qquad (1 \leqq i \leqq m) \tag{1.3}$$

have a joint multivariate t-distribution with the associated common correlation $\rho = 0$ and degrees of freedom (d.f.) ν. In Sections 1.1 and 1.2 we further expound on these two procedures and also discuss their one-sided analogs, which are based on the distribution of $\max T_i$ (known as the Studentized maximum distribution). A numerical example is given to illustrate these procedures.

1.1 Single-Step Procedures

Simultaneous $(1-\alpha)$-level two-sided confidence intervals for $\mathbf{l}_i'\boldsymbol{\theta}$ $(1 \leqq i \leqq m)$ follow from (2.7) of Chapter 2 and are given by

$$\mathbf{l}_i'\boldsymbol{\theta} \in [\mathbf{l}_i'\hat{\boldsymbol{\theta}} \pm |M|_{m,\nu}^{(\alpha)} S\sqrt{\mathbf{l}_i'\mathbf{V}\mathbf{l}_i}] \qquad (1 \leqq i \leqq m)\,; \tag{1.4}$$

here $|M|_{m,\nu}^{(\alpha)}$ is the upper α point of the Studentized maximum modulus distribution with parameter m and d.f. ν. For batch comparisons in a one-way layout (1.4) takes the form

$$\theta_i \in [\bar{Y}_i \pm |M|_{k,\nu}^{(\alpha)} S/\sqrt{n_i}] \qquad (1 \leqq i \leqq k)\,,$$

and for orthogonal contrasts in a balanced one-way layout (1.4) takes the form

$$\sum_{j=1}^{k} c_{ij}\theta_j \in \left[\sum_{j=1}^{k} c_{ij}\bar{Y}_j \pm |M|_{m,\nu}^{(\alpha)} S\sqrt{\frac{1}{n}\sum_{j=1}^{k} c_{ij}^2}\right] \qquad (1 \leqq i \leqq m)\,.$$

As noted in Chapter 2, these intervals can be used to test hypotheses on $\mathbf{l}_i'\boldsymbol{\theta}$, say, $H_{0i}: \mathbf{l}_i'\boldsymbol{\theta} = 0$ versus $H_{1i}: \mathbf{l}_i'\boldsymbol{\theta} \neq 0$ $(1 \leqq i \leqq m)$. If any H_{0i} is rejected, then one can classify the sign of $\mathbf{l}_i'\boldsymbol{\theta}$ depending on the sign of $\mathbf{l}_i'\hat{\boldsymbol{\theta}}$. This single-step test procedure controls the Type I and Type III familywise error rate (FWE) at level α.

The lower one-sided analog of (1.4) is given by

$$\mathbf{l}_i'\boldsymbol{\theta} \geqq \mathbf{l}_i'\hat{\boldsymbol{\theta}} - M_{m,\nu}^{(\alpha)} S\sqrt{\mathbf{l}_i'\mathbf{V}\mathbf{l}_i} \qquad (1 \leqq i \leqq m) \tag{1.5}$$

where $M_{m,\nu}^{(\alpha)}$ is the upper α point of the Studentized maximum distribution with parameter m and d.f. ν. If upper confidence bounds are required for a subset of the $\mathbf{l}_i'\boldsymbol{\theta}$'s and lower confidence bounds are required for the rest, then the same critical constant $M_{m,\nu}^{(\alpha)}$ can be used to guarantee that the joint confidence level $= 1 - \alpha$. The subset for which one type of confidence bounds are required may be fixed in advance or selected based on the data. The latter option enables us to make directional decisions about the $\mathbf{l}_i'\boldsymbol{\theta}$'s ($\mathbf{l}_i'\boldsymbol{\theta} > 0$ or < 0 depending on whether $\mathbf{l}_i'\hat{\boldsymbol{\theta}} > M_{m,\nu}^{(\alpha)} S\sqrt{\mathbf{l}_i'\mathbf{V}\mathbf{l}_i}$ or $< -M_{m,\nu}^{(\alpha)} S\sqrt{\mathbf{l}_i'\mathbf{V}\mathbf{l}_i}$) with control of the Type III FWE (but not the Type I FWE) at level α. The resulting procedure is a special case of the multiple three-decision procedure given in the paragraph following Theorem 2.3 of Chapter 2.

Example 1.1. The data for this example are taken from Cochran and Cox (1957, Section 5.24a). The same data were used by Bechhofer and Dunnett (1982) to demonstrate Tukey's Studentized maximum modulus procedure (1.4) for orthogonal constrasts. Here we use them to demonstrate, in addition, the directional decision procedure based on the Studentized maximum distribution.

The data pertain to the effects of four fertilizers (M = manure, N = nitrogen, P = phosphorous, K = potassium), each at two levels, on the yield of grass. The 2^4 factorial combinations of levels are randomized in each of $n = 4$ blocks. There are 15 orthogonal contrasts here consisting of the four main effects (M, N, P, K), the six two-factor interactions (MN, MP, MK, NP, NK, PK), four three-factor interactions (MNP, MNK, MPK, NPK) and one four-factor interaction (MNPK). Assuming no interaction between treatments and blocks, the mean square error (MSE) estimator of σ^2 is $S^2 = 90.5$ based on 45 d.f. and the estimated standard error of each estimated effect is $\sqrt{S^2/16} = 2.378$.

The 99% and 90% simultaneous confidence intervals using (1.4) are given in Table 1.1. In these calculations $|M|_{15,45}^{(0.01)} = 3.65$ and $|M|_{15,45}^{(0.10)} = 2.81$. Thus the allowances $|M|_{k,\nu}^{(\alpha)}\sqrt{S^2/16}$ for 99% and 90% intervals are given by $3.65 \times 2.378 = 8.68$ and $2.81 \times 2.378 = 6.68$, respectively. The t-statistics associated with the estimated effects are also given in the same table. The t-statistics that are significant at $\alpha = 0.01$ (i.e., $|t| > 3.65$) are

TABLE 1.1. Simultaneous Confidence Intervals for Effects in a 2^4 Factorial Experiment

Type	Effect	Confidence Interval $1-\alpha=0.99$	Confidence Interval $1-\alpha=0.90$	t-Statistic[a]
Main effect	M	18.0 ± 8.68	18.0 ± 6.68	7.57**
	N	21.3 ± 8.68	21.3 ± 6.68	8.96**
	P	5.5 ± 8.68	5.5 ± 6.68	2.31
	K	24.1 ± 8.68	24.1 ± 6.68	10.13**
Two-factor interactions	MN	3.2 ± 8.68	3.2 ± 6.68	1.35
	MP	-0.3 ± 8.68	-0.3 ± 6.68	−0.13
	MK	-7.5 ± 8.68	-7.5 ± 6.68	−3.15*
	NP	3.5 ± 8.68	3.5 ± 6.68	1.47
	NK	10.9 ± 8.68	10.9 ± 6.68	4.58**
	PK	3.2 ± 8.68	3.2 ± 6.68	1.35
Three-factor interactions	MNP	-1.4 ± 8.68	-1.4 ± 6.68	−0.59
	MNK	-8.5 ± 8.68	-8.5 ± 6.68	−3.57*
	MPK	0.8 ± 8.68	0.8 ± 6.68	0.34
	NPK	0.5 ± 8.68	0.5 ± 6.68	0.21
Four-factor interactions	MNPK	-1.6 ± 8.68	-1.6 ± 6.68	−0.67

[a]A double asterisk indicates significance at $\alpha = 0.01$. A single asterisk indicates significance at $\alpha = 0.10$.

marked with double asterisks and the ones that are significant at $\alpha = 0.10$ (i.e., $|t| > 2.81$) are marked with single asterisks.

If the main interest is in determining the signs of the effects (i.e., if we are concerned with protection only against Type III errors), then we can use the more powerful directional decision procedure mentioned earlier. The critical constant needed for this procedure for $\alpha = 0.10$ is $M_{15,45}^{(0.10)} = 2.55$. Comparing this value with the t-statistics we conclude that the effects M, N, K, and NK have positive signs and the effects MK and MNK have negative signs. In this instance, these are the same effects for which we are able to classify the signs by using the less powerful Studentized maximum modulus procedure. □

1.2 Step-Down Procedures

A step-down procedure for testing hypotheses on orthogonal comparisons against two-sided alternatives is given in Example 4.1 of Chapter 2. A step-down procedure for one-sided alternatives is similar to that given in Example 4.2 of Chapter 2. Here we have $\rho = 0$ and hence we use the critical constants $T^{(\alpha)}_{i,\nu,\{0\}} = M^{(\alpha)}_{i,\nu}$ in successive steps of testing ($1 \leqq i \leqq m$). The T_i-statistics are based on the pivotal r.v.'s (1.3).

From Theorem 4.1 of Chapter 2 it follows that both of these procedures control the Type I FWE at level α, these being the shortcut versions of the corresponding closed testing procedures. We also saw in Section 4.2.3 of Chapter 2 that the two-sided step-down procedure can be accompanied by directional decisions on those $\mathbf{l}_i'\boldsymbol{\theta}$'s for which the null hypotheses are rejected, without exceeding the designated level α for the Type I and Type III FWE (Holm 1979b).

By using the method of Kim, Stefánsson, and Hsu (1987) one can construct $(1-\alpha)$-level simultaneous confidence intervals associated with the one-sided step-down procedure (the corresponding analog for the two-sided procedure is not available). These intervals are given by (analogous to (4.7) of Chapter 2)

$$\mathbf{l}_i'\boldsymbol{\theta} \geqq [\mathbf{l}_i'\hat{\boldsymbol{\theta}} - M^{(\alpha)}_{p,\nu} S\sqrt{\mathbf{l}_i'\mathbf{V}\mathbf{l}_i}]^- \qquad (1 \leqq i \leqq m) \tag{1.6}$$

where $x^- = \min(x, 0)$ and p is such that $T_{(i)} > M^{(\alpha)}_{i,\nu}$ for $i = m, m-1, \ldots, p+1$ and $T_{(p)} \leqq M^{(\alpha)}_{p,\nu}$; here $T_{(1)} \leqq T_{(2)} \leqq \cdots \leqq T_{(m)}$ are the ordered T_i-statistics. One can similarly obtain upper one-sided confidence bounds.

Example 1.1 **(*Continued*).** Suppose that we only wish to test the significances of the various factorial effects. In that case it is preferable to use the two-sided step-down procedure. Clearly for any given α, all the effects that were found significant by the single-step procedure will also be found significant by the step-down procedure; the latter may find some additional ones significant. Thus for $\alpha = 0.01$, the latter will also find the effects M, N, K, and NK significant. The next effect in the order of magnitude is the three-factor interaction MNK, which has the $|t|$-statistic equal to 3.57. Comparing this with $|M|^{(0.01)}_{11,45} = 3.55$ we find that this effect is also significant. The next ordered effect is the two-factor interaction MK, which has the $|t|$-statistic equal to 3.15. Comparing this with $|M|^{(0.01)}_{10,45} = 3.51$ we find that MK is not significant at $\alpha = 0.01$. Therefore all the remaining effects are concluded to be not significant at $\alpha = 0.01$. For $\alpha = 0.10$, the step-down procedure finds the same effects significant that were found by the single-step procedure. □

2 COMPARISONS WITH A CONTROL

In many experimental situations a standard of comparison is provided by a control group with which different treatment groups are compared. For example, in a clinical trial a control group consists of patients treated with a standard existing therapy or no therapy (placebo), and the treatment groups consist of patients treated with new therapies. Often the new treatments are of interest only if they are shown to be "better" than the control. Thus comparisons with the control can be used as a screening device to eliminate noncontending treatments.

If the goal is to select the treatments that are "better" than the control, then a one-sided procedure is required. Sometimes it is desired to identify both sets of treatments, those that are "better" than the control and those that are "worse" than the control. In this case, a two-sided procedure with directional decisions is required. (For an alternative approach to this problem see Section 1.2 of Chapter 6.) We now proceed to study single-step and step-down procedures for these goals.

2.1 Single-Step Procedures

2.1.1 Exact Procedures for Balanced or Unbalanced Designs

In Example 2.3 of Chapter 2 we gave Dunnett's (1955) one-sided and two-sided simultaneous confidence intervals for $\theta_i - \theta_k$ $(1 \leqq i \leqq k-1)$ for a one-way layout where the kth treatment is regarded as a control. In the present section we extend these to general balanced or unbalanced designs. Thus consider the setting of Section 1 of Chapter 2. Let $\hat{\boldsymbol{\theta}} \sim N(\boldsymbol{\theta}, \sigma^2\mathbf{V})$ and $S^2 \sim \sigma^2\chi^2_\nu/\nu$ be the usual estimates of the vector of treatment effects $\boldsymbol{\theta}$ and the error variance σ^2, respectively.

The parameters of interest are $\gamma_i = \mathbf{l}_i'\boldsymbol{\theta} = \theta_i - \theta_k$ where $\mathbf{l}_i$ is a k-vector with $+1$ in the ith place, -1 in the kth place, and zero everywhere else $(1 \leqq i \leqq k-1)$. From (2.10) of Chapter 2 we obtain the following $(1-\alpha)$-level simultaneous one-sided confidence intervals for $\theta_i - \theta_k$:

$$\theta_i - \theta_k \geqq \hat{\theta}_i - \hat{\theta}_k - T^{(\alpha)}_{k-1,\nu,\{\rho_{ij}\}} S\sqrt{v_{ii} + v_{kk} - 2v_{ik}} \qquad (1 \leqq i \leqq k-1) \tag{2.1}$$

where by using (2.6) of Chapter 2, the correlation coefficient ρ_{ij} between $\hat{\theta}_i - \hat{\theta}_k$ and $\hat{\theta}_j - \hat{\theta}_k$ is found to be

$$\rho_{ij} = \frac{v_{ij} - v_{ik} - v_{jk} + v_{kk}}{\{(v_{ii} + v_{kk} - 2v_{ik})(v_{jj} + v_{kk} - 2v_{jk})\}^{1/2}} \qquad (1 \leqq i \neq j \leqq k-1)\,. \tag{2.2}$$

The corresponding two-sided intervals (see (2.7) of Chapter 2) are given by

$$\theta_i - \theta_k \in [\hat{\theta}_i - \hat{\theta}_k \pm |T|^{(\alpha)}_{k-1,\nu,\{\rho_{ij}\}} S\sqrt{v_{ii} + v_{kk} - 2v_{ik}}] \qquad (1 \leqq i \leqq k-1). \tag{2.3}$$

If $\rho_{ij} = \rho$ for all $i \neq j$, then we refer to the design as *balanced with respect to treatments*. The critical points $T^{(\alpha)}_{k-1,\nu,\{\rho_{ij}\}}$ and $|T|^{(\alpha)}_{k-1,\nu,\{\rho_{ij}\}}$ are tabulated only for this equicorrelated case. For unbalanced designs with arbitrary ρ_{ij}'s, it is clearly impractical to tabulate these critical points. Even a computer program to calculate these critical points would be prohibitively costly to run for moderate to large k because of the necessity to repeatedly evaluate $(k-1)$-dimensional integrals of the multivariate t density function. However, for certain special correlation structures these $(k-1)$-dimensional integrals can be reduced to lower dimensional iterated integrals, thus simplifying the computational task considerably. We first illustrate this in the simplest case of an unbalanced one-way layout.

From Example 2.3 of Chapter 2 we see that for an unbalanced one-way layout, ρ_{ij} is given by

$$\rho_{ij} = \lambda_i \lambda_j \qquad (1 \leqq i \neq j \leqq k-1) \tag{2.4}$$

where

$$\lambda_i = \left(\frac{n_i}{n_i + n_k}\right)^{1/2} \qquad (1 \leqq i \leqq k-1). \tag{2.5}$$

From (1.1a) and (1.1b) of Appendix 3 we see that when the ρ_{ij}'s have such a product structure, the cumulative distribution functions (c.d.f.'s) of $\max_{1 \leqq i \leqq k-1} T_i$ and $\max_{1 \leqq i \leqq k-1} |T_i|$ can be written as

$$\int_0^\infty \int_{-\infty}^\infty \prod_{i=1}^{k-1} \Phi\left[\frac{\lambda_i z + tu}{(1-\lambda_i^2)^{1/2}}\right] d\Phi(z)\, dF_\nu(u) \tag{2.6}$$

and

$$\int_0^\infty \int_{-\infty}^\infty \prod_{i=1}^{k-1} \left\{\Phi\left[\frac{\lambda_i z + tu}{(1-\lambda_i^2)^{1/2}}\right] - \Phi\left[\frac{\lambda_i z - tu}{(1-\lambda_i^2)^{1/2}}\right]\right\} d\Phi(z)\, dF_\nu(u), \tag{2.7}$$

respectively. Here $\Phi(\cdot)$ denotes the standard normal c.d.f. and $F_\nu(\cdot)$ denotes the c.d.f. of a $\sqrt{\chi^2_\nu/\nu}$ r.v. The desired critical points are obtained by setting (2.6) and (2.7) equal to $1-\alpha$ and solving for t. Computer programs to evaluate these integrals are given in Dunnett (1984).

When $k = 3$, there are only two comparisons, namely, $\theta_1 - \theta_3$ and $\theta_2 - \theta_3$. Therefore the critical point needed for simultaneously making these two comparisons involves only one correlation coefficient, ρ_{12}, which depends on the particular design. Thus the iterated integral representations given above to evaluate the exact critical points are not necessary. We illustrate this case with the following example.

Example 2.1. The data in Table 2.1 are taken from Dunnett (1955). The data pertain to blood count measurements on three groups of animals, one of which served as the control while the other two were treated with different drugs. Due to accidental losses, the numbers of animals in the three groups were unequal. The pooled MSE is computed to be $S^2 = 1.3805$ ($S = 1.175$) with d.f. $\nu = 12$.

The correlation coefficient between $\hat{\theta}_A - \hat{\theta}_{\text{control}}$ and $\hat{\theta}_B - \hat{\theta}_{\text{control}}$ is given by

$$\rho_{12} = \left(\frac{n_1}{n_1 + n_3}\right)^{1/2} \left(\frac{n_2}{n_2 + n_3}\right)^{1/2} = \sqrt{\frac{4}{10} \times \frac{5}{11}} = 0.4264 .$$

We can linearly interpolate in $1/(1 - \rho)$ in Tables 4 and 5 in Appendix 3 to obtain the following approximate upper 5% critical points: $T^{(0.05)}_{2,12,0.4264} \cong 2.123$ and $|T|^{(0.05)}_{2,12,0.4264} \cong 2.513$. (The corresponding exact values are 2.1211 and 2.5135, respectively.) Using the interpolated values we calculate 95% one-sided and two-sided intervals for the differences between the drugs and the control as follows.

95% One-Sided Intervals

$$\begin{aligned} \theta_A - \theta_{\text{control}} &\geqq 8.90 - 8.25 - 2.123 \times 1.175 \times \sqrt{\tfrac{1}{6} + \tfrac{1}{4}} \\ &= 0.65 - 1.61 = -0.96 . \end{aligned}$$

TABLE 2.1. Blood Counts of Animals (in Millions of Cells per Cubic Millimeter)

	Drug *A*	Drug *B*	Control
	9.76	12.80	7.40
	8.80	9.68	8.50
	7.68	12.16	7.20
	9.36	9.20	8.24
		10.55	9.84
			8.32
$\bar{Y}_i$	8.90	10.88	8.25
n_i	4	5	6

Source: Dunnett (1955).

$$\theta_B - \theta_{\text{control}} \geqq 10.88 - 8.25 - 2.123 \times 1.175 \times \sqrt{\tfrac{1}{6} + \tfrac{1}{5}}$$
$$= 2.63 - 1.51 = 1.12.$$

If upper limits are desired instead of lower limits, then they can be calculated in an analogous manner.

95% Two-Sided Intervals

$$\theta_A - \theta_{\text{control}} \in [8.90 - 8.25 \pm 2.513 \times 1.175 \times \sqrt{\tfrac{1}{6} + \tfrac{1}{4}}]$$
$$= [-1.26, 2.56].$$

$$\theta_B - \theta_{\text{control}} \in [10.88 - 8.25 \pm 2.513 \times 1.175 \times \sqrt{\tfrac{1}{6} + \tfrac{1}{5}}]$$
$$= [0.84, 4.42].$$

Thus only for drug B is the rise in the blood count statistically significant. □

In general, for any unbalanced design the exact critical points $T^{(\alpha)}_{k-1,\nu,\{\rho_{ij}\}}$ and $|T|^{(\alpha)}_{k-1,\nu,\{\rho_{ij}\}}$ can be evaluated using $(m+1)$-variate iterated integrals (where $m+1 < k-1$) if the ρ_{ij}'s can be expressed as

$$\rho_{ij} = \sum_{h=1}^{m} \lambda_{hi}\lambda_{hj} \qquad (1 \leqq i \neq j \leqq k-1) \tag{2.8}$$

where the λ_{hi}'s satisfy

$$\sum_{h=1}^{m} \lambda_{hi}^2 < 1 \qquad (1 \leqq i \leqq k-1). \tag{2.9}$$

If $T_1, T_2, \ldots, T_{k-1}$ have a joint t-distribution with this associated correlation structure, then the T_i's can be expressed as

$$T_i = \frac{\left\{\sum_{h=1}^{m} \lambda_{hi} Z_h + \left(1 - \sum_{h=1}^{m} \lambda_{hi}^2\right)^{1/2} Z_{m+i}\right\}}{U} \qquad (1 \leqq i \leqq k-1)$$

where the Z_i's $(1 \leqq i \leqq m+k-1)$ are independent standard normal r.v.'s and U is distributed independently as a $\sqrt{\chi^2_\nu/\nu}$ r.v. The probability distributions of $\max T_i$ and $\max |T_i|$ can now be evaluated in a manner similar to (2.6) and (2.7), respectively, by conditioning on U and $Z_1, \ldots, Z_m$, which results in an $(m+1)$-dimensional iterated integral in the unconditional distribution. For example, the c.d.f. of $\max T_i$ can be written as

$$\int_0^\infty \int_{-\infty}^\infty \cdots \int_{-\infty}^\infty \prod_{i=1}^{k-1} \Phi\left[\frac{tu - \sum_{h=1}^m \lambda_{hi} z_h}{\left(1 - \sum_{h=1}^m \lambda_{hi}^2\right)^{1/2}}\right] d\Phi(z_1)\cdots d\Phi(z_m)\, dF_\nu(u) . \tag{2.10}$$

Clearly, m should be small for this procedure to be computationally feasible.

Example 2.2. Consider a one-way layout with a fixed covariate. This design was discussed in Example 1.2 of Chapter 2. By using (2.2) above and (1.15) of Chapter 2 we obtain

$$\rho_{ij} = \text{corr}(\hat\theta_i - \hat\theta_k, \hat\theta_j - \hat\theta_k) = \frac{a_i a_j - a_i a_k - a_j a_k + a_k^2 + 1/n_k}{b_i b_j} \quad (1 \leqq i \neq j \leqq k-1) \tag{2.11}$$

where

$$a_i = \frac{(\bar X_{i.} - \bar X_{..})}{S_{XX}^{1/2}} \quad \text{and} \quad b_i = \left\{\frac{1}{n_i} + \frac{1}{n_k} + (a_i - a_k)^2\right\}^{1/2} \quad (1 \leqq i \leqq k) .$$

If we choose

$$\lambda_{1i} = \frac{a_i - a_k}{b_i} \quad \text{and} \quad \lambda_{2i} = \frac{1}{b_i\sqrt{n_k}} \quad (1 \leqq i \leqq k-1) ,$$

then it is easy to see that (2.11) can be expressed in the form (2.8) with $m = 2$. Also condition (2.9) is clearly satisfied since $\lambda_{1i}^2 + \lambda_{2i}^2 = 1 - 1/n_i b_i^2 < 1$. Thus the c.d.f.'s of max T_i and max $|T_i|$ can be written as trivariate integrals similar to (2.10). □

2.1.2 *Approximate Procedures for Unbalanced Designs*

For many unbalanced designs conditions (2.8) and (2.9) are not satisfied for a small value of m and therefore a substantial reduction in the dimensionality of the integrals as in (2.10) is not possible. Even if (2.8) and (2.9) are satisfied, a computer program for evaluating the reduced dimensional integrals may not be available. In such cases, it would be desirable to find suitable approximations to the exact critical points that can be readily obtained from the available tables.

Natural approximations to the critical points $T^{(\alpha)}_{k-1,\nu,\{\rho_{ij}\}}$ or $|T|^{(\alpha)}_{k-1,\nu,\{\rho_{ij}\}}$ are provided by replacing the ρ_{ij}'s by some common value ρ since the tables are widely available for this equicorrelated case. For $k-1=3$, Dutt et al. (1976) suggested using $\rho = \bar{n}_k/(n_k + \bar{n})$ where $\bar{n} = \sum_{i=1}^{k-1} n_i/(k-1)$. However, numerical results show that this approximation is generally liberal (see Example 2.3). By using the inequalities of Slepian (1962) and Šidák (1967), conservative upper bounds on these critical points are provided by using $\rho = \min \rho_{ij}$ if $\min \rho_{ij} > -1/(k-2)$. Less conservative approximations would result if ρ is chosen to be a suitable average of the ρ_{ij}'s. Dunnett (1985) considered this approach in the context of unbalanced one-way layouts. By noting that in this case the ρ_{ij}'s possess the product structure (2.4), he suggested the use of the geometric mean of the ρ_{ij}'s, which is given by

$$\tilde{\rho} = \tilde{\lambda}^2 = \left\{ \prod_{i=1}^{k-1} \lambda_i \right\}^{2/(k-1)}$$

Observe that for $k=3$ the geometric mean approximation is in fact exact since $\tilde{\rho} = \rho_{12} = \lambda_1\lambda_2$ where the λ_i's are given by (2.5). Dunnett (1985) performed an extensive numerical study of the resulting approximations $T^{(\alpha)}_{k-1,\nu,\tilde{\rho}}$ and $|T|^{(\alpha)}_{k-1,\nu,\tilde{\rho}}$ for $k>3$, and found that they are always conservative, that is, they provide upper bounds on the exact critical points $T^{(\alpha)}_{k-1,\nu,\{\rho_{ij}\}}$ and $|T|^{(\alpha)}_{k-1,\nu,\{\rho_{ij}\}}$, respectively.

Dunnett (1985) also investigated numerically the use of the arithmetic mean $\bar{\rho}$. Because of the inequality $\bar{\rho} \geqq \tilde{\rho}$, it follows (by using the inequalities of Slepian 1962 and Šidák 1967, respectively) that $T^{(\alpha)}_{k-1,\nu,\bar{\rho}}$ and $|T|^{(\alpha)}_{k-1,\nu,\bar{\rho}}$ will be smaller than the corresponding critical points obtained by using $\tilde{\rho}$. Dunnett's numerical results showed that the former still are upper bounds on the exact critical points and thus provide sharper (less conservative) approximations than $T^{(\alpha)}_{k-1,\nu,\tilde{\rho}}$ and $|T|^{(\alpha)}_{k-1,\nu,\tilde{\rho}}$, respectively. The conservatism of the approximations based on either $\bar{\rho}$ or $\tilde{\rho}$ has not been analytically established yet.

Another possible approximation is to use $\rho = \bar{\lambda}^2$ where $\bar{\lambda}$ is the arithmetic mean of the λ_i's. It is easy to see that, for $k=3$, this gives a liberal approximation since $\rho = \{(\lambda_1 + \lambda_2)/2\}^2 > \lambda_1\lambda_2 = \rho_{12}$ if $\lambda_1 \neq \lambda_2$. For $k>3$ it is not known how this approximation performs.

Example 2.3. Dutt et al. (1976) gave an example in which three treatment groups are compared with a control in a one-way layout. The

TABLE 2.2. Exact and Approximate Values of One-Sided and Two-Sided Multivariate t Critical Points

Critical Point	Exact Value	Approximate Values for Different Choices of $\rho_{ij} = \rho$			
		$\rho = \frac{\bar{n}}{n_k + \bar{n}} = 0.5091$	$\rho = \tilde{\rho} = 0.4824$	$\rho = \bar{\rho} = 0.4852$	$\rho = \bar{\lambda}^2 = 0.4877$
$T^{(0.05)}_{3,33,\{\rho_{ij}\}}$	2.1419	2.1369	2.1437	2.1430	2.1424
$\|T\|^{(0.05)}_{3,33,\{\rho_{ij}\}}$	2.4639	2.4599	2.4656	2.4650	2.4645

sample sizes for the treatment groups are $n_1 = 12$, $n_2 = 5$, and $n_3 = 11$, and that for the control group is $n_4 = 9$. Using (2.5) we calculate $\lambda_1 = 0.7559$, $\lambda_2 = 0.5976$, and $\lambda_3 = 0.7416$. Then from (2.4) we get the following correlation matrix:

$$\{\rho_{ij}\} = \begin{bmatrix} 1 & 0.4517 & 0.5606 \\ & 1 & 0.4432 \\ & & 1 \end{bmatrix}.$$

The error d.f. ν equals 33.

The exact values and various approximations to the upper 5% one-sided and two-sided critical points, $T^{(0.05)}_{3,33,\{\rho_{ij}\}}$ and $|T|^{(0.05)}_{3,33,\{\rho_{ij}\}}$, are given in Table 2.2.

Notice that all of the approximations are quite close to the corresponding exact values. The Dutt et al. approximation is liberal; the other three are conservative. Although the $\rho = \bar{\lambda}^2$ approximation is the sharpest of the latter three in this example, because of its known liberal nature for $k = 3$, it cannot be recommended for general use. Based on the current evidence, our choice is to use $\rho = \bar{\rho}$. □

2.1.3 Extensions and Other Remarks

In Chapter 6 we consider the problem of optimally allocating a given number of observations among the control and each one of the treatments when using a completely randomized design. One of the results stated there is that the allocation $n_k = n\sqrt{k-1}$ is nearly "optimal" where n is the common number of observations on each one of the treatments and n_k is the number of observations on the control. This allocation minimizes the common variance of the $(\bar{Y}_i - \bar{Y}_k)$'s $(1 \leq i \leq k-1)$.

We next discuss an extension of the simultaneous confidence intervals (2.3) to the family of all contrasts. This extension, which is similar to the extension of the Tukey (T) procedure from the family of pairwise comparisons to the family of all contrasts (see Section 2.1.1 of Chapter

3), was proposed by Shaffer (1977) and is based on the following algebraic lemma.

Lemma 2.1 (Shaffer 1977). Let $x_1, x_2, \ldots, x_k$ be any real numbers and let $\xi_1, \xi_2, \ldots, \xi_{k-1}$ be any nonnegative numbers. Then

$$|x_i - x_k| \leq \xi_i \ (1 \leq i \leq k-1) \Leftrightarrow \left|\sum_{i=1}^{k} c_i x_i\right| \leq \sum_{i=1}^{k-1} |c_i| \xi_i \quad \forall \mathbf{c} \in \mathbf{C}^k . \tag{2.12}$$

□

(Shaffer gives the result for the case $\xi_1 = \xi_2 = \cdots = \xi_{k-1}$, but it is trivial to extend her argument to the more general case given here.)

Now from (2.3) we know that with probability $1 - \alpha$,

$$|(\hat{\theta}_i - \theta_i) - (\hat{\theta}_k - \theta_k)| \leq |T|^{(\alpha)}_{k-1,\nu,\{\rho_{ij}\}} S\sqrt{v_{ii} + v_{kk} - 2v_{ik}} \qquad (1 \leq i \leq k-1)$$

where the ρ_{ij}'s are given by (2.2). Letting $x_i = \hat{\theta}_i - \theta_i$ $(1 \leq i \leq k)$, $\xi_i = |T|^{(\alpha)}_{k-1,\nu,\{\rho_{ij}\}} S\sqrt{v_{ii} + v_{kk} - 2v_{ik}}$ $(1 \leq i \leq k-1)$, and using (2.12) we obtain the following $100(1-\alpha)\%$ simultaneous confidence intervals for all contrasts:

$$\sum_{i=1}^{k} c_i \theta_i \in \left[\sum_{i=1}^{k} c_i \hat{\theta}_i \pm |T|^{(\alpha)}_{k-1,\nu,\{\rho_{ij}\}} S \sum_{i=1}^{k} |c_i| \sqrt{v_{ii} + v_{kk} - 2v_{ik}}\right] \qquad \forall \mathbf{c} \in \mathbf{C}^k . \tag{2.13}$$

Shaffer (1977) made a comparative study of these intervals with the Scheffé intervals (see (1.2) of Chapter 3) and the Tukey intervals (see (2.2) of Chapter 3) for the one-way layout case when either $n_1 = \cdots = n_k$ or $n_1 = \cdots = n_{k-1} = n_k/\sqrt{k-1}$. (In the latter case the Tukey–Kramer (TK) intervals are used instead of the Tukey intervals.) Generally speaking, the intervals (2.13) are competitive only when the comparison is between the control on the one hand and only a few treatments on the other hand. If a comparison does not involve the control, then the interval (2.13) is excessively wide.

2.2 Step-Down Procedures

A shortcut form of the closed testing procedure of Marcus, Peritz, and Gabriel (1976) for testing one-sided hypotheses on $\theta_i - \theta_k$ $(1 \leq i \leq k-1)$

was given in Example 4.2 of Chapter 2. That step-down procedure is applicable when the correlations ρ_{ij} are all equal, and in that case it clearly provides a more powerful alternative than the corresponding single-step procedure. A two-sided step-down procedure can be constructed in an analogous manner.

When the ρ_{ij}'s are not all equal, one can use the sequentially rejective Bonferroni procedure of Section 4.2.2 of Chapter 2, which provides conservative control of the FWE. For testing one-sided hypotheses, a slightly less conservative procedure is obtained if at the lth step of testing $(l = 1, 2, \ldots, k-2)$, we use the critical constant $M_{k-l,\nu}^{(\alpha)}$ instead of $T_{\nu}^{(\alpha/(k-l))}$ used by the sequentially rejective Bonferroni procedure. For testing two-sided hypotheses one can similarly use the critical constant $|M|_{k-l,\nu}^{(\alpha)}$ in place of $T_{\nu}^{(\alpha/2(k-l))}$.

Simultaneous confidence intervals for $\theta_i - \theta_k$ $(1 \leqq i \leqq k-1)$ derived using the method of Kim, Stefánsson, and Hsu (1987) from the one-sided step-down test procedure were given in Example 4.3 of Chapter 2 for the equicorrelated case. When the ρ_{ij}'s are not equal and one of the aforementioned conservative procedures is used, an obvious modification of the critical constant in (4.7) of Chapter 2 gives the corresponding simultaneous confidence intervals. It is not as yet known how to derive two-sided simultaneous confidence intervals associated with the corresponding step-down procedures.

Example 2.4. Table 2.3 gives some data on the uterine weights of mice (this variable being used as a measure of estrogenic activity) obtained from an estrogen assay of a control solution and six test solutions (referred to as treatments) that had been subjected to an *in vitro* activation technique. These data are taken from Steel and Torrie (1980, p. 144). The design consists of a one-way layout with four mice per group.

TABLE 2.3. Uterine Weights of Mice (in mg)

	Solutions						
	1	2	3	4	5	6	Control
	84.4	64.4	75.2	88.4	56.4	65.6	89.8
	116.0	79.8	62.4	90.2	83.2	79.4	93.8
	84.0	88.0	62.4	73.2	90.4	65.6	88.4
	68.6	69.4	73.8	87.8	85.6	70.2	112.6
$\bar{Y}_i$	88.25	75.40	68.45	84.90	78.90	70.20	96.15

Source: Steel and Torrie (1980).

It is desired to identify those treatments that significantly lower the estrogenic activity as compared to the control. For this purpose we apply the one-sided step-down test procedure given in Example 4.2 of Chapter 2. The pooled MSE for these data is $S^2 = 145.84$ with $\nu = 21$ d.f. The t-statistics for testing the differences between the control mean and the treatment means are shown in Table 2.4.

The ordered t-statistics are to be compared with the critical points $T^{(\alpha)}_{p,\nu,\rho}$ where $\rho = \frac{1}{2}$ (since all groups have the same sample size), $\nu = 21$, $p = 6, 5, \ldots, 1$, and α is chosen to be 0.10. These critical points are listed in Table 2.5 for convenience.

Since the largest t-statistic, 3.244, exceeds $T^{(0.10)}_{6,21,1/2} = 2.08$ we conclude that treatment 3 lowers the estrogenic activity with respect to the control. Comparing the next largest t-statistic, 3.039, with $T^{(0.10)}_{5,21,1/2} = 2.01$ we conclude that treatment 6 also lowers the estrogenic activity with respect to the control. Proceeding in this manner we find that the same conclusion is drawn for treatments 2 and 5 but not for treatments 1 and 4. If we use the corresponding single-step test procedure, then all the t-statistics will be compared with the common critical point $T^{(0.10)}_{6,21,1/2} = 2.08$, and thus significant results will be found only for treatments 3, 6, and 2.

Associated with this step-down procedure there are simultaneous lower one-sided 90% confidence limits on the differences between the treatment and control means, which we now compute. The lower confidence limits are given by (4.7) of Chapter 2:

$$\theta_k - \theta_i \geqq \left[\bar{Y}_k - \bar{Y}_i - T^{(\alpha)}_{p,\nu,1/2} S\sqrt{\frac{2}{n}}\right]^- \qquad (1 \leqq i \leqq k-1)$$

where $x^- = \min(x, 0)$ and p is the index of the largest ordered t-statistic

TABLE 2.4. t-Statistics for the Differences in Means Between the Control and the Treatments

Treatments					
1	2	3	4	5	6
0.925	2.430	3.244	1.317	2.020	3.039

TABLE 2.5. Critical Points

p	1	2	3	4	5	6
$T^{(0.10)}_{p,21,1/2}$	1.32	1.64	1.81	1.92	2.01	2.08

that is not found statistically significant by the step-down test procedure. In the present example $p = 2$ and therefore $T^{(\alpha)}_{p,\nu,1/2} S\sqrt{2/n} = 1.64 \times 12.08 \times \sqrt{2/4} = 14.00$. The lower confidence limit is zero for $\theta_{\text{control}} - \theta_i$ for $i = 3, 6, 2$ and 5, that is, for those mean differences that are concluded to be positive by the step-down test procedure. The lower confidence limits for $\theta_{\text{control}} - \theta_4$ and $\theta_{\text{control}} - \theta_1$ are given by $96.15 - 84.90 - 14.00 = -2.75$ and $96.15 - 88.25 - 14.00 = -6.10$, respectively. □

3 COMPARISONS WITH THE "BEST" TREATMENT

In the preceding section we considered the problem of comparing the treatments under study with a control treatment that is used as a benchmark. In some applications the relevant benchmark is the (unknown) "best" treatment, that is, the treatment having the largest θ-value. The parameters of interest in this case are $\theta_i - \max_j \theta_j$, or alternatively $\theta_i - \max_{j \neq i} \theta_j$ $(1 \leqq i \leqq k)$. Notice that these are *nonlinear* functions of $\boldsymbol{\theta}$; all of the families that we have dealt with thus far have involved *linear* parametric functions of $\boldsymbol{\theta}$. Hsu (1981) constructed simultaneous upper one-sided confidence intervals for $\theta_i - \max_j \theta_j$. Later Hsu (1984a) obtained simultaneous two-sided confidence intervals for $\theta_i - \max_{j \neq i} \theta_j$ that imply the results of his 1981 paper.

In Section 3.1 we give Hsu's (1984a) basic procedure for the case of a balanced one-way layout. We also discuss how his procedure provides stronger inferences than those inherent in the indifference-zone selection procedure of Bechhofer (1954) and the subset selection procedure of Gupta (1956, 1965). In Section 3.2 we note some extensions of the basic procedure, particularly to unbalanced one-way layouts.

3.1 The Basic Procedure

We use the notation of Example 1.1 of Chapter 2 throughout. The following additional notation is used in the proof of the following theorem. Let $\theta_{(1)} \leqq \theta_{(2)} \leqq \cdots \leqq \theta_{(k)}$ denote the ordered treatment means and let $E\bar{Y}_{(i)} = \theta_{(i)}$, $i = 1, 2, \ldots, k$ (i.e., $\bar{Y}_{(i)}$ is the sample mean of the treatment having the ith smallest θ-value, $\theta_{(i)}$).

Theorem 3.1 (Hsu 1984a). Simultaneous $(1-\alpha)$-level confidence intervals for $\theta_i - \max_{j \neq i} \theta_j$ are given by

$$\theta_i - \max_{j \neq i} \theta_j \in [(\bar{Y}_i - \max_{j \neq i} \bar{Y}_j - d)^-, (\bar{Y}_i - \max_{j \neq i} \bar{Y}_j + d)^+] \qquad (1 \leqq i \leqq k) \tag{3.1}$$

where

$$d = T^{(\alpha)}_{k-1,\nu,1/2} S \sqrt{\frac{2}{n}} ,$$

$x^- = \min(x, 0)$, and $x^+ = \max(x, 0)$.

Proof. Define the events

$$E = \{\bar{Y}_{(k)} - \bar{Y}_{(i)} - (\theta_{(k)} - \theta_{(i)}) > -d \ (1 \leqq i \leqq k-1)\} ,$$

$$E_1 = \{\theta_i - \max_{j \neq i} \theta_j \leqq (\bar{Y}_i - \max_{j \neq i} \bar{Y}_j + d)^+ \ (1 \leqq i \leqq k)\} ,$$

and

$$E_2 = \{\theta_i - \max_{j \neq i} \theta_j \geqq (\bar{Y}_i - \max_{j \neq i} \bar{Y}_j - d)^- \ (1 \leqq i \leqq k)\} .$$

Clearly, $\Pr(E) = 1 - \alpha$ for all $\boldsymbol{\theta}$ and σ^2. We show that $E \subseteq E_1 \cap E_2$ whence the theorem follows. To simplify the notation put $L_i = (\bar{Y}_i - \max_{j \neq i} \bar{Y}_j - d)^-$ and $U_i = (\bar{Y}_i - \max_{j \neq i} \bar{Y}_j + d)^+$ $(1 \leqq i \leqq k)$. Also let $L_{ij} = (\bar{Y}_i - \bar{Y}_j - d)^-$ and note that $L_i = \min_{j \neq i} L_{ij}$.

We first show that $E \subseteq E_1$:

$$\begin{aligned} E &\subseteq \{\bar{Y}_{(k)} - \bar{Y}_{(i)} - (\theta_{(k)} - \theta_{(k-1)}) > -d \ \forall i \neq k\} \\ &= \{\theta_{(k)} - \theta_{(k-1)} \leqq \bar{Y}_{(k)} - \bar{Y}_i + d \ \forall i \neq (k);\ \theta_i - \theta_{(k-1)} \leqq 0 \ \forall i \neq (k)\} \\ &\subseteq \{\theta_i - \theta_{(k-1)} \leqq U_i \ \forall i\} \\ &= \{\theta_i - \max_{j \neq i} \theta_j \leqq U_i \ \forall i\} \end{aligned}$$

where the last equality follows because all the U_i's are $\geqq 0$.

We next show that $E \subseteq E_2$:

$$\begin{aligned} E &\subseteq \{\theta_{(k)} - \theta_{(i)} \leqq (\bar{Y}_{(k)} - \bar{Y}_{(i)} + d)^+ \ \forall i \neq k\} \\ &= \{(\bar{Y}_i - \bar{Y}_{(k)} - d)^- \leqq \theta_i - \theta_{(k)} \ \forall i \neq (k);\ \theta_i - \theta_{(k)} = 0 \text{ for } i = (k)\} \\ &\subseteq \{\min_{j \neq i} L_{ij} \leqq \theta_i - \theta_{(k)} \ \forall i\} \\ &= \{L_i \leqq \theta_i - \max_{j \neq i} \theta_j \ \forall i\} \end{aligned}$$

where the last equality follows because all the L_i's are $\leqq 0$. □

If we set all the U_i's $= +\infty$, then we obtain lower one-sided confidence bounds on $\theta_i - \max_{j \neq i} \theta_j$ $(1 \leqq i \leqq k)$. Because these lower confidence bounds are constrained to be nonpositive, they also provide lower confidence bounds on $\theta_i - \theta_{(k)}$ $(1 \leqq i \leqq k)$, which were originally derived by Hsu (1981).

Two-sided simultaneous confidence intervals on $\theta_i - \theta_{(k)}$ were derived by Edwards and Hsu (1983). Their approach is based on the idea of regarding each treatment as a known control, and adapting the set of all resulting two-sided intervals for comparisons with known controls to obtain the following intervals for comparisons with the unknown "best":

$$\theta_{(k)} - \theta_i \in \left[\left(\min_{j \in \mathcal{S}} \bar{Y}_j - \bar{Y}_i - \xi_1 S \sqrt{\frac{2}{n}} \right)^+ , \left(\max_{j \neq i} \bar{Y}_j - \bar{Y}_i + \xi_2 S \sqrt{\frac{2}{n}} \right)^+ \right] \quad (1 \leqq i \leqq k) \qquad (3.2)$$

where

$$\mathcal{S} = \left\{ i : \bar{Y}_i \geqq \max \bar{Y}_j - \xi_1 S \sqrt{\frac{2}{n}} \right\} .$$

To guarantee that the joint confidence level $\geqq 1 - \alpha$, the critical constants ξ_1 and ξ_2 must be chosen so as to satisfy

$$\Pr\{-\xi_1 \leqq T_i \leqq \xi_2 \ (1 \leqq i \leqq k-1)\} = 1 - \alpha$$

where $T_1, T_2, \ldots, T_{k-1}$ have a joint $(k-1)$-variate t-distribution with ν d.f. and the associated common correlation $\rho = \frac{1}{2}$. If we set $\xi_1 = +\infty$ in (3.2), then Hsu's (1981) upper confidence bounds on $\theta_{(k)} - \theta_i$ $(1 \leqq i \leqq k-1)$ are obtained. If we set $\xi_1 = \xi_2 = \xi$ (say), then $\xi = |T|^{(\alpha)}_{k-1,\nu,1/2}$. Notice that although both (3.1) and (3.2) are two-sided intervals, the former are based on one-sided critical points, while the latter are based on two-sided critical points. However, the former give constrained (because all the L_i's are $\leqq 0$ and all the U_i's are $\geqq 0$) intervals on $\theta_i - \max_{j \neq i} \theta_j$ $(1 \leqq i \leqq k)$ while it can be shown that the latter imply unconstrained intervals on $\theta_i - \max_{j \neq i} \theta_j$ $(1 \leqq i \leqq k)$. Hsu (1985) has given a method for removing the nonpositivity constraint on the L_i's that requires the use of a slightly larger critical value than d. This enables one to obtain a positive lower bound on the amount by which the treatment that appears to be the "best" (i.e., which produces the largest $\bar{Y}_i$) is better than the best of the remaining treatments. Kim, Stefánsson, and Hsu (1987) have given a unified approach to deriving both sets of confidence intervals by inverting appropriate one-sided and two-sided tests for comparisons with a control.

We next note connections between the intervals (3.1) and some results from the ranking and selection theory. Gupta (1956, 1965) showed that if one selects a subset $\mathscr{S}$ of treatments according to the rule

$$\mathscr{S} = \left\{ i : \bar{Y}_i > \max_{j \neq i} \bar{Y}_j - T^{(\alpha)}_{k,\nu,1/2} S \sqrt{\frac{2}{n}} \right\}, \tag{3.3}$$

then $\mathscr{S}$ contains the "best" treatment with probability at least $1 - \alpha$ for all $\boldsymbol{\theta}$ and σ^2; thus $\mathscr{S}$ is a $(1-\alpha)$-level confidence set for the "best" treatment. Note that (3.3) is equivalent to selecting treatment i if $U_i > 0$ $(1 \leqq i \leqq k)$. Hsu (1984a) pointed out that one can make the confidence statement (3.1) and the statement $\{(k) \in \mathscr{S}\}$ simultaneously with confidence level $1 - \alpha$ since both follow from the same event E, which has the probability $1 - \alpha$.

Whereas the connection with the subset selection is through the upper bounds U_i, the connection with the indifference-zone selection is through the lower bounds L_i. Suppose we define treatment i as "good" if $\theta_i - \theta_{(k)} \geqq -\delta^*$ $(1 \leqq i \leqq k)$ where $\delta^* > 0$ is a prespecified constant. Then it can be shown that all treatments with $L_i \geqq -\delta^*$ are "good" with probability at least $1 - \alpha$. This is the confidence statement implied by Desu's (1970) procedure (extended to the unknown σ^2 case). If treatment $[k]$ corresponds to $\bar{Y}_{[k]} = \max_j \bar{Y}_j$, then Fabian's (1962) result (extended to the unknown σ^2 case) states that with probability $1 - \alpha$, $\theta_{[k]} - \theta_{(k)} \geqq L_{[k]} \geqq -\delta^*$. Note that Hsu's procedure is an improvement over Desu's and Fabian's procedures because it provides lower confidence bounds (which are sharper than those of Desu's) on *all* $\theta_i - \theta_{(k)}$ (while Fabian's bound is only on $\theta_{[k]} - \theta_{(k)}$) at the same confidence level $1 - \alpha$.

By letting $\delta^* \to 0$ we can choose treatment i as the "best" if $L_i = 0$, that is, if $\bar{Y}_i \geqq \max_{j \neq i} \bar{Y}_j + d$. Clearly, at most one treatment will satisfy this condition. For $\delta^* > 0$, if σ^2 is known, then $n \geqq 2(Z^{(\alpha)}_{k-1,1/2}\sigma/\delta^*)^2$ will guarantee that at least one treatment satisfies $L_i \geqq -\delta^*$. If σ^2 is unknown, then at least two stages of sampling (as in Stein 1945) are required to guarantee that at least one treatment satisfies $L_i \geqq -\delta^*$.

Example 3.1. Consider the barley data (Duncan 1955) given in Example 1.1 of Chapter 3. Suppose that we wish to construct 90% simultaneous confidence intervals for $\theta_i - \max_{j \neq i} \theta_j$ $(1 \leqq i \leqq k)$. For this purpose we need the value of the critical point $T^{(0.10)}_{k-1,\nu,1/2}$. For $k = 7$ and $\nu = 30$ it is found to be 2.05 from Table 4 in Appendix 3. The $\bar{Y}_i$'s (each based on $n = 6$ observations) are given in Table 3.1, and $S = 8.924$. Thus the "allowance" $d = 2.05 \times 8.924 \times \sqrt{2/6} = 10.56$. The desired intervals can now be computed by using (3.1) and are shown in Table 3.1.

TABLE 3.1. 90% Simultaneous Confidence Intervals for $\theta_i - \max_{j \neq i} \theta_j$

Variety	$\bar{Y}_i$	$[L_i, U_i]$
A	49.6	[−32.26, 0.00]
B	71.2	[−10.66, 10.46]
C	67.6	[−14.26, 6.86]
D	61.5	[−20.36, 0.76]
E	71.3	[−10.46, 10.66]
F	58.1	[−23.76, 0.00]
G	61.0	[−20.86, 0.26]

From these confidence intervals we see that varieties A and F can be eliminated as being not in contention for the "best" since the upper bounds of their intervals are zero, indicating that there is another variety (namely E) that is "better." Note that no variety can be selected as the "best" at $\alpha = 0.10$ since all the lower bounds are negative. □

3.2 Extensions

We first consider an extension of the basic procedure to unbalanced one-way layouts. In this case, simultaneous $(1-\alpha)$-level confidence intervals for $\theta_i - \max_{j \neq i} \theta_j$ are given by $[L_i, U_i]$ $(1 \leqq i \leqq k)$ where

$$L_i = \left\{ \min_{\substack{j \in \mathcal{S} \\ j \neq i}} \left[\bar{Y}_i - \bar{Y}_j - T^{(\alpha)}_{k-1,\nu,\mathbf{R}_j} S \left(\frac{1}{n_i} + \frac{1}{n_j} \right)^{1/2} \right] \right\}^- \quad (1 \leqq i \leqq k),$$

$$U_i = \left\{ \min_{j \neq i} \left[\bar{Y}_i - \bar{Y}_j + T^{(\alpha)}_{k-1,\nu,\mathbf{R}_i} S \left(\frac{1}{n_i} + \frac{1}{n_j} \right)^{1/2} \right] \right\}^+ \quad (1 \leqq i \leqq k), \tag{3.4}$$

and

$$\mathcal{S} = \left\{ i : \min_{j \neq i} \left[\bar{Y}_i - \bar{Y}_j + T^{(\alpha)}_{k-1,\nu,\mathbf{R}_i} S \left(\frac{1}{n_i} + \frac{1}{n_j} \right)^{1/2} \right] \geqq 0 \right\}. \tag{3.5}$$

In these expressions $T^{(\alpha)}_{k-1,\nu,\mathbf{R}_l}$ is the upper α point of max $(T_1, \ldots, T_{l-1}, T_{l+1}, \ldots, T_k)$ where the T_i's, $i \neq l$, have a joint $(k-1)$-variate t-distribution with ν d.f. and the associated correlation matrix

$\mathbf{R}_l$, whose off-diagonal elements are given by

$$\rho_{ij}^{(l)} = \left(\frac{n_i}{n_i + n_l}\right)^{1/2}\left(\frac{n_j}{n_j + n_l}\right)^{1/2} \qquad (i \neq j \neq l, 1 \leqq i, j, l \leqq k) .$$

This extension in slightly modified forms is given for the two-sided case in Edwards and Hsu (1983) and for the one-sided case in Hsu (1984b).

If we set $n_i = n$ for all i, then the confidence limits (3.4) reduce to (3.1). Also notice that in the unbalanced case, the minimum in the formula for L_i is taken over treatments in set $\mathcal{S}$; in the balanced case this modification is immaterial.

We have confined our discussion here only to one-way layouts. But some of the procedures can be extended to more complex designs, for example, block designs (Hsu 1982) and split-plot and split-block designs (Federer and McCulloch 1984).

Another extension of the basic procedure proposd by Hsu (1984b) is the use of the so-called R- and S-values. The R-value for a treatment is the smallest α for which that treatment can be rejected, that is, not included in set $\mathcal{S}$. For the treatment that appears to be the "best" (this is the treatment yielding the largest L_i, which in the balanced case is the same as the one yielding the largest $\bar{Y}_i$), the S-value is the smallest α for which that treatment can be selected as the "best." The R- and S-values are informative in the same way that the P-values are informative in significance testing.

The R-value for treatment i is given by (Hsu 1984b)

$$R_i = 1 - \Pr\{T_j \leqq t_i \ \forall j \neq i\} \qquad (1 \leqq i \leqq k)$$

where

$$t_i = \max_{j \neq i} \frac{\bar{Y}_j - \bar{Y}_i}{S\sqrt{1/n_i + 1/n_j}} \qquad (1 \leqq i \leqq k)$$

is the observed value, and the joint distribution of $T_1, \ldots, T_{l-1}, T_{l+1}, \ldots, T_k$ is the same as that encountered following (3.5). By using (2.6) we can write R_i as

$$R_i = 1 - \int_0^{\infty}\int_{-\infty}^{\infty} \prod_{\substack{j=1\\ j\neq i}}^{k} \Phi\left[\left(\frac{n_j}{n_i}\right)^{1/2} z + t_i u\left(1 + \frac{n_j}{n_i}\right)^{1/2}\right] d\Phi(z)\, dF_\nu(u)$$

$$(1 \leqq i \leqq k) .$$

If the treatment that appears to be the "best" is denoted by $[k]$, then clearly $S_{[k]} = \max_{i \neq [k]} R_i$.

The computations of the confidence limits (3.4) and of the R- and S-values are quite formidable. Hsu has developed a computer program for this purpose (which also deals with the balanced case) called RSMCB (for Ranking and Selection and Multiple Comparisons with the Best) that is now available in version 5 of the SAS package (see Aubuchon, Gupta, and Hsu 1986). The following example is taken from Gupta and Hsu (1984), which gives a more detailed description of this program.

Example 3.2. An experiment was performed to compare seven brands of filters for their ability to reveal the microorganism fecal coliform in river water. Three samples of each brand of filter were selected and 100 ml of sample water was poured through each. After 24 hours of incubation, the number of colonies of the target microorganism on each filter was recorded. Two samples were damaged because of colony blurring and spreading, resulting in missing data. The final results are shown in Table 3.2. For this data set the pooled sample variance is $S^2 = 411.2$ with $\nu = 12$ d.f. For $1 - \alpha = 0.95$, the PROC RSMCB gave the results shown in Table 3.3.

The upper bounds for brands 1, 2, 3, 4, and 6 are zero, indicating that some other brand is "better" than these. This result is confirmed by the R-values for these brands, all of which are less than the designated level $\alpha = 0.05$. Thus only brands 5 and 7 will be selected by the subset selection rule (3.3). Note that brand 5 cannot be selected as the "best" at $\alpha = 0.05$ since the corresponding lower bound is less than zero. This is also confirmed by its S-value $= 0.1007$, which is greater than 0.05.

TABLE 3.2. Counts of Colonies on Different Brands of Filters

Brand	Counts			$\bar{Y}_i$
1	69	122	95	95.3
2	118	154	102	124.7
3	171	132	182	161.7
4	122	119	—	120.5
5	204	225	190	206.3
6	140	130	127	132.3
7	170	165	—	167.5

Source: Gupta and Hsu (1984).

TABLE 3.3. 95% Simultaneous Confidence Intervals for $\theta_i - \max_{j \neq i} \theta_j$

Brand	$[L_i, U_i]$	R_i	S_i
1	[−153.94, 0.00]	0.0001	—
2	[−124.61, 0.00]	0.0009	—
3	[−87.61, 0.00]	0.0418	—
4	[−133.84, 0.00]	0.0013	—
5	[−7.95, 86.84]	—	0.1007
6	[−116.94, 0.00]	0.0019	—
7	[−86.84, 7.95]	0.1007	—

□

4 TWO MISCELLANEOUS FAMILIES

4.1 Comparisons with the Average

The last two sections dealt with the problems of comparing the treatments with different benchmarks; in Section 2 the benchmark was a given control, while in Section 3 the benchmark was the (unknown) "best" treatment. Here we consider the problem of comparing the treatment means θ_i with their average $\bar{\theta} = \Sigma_{i=1}^{k} \theta_i/k$, which serves as the benchmark. The family of interest is thus the deviations $\theta_i - \bar{\theta}$ $(1 \leqq i \leqq k)$ that measure how much "better" or "worse" each treatment is compared to their average. Notice that each deviation $\theta_i - \bar{\theta}$ is a contrast $\mathbf{c}_i'\boldsymbol{\theta}$ where $\mathbf{c}_i$ has the ith entry equal to $(k-1)/k$ and the remaining $(k-1)$ entries equal to $-1/k$ each. This is thus a finite family of linear parametric functions and we can apply the results of Section 2.1.1 of Chapter 2.

To keep the discussion simple, throughout this section we confine our attention to the setting of a balanced one-way layout. In that case the LS estimator of $\theta_i - \bar{\theta}$ is $\bar{Y}_i - \bar{\bar{Y}}$ $(1 \leqq i \leqq k)$ where $\bar{\bar{Y}} = \Sigma_{i=1}^{k} \bar{Y}_i/k$. These estimators have a singular multivariate normal distribution with means $\theta_i - \bar{\theta}$, common variance $\sigma^2(k-1)/kn$, and common correlation $\rho = -1/(k-1)$. The pivotal r.v.'s

$$T_i = \sqrt{\frac{kn}{k-1}}\left(\frac{\bar{Y}_i - \bar{\bar{Y}} - \theta_i + \bar{\theta}}{S}\right) \qquad (1 \leqq i \leqq k)$$

have a multivariate t-distribution with ν d.f. and the associated common correlation $\rho = -1/(k-1)$. Thus $100(1-\alpha)\%$ one-sided simultaneous confidence intervals for $\theta_i - \bar{\theta}$ $(1 \leqq i \leqq k)$ will have the form

$$\theta_i - \bar{\theta} \leqq \bar{Y}_i - \bar{\bar{Y}} + T^{(\alpha)}_{k,\nu,-1/(k-1)} S \sqrt{\frac{k-1}{kn}} \qquad (1 \leqq i \leqq k), \tag{4.1}$$

and the corresponding two-sided intervals will have the form

$$\theta_i - \bar{\theta} \in \left[\bar{Y}_i - \bar{\bar{Y}} \pm |T|^{(\alpha)}_{k,\nu,-1/(k-1)} S \sqrt{\frac{k-1}{kn}} \right] \qquad (1 \leqq i \leqq k). \tag{4.2}$$

The intervals (4.2) can be extended to the family of all linear combinations $\Sigma_{i=1}^k l_i(\theta_i - \bar{\theta})$. It is clear that the resulting family is the family of all contrasts $\Sigma_{i=1}^k c_i\theta_i$ since the contrast vectors corresponding to $\theta_i - \bar{\theta}$ $(1 \leqq i \leqq k-1)$ form a basis for the space of all contrasts. This is also seen from the fact that $\Sigma_{i=1}^k l_i(\theta_i - \bar{\theta}) = \Sigma_{i=1}^k c_i\theta_i$ with $c_i = l_i - \bar{l}$ $(1 \leqq i \leqq k)$. The extension is based on the following algebraic lemma.

Lemma 4.1. Let $(x_1, x_2, \ldots, x_k)'$ and $(l_1, l_2, \ldots, l_k)'$ be any real vectors in $\mathbb{R}^k$ and let $\xi > 0$. Then

$$|x_i - \bar{x}| \leqq \xi \ \forall i \Rightarrow \left| \sum_{i=1}^k l_i(x_i - \bar{x}) \right| \leqq \xi \sum_{i=1}^k |l_i - \bar{l}| .$$

Proof.

$$\begin{aligned} \left| \sum_{i=1}^k l_i(x_i - \bar{x}) \right| &= \left| \sum_{i=1}^k (l_i - \bar{l})(x_i - \bar{x}) \right| \\ &\leqq \max_{1 \leqq i \leqq k} |x_i - \bar{x}| \cdot \sum_{i=1}^k |l_i - \bar{l}| \\ &\leqq \xi \sum_{i=1}^k |l_i - \bar{l}| . \end{aligned}$$

□

By letting $x_i = \bar{Y}_i - \theta_i$ $(1 \leqq i \leqq k)$ and $\xi = |T|^{(\alpha)}_{k,\nu,-1/(k-1)} S\sqrt{(k-1)/kn}$ in this lemma and using (4.2), we obtain the following $100(1-\alpha)\%$ simultaneous confidence intervals for all $\Sigma_{i=1}^k l_i(\theta_i - \bar{\theta})$:

$$\sum_{i=1}^k l_i(\theta_i - \bar{\theta}) \in \left[\sum_{i=1}^k l_i(\bar{Y}_i - \bar{\bar{Y}}) \pm |T|^{(\alpha)}_{k,\nu,-1/(k-1)} S \sqrt{\frac{k-1}{kn}} \sum_{i=1}^k |l_i - \bar{l}| \right] \quad \forall l \in \mathbb{R}^k .$$

The intervals (4.1) and (4.2) can be used to identify "discrepant" treatments in a balanced one-way layout. For example, suppose that the discrepancy $\theta_i - \bar{\theta}$ can be in either direction. Then any treatment i satisfying

$$|T_i| = \sqrt{\frac{kn}{k-1}} \cdot \frac{|\bar{Y}_i - \bar{\bar{Y}}|}{S} > |T|^{(\alpha)}_{k,\nu,-1/(k-1)}$$

can be regarded as discrepant at FWE $= \alpha$. This is a single-step test procedure for the family of hypotheses $\{H_{0i}: \theta_i - \bar{\theta} = 0\ (1 \leqq i \leqq k)\}$ corresponding to the two-sided confidence intervals (4.2). In the case of assertions of discrepancies, one can also make directional decisions about the signs of discrepancies in the usual manner. The Type I and Type III FWE for this single-step procedure is controlled at level α.

When only the identification of discrepant treatments is of interest (and not the estimates of the discrepancies), a step-down test procedure as in Example 4.1 of Chapter 2 can be used. This procedure compares the ordered $|T_i|$-statistics, $|T|_{(k)} \geqq |T|_{(k-1)} \geqq \cdots \geqq |T|_{(1)}$ with successively smaller critical points $|T|^{(\alpha)}_{p,\nu,-1/(k-1)}$ $(p = k, k-1, \ldots)$.

Procedures similar to those given for detecting discrepant treatments can be used in the closely related problem of detecting outlier observations in a single normal sample. This problem was studied by Nair (1948), who considered the one-sided case, and Halperin et al. (1955), who considered the two-sided case. Nair (1948, 1952) tabulated the one-sided critical points $((k-1)/k)^{1/2}\ T^{(\alpha)}_{k,\nu,-1/(k-1)}$, which were later corrected by David (1956) and further supplemented by Pillai and Tienzo (1959) and Pillai (1959). Upper and lower bounds (based on the first order and second order Bonferroni inequalities, respectively) on the two-sided critical points $((k-1)/k)^{1/2}|T|^{(\alpha)}_{k,\nu,-1/(k-1)}$ were tabled by Halperin et al. (1955). Thus the values of the critical points needed for computing (4.1) and (4.2) are available in the literature.

4.2 Successive Comparisons among Ordered Means

In some applications the means θ_i of interest are ordered in a natural way (e.g., according to the time sequence of collection of the corresponding samples) and it is desired to compare the adjacent pairs of means. The objective in making such successive comparisons may be to detect when changes in the means occurred. Such comparisons may also be of interest when the θ_i's are the mean responses for increasing dose-levels of some drug and they are not necessarily monotone (See Chapter 10, Section 4 for the problem of multiple comparisons among the θ_i's when they are monotonically ordered.)

The family of parametric functions of interest in these problems is $\{\theta_{i+1} - \theta_i\ (1 \leqq i \leqq k-1)\}$. This is a finite family of linear parametric functions and we can apply the results of Section 2.1.1 of Chapter 2 to obtain simultaneous confidence intervals and hypotheses tests on $\theta_{i+1} - \theta_i$. Hochberg and Marcus (1978) consider a related design problem, which is discussed in Chapter 6.

The LS estimators of $\theta_{i+1} - \theta_i$ are $\bar{Y}_{i+1} - \bar{Y}_i\ (1 \leqq i \leqq k-1)$, which follow a $(k-1)$-variate normal distribution with means $\theta_{i+1} - \theta_i$, common variance $2\sigma^2/n$, and correlations

$$\rho_{ij} = \begin{cases} -\frac{1}{2} & \text{if } |i-j| = 1 \\ 0 & \text{if } |i-j| > 1 \end{cases} \qquad (1 \leqq i \neq j \leqq k-1)\,. \tag{4.3}$$

Simultaneous $100(1-\alpha)\%$ upper one-sided confidence intervals for $\theta_{i+1} - \theta_i$ have the form

$$\theta_{i+1} - \theta_i \leqq \bar{Y}_{i+1} - \bar{Y}_i + T^{(\alpha)}_{k-1,\nu,\{\rho_{ij}\}} S\sqrt{\frac{2}{n}} \qquad (1 \leqq i \leqq k-1) \tag{4.4}$$

and the corresponding two-sided confidence intervals have the form

$$\theta_{i+1} - \theta_i \in \left[\bar{Y}_{i+1} - \bar{Y}_i \pm |T|^{(\alpha)}_{k-1,\nu,\{\rho_{ij}\}} S\sqrt{\frac{2}{n}}\right] \qquad (1 \leqq i \leqq k-1) \tag{4.5}$$

where the ρ_{ij}'s are given by (4.3).

The exact values of the critical points $T^{(\alpha)}_{k-1,\nu,\{\rho_{ij}\}}$ and $|T|^{(\alpha)}_{k-1,\nu,\{\rho_{ij}\}}$ have not been tabulated. Cox et al. (1980) proposed using $|M|^{(\alpha)}_{k-1,\nu}$ in place of $|T|^{(\alpha)}_{k-1,\nu,\{\rho_{ij}\}}$, thus providing a conservative approximation. They also evaluated the second order Bonferroni lower bound on $|T|^{(\alpha)}_{k-1,\nu,\{\rho_{ij}\}}$. The numerical results in their Table 19 show that the two bounds are fairly close, and thus $|M|^{(\alpha)}_{k-1,\nu}$ is a good approximation. An analogous conservative approximation to $T^{(\alpha)}_{k-1,\nu,\{\rho_{ij}\}}$ is $M^{(\alpha)}_{k-1,\nu}$.

One-sided and two-sided step-down test procedures for tests on $\theta_{i+1} - \theta_i\ (1 \leqq i \leqq k-1)$ can be constructed using the methods of Section 4.2 of Chapter 2. For conservative control of the FWE one can use the critical constants $M^{(\alpha)}_{p,\nu}$ in the one-sided case, and $|M|^{(\alpha)}_{p,\nu}$ in the two-sided case $(1 \leqq p \leqq k-1)$. Sequentially rejective Bonferroni procedures of the type discussed in Section 4.2.2 of Chapter 2 would be even more conservative.

CHAPTER 6

Designing Experiments for Multiple Comparisons

Most of the work in multiple comparisons has focused on the inferential aspects of the procedures and on the control of Type I error rates. This is, of course, consistent with the way the problem of significance testing is generally approached. However, in many applications it is of interest to control a suitable Type II error rate under some specified nonnull configurations. An analog of this latter requirement in the case of simultaneous confidence estimation is the specification of prescribed "widths" for the intervals. When such requirements are specified by an experimenter, they should be taken into account in *designing* the experiment, specifically in the determination of sample sizes. The present chapter is devoted to such design considerations arising in multiple comparison problems.

There are essentially two different formulations used in specifying design requirements in multiple comparison problems:

(i) An indifference-zone type formulation, which involves specifying two "threshold" constants $\delta_1 < \delta_2$ such that all contrasts with values no greater than δ_1 are to be classified as "small," while those with values no less than δ_2 are to be classified as "large." Any contrast with a value between these two limits (the so-called "indifference-zone") can be classified either way. In this formulation the familywise error rate (FWE) corresponds to the probability of any misclassification. It is desired to control this probability at or below a designated level α for all parameter configurations. This formulation owes its origin to Bechhofer's (1954) indifference-zone approach to selection problems.

(ii) In the second formulation the focus is on simultaneous confidence estimation. It is desired to construct simultaneous confidence intervals with specified joint confidence level $1-\alpha$ for a family of parametric functions of interest such that the intervals have prescribed "widths" (or "allowances").

These formulations are illustrated in Section 1 for a one-way layout setting under the assumption that the error variance σ^2 is known. This assumption enables us to use a single-stage sampling procedure for controlling the designated error rate. The families considered are: (i) all pairwise comparisons, (ii) comparisons with a control, (iii) comparisons with the "best," and (iv) comparisons between successive treatments in a set of *a priori* ordered treatments. In practice, σ^2 is usually not known accurately. If it is still desired to use a single-stage procedure, then two options are available for specifying design requirements: (i) If a conservative upper bound on σ^2 is known, then it can be used for design purposes. (However, the analysis should be done using an estimate of σ^2 obtained from the data.) (ii) Alternatively, instead of specifying the thresholds or allowances in absolute units, they can be specified as multiples of σ. Thus, for example, in the indifference-zone formulation, instead of directly specifying δ_1 and δ_2, one could specify $\Delta_1 = \delta_1/\sigma$ and $\Delta_2 = \delta_2/\sigma$, respectively. If δ_1 and δ_2 are specified in absolute units and if option (i) is not available, then the design requirements cannot be met by using a single-stage sampling procedure. In Section 2 we describe some two-stage procedures for the aforementioned four families when σ^2 is unknown.* Finally, in Section 3 a class of incomplete block designs for comparing treatments with a control is discussed. Illustrative examples are given for representative problems.

* As this book was going to press, Hsu (1986) sent us a preprint of a paper in which he proposed an alternative approach (conceptually similar to the one proposed by Tukey 1953, Chapter 18) to sample size determination in the case of simultaneous confidence estimation when σ^2 is unknown and a single-stage sampling procedure is to be used. In this approach, in addition to the usual requirement that the specified joint confidence level be $\geqq 1-\alpha$, it is also required that the probability of the joint occurrence of the event of simultaneous coverage ("correct" inference) *and* the event that the common random width of the confidence intervals is $\leqq 2\delta$ ("useful" inference) be at least $1-\beta$; here $1-\beta < 1-\alpha$ and $\delta > 0$ are preassigned constants. Note that in this approach it is not required that the probability of "useful" inference be unity as is required in the approaches described in Sections 1 and 2. It is to meet this latter more stringent requirement that σ^2 must be known if a single-stage procedure is to be used; if σ^2 is unknown then a two-stage (or more generally a sequential) procedure must be used. For a given family (e.g., the family of all pairwise comparisons), instead of determining the common sample size n per treatment to satisfy a specified requirement concerning the joint probability of "correct" and "useful" inferences, Hsu recommends studying that probability as a function of n for specified $1-\alpha$ and δ/σ.

1 SINGLE-STAGE PROCEDURES

1.1 Pairwise Comparisons

We consider the one-way layout model of Example 1.1 of Chapter 2 but assume the balanced case: $n_1 = \cdots = n_k = n$ (say). For design purposes we assume throughout Section 1 that the common error variance σ^2 is known. For this setting Reading (1975) proposed an indifference-zone formulation for the problem of all pairwise comparisons. In this formulation two positive constants δ_1 and δ_2 are specified such that $\delta_1 < \delta_2$. The pairwise difference $\theta_i - \theta_j$ is regarded as "small" if $|\theta_i - \theta_j| \leqq \delta_1$ and as "large" if $|\theta_i - \theta_j| \geqq \delta_2$; the corresponding treatments are regarded as "practically equal" and "different," respectively $(1 \leqq i < j \leqq k)$. Each pairwise difference is to be classified as "small" or "large." Any decision with regard to the pairwise difference lying in the "indifference-zone," $\delta_1 < |\theta_i - \theta_j| < \delta_2$, is regarded as correct, but if any "small" or "large" difference is misclassified, then the overall classification is regarded as erroneous. For the pairwise differences that are declared "large," the sign of the difference is also part of the decision, that is, if a "large" $|\theta_i - \theta_j|$ is correctly classified but it is concluded that $\theta_i > \theta_j$ when in fact $\theta_i < \theta_j$, then the classification is regarded as erroneous. Note that misclassifying a "small" (respectively "large") difference corresponds to a Type I (respectively, Type II) error and the misclassification of the sign of a correctly classified "large" difference corresponds to a Type III error. It is desired to control the probability of an erroneous classification (which corresponds to the FWE combining Type I, Type II, and Type III errors) at or below a designated level α $(0 < \alpha < 1 - (\frac{1}{2})^{k^*}$ where $k^* = \binom{k}{2})$ for all values of the θ_i's.

Reading proposed the following procedure: Take n independent observations on each treatment and compute the sample means $\bar{Y}_i$ $(1 \leqq i \leqq k)$. Classify the pairwise difference $|\theta_i - \theta_j|$ as "small" if $|\bar{Y}_i - \bar{Y}_j| \leqq \xi$ and as "large" if $|\bar{Y}_i - \bar{Y}_j| > \xi$; in the latter case also conclude that $\theta_i > \theta_j$ or $\theta_i < \theta_j$ depending on whether $\bar{Y}_i > \bar{Y}_j$ or $\bar{Y}_i < \bar{Y}_j$ $(1 \leqq i \neq j \leqq k)$. Here the critical constant $\xi > 0$ and the minimum sample size n are to be determined so as to satisfy the specified FWE requirement.

The crucial step in obtaining the required values of ξ and n is the determination of the "least favorable" configuration of the θ_i's that maximizes the FWE (i.e., the probability of misclassification) for given (k, σ) and specified (δ_1, δ_2). Reading obtained an exact solution to this problem for $k = 2$. For $k \geqq 3$ he obtained only an approximate solution by determining the least favorable configuration of the θ_i's for a Bonferroni upper bound on the exact probability of misclassification. Using the expressions thus obtained Reading gave tables of constants from which ξ and n can be determined for given k and σ^2, and specified δ_1, δ_2, and α. When these values of ξ and n are used in the procedure described above, the FWE is controlled at or below α for all values of the θ_i's. The use of Reading's table is illustrated in Example 1.1.

In the second design formulation one specifies a (common) fixed allowance $A > 0$ for simultaneous confidence intervals on all pairwise differences $\theta_i - \theta_j$, that is, one requires, say, $100(1-\alpha)\%$ simultaneous confidence intervals of the following form:

$$\theta_i - \theta_j \in [\bar{Y}_i - \bar{Y}_j \pm A] \qquad (1 \leqq i < j \leqq k) .$$

In this case it is easy to see (using equation (2.1) of Chapter 3 for the Tukey intervals) that the common sample size n per treatment is given by

$$n = \left\lfloor \left(\frac{Q_{k,\infty}^{(\alpha)} \sigma}{A} \right)^2 \right\rfloor$$

where $\lfloor x \rfloor$ denotes the smallest integer $\geqq x$.

Example 1.1. Suppose that $k = 4$ treatments are to be compared with each other and each has a standard deviation $\sigma = 0.5$. Two treatments are considered different if their means are apart by at least $\delta_2 = 1$ unit while they are considered practically equal if their means are apart by no more than $\delta_1 = 0.2$ units. A correct classification of all pairwise differences is to be guaranteed with probability at least $1 - \alpha = 0.95$. We wish to determine the smallest sample size n per treatment and the associated critical constant ξ that will guarantee the specified design requirement when used in Reading's (1975) procedure.

From Reading's (1975) Table 3 (entering the row corresponding to the ratio $\delta_1/\delta_2 = 0.2$) we find that $n(\delta_2/\sigma)^2 = 68.72$ and $\xi/\delta_2 = 0.584$. Therefore $n = \lfloor 68.72(0.5)^2 \rfloor = 18$ and $\xi = 0.584$. If after taking a random sample of size 18 from each treatment we obtain the sample means $\bar{Y}_1 = 5.5$, $\bar{Y}_2 = 6.0$, $\bar{Y}_3 = 7.0$, and $\bar{Y}_4 = 7.5$, then treatments in the pairs (1, 2) and (3, 4) can be classified as practically equal; treatments in all other pairs can be classified as different. This result can be depicted graphically as in the case of the step-down range procedure of Section 2.1 of Chapter 4. For every pair of treatments that are classified as different we can make a directional decision in the usual manner. □

1.2 Comparisons with a Control

Now consider the setting of Chapter 5, Section 2 where the kth treatment is regarded as a control with which the first $k - 1$ treatments are to be compared. Tong (1969) considered the problem of classifying the treatments as "good" or "bad" relative to the control using a formulation analogous to that in the previous section for the pairwise comparisons problem. In this formulation two constants δ_1 and δ_2 are specified such that $\delta_1 < \delta_2$; note that these constants need not be positive. Let

$$G = \{i : \theta_i \geqq \theta_k + \delta_2, 1 \leqq i \leqq k-1\},$$
$$B = \{i : \theta_i \leqq \theta_k + \delta_1, 1 \leqq i \leqq k-1\},$$

and

$$I = \{i : \theta_k + \delta_1 < \theta_i < \theta_k + \delta_2, 1 \leqq i \leqq k-1\}$$

be the index sets of "good," "bad," and "indifferent" treatments, respectively, relative to the control. Each treatment is to be classified either as "good" or "bad." Any classification that misclassifies at least one "good" or "bad" treatment is regarded as erroneous, but "indifferent" treatments can be classified either way without affecting the correctness of the overall classification. The FWE is to be controlled at or below a designated level α $(0 < \alpha < 1 - (\frac{1}{2})^{k-1})$ for all values of the θ_i's.

Tong proposed the following procedure for this purpose: Take n independent observations on each treatment and the control, and compute the sample means $\bar{Y}_i$ $(1 \leqq i \leqq k)$. Classify treatment i as "good" if $\bar{Y}_i - \bar{Y}_k \geqq \xi$ and as "bad" if $\bar{Y}_i - \bar{Y}_k < \xi$ $(1 \leqq i \leqq k-1)$ where ξ and n are to be determined so as to satisfy the FWE requirement. Tong made the choice $\xi = (\delta_1 + \delta_2)/2$; thus it only remains to determine the smallest n.

To find the least favorable configuration of the θ_i's for this procedure, first note that it suffices to consider the case where there are no "indifferent" treatments. Furthermore, for fixed sets G and B with cardinalities $g \geqq 0$ and $b \geqq 0$ (with $g + b = k-1$), respectively, it suffices to consider the configurations $\theta_i = \theta_k + \delta_2$ for all $i \in G$ and $\theta_i = \theta_k + \delta_1$ for all $i \in B$. Then the probability of a correct classification can be written as

$$\Pr\left\{\bar{Y}_i - \bar{Y}_k \geqq \frac{\delta_1 + \delta_2}{2}\ \forall i \in G,\ \bar{Y}_i - \bar{Y}_k < \frac{\delta_1 + \delta_2}{2}\ \forall i \in B\right\}$$
$$= \Pr\left\{Z_i < \frac{\delta_2 - \delta_1}{2\sigma}\sqrt{\frac{n}{2}}\ (1 \leqq i \leqq k-1)\right\} \tag{1.1}$$

where the Z_i's have a $(k-1)$-variate normal distribution with zero means, unit variances, and

$$\operatorname{corr}(Z_i, Z_j) = \begin{cases} \frac{1}{2} & 1 \leqq i \neq j \leqq g,\ g+1 \leqq i \neq j \leqq k-1 \\ -\frac{1}{2} & 1 \leqq i \leqq g,\ g+1 \leqq j \leqq k-1 . \end{cases} \tag{1.2}$$

Tong showed that (1.1) is minimized with respect to unknown g when $g = (k-1)/2$ if k is odd and $g = k/2$ if k is even. For this configuration he tabulated the solution $\lambda = \{(\delta_2 - \delta_1)/2\sigma\}\sqrt{n/2}$ to the equation obtained by setting (1.1) equal to $1-\alpha$. This table can be used for determining the common sample size n necessary for given k and σ^2, and specified δ_1, δ_2, and α.

Since the control plays a special role in that each treatment is compared to it, one would expect that more observations should be allocated to it than to each one of the treatments. Sobel and Tong (1971) formulated this design problem as follows: Let n be the common number of observations on each treatment and let n_k be the number of observations on the control. Let $\xi = (\delta_1 + \delta_2)/2$ as in Tong's (1969) procedure. Find the "optimal" values of n and n_k that minimize the total sample size $N = n_k + (k-1)n$ subject to the condition that FWE $\leqq \alpha$ (i.e., the probability of a correct classification $\geqq 1-\alpha$) for all values of the θ_i's. Sobel and Tong derived the equations necessary for determining the optimal values of n_k and n. They also studied the asymptotic ($N \to \infty$) behavior of the solution. But they did not provide tables for implementing their solution for finite values of N. Tamhane (1987) considered the problem of simultaneous optimal choice of n_k, n, *and* ξ. He showed that for k odd, Tong's choice $\xi = (\delta_1 + \delta_2)/2$ is optimal, but for k even, a different choice of ξ is optimal. This optimal choice can be numerically determined, along with n_k and n, by solving a set of simultaneous equations. He gave tables for determining the optimal values of ξ, n_k, and n. In the same article he also extended Sobel and Tong's (1971) asymptotic result that for $N = n_k + (k-1)n$ approaching infinity, the square-root allocation rule, $n_k = n\sqrt{k-1}$ (see Section 2.1.3 of Chapter 5), recommended by Dunnett (1955), is optimal. Both of these articles regard n and n_k as nonnegative continuous variables, or equivalently, $\gamma = n_k/N$ as a continuous variable taking values in the interval (0, 1). If N is large, then this continuous approximation is quite good.

Example 1.2. Consider the same setting as in Example 1.1 but now suppose that treatment 4 is a control with which the first three treatments are to be compared. If $\theta_i - \theta_4 \leqq \delta_1 = 0.2$, then the ith treatment is regarded as "bad" relative to the control and if $\theta_i - \theta_4 \geqq \delta_2 = 1.0$, then the ith treatment is regarded as "good" relative to the control ($i = 1, 2, 3$). A correct classification of the three treatments is desired with probability at least $1 - \alpha = 0.90$. We wish to find the smallest total sample size N and the associated values of n, n_k, and ξ that when used in the given classification procedure will guarantee the specified design requirement.

From Table 2 in Tamhane (1987) we obtain the optimal values of

$b = (\delta_2 - \delta_1)\sqrt{N}/2\sigma$, $c = \sqrt{n/n_4}$, and $d = \xi\sqrt{N}/\sigma$ to be 4.8826, 0.7684, and 14.1926, respectively. It should be noted that d is tabulated for $\gamma = \delta_2/\delta_1 = 2$ from which the optimal d for any other γ can be obtained by the formula $d + b(2-\gamma)/(\gamma - 1)$.

From the tabulated value of b, we first calculate

$$N = \left\lceil \left(\frac{4.8826 \times 2 \times 0.5}{1 - 0.2}\right)^2 \right\rceil = 38 .$$

If we allocate 24 of these 38 observations to the three treatments with $n = 8$ per treatment and the remaining $n_4 = 14$ observations to the control, then we get $\sqrt{n/n_4} = 0.7559$, which is quite close to the optimal value of c. Notice that the square root allocation rule gives $c = \sqrt{n/n_4} = (\frac{1}{3})^{1/4} = 0.7598$. Next we have $\gamma = 1/0.2 = 5$ and therefore the optimal $d = 14.1926 + 4.8826(-3)/4 = 10.5307$, from which we get the optimal $\xi = 10.5307 \times 0.5/\sqrt{38} = 0.8542$.

If we use Tong's (1969) original procedure, then the required critical constant is $\xi = (1 + 0.2)/2 = 0.6$. The tabulated $\lambda = \{(\delta_2 - \delta_1)/2\sigma\}\sqrt{n/2}$ in this case equals 1.8004 from Tong's tables, which gives n, the common sample size for each treatment and the control, equal to $\lfloor 2(2 \times 0.5 \times 1.8004/0.8)^2 \rfloor = 11$. The total sample size required by Tong's procedure is thus $N = 4 \times 11 = 44$. Compare this with $N = 38$ obtained using the optimal procedure and notice the saving. Larger savings are possible for larger k and smaller $(\delta_2 - \delta_1)/\sigma$. □

Bechhofer (1969) considered a related design problem from the viewpoint of simultaneous one-sided confidence interval estimation of the differences $\theta_i - \theta_k$ $(1 \leqq i \leqq k-1)$. Bechhofer and Nocturne (1972) considered the analogous problem for simultaneous two-sided interval estimation. In these formulations it is assumed that the experimenter can prespecify a common "allowance" $A > 0$ for each interval estimate and a joint confidence level $1 - \alpha$. The design problem is then for given k and σ^2, and specified $A > 0$ and α, to determine both the smallest total sample size $N = n_k + n(k-1)$ and the associated optimal allocation (n_k, n), so as to guarantee

$$\Pr\{\theta_i - \theta_k \leqq \bar{Y}_i - \bar{Y}_k + A \ (1 \leqq i \leqq k-1)\} \geqq 1 - \alpha \tag{1.3}$$

for one-sided interval estimates, and

$$\Pr\{\theta_i - \theta_k \in [\bar{Y}_i - \bar{Y}_k \pm A] \ (1 \leqq i \leqq k-1)\} \geqq 1 - \alpha \tag{1.4}$$

for two-sided interval estimates.

Solutions to these optimal allocation problems were given by Bechhofer (1969) (for one-sided intervals) and Bechhofer and Nocturne (1972) (for two-sided intervals). Letting $\gamma = n_k/N$ and $\lambda = A\sqrt{N}/\sigma$, the left hand side of (1.3) can be written as

$$\int_{-\infty}^{\infty} \Phi^{k-1}\left[\left(\frac{x}{\sqrt{\gamma}} + \lambda\right)\left(\frac{1-\gamma}{k-1}\right)^{1/2}\right] d\Phi(x) \tag{1.5}$$

and that of (1.4) can be written as

$$\int_{-\infty}^{\infty} \left\{\Phi\left[\left(\frac{x}{\sqrt{\gamma}} + \lambda\right)\left(\frac{1-\gamma}{k-1}\right)^{1/2}\right] - \Phi\left[\left(\frac{x}{\sqrt{\gamma}} - \lambda\right)\left(\frac{1-\gamma}{k-1}\right)^{1/2}\right]\right\}^{k-1} d\Phi(x)\,, \tag{1.6}$$

where $\Phi(\cdot)$ denotes the standard normal distribution function. Note that these integrals depend on A, N, and σ only through a single unitless quantity λ.

First consider the problem of determining the optimal proportion γ of observations on the control for given $\lambda > 0$. As mentioned before, it is convenient to regard γ as a continuous variable. The optimal γ in the one-sided case (respectively, two-sided case) is the value that maximizes (1.5) (respectively, (1.6)). The optimal γ can be obtained by solving the equation obtained by setting the partial derivative of (1.5) (respectively, (1.6)) with respect to γ equal to zero. To determine the smallest N, or equivalently the smallest λ that guarantees (1.3) (respectively, (1.4)), it is necessary to solve two simultaneous equations for γ and λ, one obtained by setting the partial derivative of (1.5) (respectively, (1.6)) with respect to γ equal to zero and the other obtained by setting (1.5) (respectively, (1.6)) equal to the specified value $1 - \alpha$. Solutions to these simultaneous equations are tabulated in Bechhofer and Tamhane (1983) for $k - 1 = 2(1)10$ and $1 - \alpha = 0.75, 0.90, 0.95$, and 0.99.

Bechhofer (1969) and Bechhofer and Nocturne (1972) also showed for their respective problems that as $\lambda \to \infty$, the optimal proportion $\hat{\gamma} \to \sqrt{k-1}/(1 + \sqrt{k-1})$ or equivalently $n_k/n \to \sqrt{k-1}$, which is the square-root allocation rule mentioned earlier.

Example 1.3. Consider the setting of Example 1.2 but now suppose that instead of classifying the treatments as "good" or "bad" relative to the control, we wish to estimate the differences $\theta_i - \theta_4$ $(1 \leq i \leq 3)$ using 90% simultaneous upper one-sided confidence intervals (1.3). Let the specified allowance $A = 0.2$. From Table 2 in Bechhofer and Tamhane (1983) we find that the optimal values of γ and λ are 0.335 and 4.819, respectively.

From the latter we calculate the minimum value of N to be $\lfloor(4.819 \times 0.5/0.2)^2\rfloor = 146$. Of these 146 observations, we allocate $n = 32$ to each one of the three treatments and the remaining $n_4 = 50$ observations to the control. This gives $\gamma = 50/146 = 0.342$, which is quite close to the optimum γ. □

1.3 Comparisons with the "Best"

This family was considered in Chapter 5, Section 3. Desu (1970) studied the design problem associated with this family from the indifference-zone viewpoint. Let δ_1 and δ_2 be two preassigned constants such that $0 < \delta_1 < \delta_2$, and let

$$G = \{i : \theta_i \geqq \theta_{[k]} - \delta_1 \ (1 \leqq i \leqq k)\}$$

and

$$B = \{i : \theta_i \leqq \theta_{[k]} - \delta_2 \ (1 \leqq i \leqq k)\}$$

be the index sets of "good" and "bad" treatments, respectively, where $\theta_{[k]} = \max \theta_i$. Notice that here the unknown "best" treatment (i.e., the treatment with mean $\theta_{[k]}$) is used as a benchmark for comparisons, while in the previous section the known control treatment was used as a benchmark. Desu considered the goal of selecting a random subset $\hat{G}$ of treatments such that $\hat{G} \cap B = \phi$, which he referred to as a *correct decision* (CD). In other words, a correct decision is made when all bad treatments are excluded from the selected subset $\hat{G}$. (Note, however, that some good treatments may also be excluded from $\hat{G}$.) The selection procedure is required to satisfy the probability requirement that $\Pr_{\boldsymbol{\theta}}\{\text{CD}\} \geqq 1 - \alpha$ for all $\boldsymbol{\theta}$. This probability requirement implies that one can make the simultaneous confidence statement

$$\Pr_{\boldsymbol{\theta}}\{\theta_{[k]} - \theta_i < \delta_2 \ \forall\, i \in \hat{G}\} \geqq 1 - \alpha \qquad \forall\, \boldsymbol{\theta}. \tag{1.7}$$

Desu proposed the following procedure: Take a random sample of size n on each treatment and compute the sample means $\bar{Y}_i$ $(1 \leqq i \leqq k)$. Let $\bar{Y}_{[k]} = \max \bar{Y}_i$. Then choose the subset

$$\hat{G} = \left\{i : \bar{Y}_i \geqq \bar{Y}_{[k]} - \left(\delta_2 - \frac{c\sigma}{\sqrt{n}}\right) (1 \leqq i \leqq k)\right\}.$$

Desu showed that to satisfy the stated probability requirement, the constant c must be equal to $Z^{(\alpha)}_{k-1,1/2}$, the upper α equicoordinate point of the standard $(k-1)$-variate equicorrelated normal distribution with common correlation $= \frac{1}{2}$. In order that $\hat{G}$ be nonempty, we must have $n \geqq (c\sigma/\delta_2)^2$, which determines n.

Hsu (1981) showed that using Desu's procedure one can make a stronger simultaneous confidence statement than (1.7), namely,

$$\Pr_{\boldsymbol{\theta}}\{\theta_{[k]} - \theta_i \leqq U_i \ (1 \leqq i \leqq k)\} \geqq 1 - \alpha \qquad \forall\, \boldsymbol{\theta} \tag{1.8}$$

where

$$U_i = \max\left\{\max_{j \neq i} \bar{Y}_j - \bar{Y}_i + \frac{c\sigma}{\sqrt{n}}, 0\right\} \qquad (1 \leqq i \leqq k)\,.$$

Note that Hsu provides confidence intervals for all differences $\theta_{[k]} - \theta_i$ from the "best." Furthermore, his upper confidence bounds $U_i \leqq \delta_2$ for $i \in \hat{G}$, and hence are sharper than those in (1.7). For a further discussion of Hsu's work we refer the reader to Section 3 of Chapter 5.

1.4 Successive Comparisons among Ordered Means

This family was discussed in Chapter 5, Section 4.2. The design problem for this family was considered by Hochberg and Marcus (1978), who proposed an indifference-zone formulation analogous to the ones given in the preceding two sections.

Let δ_1 and δ_2 be prespecified constants such that $\delta_1 < \delta_2$. The ith increment $\Delta_i = \theta_{i+1} - \theta_i$ is regarded as "small" if $\Delta_i \leqq \delta_1$ and "large" if $\Delta_i \geqq \delta_2$ $(1 \leqq i \leqq k-1)$. Each one of the $k-1$ successive differences is to be classified as "small" or "large." An erroneous classification is defined as one that misclassifies at least one "small" or "large" Δ_i. Any Δ_i lying in the indifference-zone (δ_1, δ_2) may be classified either way without affecting the correctness of the overall classification. The FWE is to be controlled at or below a designated level α $(0 < \alpha < 1 - (\frac{1}{2})^{k-1})$ for all values of the θ_i's.

Hochberg and Marcus (1978) proposed the following procedure for this purpose: Let $\bar{Y}_i$ be the sample mean based on n independent observations $(1 \leqq i \leqq k)$. Classify the ith successive difference, $\Delta_i = \theta_{i+1} - \theta_i$, as "small" if $\bar{Y}_{i+1} - \bar{Y}_i \leqq \xi$, and as "large" if $\bar{Y}_{i+1} - \bar{Y}_i > \xi$ $(1 \leqq i \leqq k-1)$. The critical constant ξ and the common sample size n are to be determined so as to satisfy the specified FWE requirement.

For determining a least favorable configuration of the θ_i's that maximizes the FWE of this procedure, it suffices to restrict consideration (as in Section 1.2) to configurations where each Δ_i equals either δ_1 or δ_2. Hochberg and Marcus (1978) showed that the configuration $\Delta_i = \delta_1$ for all

i or $\Delta_i = \delta_2$ for all i maximizes the FWE and the "optimal" value of ξ that minimizes the corresponding maximum FWE is $(\delta_1 + \delta_2)/2$. They also showed that for this choice of ξ, the smallest common sample size n that controls the FWE in the least favorable configuration (and hence for all values of the θ_i's) is given by $n = \lfloor \{2\lambda\sigma/(\delta_2 - \delta_1)\}^2 \rfloor$ where λ is the solution to the equation

$$\Pr\{\max_{1 \leqq i \leqq k-1} Z_i \leqq \lambda\} = 1 - \alpha\,; \tag{1.9}$$

here the Z_i's are jointly distributed standard normal random variables (r.v.'s) with the correlation structure

$$\operatorname{corr}(Z_i, Z_j) = \begin{cases} -\frac{1}{2} & \text{if } |i-j| = 1 \\ 0 & \text{if } |i-j| > 1\,. \end{cases}$$

Unfortunately for this correlation structure the left hand side of (1.9), which gives the probability of a correct classification under a least favorable configuration, cannot be expressed in a form convenient for computation. Hochberg and Marcus (1978) obtained the following lower bound on it by using Kounias's (1968) improved Bonferroni inequality:

$$1 - (k-1)\Pr\{Z_1 > \lambda\} + \Pr\{Z_1 > \lambda, Z_2 > \lambda\} + (k-3)(\Pr\{Z_1 > \lambda\})^2\,. \tag{1.10}$$

This lower bound is exact for $k = 3$. Hochberg and Marcus (1978) tabulated the solutions λ to the equation obtained by setting (1.10) equal to $1 - \alpha$ for selected k and α. This table can be used for determining a conservative sample size n (exact for $k = 3$) that meets the specified FWE requirement.

2 TWO-STAGE PROCEDURES

In Section 1 we assumed that σ^2 is known. When σ^2 is unknown, the probability of a Type II error at a specified nonnull alternative cannot be controlled at a designated level (or equivalently simultaneous confidence intervals of prespecified widths cannot be constructed) using a single-stage procedure. Sequential procedures are needed for such purposes. The simplest types of sequential procedures are two-stage procedures. Such two-stage procedures are based on the following basic result.

Theorem 2.1 (Stein 1945). Let Y_i $(i = 1, 2, \ldots)$ be independent $N(\mu, \sigma^2)$ r.v.'s. Let S^2 be an unbiased estimate of σ^2 that is distributed as $\sigma^2\chi^2_\nu/\nu$ independently of $\Sigma_{i=1}^n Y_i$ and $Y_{n+1}, Y_{n+2}, \ldots$, where n is some fixed positive integer. Let $N \geqq n$ be a positive integer valued function of S^2. Denote $\bar{Y} = \Sigma_{i=1}^N Y_i/N$. Then $(\bar{Y} - \mu)\sqrt{N}/S$ is distributed as a Student's t r.v. with ν degrees of freedom (d.f.). □

2.1 Pairwise Comparisons

Consider the balanced one-way layout model of Section 1.1. Suppose that simultaneous confidence intervals with confidence coefficient $1 - \alpha$ and prespecified allowance $A > 0$ are desired for all pairwise differences $\theta_i - \theta_j$ $(1 \leqq i < j \leqq k)$. For this problem Hochberg and Lachenbruch (1976) proposed a two-stage procedure that is based on the result of Theorem 2.1. This procedure operates as follows.

Stage 1. Take $n \geqq 2$ independent observations from each of the k treatments and compute the usual pooled unbiased estimate S^2 of σ^2 based on $\nu = k(n-1)$ d.f. Let

$$N = \max\left\{\left\lfloor \frac{S^2}{\zeta^2} \right\rfloor, n\right\} \tag{2.1}$$

where $\zeta = A/Q_{k,\nu}^{(\alpha)}$ and $Q_{k,\nu}^{(\alpha)}$ is the upper α point of the Studentized range distribution with parameter k and d.f. ν.

Stage 2. In the second stage take $N - n$ additional independent observations from each of the k treatments and compute the overall sample means $\bar{Y}_i = \Sigma_{j=1}^N Y_{ij}/N$ $(1 \leqq i \leqq k)$. Then $100(1-\alpha)\%$ simultaneous confidence intervals for all pairwise differences $\theta_i - \theta_j$ are given by

$$\theta_i - \theta_j \in [\bar{Y}_i - \bar{Y}_j \pm A] \qquad (1 \leqq i < j \leqq k). \tag{2.2}$$

These confidence intervals can be extended to the family of all contrasts by using Lemma 2.1 of Chapter 3. Thus $100(1-\alpha)\%$ simultaneous confidence intervals for all contrasts $\Sigma_{i=1}^k c_i\theta_i$ are given by

$$\sum_{i=1}^k c_i\theta_i \in \left[\sum_{i=1}^k c_i\bar{Y}_i \pm \frac{A}{2}\sum_{i=1}^k |c_i|\right] \qquad \forall\, \mathbf{c} \in \mathbb{C}^k .$$

The confidence intervals (2.2) with N given by (2.1) follow from the result (which follows from Theorem 2.1) that $\sqrt{N}/S$ times the range of

$\bar{Y}_1 - \theta_1, \ldots, \bar{Y}_k - \theta_k$ has the Studentized range distribution with parameter k and d.f. ν. Hochberg and Lachenbruch (1976) also gave a two-stage procedure for testing $H_0: \theta_1 = \theta_2 = \cdots = \theta_k$ using the Studentized range statistic $\sqrt{N} \max_{1 \leqq i < j \leqq k} |\bar{Y}_i - \bar{Y}_j| / S$. This procedure guarantees specified power whenever $\max_{1 \leqq i < j \leqq k} |\theta_i - \theta_j| \geqq \delta$ where $\delta > 0$ is a specified constant.

Example 2.1. As in Example 1.1 suppose that pairwise comparisons are to be made among $k = 4$ treatments, but now assume that σ^2 is unknown and simultaneous confidence intervals are of interest. Let $1 - \alpha = 0.95$ and $A = 0.4$. If $n = 7$ observations are taken from each treatment in the first stage, then $\nu = k(n-1) = 24$. Thus $\zeta = 0.4/3.90 = 0.1026$ where $Q_{4,24}^{(0.05)} = 3.90$ is taken from Table 8 in Appendix 3. If the first stage data yield the pooled estimate $S = 0.5$ (the value of σ used in Example 1.1), then

$$N = \max\left\{\left[\left(\frac{0.5}{0.1026}\right)^2\right], 7\right\} = 24 ,$$

and therefore 17 observations must be taken on each treatment in the second stage. □

2.2 Comparisons with a Control

Consider the setup of Section 1.2. For the problem of classification of treatments with respect to a control, Tong (1969) proposed a two-stage procedure similar to the one described in Section 1.2. For Tong's procedure the constant ζ in (2.1) equals $(\delta_2 - \delta_1)/(2\sqrt{2}T_{k-1,\nu,\mathbf{R}}^{(\alpha)})$; here $T_{k-1,\nu,\mathbf{R}}^{(\alpha)}$ is the upper α equicoordinate point of a $(k-1)$-variate t-distribution with ν d.f. and associated correlation matrix $\mathbf{R}$ whose entries are given by (1.2) with $g = (k-1)/2$ if k is odd and $g = k/2$ if k is even. A table of these equicoordinate points is given by Tong. Having obtained the overall sample means $\bar{Y}_i$, the treatments are classified in the same way as in the case of the single-stage procedure of Section 1.2; thus treatment i is classified as "good" with respect to the control treatment k if $\bar{Y}_i - \bar{Y}_k \geqq \xi$ and as "bad" if $\bar{Y}_i - \bar{Y}_k < \xi$ $(1 \leqq i \leqq k-1)$ where $\xi = (\delta_1 + \delta_2)/2$. That this procedure controls the FWE at or below α for all values of the θ_i's follows from the fact that the vector

$$\frac{\sqrt{N}}{S}(\bar{Y}_1 - \bar{Y}_k - \theta_1 + \theta_k, \ldots, \bar{Y}_{k-1} - \bar{Y}_k - \theta_{k-1} + \theta_k)$$

has the $(k-1)$-variate t-distribution referred to above but with arbitrary (unknown) g ($0 \leqq g \leqq k-1$), and $g = k/2$ (respectively, $g = (k-1)/2$) is the least favorable value of g for k even (respectively, k odd).

This same approach can be used for constructing two-stage procedures to obtain simultaneous confidence intervals for $\theta_i - \theta_k$ ($1 \leqq i \leqq k-1$) having a common prescribed width $2A$. We omit the details. Similarly a two-stage procedure can be constructed that uses a fixed critical constant for determining whether each difference between the sample means of successively ordered treatments is "small" or "large." The details can be found in the Hochberg and Marcus (1978) article.

3 INCOMPLETE BLOCK DESIGNS FOR COMPARING TREATMENTS WITH A CONTROL

Blocking is a standard technique used to improve the precision of a comparative experiment. When all pairwise comparisons between treatment means or specified orthogonal contrasts are of equal interest, symmetrical block designs are the natural choice. In particular, if the size of each block is the same and this common size is less than the number of treatments (i.e., the blocks are incomplete) then balanced incomplete block (BIB) designs have certain optimality properties and should be used (Kiefer 1958).

Still restricting our attention to the incomplete block setting we now consider the question of appropriate designs to use for the comparisons with a control problem. Cox (1958, p. 238) noted the inappropriateness of BIB designs between all treatments (including the control) for this purpose in view of the special role played by the control treatment. Cox suggested a design in which the same number of plots in each block are allocated to the control, and the test treatments are arranged in a BIB design in the remaining plots of the blocks. Bechhofer and Tamhane (1981) proposed a general class of designs called *balanced treatment incomplete block* (*BTIB*) designs; Cox's design is a member of this class. To define this class we first introduce some notation.

Let the treatments be indexed $1, 2, \ldots, k$ with k denoting the control treatment and $1, 2, \ldots, k-1$ denoting the $k-1$ test treatments. Let p denote the common size of each block and b, the number of blocks. Bechhofer and Tamhane (1981) considered the following additive linear model for an observation Y_{ijh} obtained when the ith treatment is assigned to the hth plot of the jth block:

$$Y_{ijh} = \theta_i + \beta_j + E_{ijh} \qquad (1 \leqq i \leqq k,\ 1 \leqq j \leqq b,\ 1 \leqq h \leqq p) \qquad (3.1)$$

where the E_{ijh}'s are independent $N(0, \sigma^2)$ random errors. Let $\hat{\theta}_i - \hat{\theta}_k$ be the LS estimate of $\theta_i - \theta_k$ $(1 \leqq i \leqq k-1)$.

For given (k, p, b) a design is referred to as a *BTIB design* if it has the property that

$$\operatorname{var}(\hat{\theta}_i - \hat{\theta}_k) = \tau^2\sigma^2 \qquad (1 \leqq i \leqq k-1) \tag{3.2}$$

and

$$\operatorname{corr}(\hat{\theta}_i - \hat{\theta}_k, \hat{\theta}_{i'} - \hat{\theta}_k) = \rho \qquad (1 \leqq i \neq i' \leqq k-1) \tag{3.3}$$

for some $\tau^2 > 0$ and ρ $(-1/(k-2) < \rho < 1)$. Here τ^2 and ρ depend, of course, on the particular design employed. Designs having this same balance property were considered earlier by Pearce (1960).

Let n_{ij} denote the number of replications of the ith treatment in the jth block $(1 \leqq i \leqq k,\ 1 \leqq j \leqq b)$. Theorem 3.1 of Bechhofer and Tamhane (1981) gives the following conditions as both necessary and sufficient for a design to be BTIB:

$$\sum_{j=1}^{b} n_{kj} n_{ij} = \lambda_0 \qquad (1 \leqq i \leqq k-1) \tag{3.4}$$

and

$$\sum_{j=1}^{b} n_{ij} n_{i'j} = \lambda_1 \qquad (1 \leqq i \neq i' \leqq k-1) \tag{3.5}$$

for some nonnegative integers λ_0, λ_1. In order that the $(\theta_i - \theta_k)$'s be estimable we must have $\lambda_0 > 0$.

The following is an example of a BTIB design for $k = 5$, $p = 3$, and $b = 7$:

$$\begin{bmatrix} 5 & 5 & 5 & 5 & 5 & 5 & 1 \\ 1 & 2 & 3 & 5 & 5 & 5 & 2 \\ 3 & 3 & 4 & 1 & 2 & 4 & 4 \end{bmatrix};$$

here the columns represent the blocks. This design has $\lambda_0 = 3$ and $\lambda_1 = 1$.

We now consider the analysis aspects of BTIB designs. Let T_i denote the sum of all observations on treatment i $(1 \leqq i \leqq k)$ and let B_j denote the sum of all observations in the jth block $(1 \leqq j \leqq b)$. Define $B_i^* = \Sigma_{j=1}^{b} n_{ij} B_j$ and $Q_i = kT_i - B_i^*$ $(1 \leqq i \leqq k)$. Bechhofer and Tamhane showed that the LS estimate of $\theta_i - \theta_k$ is given by

$$\hat{\theta}_i - \hat{\theta}_k = \frac{\lambda_0 Q_i - \lambda_1 Q_k}{\lambda_0[\lambda_0 + (k-1)\lambda_1]} \qquad (1 \leqq i \leqq k-1)\,. \tag{3.6}$$

Also τ^2 of (3.2) is given by

$$\tau^2 = \frac{p(\lambda_0 + \lambda_1)}{\lambda_0[\lambda_0 + (k-1)\lambda_1]} \tag{3.7}$$

and ρ of (3.3) is given by

$$\rho = \frac{\lambda_1}{\lambda_0 + \lambda_1}. \tag{3.8}$$

The analysis of variance table for a BTIB design is as shown in Table 3.1. In this table G denotes the sum of all observations, and H denotes the sum of the squares of all observations. An unbiased estimate of σ^2 is given by $S^2 = MS_{\text{error}} = SS_{\text{error}}/(bp - b - k + 1)$ with $\nu = bp - b - k + 1$ d.f.

The estimates $\hat{\theta}_i - \hat{\theta}_k$ along with the estimate S^2 of σ^2 can be used in Dunnett's procedure (see Section 2 of Chapter 5) to obtain the following $100(1-\alpha)\%$ simultaneous one-sided confidence intervals:

$$\theta_i - \theta_k \leqq \hat{\theta}_i - \hat{\theta}_k + T^{(\alpha)}_{k-1,\nu,\rho}\tau S \qquad (1 \leqq i \leqq k-1). \tag{3.9}$$

The corresponding two-sided intervals are given by

$$\theta_i - \theta_k \in [\hat{\theta}_i - \hat{\theta}_k \pm |T|^{(\alpha)}_{k-1,\nu,\rho}\tau S] \qquad (1 \leqq i \leqq k-1). \tag{3.10}$$

In (3.9), $T^{(\alpha)}_{k-1,\nu,\rho}$ (respectively, in (3.10), $|T|^{(\alpha)}_{k-1,\nu,\rho}$) is the upper α point

TABLE 3.1. Analysis of Variance for a BTIB Design

Source	Sum of Squares (SS)	d.f.
Treatments (adjusted for blocks)	$\frac{1}{p[\lambda_0 + (k-1)\lambda_1]}\left\{\frac{\lambda_1}{\lambda_0} Q_k^2 + \sum_{i=1}^{k-1} Q_i^2\right\}$	$k-1$
Blocks (unadjusted)	$\frac{1}{p}\sum_{j=1}^{b} B_j^2 - \frac{G^2}{bp}$	$b-1$
Error	By subtraction	$bp - b - k + 1$
Total	$H - \frac{G^2}{bp}$	$bp - 1$

of the distribution of the maximum of the components (respectively, absolute values of the components) of a $(k-1)$-variate equicorrelated t-distribution with ν d.f. and common correlation ρ given by (3.8).

Bechhofer and Tamhane formulated the problem of determining an "optimal" BTIB design in the following way: Suppose that k, p, and b are given and σ^2 is assumed to be known for design purposes. Further suppose that a common allowance $A > 0$ is specified and it is desired to have simultaneous one-sided confidence intervals of the form

$$\theta_i - \theta_k \leqq \hat{\theta}_i - \hat{\theta}_k + A \qquad (1 \leqq i \leqq k-1), \tag{3.11}$$

or simultaneous two-sided confidence intervals of the form

$$\theta_i - \theta_k \in [\hat{\theta}_i - \hat{\theta}_k \pm A] \qquad (1 \leqq i \leqq k-1). \tag{3.12}$$

An "optimal" BTIB design for the one-sided (respectively, two-sided) case maximizes the joint confidence coefficient associated with (3.11) (respectively, (3.12)). Tables of such optimal designs for $k-1 = 2(1)6$ and $p = 2, 3$ are given in Bechhofer and Tamhane (1985). The following example is taken from this reference.

Example 3.1. Suppose that an experimenter wishes to compare two treatments (labeled 1 and 2) with a control (labeled 3) in blocks of size $p = 2$. Further suppose that the experimenter wishes to make simultaneous 95% one-sided confidence statements about $\theta_i - \theta_3$ $(i = 1, 2)$ and that for design purposes a common fixed standardized allowance $A/\sigma = 1$ is specified. From Table OPT1.2.2 in Bechhofer and Tamhane (1985) we find that for $A/\sigma = 1.0$, the smallest number of blocks required to guarantee a joint confidence coefficient of 0.95 is $b = 15$ and the actual achieved joint confidence coefficient is 0.9568. The corresponding optimal design requires six copies of design $D_1 = \{\begin{smallmatrix} 0 & 0 \\ 1 & 2 \end{smallmatrix}\}$ and three copies of design $D_2 = \{\begin{smallmatrix} 1 \\ 2 \end{smallmatrix}\}$. Thus out of a total of 30 experimental units, 12 are allocated to the control and 9 are allocated to each one of the two treatments.

This optimal design calculation assumes that σ^2 is known. After the data are collected using this design, the experimenter must estimate σ^2 by $S^2 = MS_{\text{error}}$ from Table 3.1 and then apply (3.9) to calculate the desired one-sided confidence intervals. For this design it is easy to verify that $\lambda_0 = 6$, $\lambda_1 = 3$ and hence $\tau^2 = 0.25$ and $\rho = \frac{1}{3}$ from (3.7) and (3.8), respectively. From Table 3.1 we see that the error d.f. $\nu = 13$. The critical constant needed for making 95% simultaneous one-sided confidence intervals is $T^{(0.05)}_{2,13,1/3}$, which approximately equals 2.116 as found by linear interpolation with respect to $1/(1-\rho)$ and $1/\nu$ in Table 4 in Appendix 3.

Thus the desired upper one-sided confidence intervals will have the form

$$\theta_i - \theta_3 \leqq \hat{\theta}_i - \hat{\theta}_3 + 2.116 \times 0.5 \times S \qquad (i = 1, 2)$$

where $\hat{\theta}_i - \hat{\theta}_3$ $(i = 1, 2)$ can be calculated using (3.6).

If two-sided confidence intervals are desired, then from Table OPT2.2.2 of Bechhofer and Tamhane (1985) the optimal design is found to consist of eight copies of D_1 and three copies of D_2 with a total of 19 blocks. The rest of the calculations are similar to those given above.

□

PART II

Procedures for Other Models and Problems, and Procedures Based on Alternative Approaches

CHAPTER 7

Procedures for One-Way Layouts with Unequal Variances

In Part I we studied multiple comparison procedures (MCPs) for fixed-effects linear models with independent, homoscedastic normal errors. These MCPs use the usual mean square error (MSE) as an estimate of the common σ^2. However, if the homoscedasticity assumption is violated, then many of these MCPs are not very robust in terms of their Type I error rates, that is, their Type I error rates differ by not insignificant amounts from their nominal values when the variances are unequal. In the present chapter we consider MCPs that are designed to take into account possibly heterogeneous variances by utilizing their separate sample estimates. We restrict to the one-way layout setting and follow the same notation as in Example 1.1 of Chapter 2, but now denote the ith treatment variance by σ_i^2 $(1 \leq i \leq k)$.

Many simulation studies have shown that the MCPs based on the usual MSE estimate are in fact liberal (i.e., their Type I error rates exceed their nominal values), particularly when large σ_i^2's are paired with small n_i's and vice versa (inverse pairing); see, for example, Keselman and Rogan (1978), Dunnett (1980b), Tamhane (1979), and Korhonen (1979). Keselman and Rogan's (1978) simulation results showed that the Tukey–Kramer (TK) procedure and the Scheffé (S) procedure are both liberal in the presence of heterogeneous variances, the latter being somewhat less so (see also Scheffé 1959, Section 10.4). Dunnett's (1980b) simulation study showed that the actual Type I familywise error rate (FWE) of the TK-procedure can be more than four times the nominal value α (for $\alpha = 0.05$) if the σ_i^2's and n_i's are inversely paired, and thus the (σ_i^2/n_i)'s are highly imbalanced. In Tamhane's (1979) simulation study the TK-procedure was not included but the Tukey–Spjøtvoll–Stoline (TSS) and

Hochberg's (1974b) GT2-procedures were included. Both of these procedures were found to have excessive Type I error rates only when the imbalance among the (σ_i^2/n_i)-values was large. However, this apparently robust behavior is due to the fact that these procedures are inherently conservative (for the case of homogeneous variances) while the TK-procedure is sharp. Korhonen (1979) studied the robustness of Tukey's T-procedure, the TSS-procedure and the Bonferroni procedure with respect to the violations of the normality, independence, and homoscedasticity assumptions. He found that these procedures are quite robust against departures from normality but not against departures from the other two assumptions. We again note that all of these articles study robustness with respect to the Type I FWE only.

This discussion shows the need for MCPs that are not based on the assumption of homogeneous variances. In Section 1 we review such single-stage MCPs. It is possible to construct such exact (having FWE = the nominal value α) single-stage MCPs for the family of all linear combinations of the θ_i's but not for the family of all contrasts (or for any subset of that family such as the family of pairwise comparisons or the family of comparisons with a control). Only approximate (having FWE $\cong \alpha$) MCPs can be used in the latter case. These approximate MCPs are suitable extensions of the available procedures for the Behrens–Fisher problem, which is a special case for $k = 2$ of the contrasts problem under variance heterogeneity. Although only approximate single-stage MCPs can be constructed for the contrasts problem, it is possible to construct exact two-stage MCPs (having FWE $= \alpha$) for the same problem. In Section 2 we discuss these two-stage MCPs. All of the procedures in Sections 1 and 2 are simultaneous confidence (and test) procedures. No stepwise test procedures have been proposed for comparing the means without assuming homogeneous error variances. In Section 3 we offer some suggestions for two step-down procedures but their details remain to be worked out.

1 SINGLE-STAGE PROCEDURES

As stated before, the MCPs discussed in this section fall into two groups. The MCPs in the first group address the family of all linear combinations of the θ_i's while those in the second group address the family of all contrasts. The MCPs in the first group are exact while those in the second group are approximate, usually conservative. Note that the contrasts problem is a generalization of the Behrens–Fisher problem for which also no exact single-stage solution is known to exist.

In the following we let $\bar{Y}_i = \sum_{j=1}^{n_i} Y_{ij}/n_i$ and $S_i^2 = \sum_{j=1}^{n_i} (Y_{ij} - \bar{Y}_i)^2/(n_i - 1)$ denote the sample mean and sample variance, respectively, for the ith treatment based on n_i observations ($1 \leqq i \leqq k$). Note that the S_i^2's are distributed independently of each other and of the $\bar{Y}_i$'s as $\sigma_i^2 \chi_{\nu_i}^2/\nu_i$ random variables (r.v.'s) where $\nu_i = n_i - 1$ ($1 \leqq i \leqq k$).

1.1 All Linear Combinations

DL-PROCEDURE. This procedure proposed by Dalal (1978) is based on the following variation of Hölder's inequality, which was earlier used for constructing MCPs by Dwass (1959).

Lemma 1.1. Let $p, q \geqq 1$ such that $1/p + 1/q = 1$. Then for any vector $\mathbf{x} \in \mathbb{R}^k$ and any $\xi \geqq 0$

$$|\mathbf{a}'\mathbf{x}| \leqq \xi\left(\sum_{i=1}^{k} |a_i|^p\right)^{1/p} \quad \forall\, \mathbf{a} \in \mathbb{R}_k \Leftrightarrow \left(\sum_{i=1}^{k} x_i^q\right)^{1/q} \leqq \xi . \tag{1.1}$$

An important special case of (1.1) for $p = 1$, $q = \infty$ gives

$$|\mathbf{a}'\mathbf{x}| \leqq \xi\left(\sum_{i=1}^{k} |a_i|\right) \quad \forall\, \mathbf{a} \in \mathbb{R}^k \Leftrightarrow \max_{1 \leqq i \leqq k} |x_i| \leqq \xi . \tag{1.2}$$

Another important special case for $p = q = 2$ gives

$$|\mathbf{a}'\mathbf{x}| \leqq \xi\left(\sum_{i=1}^{k} a_i^2\right)^{1/2} \quad \forall\, \mathbf{a} \in \mathbb{R}^k \Leftrightarrow \left(\sum_{i=1}^{k} x_i^2\right) \leqq \xi^2 . \tag{1.3}$$

□

Let $D_{q;\,\nu_1,\ldots,\nu_k}^{(\alpha)}$ be the upper α point of the distribution of

$$D_{q;\,\nu_1,\ldots,\nu_k} = \left(\sum_{i=1}^{k} T_{\nu_i}^q\right)^{1/q}$$

where T_{ν_i} is a Student's t r.v. with ν_i degrees of freedom (d.f.) ($1 \leqq i \leqq k$) and the T_{ν_i}'s are independently distributed. We now show that exact $(1-\alpha)$-level simultaneous confidence intervals for all linear combinations $\sum_{i=1}^{k} a_i\theta_i$ are given by

$$\sum_{i=1}^{k} a_i\theta_i \in \left[\sum_{i=1}^{k} a_i\bar{Y}_i \pm D_{q;\,\nu_1,\ldots,\nu_k}^{(\alpha)} \left\{\sum_{i=1}^{k} \left(\frac{|a_i|S_i}{\sqrt{n_i}}\right)^p\right\}^{1/p}\right] \quad \forall\, \mathbf{a} \in \mathbb{R}^k \tag{1.4}$$

where $p, q \geqq 0$ are such that $1/p + 1/q = 1$. On letting $b_i = a_iS_i/\sqrt{n_i}$

$(1 \leq i \leq k)$ we have

$$\Pr\left\{\sum_{i=1}^{k} a_i\theta_i \in \left[\sum_{i=1}^{k} a_i\bar{Y}_i \pm D^{(\alpha)}_{q;\,\nu_1,\ldots,\nu_k}\left\{\sum_{i=1}^{k}\left(\frac{|a_i|S_i}{\sqrt{n_i}}\right)^p\right\}^{1/p}\right] \ \forall\, \mathbf{a}\in\mathbb{R}^k\right\}$$

$$= \Pr\left\{\left|\sum_{i=1}^{k} b_i\left(\frac{\bar{Y}_i-\theta_i}{S_i/\sqrt{n_i}}\right)\right| \leq D^{(\alpha)}_{q;\,\nu_1,\ldots,\nu_k}\left(\sum_{i=1}^{k}|b_i|^p\right)^{1/p} \ \forall\, \mathbf{b}\in\mathbb{R}^k\right\}$$

$$= \Pr\left\{\left[\sum_{i=1}^{k}\left(\frac{\bar{Y}_i-\theta_i}{S_i/\sqrt{n_i}}\right)^q\right]^{1/q} \leq D^{(\alpha)}_{q;\,\nu_1,\ldots,\nu_k}\right\} \qquad \text{(using (1.1))}$$

$$= \Pr\left\{\left(\sum_{i=1}^{k} T^q_{\nu_i}\right)^{1/q} \leq D^{(\alpha)}_{q;\,\nu_1,\ldots,\nu_k}\right\}$$

$$= 1-\alpha\,.$$

For $p = 1$, $q = \infty$, it follows that $D_{\infty;\,\nu_1,\ldots,\nu_k} = \max_{1\leq i\leq k}|T_{\nu_i}|$ and thus $D^{(\alpha)}_{\infty;\,\nu_1,\ldots,\nu_k}$ is the solution in D of the equation

$$\prod_{i=1}^{k}\{2F_{\nu_i}(D)-1\} = 1-\alpha \tag{1.5}$$

where $F_\nu(\cdot)$ denotes the distribution function of a T_ν r.v.

For $p = 1$, $q = \infty$, a modified version of the DL-procedure, which is easier to implement in practice was proposed by Tamhane (1979). This modification of (1.4) results in the following exact $(1-\alpha)$-level simultaneous confidence intervals:

$$\sum_{i=1}^{k} a_i\theta_i \in \left[\sum_{i=1}^{k} a_i\bar{Y}_i \pm \sum_{i=1}^{k}\left\{T^{(\alpha^*)}_{\nu_i}|a_i|\frac{S_i}{\sqrt{n_i}}\right\}\right] \quad \forall\, \mathbf{a}\in\mathbb{R}^k \tag{1.6}$$

where $T^{(\alpha^*)}_\nu$ is the upper α^* point of Student's t distribution with ν d.f. and $\alpha^* = \frac{1}{2}\{1-(1-\alpha)^{1/k}\}$. The $T^{(\alpha^*)}_{\nu_i}$'s are easier to obtain (e.g., from the tables of Games 1977 or by interpolating in Table 1 in Appendix 3) than solving equation (1.5). The original DL-procedure and the modified DL-procedure are identical when the ν_i's are equal and also when the ν_i's$\rightarrow\infty$ (the known variances case). For other situations the performances of the modified DL-procedure and the original DL-procedure are comparable on the average (Tamhane 1979). Thus it does not seem necessary to consider the original DL-procedure (for $p = 1$, $q = \infty$) separately.

For $p = q = 2$ the DL-procedure was proposed earlier by Spjøtvoll (1972b). He gave the following approximation to $D^{(\alpha)}_{2;\,\nu_1,\ldots,\nu_k}$ obtained by matching the first two moments of $D_{2;\nu_1,\ldots,\nu_k}$ with those of a scaled F r.v.:

$$D^{(\alpha)}_{2;\nu_1,\ldots,\nu_k} \cong (aF^{(\alpha)}_{k,b})^{1/2} \tag{1.7}$$

where

$$b = \frac{(k-2)\left[\sum_{i=1}^{k}\{\nu_i/(\nu_i-2)\}\right]^2 + 4k\sum_{i=1}^{k}\{\nu_i^2(\nu_i-1)/(\nu_i-2)^2(\nu_i-4)\}}{k\sum_{i=1}^{k}\{\nu_i^2(\nu_i-1)/(\nu_i-2)^2(\nu_i-4)\} - \left[\sum_{i=1}^{k}\{\nu_i/(\nu_i-2)\}\right]^2}, \tag{1.8}$$

and

$$a = \left(1-\frac{2}{b}\right)\sum_{i=1}^{k}\left\{\frac{\nu_i}{\nu_i-2}\right\}. \tag{1.9}$$

We refer to this procedure as the *SP-procedure*.

H1-PROCEDURE. This procedure was proposed by Hochberg (1976a). Let $R'^{(\alpha)}_{\nu_1,\ldots,\nu_k}$ denote the upper α point of the augmented range $R'_{\nu_1,\ldots,\nu_k}$ of k independent t r.v.'s $T_{\nu_1}, T_{\nu_2}, \ldots, T_{\nu_k}$, that is,

$$R'_{\nu_1,\ldots,\nu_k} = \max\{\max_{1\le i<j\le k}|T_{\nu_i} - T_{\nu_j}|, \max_{1\le i\le k}|T_{\nu_i}|\}.$$

Then exact $(1-\alpha)$-level simultaneous confidence intervals for all linear combinations $\Sigma_{i=1}^{k} a_i\theta_i$ are given by

$$\sum_{i=1}^{k} a_i\theta_i \in \left[\sum_{i=1}^{k} a_i\bar{Y}_i \pm R'^{(\alpha)}_{\nu_1,\ldots,\nu_k} M(\mathbf{b})\right] \quad \forall\, \mathbf{a} \in \mathbb{R}^k \tag{1.10}$$

where $\mathbf{b} = (b_1, \ldots, b_k)'$,

$$b_i = \frac{a_i S_i}{\sqrt{n_i}}, \qquad M(\mathbf{b}) = \max\left\{\sum_{i=1}^{k} b_i^+, \sum_{i=1}^{k} b_i^-\right\},$$

$$b_i^+ = \max(b_i, 0), \qquad \text{and} \qquad b_i^- = \max(-b_i, 0),$$

$i = 1, 2, \ldots, k$.

That the intervals (1.10) have exact $1-\alpha$ joint confidence level follows from the probability calculation below:

$$\Pr\left\{\sum_{i=1}^{k} a_i\theta_i \in \left[\sum_{i=1}^{k} a_i\bar{Y}_i \pm R'^{(\alpha)}_{\nu_1,\ldots,\nu_k} M(\mathbf{b})\right] \ \forall\, \mathbf{a} \in \mathbb{R}^k\right\}$$

$$= \Pr\left\{\frac{\left|\sum_{i=1}^{k} b_i(\bar{Y}_i - \theta_i)/[S_i/\sqrt{n_i}]\right|}{M(\mathbf{b})} \leqq R'^{(\alpha)}_{\nu_1,\ldots,\nu_k} \ \forall\, \mathbf{b} \in \mathbb{R}^k\right\}$$

$$= \Pr\left\{\frac{\left|\sum_{i=1}^{k} b_i T_{\nu_i}\right|}{M(\mathbf{b})} \leqq R'^{(\alpha)}_{\nu_1,\ldots,\nu_k} \ \forall\, \mathbf{b} \in \mathbb{R}^k\right\}$$

$$= \Pr\{R'_{\nu_1,\ldots,\nu_k} \leqq R'^{(\alpha)}_{\nu_1,\ldots,\nu_k}\} \qquad \text{(using Lemma 2.2 of Chapter 3)}$$

$$= 1 - \alpha\,.$$

Tables of $R'^{(\alpha)}_{\nu_1,\ldots,\nu_k}$ are not available. If the ν_i's are large, then $R'^{(\alpha)}_{\nu_1,\ldots,\nu_k}$ may be approximated by the upper α point of the augmented range of k independent standard normal variables; the latter may be obtained from Stoline's (1978) tables of the Studentized augmented range distribution with d.f. equal to ∞. Hochberg (1976a) has given a short table of the upper α points of the range of k independent Student t variates each with ν d.f. for $\alpha = 0.05$ and 0.10; these critical points (denoted by $R^{(\alpha)}_{\nu,\ldots,\nu}$) provide close approximations (lower bounds) to $R'^{(\alpha)}_{\nu_1,\ldots,\nu_k}$ when $\nu_1 = \cdots = \nu_k = \nu$.

1.2 All Contrasts

1.2.1 The Special Case k = 2 (The Behrens–Fisher Problem)

The Behrens–Fisher problem involves making an inference on $\theta_1 - \theta_2$ when σ_1^2 and σ_2^2 are unknown and unequal. Thus it is a special case of the contrasts problem for $k = 2$. No exact single-stage solution (i.e., an exact α-level single-stage test or a $(1-\alpha)$-level single-stage confidence interval for $\theta_1 - \theta_2$) is available for this problem. However, a variety of approximate solutions have been proposed; for a review, see Lee and Gurland (1975). Perhaps the most popular among these is the one proposed by Welch (1938), which involves approximating the distribution of $\{\bar{Y}_1 - \bar{Y}_2 - (\theta_1 - \theta_2)\}/(S_1^2/n_1 + S_2^2/n_2)^{1/2}$ by Student's t-distribution with estimated (random) d.f. $\hat{\nu}$ given by

$$\hat{\nu} = \frac{(S_1^2/n_1 + S_2^2/n_2)^2}{\{S_1^4/n_1^2(n_1 - 1) + S_2^4/n_2^2(n_2 - 1)\}}\,. \tag{1.11}$$

Wang (1971) made a numerical study of the Welch approximate solution and found that it controls the Type I error rate fairly closely over a wide range of σ_1^2/σ_2^2 values as long as $n_1, n_2 \geqq 6$. Many of the single-stage MCPs for the contrasts problem are extensions of this approximate solution.

1.2.2 The General Case $k \geqq 2$

All of the following procedures are of Tukey-type (except the Brown and Forsythe 1974a procedure, which is of Scheffé-type). In a Tukey-type procedure simultaneous confidence intervals for all pairwise contrasts are first obtained having the following form:

$$\theta_i - \theta_j \in \left[\bar{Y}_i - \bar{Y}_j \pm \xi_{ij}^{(\alpha)} \left(\frac{S_i^2}{n_i} + \frac{S_j^2}{n_j} \right)^{1/2} \right] \quad (1 \leqq i < j \leqq k). \tag{1.12}$$

These intervals can then be extended to all contrasts using Lemma 2.1 of Chapter 3 as follows:

$$\sum_{i=1}^{k} c_i \theta_i \in \left[\sum_{i=1}^{k} c_i \bar{Y}_i \pm \sum_{i=1}^{k} \sum_{j=1}^{k} c_i^+ c_j^- W_{ij}^{(\alpha)} \Big/ \left(\frac{1}{2} \sum_{i=1}^{k} |c_i| \right) \right] \quad \forall \mathbf{c} \in \mathbb{C}^k \tag{1.13}$$

where $\mathbb{C}^k$ is the k-dimensional contrast space $\{\mathbf{c} \in \mathbb{R}^k : \Sigma_{i=1}^k c_i = 0\}$ and

$$W_{ij}^{(\alpha)} = \xi_{ij}^{(\alpha)} \left(\frac{S_i^2}{n_i} + \frac{S_j^2}{n_j} \right)^{1/2} \quad (1 \leqq i < j \leqq k).$$

In (1.12), the $\xi_{ij}^{(\alpha)}$'s are suitably chosen so that the joint confidence level is approximately $1 - \alpha$. We now discuss various procedures that have been proposed for choosing the $\xi_{ij}^{(\alpha)}$'s.

H2-PROCEDURE. In this procedure proposed by Hochberg (1976a) the $\xi_{ij}^{(\alpha)}$'s are set equal to a common value (say) $\xi^{(\alpha)}$ that is determined by using the Welch approximate solution for $k = 2$ and the Bonferroni inequality. This results in the following equation for determining $\xi^{(\alpha)}$:

$$\sum_{1 \leqq i < j \leqq k} \Pr\{|T_{ij}| > \xi^{(\alpha)}\} = \alpha \tag{1.14}$$

where each $T_{ij} = (\bar{Y}_i - \bar{Y}_j)/(S_i^2/n_i + S_j^2/n_j)^{1/2}$ is approximately distributed as a Student's t r.v. with

$$\hat{\nu}_{ij} = \frac{(S_i^2/n_i + S_j^2/n_j)^2}{\{S_i^4/n_i^2(n_i - 1) + S_j^4/n_j^2(n_j - 1)\}} \tag{1.15}$$

d.f. in analogy with (1.11) ($1 \leqq i < j \leqq k$).

Because of the difficulty in numerically solving (1.14), Tamhane (1979) proposed a modified version of the H2-procedure (also proposed earlier independently by Ury and Wiggins 1971) that uses $\xi_{ij}^{(\alpha)} = T_{\hat{\nu}_{ij}}^{(\alpha')}$ where $\alpha' = \alpha/2k^*$ and $k^* = \binom{k}{2}$. Clearly, the $T_{\hat{\nu}_{ij}}^{(\alpha')}$'s are much easier to obtain (e.g., from the tables of Bailey 1977) than solving (1.14). This modified H2-procedure also has the advantage of having a constant per-comparison error rate (PCE). As in the case of the DL-procedure, the average performances of the original and the modified versions of the H2-procedure are comparable and hence the original H2-procedure may be dropped from consideration.

GH- AND C-PROCEDURES. The GH-procedure proposed by Games and Howell (1976) can be regarded as an extension of the TK-procedure to the case of unequal variances; the extension is achieved by using the Welch approximate solution. Thus the GH-procedure uses $\xi_{ij}^{(\alpha)} = Q_{k,\hat{\nu}_{ij}}^{(\alpha)}/\sqrt{2}$ where $\hat{\nu}_{ij}$ is given by (1.15).

In view of the liberal nature of this procedure (see the discussion in Section 1.4), Dunnett (1980b) proposed instead using

$$\xi_{ij}^{(\alpha)} = \frac{Q_{k,\nu_i}^{(\alpha)}(S_i^2/n_i) + Q_{k,\nu_j}^{(\alpha)}(S_j^2/n_j)}{\sqrt{2}(S_i^2/n_i + S_j^2/n_j)} \quad (1 \leqq i < j \leqq k)\,, \tag{1.16}$$

which for $k = 2$ reduces to Cochran's (1964) solution to the Behrens–Fisher problem. Dunnett refers to this procedure as the *C-procedure*. We note that the GH- and C-procedures become identical when all the ν_i's$\rightarrow \infty$. When all the n_i's are equal (to n, say), $\xi_{ij}^{(\alpha)} = Q_{k,n-1}^{(\alpha)}/\sqrt{2}$ for the C-procedure, which is strictly larger (with probability 1) than all the $\xi_{ij}^{(\alpha)} = Q_{k,\hat{\nu}_{ij}}^{(\alpha)}/\sqrt{2}$ for the GH-procedure where the $\hat{\nu}_{ij}$'s are given by (1.15).

T2- AND T3-PROCEDURES. The T2-procedure proposed by Tamhane (1977) employs Šidák's (1967) multiplicative inequality (instead of the Bonferroni inequality used in the H2-procedure) in conjunction with the Welch approximate solution; this yields $\xi_{ij}^{(\alpha)} = T_{\hat{\nu}_{ij}}^{(\alpha'')}$ where $\alpha'' = \frac{1}{2}\{1 - (1-\alpha)^{1/k^*}\}$. Because the Šidák inequality is sharper than the Bonferroni inequality ($T_{\hat{\nu}_{ij}}^{(\alpha'')} < T_{ij}^{(\alpha')}$ for $k > 2$), the T2 – procedure always provides narrower confidence intervals than the H2-procedure.

In Tamhane's (1979) simulation study the T2-procedure was found to be conservative; therefore the modified H2-procedure would be even more conservative and may be dropped from consideration. To reduce the conservatism of the T2-procedure, Dunnett (1980b) proposed a modification based on the Kimball (1951) inequality, which leads to the choice $\xi_{ij}^{(\alpha)} = |M|_{k^*, \hat{\nu}_{ij}}^{(\alpha)}$. For $k > 2$ this critical constant is smaller than the critical constant $T_{\hat{\nu}_{ij}}^{(\alpha'')}$ used in the T2-procedure. Dunnett refers to this modified procedure as the *T3-procedure*. Note that T2 and T3 become identical when all the ν_i's$\rightarrow \infty$.

BF-PROCEDURE. This procedure proposed by Brown and Forsythe (1974a) uses Scheffé's projection method to obtain the following approximate $(1-\alpha)$-level simultaneous confidence intervals for all contrasts $\sum_{i=1}^{k} c_i\theta_i$:

$$\sum_{i=1}^{k} c_i\theta_i \in \left[\sum_{i=1}^{k} c_i\bar{Y}_i \pm \{(k-1)F_{k-1,\hat{\nu}_c}^{(\alpha)}\}^{1/2}\left\{\sum_{i=1}^{k} \frac{c_i^2 S_i^2}{n_i}\right\}^{1/2}\right] \quad \forall \mathbf{c} \in \mathbb{C}^k \tag{1.17}$$

where

$$\hat{\nu}_c = \frac{\left(\sum_{i=1}^{k} [c_i^2 S_i^2/n_i]\right)^2}{\left\{\sum_{i=1}^{k} [c_i^4 S_i^4/n_i^2(n_i-1)]\right\}} . \tag{1.18}$$

For pairwise contrasts (1.17) reduces to the general form (1.12) with $\xi_{ij}^{(\alpha)} = \{(k-1)F_{k-1,\hat{\nu}_{ij}}^{(\alpha)}\}^{1/2}$. Closely related procedures have been proposed by Naik (1967).

Example 1.1. The data in Table 1.1 on the amounts of different types of fat absorbed by doughnuts during cooking are taken from Snedecor and Cochran (1976, p. 259). We assume that the one-way layout model holds for these data.

Suppose that it is desired to make all pairwise comparisons between the four types of fat. If the variances σ_i^2 are not assumed to be *a priori* equal, then a possible approach is to first perform a preliminary test of homogeneity on the variances using a test such as the Bartlett test. If the test does not reject the null hypothesis, then one may wish to assume equal variances and use an appropriate MCP (e.g., the T-procedure in the present example). If the test does reject the null hypothesis, then one may wish to use an MCP such as the T3-procedure, which is designed for

TABLE 1.1. Fat Absorbed per Batch in Grams

	Fat			
	1	2	3	4
	164	178	175	155
	172	191	193	166
	168	197	178	149
	177	182	171	164
	156	185	163	170
	195	177	176	168
$\bar{Y}_i$	172	185	176	162
S_i^2	178	60	98	68
n_i	6	6	6	6
ν_i	5	5	5	5
$S_i/\sqrt{n_i}$	5.45	3.16	4.04	3.37

Source: Snedecor and Cochran (1976).

unequal variances. However, the properties of such a composite procedure involving a preliminary test on the variances are unknown. If the preliminary test is not very powerful, then the null hypothesis will be accepted with high probability even when the variances are unequal, resulting in a "wrong" MCP being used, which will have excessive Type I error rates.

If a guaranteed protection against the Type I FWE is desired, then one may wish to use an MCP for unequal variances without performing any preliminary test on the variances. If the variances are in fact equal, then such an MCP may not be as powerful as an MCP that is based on the homogeneous variances assumption.

In the present example the Bartlett test for the equality of variances leads to the retention of the null hypothesis (P-value>0.50). However, it is known that for small sample sizes this test is not very powerful. Therefore we may wish to ignore this nonsignificant test outcome and apply one of the MCPs for unequal variances.

For the family of pairwise comparisons we only illustrate the use of T3-, GH- and C-procedures since other procedures are not competitive in this case. We use $\alpha = 0.05$.

T3-Procedure. Using (1.15) we calculate $\hat{\nu}_{12} = 8.03$, $\hat{\nu}_{13} = 9.22$, $\hat{\nu}_{14} = 8.33$, $\hat{\nu}_{23} = 9.45$, $\hat{\nu}_{24} = 9.96$, and $\hat{\nu}_{34} = 9.68$. The critical constants $|M|_{k^*, \hat{\nu}_{ij}}^{(\alpha)}$ for $\alpha = 0.05$ and $k^* = \binom{4}{2} = 6$ are obtained by linear interpolation in $1/\nu$ from Table 7 of Appendix 3. Using these critical constants, simultaneous

95% confidence intervals for all pairwise differences are computed as follows:

$$\theta_1 - \theta_2 : 172 - 185 \pm 3.361\left(\frac{178}{6} + \frac{60}{6}\right)^{1/2} = -13 \pm 21.17$$

$$\theta_1 - \theta_3 : 172 - 176 \pm 3.255\left(\frac{178}{6} + \frac{98}{6}\right)^{1/2} = -4 \pm 22.08$$

$$\theta_1 - \theta_4 : 172 - 162 \pm 3.332\left(\frac{178}{6} + \frac{68}{6}\right)^{1/2} = 10 \pm 21.34$$

$$\theta_2 - \theta_3 : 185 - 176 \pm 3.237\left(\frac{60}{6} + \frac{98}{6}\right)^{1/2} = 9 \pm 16.61$$

$$\theta_2 - \theta_4 : 185 - 162 \pm 3.202\left(\frac{60}{6} + \frac{68}{6}\right)^{1/2} = 23 \pm 14.79$$

$$\theta_3 - \theta_4 : 176 - 162 \pm 3.221\left(\frac{98}{6} + \frac{68}{6}\right)^{1/2} = 14 \pm 16.94 .$$

Thus only the difference $\theta_2 - \theta_4$ is found significant using this procedure.

GH-Procedure. For this procedure the critical constants are $Q^{(\alpha)}_{k,\hat{\nu}_{ij}}/\sqrt{2}$ where the $Q^{(\alpha)}_{k,\hat{\nu}_{ij}}$'s are obtained by linear interpolation in $1/\nu$ in Table 8 of Appendix 3. Using these critical constants, simultaneous 95% confidence intervals for all pairwise differences are calculated as follows:

$$\theta_1 - \theta_2 : 172 - 185 \pm \left(\frac{4.525}{\sqrt{2}}\right)\left(\frac{178}{6} + \frac{60}{6}\right)^{1/2} = -13 \pm 20.15$$

$$\theta_1 - \theta_3 : 172 - 176 \pm \left(\frac{4.395}{\sqrt{2}}\right)\left(\frac{178}{6} + \frac{98}{6}\right)^{1/2} = -4 \pm 21.08$$

$$\theta_1 - \theta_4 : 172 - 162 \pm \left(\frac{4.488}{\sqrt{2}}\right)\left(\frac{178}{6} + \frac{68}{6}\right)^{1/2} = 10 \pm 20.32$$

$$\theta_2 - \theta_3 : 185 - 176 \pm \left(\frac{4.373}{\sqrt{2}}\right)\left(\frac{60}{6} + \frac{98}{6}\right)^{1/2} = 9 \pm 15.87$$

$$\theta_2 - \theta_4 : 185 - 162 \pm \left(\frac{4.330}{\sqrt{2}}\right)\left(\frac{60}{6} + \frac{68}{6}\right)^{1/2} = 23 \pm 14.14$$

$$\theta_3 - \theta_4 : 176 - 162 \pm \left(\frac{4.353}{\sqrt{2}}\right)\left(\frac{98}{6} + \frac{68}{6}\right)^{1/2} = 14 \pm 16.19 .$$

Once again, only the difference $\theta_2 - \theta_4$ comes out significant. Note that because $Q^{(\alpha)}_{k,\hat{\nu}_{ij}}/\sqrt{2} < |M|^{(\alpha)}_{k,\hat{\nu}_{ij}}$ for $k > 2$, the intervals obtained using the GH-procedure are uniformly shorter than those obtained using the T3-procedure. However, as noted earlier, the GH-procedure may be liberal.

C-Procedure. Since all the ν_i's are equal, as noted in the discussion of the C-procedure, the critical constants $\xi^{(\alpha)}_{ij}$ are the same for all pairwise comparisons and this common value equals $Q^{(0.05)}_{4,5}/\sqrt{2} = 5.22/\sqrt{2}$. This will yield intervals that are uniformly longer than those given by either the T3- or the GH-procedure.

Now let us suppose that after looking at the data, the experimenter decided to estimate the contrast $(\theta_2 + \theta_3)/2 - (\theta_1 + \theta_4)/2$ because it corresponds to the comparison between the average for treatments with high sample means with that for treatments with low sample means. For the T3-procedure a 95% confidence interval for this contrast using (1.13) is

$$\frac{185 + 176}{2} - \frac{172 + 162}{2} \pm (20.15 + 21.05 + 14.14 + 16.19) = 13.5 \pm 71.53\,.$$

Let us next illustrate the use of the procedures for all linear combinations and also the use of the BF-procedure to estimate this contrast.

DL-Procedure. For $p = 1$, $q = \infty$, we have $D^{(\alpha)}_{\infty;\nu,\ldots,\nu} = T^{(\alpha^*)}_{\nu} = 3.791$ for $\nu = 5$ and $\alpha = 0.05$ ($\alpha^* = \frac{1}{2}\{1 - (1-\alpha)^{1/4}\} = 0.00637$) using the tables of Games (1977) or by interpolating in Table 1 of Appendix 3. Thus from (1.4) we obtain the following 95% confidence interval for the contrast $(\theta_2 + \theta_3)/2 - (\theta_1 + \theta_4)/2$:

$$\frac{185 + 176}{2} - \frac{172 + 162}{2} \pm 3.791(5.45 + 3.16 + 4.04 + 3.37) = 13.5 \pm 60.73\,.$$

SP-Procedure. We use the approximation (1.7) where for $\nu_1 = \cdots = \nu_k = 5$ we get $a = 40/9$ and $b = 6$ from (1.9) and (1.8), respectively. Thus $D^{(0.05)}_{2;5,\ldots,5} \cong ((40/9)F^{(0.05)}_{4,6})^{1/2} = ((40/9) \times 4.53)^{1/2} = 4.49$. Then from (1.4) we obtain the following 95% confidence interval for the contrast $(\theta_2 + \theta_3)/2 - (\theta_1 + \theta_4)/2$:

$$\frac{185 + 176}{2} - \frac{172 + 162}{2} \pm 4.49\left(\frac{178}{6} + \frac{60}{6} + \frac{98}{6} + \frac{68}{6}\right)^{1/2} = 13.5 \pm 36.84\,.$$

H1-Procedure. To implement this procedure we require the value of the critical point $R'^{(\alpha)}_{\nu,\ldots,\nu}$, which is not tabulated. Instead we use the lower bound $R^{(\alpha)}_{\nu,\ldots,\nu}$, which equals 5.05 for $k = 4$, $\nu = 5$, and $\alpha = 0.05$ from the table provided by Hochberg (1976a); note that this results in a slightly shorter (and possibly liberal) interval. From (1.10) we obtain the following 95% confidence interval for the contrast $(\theta_2 + \theta_3)/2 - (\theta_1 + \theta_4)/2$:

$$\frac{185 + 176}{2} - \frac{172 + 162}{2} \pm 5.05 \max(3.16 + 4.04, 5.45 + 3.37)$$
$$= 13.5 \pm 44.54 .$$

BF-Procedure. From (1.18) we calculate $\hat{\nu}_{\mathbf{c}} = 16.48$ for $\mathbf{c} = (-0.5, +0.5, +0.5, -0.5)'$. By linear interpolation in $1/\nu$ in the F-table we obtain $F^{(0.05)}_{3,\,16.48} \cong 3.22$. Thus from (1.17) we obtain the following 95% confidence interval for the contrast $(\theta_2 + \theta_3)/2 - (\theta_1 + \theta_4)/2$:

$$\frac{185 + 176}{2} - \frac{172 + 162}{2} \pm (3 \times 3.22)^{1/2}\left(\frac{178}{6} + \frac{60}{6} + \frac{98}{6} + \frac{68}{6}\right)^{1/2}$$
$$= 13.5 \pm 25.50 .$$

We see that the BF-procedure provides the shortest interval and the SP-procedure the next shortest for the contrast of interest in this example. This holds generally for higher order contrasts. Also note that the T3-procedure provides the longest interval for this high order contrast, although it generally provides the shortest intervals for pairwise contrasts. □

1.3 Comparisons with a Control

If the kth treatment is a control and simultaneous confidence intervals are desired for contrasts $\theta_i - \theta_k$ $(1 \leqq i \leqq k-1)$, then some of the procedures described above can be readily modified to address this family. For simultaneous two-sided intervals on $\theta_i - \theta_k$ $(1 \leqq i \leqq k-1)$, the H2-, T2- and T3-procedures can be modified by simply replacing $k^* = \binom{k}{2}$ by $k-1$, which is the new number of comparisons. For obtaining simultaneous one-sided intervals on $\theta_i - \theta_k$ $(1 \leqq i \leqq k-1)$, in addition to this modification, one must use $2\alpha'$ and $2\alpha''$ in the H2- and T2-procedures, respectively. In the case of the T3-procedure, the critical constant $|M|^{(\alpha)}_{k-1,\,\hat{\nu}_{ik}}$ must be replaced by the corresponding critical constant $M^{(\alpha)}_{k-1,\,\hat{\nu}_{ik}}$ from the Studentized maximum distribution.

1.4 A Comparison of Procedures

We first consider the MCPs that address the family of all linear combinations of the θ_i's. It was shown in Tamhane's (1979) simulation study that the DL- and H1-procedures are highly conservative not only for pairwise contrasts but also for higher order contrasts. That simulation study also showed that the SP-procedure, although very conservative for pairwise contrasts, offers intervals for higher order contrasts that are shorter than those yielded by all other procedures except the BF-procedure. The conservatism of the DL- and H1-procedures even for higher order contrasts would seem to rule out their use in most applications. The SP-procedure also cannot be recommended based only on its second best performance for higher order contrasts.

For pairwise contrasts, the only contenders are the T3-, C- and GH-procedures. The GH-procedure yields shorter intervals than the T3-procedure because of the inequality $|M|^{(\alpha)}_{k^*, \hat{\nu}_{ij}} > Q^{(\alpha)}_{k, \hat{\nu}_{ij}}/\sqrt{2}$ for $k > 2$. However, as alluded to earlier, the GH-procedure can sometimes be liberal; see the simulation results of Keselman and Rogan (1978), Tamhane (1979), and Dunnett (1980b). Thus narrower confidence intervals (higher power) associated with the GH-procedure are obtained at the expense of occasionally excessive Type I FWE. Dunnett's (1980b) results indicate that the GH-procedure is most liberal when the variances of the sample means, σ_i^2/n_i, are approximately equal.

While the GH-procedure is sometimes liberal, the C- and T3-procedures are found to be always conservative (and hence the T2-procedure, too). Between the T3- and C-procedures, the former gives better performance for pairwise contrasts when the ν_i's are small, while the latter gives better performance when the ν_i's are moderately large.

The BF-procedure is very conservative for pairwise contrasts but for higher order contrasts it gives the best performance among all procedures.

To summarize, for pairwise and lower order contrasts the T3-procedure is recommended when the sample sizes are small while the C-procedure is recommended when the sample sizes are large. It should be mentioned that when the sample sizes are not too small, these procedures do not lose much in terms of power if, unknown to the experimenter, the variances are in fact equal or nearly equal. The GH-procedure may be used in these same situations with the benefit of narrower confidence intervals if slightly excessive Type I FWE can be tolerated. For higher order contrasts we recommend the BF-procedure.

2 TWO-STAGE PROCEDURES

2.1 Preliminaries

As noted earlier, none of the single-stage MCPs for the contrasts problem are exact. It is, however, possible to construct exact MCPs for this problem using a two-stage sampling scheme due to Stein (1945). An important feature of the resulting two-stage MCPs is that they yield *fixed* width (which can be prespecified by the experimenter) simultaneous confidence intervals for the contrasts of interest, as opposed to the *random* width intervals associated with single-stage MCPs. Such a use of Stein-type procedures was made earlier in Chapter 6, where the error variances were assumed to be unknown but equal. We need the following analog of Theorem 2.1 of Chapter 6, which is suitable for unequal variances.

Theorem 2.1 (Stein 1945). Let the Y_i's ($i = 1, 2, \ldots$) be independent and identically distributed $N(\mu, \sigma^2)$ r.v.'s. Let S^2 be an unbiased estimate of σ^2 that is distributed as $\sigma^2\chi^2_\nu/\nu$ independently of $\Sigma_{i=1}^n Y_i$ and $Y_{n+1}, Y_{n+2}, \ldots$ where n is some fixed positive integer. Let

$$N = \max\left\{\left\lfloor \frac{S^2}{\xi^2} \right\rfloor, n+1\right\}$$

where $\xi > 0$ is an arbitrary constant and $\lfloor x \rfloor$ denotes the smallest integer $\geqq x$. Then there exist real numbers $l_1, l_2, \ldots, l_N$ satisfying

$$\sum_{i=1}^N l_i = 1\,, \qquad l_1 = l_2 = \cdots = l_n\,, \qquad \text{and} \qquad S^2 \sum_{i=1}^N l_i^2 = \xi^2$$

such that

$$\frac{\left(\sum_{i=1}^N l_i Y_i - \mu\right)}{\xi} = \frac{\left(\sum_{i=1}^N l_i Y_i - \mu\right)}{\left(S^2 \sum_{i=1}^N l_i^2\right)^{1/2}}$$

has Student's t-distribution with ν d.f. □

The use of this result in providing a solution to the Behrens–Fisher problem was first made by Chapman (1950) and later by Ghosh (1975). In multiple comparison problems involving variance heterogeneity this result was used by Dudewicz and Dalal (1975); see also Healy (1956).

Two-stage MCPs can be based on ordinary sample means or generalized sample means (see (2.2b) below). In the former case Theorem 2.1 of Chapter 6 is used to claim the desired Student's t-distribution result, while in the latter case Theorem 2.1 of the present chapter is used. The sampling scheme for any two-stage MCP based either on sample means or on generalized sample means is basically the same. It is described here for convenience, and is not repeated for each individual MCP.

Two-Stage Sampling Scheme

Stage 1. Take a random sample of size n_i from the ith treatment and calculate the sample variance S_i^2 based on $\nu_i = n_i - 1$ d.f. $(1 \leqq i \leqq k)$. Determine the total sample size N_i to be taken on the ith treatment using the formula

$$N_i = \max\left\{n_i, \left\lfloor \frac{S_i^2}{\xi^2} \right\rfloor\right\} \quad \text{(for a sample mean MCP)} \tag{2.1a}$$

or

$$N_i = \max\left\{n_i + 1, \left\lfloor \frac{S_i^2}{\xi^2} \right\rfloor\right\} \quad \text{(for a generalized sample mean MCP)}\,. \tag{2.1b}$$

Here $\xi > 0$ is a predetermined constant that depends on the MCP to be employed, the joint confidence level $1 - \alpha$, the d.f. ν_i, and the desired fixed width associated with each interval.

Stage 2. Take an additional random sample of size $N_i - n_i$ from the ith treatment and calculate for $1 \leqq i \leqq k$,

$$\bar{Y}_i = \frac{\sum_{j=1}^{N_i} Y_{ij}}{N_i} \quad \text{(for a sample mean MCP)} \tag{2.2a}$$

or

$$\tilde{Y}_i = \sum_{j=1}^{N_i} l_{ij} Y_{ij} \quad \text{(for a generalized sample mean MCP)} \tag{2.2b}$$

where the l_{ij}'s satisfy

$$\sum_{j=1}^{N_i} l_{ij} = 1\,, \qquad l_{i1} = l_{i2} = \cdots = l_{in_i}\,,$$

$$S_i^2 \sum_{j=1}^{N_i} l_{ij}^2 = \xi^2 \qquad (1 \leqq i \leqq k)\,.$$

Use the sample means $\bar{Y}_i$ or the generalized sample means $\tilde{Y}_i$ in the appropriate MCP.

Dudewicz, Ramberg, and Chen (1975) have recommended the following choice of the l_{ij}'s for generalized sample mean MCPs:

$$l_{i1} = \cdots = l_{in_i} = l_i \text{ (say)},$$

$$l_{i,n_i+1} = \cdots = l_{iN_i} = \left(\frac{1 - n_i l_i}{N_i - n_i}\right),$$

where

$$l_i = \frac{1}{N_i}\left\{1 + \left[1 - \frac{N_i}{n_i}\left(1 - \frac{N_i - n_i}{(S_i/\xi)^2}\right)\right]^{1/2}\right\}. \tag{2.3}$$

With this choice the calculation of the generalized sample mean $\tilde{Y}_i$ given by (2.2b) simplifies to

$$\tilde{Y}_i = l_i n_i \bar{Y}_i^{(1)} + (1 - l_i n_i)\bar{Y}_i^{(2)} \tag{2.4}$$

where for the ith treatment, $\bar{Y}_i^{(1)}$ is the sample mean based on n_i observations from the first stage and $\bar{Y}_i^{(2)}$ is the sample mean based on $N_i - n_i$ observations from the second stage $(1 \leq i \leq k)$.

2.2 All Linear Combinations

2.2.1 *Procedures Based on Sample Means*

Let $D^{(\alpha)}_{2;\,\nu_1,\ldots,\nu_k}$ be as defined in Section 1.1. Then simultaneous confidence intervals for all linear combinations $\Sigma_{i=1}^k a_i\theta_i$ with confidence level at least $1 - \alpha$ are given by

$$\sum_{i=1}^k a_i\theta_i \in \left[\sum_{i=1}^k a_i\tilde{Y}_i \pm \xi D^{(\alpha)}_{2;\,\nu_1,\ldots,\nu_k}\left(\sum_{i=1}^k a_i^2\right)^{1/2}\right] \quad \forall\, \mathbf{a} \in \mathbb{R}^k. \tag{2.5}$$

If the a_i's are normalized so that $\Sigma_{i=1}^k a_i^2 = 1$, then to obtain a common fixed width $= 2W$ for all intervals (2.5), we choose $\xi = W/D^{(\alpha)}_{2;\,\nu_1,\ldots,\nu_k}$ in (2.1a) for the two-stage sampling scheme. An approximation to $D^{(\alpha)}_{2;\,\nu_1,\ldots,\nu_k}$ is given by (1.7); if $n_1 = \cdots = n_k = n$ (say), then a and b in that approximation simplify to

$$a = \frac{k(n-1)(b-2)}{b(n-3)}, \qquad b = \frac{(k+2)n - 5k + 2}{3}.$$

That the intervals (2.5) have a joint confidence level at least $1-\alpha$ follows because

$$\Pr\left\{\sum_{i=1}^{k} a_i\theta_i \in \left[\sum_{i=1}^{k} a_i\bar{Y}_i \pm \xi D^{(\alpha)}_{2;\,\nu_1,\ldots,\nu_k}\left(\sum_{i=1}^{k} a_i^2\right)^{1/2}\right] \ \forall\, \mathbf{a}\in\mathbb{R}^k\right\}$$

$$\geqq \Pr\left\{\left|\sum_{i=1}^{k} a_i\left(\frac{\bar{Y}_i-\theta_i}{S_i/\sqrt{N_i}}\right)\right| \leqq D^{(\alpha)}_{2;\,\nu_1,\ldots,\nu_k}\left(\sum_{i=1}^{k} a_i^2\right)^{1/2} \ \forall\, \mathbf{a}\in\mathbb{R}^k\right\}$$

(since $S_i/\sqrt{N_i} \leqq \xi \ \forall\, i$ from (2.1))

$$= \Pr\left\{\left|\sum_{i=1}^{k} a_i T_{\nu_i}\right| \leqq D^{(\alpha)}_{2;\,\nu_1,\ldots,\nu_k}\left(\sum_{i=1}^{k} a_i^2\right)^{1/2} \ \forall\, \mathbf{a}\in\mathbb{R}^k\right\}$$

$$= \Pr\left\{\left(\sum_{i=1}^{k} T_{\nu_i}^2\right)^{1/2} \leqq D^{(\alpha)}_{2;\,\nu_1,\ldots,\nu_k}\right\} \qquad \text{(using (1.3))}$$

$$= 1-\alpha\,.$$

Hochberg (1975b) proposed a Tukey-type two-stage procedure based on Lemma 2.2 of Chapter 3 that yields the following simultaneous confidence intervals for all linear combinations $\Sigma_{i=1}^{k} a_i\theta_i$:

$$\sum_{i=1}^{k} a_i\theta_i \in \left[\sum_{i=1}^{k} a_i\bar{Y}_i \pm \xi R'^{(\alpha)}_{\nu_1,\ldots,\nu_k} M(\mathbf{b})\right] \qquad \forall\, \mathbf{a}\in\mathbb{R}^k \tag{2.6}$$

where $b_i = a_iS_i/\sqrt{N_i}$ $(1 \leqq i \leqq k)$ and $M(\mathbf{b})$ and $R'^{(\alpha)}_{\nu_1,\ldots,\nu_k}$ are as defined in Section 1.1 in the context of the H1-procedure. Comments on how to approximate $R'^{(\alpha)}_{\nu_1,\ldots,\nu_k}$ are also provided in Section 1.1.

2.2.2 *Procedures Based on Generalized Sample Means*

One can use the generalized sample means $\tilde{Y}_i$ in place of the sample means $\bar{Y}_i$ in (2.6) and obtain simultaneous confidence intervals for all linear combinations $\Sigma_{i=1}^{k} a_i\theta_i$ with a joint confidence level exactly equal to $1-\alpha$. This procedure was proposed by Tamhane (1977).

Bishop (1979) proposed a related MCP for the family of all linear combinations of the treatment effects α_i where $\alpha_i = \theta_i - (1/k)\,\Sigma_{i=1}^{k}\theta_i$ $(1 \leqq i \leqq k)$, which is equivalent to the family of all contrasts of the θ_i's. Let $V^{(\alpha)}_{\nu_1,\nu_2,\ldots,\nu_k}$ denote the upper α point of the distribution of

$$V_{\nu_1,\nu_2,\ldots,\nu_k} = \left\{\sum_{i=1}^{k}(T_{\nu_i}-\bar{T})^2\right\}^{1/2}$$

where the T_{ν_i}'s are independent Student t r.v.'s $(1 \leqq i \leqq k)$ and $\bar{T} =$

$(1/k)\sum_{i=1}^{k} T_{\nu_i}$. Then exact $(1-\alpha)$-level simultaneous confidence intervals for all linear combinations $\sum_{i=1}^{k} a_i\alpha_i$ are given by

$$\sum_{i=1}^{k} a_i\alpha_i \in \left[\sum_{i=1}^{k} a_i(\tilde{Y}_i - \bar{\tilde{Y}}) \pm \xi V_{\nu_1,\ldots,\nu_k}^{(\alpha)}\left(\sum_{i=1}^{k} a_i^2\right)^{1/2}\right] \quad \forall \mathbf{a} \in \mathbb{R}^k \tag{2.7}$$

where $\bar{\tilde{Y}} = (1/k)\sum_{i=1}^{k} \tilde{Y}_i$. If the a_i's are normalized so that $\sum_{i=1}^{k} a_i^2 = 1$, then to obtain a common fixed width $= 2W$ for all intervals (2.7), we choose $\xi = W/V_{\nu_1,\ldots,\nu_k}^{(\alpha)}$ in (2.1b) in the two-stage sampling scheme.

That the intervals (2.7) have a joint confidence level exactly equal to $1-\alpha$ follows because

$$\Pr\left\{\sum_{i=1}^{k} a_i\alpha_i \in \left[\sum_{i=1}^{k} a_i(\tilde{Y}_i - \bar{\tilde{Y}}) \pm \xi V_{\nu_1,\ldots,\nu_k}^{(\alpha)}\left(\sum_{i=1}^{k} a_i^2\right)^{1/2}\right] \ \forall \mathbf{a} \in \mathbb{R}^k\right\}$$

$$= \Pr\left\{\left|\frac{\sum_{i=1}^{k} a_i(\tilde{Y}_i - \bar{\tilde{Y}} - \alpha_i)}{\xi}\right| \le V_{\nu_1,\ldots,\nu_k}^{(\alpha)}\left(\sum_{i=1}^{k} a_i^2\right)^{1/2} \ \forall \mathbf{a} \in \mathbb{R}^k\right\}$$

$$= \Pr\left\{\left|\sum_{i=1}^{k} a_i(T_{\nu_i} - \bar{T})\right| \le V_{\nu_1,\ldots,\nu_k}^{(\alpha)}\left(\sum_{i=1}^{k} a_i^2\right)^{1/2} \ \forall \mathbf{a} \in \mathbb{R}^k\right\}$$

$$= \Pr\left\{\left[\sum_{i=1}^{k} (T_{\nu_i} - \bar{T})^2\right]^{1/2} \le V_{\nu_1,\ldots,\nu_k}^{(\alpha)}\right\} \quad \text{(using (1.3))}$$

$$= 1 - \alpha\,.$$

Tables of $(V_{\nu_1,\ldots,\nu_k}^{(\alpha)})^2$ for selected values of $\nu_1 = \cdots = \nu_k = \nu$ (say) and α have been given by Bishop et al. (1978). For other cases the following approximation suggested by Hochberg (1975b) can be used:

$$V_{\nu_1,\ldots,\nu_k}^{(\alpha)} \cong (aF_{a,b}^{(\alpha)})^{1/2}$$

where a and b are obtained by matching the first two moments of $V_{\nu_1,\ldots,\nu_k}^2$ with those of $aF_{a,b}$. When $\nu_1 = \cdots = \nu_k = \nu$, this results in the following equations for a and b:

$$\frac{ab}{b-2} = (k-1)\frac{\nu}{\nu-2}$$

and

$$\frac{b^2a(a+2)}{(b-2)(b-4)} = \frac{\nu^2(k-1)}{(\nu-2)k}\left[\frac{3(k-1)}{\nu-4} + \frac{k^2-2k+3}{\nu-2}\right].$$

2.3 All Contrasts

2.3.1 Procedures Based on Sample Means

Hochberg (1975b) proposed an MCP based on sample means that gives the following simultaneous confidence intervals for all contrasts $\Sigma_{i=1}^{k} c_i\theta_i$ with joint confidence level at least $1-\alpha$:

$$\sum_{i=1}^{k} c_i\theta_i \in \left[\sum_{i=1}^{k} c_i\bar{Y}_i \pm \xi V_{\nu_1,\ldots,\nu_k}^{(\alpha)}\left(\sum_{i=1}^{k} c_i^2\right)^{1/2}\right] \quad \forall \mathbf{c} \in \mathbb{C}^k \tag{2.8}$$

where $V_{\nu_1,\ldots,\nu_k}^{(\alpha)}$ is as defined in the preceding section. If the c_i's are normalized so that $\Sigma_{i=1}^{k} c_i^2 = 2$, then to obtain a common fixed width $= 2W$ for all intervals (2.8) we choose $\xi = W/\sqrt{2}V_{\nu_1,\ldots,\nu_k}^{(\alpha)}$ in (2.1a) in the two-stage sampling scheme.

The following probability calculation shows that the intervals (2.8) have a joint confidence level at least $1-\alpha$:

$$\begin{aligned}
&\Pr\left\{\sum_{i=1}^{k} c_i\theta_i \in \left[\sum_{i=1}^{k} c_i\bar{Y}_i \pm \xi V_{\nu_1,\ldots,\nu_k}^{(\alpha)}\left(\sum_{i=1}^{k} c_i^2\right)^{1/2}\right] \forall \mathbf{c} \in \mathbb{C}^k\right\} \\
&\geqq \Pr\left\{\left|\sum_{i=1}^{k} c_i\left(\frac{\bar{Y}_i - \theta_i}{S_i/\sqrt{N_i}}\right)\right| \leqq V_{\nu_1,\ldots,\nu_k}^{(\alpha)}\left(\sum_{i=1}^{k} c_i^2\right)^{1/2} \forall \mathbf{c} \in \mathbb{C}^k\right\} \\
&\qquad (\text{since } S_i/\sqrt{N_i} \leqq \xi \ \forall i \text{ from (2.1a)}) \\
&= \Pr\{|\mathbf{c}'\mathbf{T}| \leqq V_{\nu_1,\ldots,\nu_k}^{(\alpha)}(\mathbf{c}'\mathbf{c})^{1/2} \ \forall \mathbf{c} \in \mathbb{C}^k\}
\end{aligned} \tag{2.9}$$

where $\mathbf{T} = (T_{\nu_1}, \ldots, T_{\nu_k})'$ and the T_{ν_i}'s are independent Student t r.v.'s. Now let $\mathbf{A} = \mathbf{I} - (1/k)\mathbf{J}$ where $\mathbf{I}$ is the identity matrix and $\mathbf{J}$ is the matrix of all 1's, both matrices being of order k. Then for any contrast $\mathbf{c} \in \mathbb{C}^k$, $\mathbf{Ac} = \mathbf{c}$ and $\mathbf{A}$ is idempotent. By the Cauchy–Schwartz inequality we have

$$|\mathbf{c}'\mathbf{T}| = |\mathbf{c}'\mathbf{A}'\mathbf{T}| \leqq (\mathbf{c}'\mathbf{c})^{1/2}(\mathbf{T}'\mathbf{A}\mathbf{T})^{1/2} = (\mathbf{c}'\mathbf{c})^{1/2}\left\{\sum_{i=1}^{k}(T_{\nu_i} - \bar{T})^2\right\}^{1/2},$$

and therefore it is readily seen that (2.9) is bounded below by $1-\alpha$.

2.3.2 Procedures Based on Generalized Sample Means

Tamhane (1977) proposed an MCP based on generalized sample means that gives the following simultaneous confidence intervals for all contrasts $\Sigma_{i=1}^{k} c_i\theta_i$ with confidence level exactly $1-\alpha$:

$$\sum_{i=1}^{k} c_i\theta_i \in \left[\sum_{i=1}^{k} c_i\tilde{Y}_i \pm \xi R_{\nu_1,\ldots,\nu_k}^{(\alpha)} \sum_{i=1}^{k} \frac{|c_i|}{2}\right] \quad \forall \mathbf{c} \in \mathbb{C}^k \tag{2.10}$$

where $R^{(\alpha)}_{\nu_1,\ldots,\nu_k}$ is the upper α point of the range of k independent Student t r.v.'s $T_{\nu_1}, T_{\nu_2}, \ldots, T_{\nu_k}$. If the c_i's are normalized so that $\Sigma_{i=1}^k |c_i| = 2$, then to obtain a common fixed width $= 2W$ for all intervals in (2.10) we choose $\xi = W/R^{(\alpha)}_{\nu_1,\ldots,\nu_k}$ in (2.1b) in the two-stage sampling scheme. As noted earlier in Section 1.1, Hochberg (1976a) has provided a short table of $R^{(\alpha)}_{\nu_1,\ldots,\nu_k}$ for the special case $\nu_1 = \cdots = \nu_k$.

That (2.10) provides exact $(1-\alpha)$-level simultaneous confidence intervals for all contrasts $\Sigma_{i=1}^k c_i\theta_i$ can be shown using arguments similar to those given before.

2.4 Comparisons with a Control

In this section we assume that treatment k is the control treatment with which the remaining $k-1$ treatments are to be compared. Thus it is desired to construct simultaneous confidence intervals (one-sided or two-sided) for $\theta_i - \theta_k$ $(1 \leq i \leq k-1)$. In the literature only MCPs based on generalized sample means have been proposed for this problem although it is not difficult to develop their sample mean analogs.

2.4.1 One-Sided Comparisons

Dudewicz and Ramberg (1972) and Dudewicz, Ramberg, and Chen (1975) proposed the following exact $(1-\alpha)$-level simultaneous one-sided confidence intervals for all differences $\theta_i - \theta_k$:

$$\theta_i - \theta_k \leq \tilde{Y}_i - \tilde{Y}_k + \xi H^{(\alpha)}_{\nu_1,\ldots,\nu_k} \qquad (1 \leq i \leq k-1) \tag{2.11}$$

where $H^{(\alpha)}_{\nu_1,\ldots,\nu_k}$ is the upper α point of the distribution of $H_{\nu_1,\ldots,\nu_k} = \max_{1\leq i\leq k-1}(T_{\nu_i} - T_{\nu_k})$ where the T_{ν_i}'s are independent Student t r.v.'s. If it is desired to have a common fixed "width" (or allowance) $= W$ for all intervals (2.11), then we use $\xi = W/H^{(\alpha)}_{\nu_1,\ldots,\nu_k}$ in (2.1b) in the two-stage sampling scheme.

It is readily seen that $H^{(\alpha)}_{\nu_1,\ldots,\nu_k}$ is the solution in H to the equation

$$\int_{-\infty}^{\infty} \prod_{i=1}^{k-1} F_{\nu_i}(x+H)\, dF_{\nu_k}(x) = 1-\alpha \tag{2.12}$$

where $F_\nu(\cdot)$ is the c.d.f. of Student's t-distribution with ν d.f. Dudewicz, Ramberg, and Chen (1975) have tabulated the solution to this equation for the special case $\nu_1 = \cdots = \nu_k = \nu$ (say) for selected values of k, α, and ν.

2.4.2 Two-Sided Comparisons

Dudewicz and Dalal (1983) proposed the following exact $(1-\alpha)$-level simultaneous two-sided confidence intervals for all differences $\theta_i - \theta_k$

$(1 \leq i \leq k-1)$:

$$\theta_i - \theta_k \in [\tilde{Y}_i - \tilde{Y}_k \pm \xi H'^{(\alpha)}_{\nu_1,\ldots,\nu_k}] \qquad (1 \leq i \leq k-1) \qquad (2.13)$$

where $H'^{(\alpha)}_{\nu_1,\ldots,\nu_k}$ is the upper α point of the distribution of $H'_{\nu_1,\ldots,\nu_k} = \max_{1 \leq i \leq k-1} |T_{\nu_i} - T_{\nu_k}|$. If it is desired to have a common fixed width $= 2W$ for all intervals (2.13), then we use $\xi = W/H'^{(\alpha)}_{\nu_1,\ldots,\nu_k}$ in (2.1b) in the two-stage sampling scheme.

It is readily seen that $H'^{(\alpha)}_{\nu_1,\ldots,\nu_k}$ is the solution in H' to the equation

$$\int_{-\infty}^{\infty} \prod_{i=1}^{k-1} \{F_{\nu_i}(x + H') - F_{\nu_i}(x - H')\} \, dF_{\nu_k}(x) = 1 - \alpha \, . \qquad (2.14)$$

Dudewicz and Dalal (1983) have tabulated the solution to this equation for the special case $\nu_1 = \cdots = \nu_k = \nu$ (say) for selected values of k, α, and ν.

Although tables of the critical constants $H^{(\alpha)}$ and $H'^{(\alpha)}$ are available only for the case where in the first stage an equal number of observations are taken on each treatment (including the control), it is in fact desirable to have unequal first stage sample sizes. This is because (i) the treatment variances σ_i^2 are unequal, and (ii) even when the σ_i^2's are equal, since the control treatment plays a special role of being the benchmark for comparisons, more observations should be taken on it than the other treatments (see Section 1.2 of Chapter 6). Bechhofer and Turnbull (1971) have shown that if the σ_i^2's are known, then an asymptotically optimal choice for a single-stage MCP for one-sided comparisons is given by

$$\frac{n_k}{n_i} = \frac{\sigma_k^2}{\sigma_i^2} \cdot \left(\sum_{j=1}^{k-1} \frac{\sigma_j^2}{\sigma_k^2} \right)^{1/2} \qquad (1 \leq i \leq k-1) \, .$$

In the case of a single-stage MCP, an alternative criterion based on minimizing the sum (or the maximum) of the variances of $\bar{Y}_i - \bar{Y}_k$ leads to the choice

$$\frac{n_k}{n_i} = \frac{\sigma_k^2}{\sigma_i^2} \cdot \sqrt{k-1} \qquad (1 \leq i \leq k-1) \, .$$

One of these choices may be used for the first stage sample sizes in the two-stage sampling scheme if some prior estimates of the σ_i^2's are available. For the given choice of the n_i's one can then compute $H^{(\alpha)}$ (for one-sided comparisons) from (2.13) or $H'^{(\alpha)}$ (for two-sided comparisons) from (2.14).

Example 2.1. The data for the following example are taken from Bishop and Dudewicz (1978). These data pertain to the effects of different solvents on the ability of the fungicide methyl-2-benzimidazole carbamate to destroy the fungus *Penicillium expansum*. The fungicide was diluted in four different solvents. In the first stage $n = 15$ samples of fungus were sprayed with each mixture and the percentages of fungus destroyed (the Y_{ij}'s) were noted. The summary statistics for the first stage are as shown in Table 2.1.

Let us suppose that it is desired to construct simultaneous confidence intervals for all pairwise contrasts at a joint confidence level of 95% with each interval having half-width $= W = 2.0$ (say).

We first illustrate the application of Hochberg's (1975b) sample means procedure (2.8) to this problem. From Bishop et al. (1978) we obtain $(V^{(0.05)}_{14,\ldots,14})^2 = 9.65$. Thus $\xi = W/\sqrt{2}V^{(0.05)}_{14,\ldots,14} = 2/\sqrt{2 \times 9.65} = 0.455$. Using (2.1a) we calculate the following total sample sizes: $N_1 = 15$, $N_2 = 16$, $N_3 = 29$, and $N_4 = 15$. Thus no additional observations are needed on solvents 1 and 4, but one and fourteen additional observations must be taken in the second stage on solvents 2 and 3, respectively. After having taken these additional observations, we calculate the cumulative sample means $\bar{Y}_i$ $(1 \leqq i \leqq k)$. Simultaneous 95% confidence intervals for all pairwise differences $\theta_i - \theta_j$ are then given by $\bar{Y}_i - \bar{Y}_j \pm 2$ $(1 \leqq i < j \leqq 4)$.

We next illustrate the application of Tamhane's (1977) generalized sample means procedure (2.10) to this simultaneous confidence estimation problem. From Table 1 in Hochberg (1976a) we obtain $R^{(0.05)}_{14,\ldots,14} \cong 4.03$ by linear interpolation in $1/\nu$. Thus $\xi = W/R^{(0.05)}_{14,\ldots,14} = 2/4.03 = 0.496$. Using (2.1b) we calculate the following total sample sizes: $N_1 = 16$, $N_2 = 16$, $N_3 = 24$, and $N_4 = 16$. Thus only one additional observation must be taken on solvents 1, 2, and 4, and nine additional observations must be taken on solvent 3. After having taken these additional observations, we calculate the second stage sample means $\bar{Y}^{(2)}_i$ and then the generalized sample means $\tilde{Y}_i$ via (2.4) where the l_i's given by (2.3) are as follows: $l_1 = 0.0775$, $l_2 = 0.0704$, $l_3 = 0.0438$, and $l_4 = 0.0950$. Simultaneous confidence intervals for all pairwise differences are then given by $\tilde{Y}_i - \tilde{Y}_j \pm 2$ $(1 \leqq i < j \leqq 4)$.

TABLE 2.1. Summary Statistics for the First Stage

Solvent	1	2	3	4
$\bar{Y}^{(1)}_i$	96.84	94.69	94.38	97.33
S^2_i	2.110	3.171	5.884	0.780

If instead of all pairwise comparisons, only comparisons with solvent 4 (say) are of interest, then the procedures of Section 2.4 can be employed. If one-sided confidence intervals are of interest, then (2.11) can be used with $H'^{(0.05)}_{14,\ldots,14} = 3.17$ from the tables given in Dudewicz, Ramberg, and Chen (1975). If two-sided confidence intervals are of interest, then (2.13) can be used with $H'^{(0.05)}_{14,\ldots,14} = 3.67$ from the tables given in Dudewicz and Dalal (1983). □

3 STEP-DOWN PROCEDURES

In this section we discuss possible modifications (to account for variance heterogeneity) of some classical step-down procedures for comparisons of means in a one-way layout. Some of these modifications were suggested by Dijkstra (1983).

3.1 Modified Fisher's Least Significant Difference Procedure

The preliminary F-test of the overall null hypothesis $H_0: \theta_1 = \theta_2 = \cdots = \theta_k$ can be replaced by one of its robust (to departures from variance homogeneity) versions. Such robust F-tests for this generalized Behrens–Fisher problem of comparing $k \geqq 2$ treatment means in the presence of variance heterogeneity have been proposed by James (1951), Welch (1947, 1951), and Brown and Forsythe (1974b). For example, Welch (1947) proposed the modified F-statistic

$$\tilde{F} = \frac{1}{(k-1)} \sum_{i=1}^{k} W_i(\bar{Y}_i - \bar{\bar{Y}})^2 \tag{3.1}$$

where

$$W_i = \frac{n_i}{S_i^2} \ (1 \leqq i \leqq k), \qquad \bar{\bar{Y}} = \frac{\sum_{i=1}^{k} W_i Y_i}{W} \quad \text{and} \quad W = \sum_{i=1}^{k} W_i . \tag{3.2}$$

The overall null hypothesis is rejected if

$$\tilde{F} > F^{(\alpha)}_{k-1,\,N-k}\left\{1 + \frac{2(k-2)}{k^2-1} \sum_{i=1}^{k} \frac{(1 - W_i/W)^2}{(n_i - 1)}\right\} \tag{3.3}$$

where $N = \Sigma_{i=1}^{k} n_i$ and $F^{(\alpha)}_{k-1,\,N-k}$ is the upper α point of the F-distribution with $k-1$ and $N-k$ d.f. (See the simulation studies by Brown and Forsythe 1974b, and Dijkstra and Werter 1981 for a performance comparison of different robust modifications of the F-test.)

As in Fisher's least significant difference (LSD) procedure, if H_0 is not rejected using (3.3) (say), then all pairwise null hypotheses $H_{ij}: \theta_i = \theta_j$ are retained without further tests. If H_0 is rejected, then each pairwise null hypothesis is tested at level α using the Welch test for the Behrens–Fisher problem discussed in Section 1.2.1. Because robust tests are used at both steps of testing, the operating characteristics of the modified LSD (under heterogeneous variances) would be expected to be similar to those of Fisher's LSD (under homogeneous variances).

3.2 Modified Newman–Keuls Type Procedures

Newman–Keuls (NK) type step-down procedures based on F-statistics (discussed in Section 1 of Chapter 4) can be readily modified by using robust F-tests such as (3.3) in place of the usual F-tests for subset homogeneity hypotheses.

Next let us discuss a step-down procedure based on Studentized range statistics. Consider the statistic

$$\max_{i,\, j \in P} \frac{|\bar{Y}_i - \bar{Y}_j|}{\sqrt{S_i^2/n_i + S_j^2/n_j}} \tag{3.4}$$

for testing the homogeneity hypothesis on subset $P \subseteq \{1, 2, \ldots, k\}$ of treatments. This is a union-intersection (UI) test statistic obtained by considering the homogeneity hypothesis on set P as the intersection of the corresponding pairwise null hypotheses, and using the usual standardized difference as the statistic for each pairwise test. The homogeneity hypothesis is rejected if (3.4) exceeds some critical constant that depends on the null distribution of (3.4) and the nominal significance level α_p where p is the cardinality of set P. The statistics (3.4) can be used in the usual step-down testing scheme wherein if the homogeneity hypothesis on any set P (with cardinality $p > 2$) is not rejected, then all subsets of P are retained as homogeneous without further tests. If the homogeneity hypothesis is rejected, then all subsets of P of size $p-1$ are tested.

This step-down procedure is completely specified once we fix the nominal significance levels α_p $(2 \leqq p \leqq k)$ and the associated critical constants with which the statistics (3.4) are to be compared. The α_p's may be chosen according to one of the specification schemes given in Section 4.3.3 of Chapter 2. To determine the associated critical constants we need to know the distribution of (3.4) under the homogeneity hypothesis H_P for set P.

In the Welch (1938) method the exact distribution of the (i, j)th standardized difference under $\theta_i = \theta_j$ is approximated by that of $t_{\hat{\nu}_{ij}}$ r.v. where $\hat{\nu}_{ij}$ is the estimated d.f. given by (1.15). Thus (3.4) is approximately distributed as the maximum over $i, j \in P$ of correlated r.v.'s $t_{\hat{\nu}_{ij}}$. No attempt has been made to get a handle on the distribution of this maximum, which is a very difficult problem. Note, however, that by using the Bonferroni inequality we can construct a nominal α_p-level UI test of H_P that rejects H_P if, for at least one $i, j \in P$,

$$\frac{|\bar{Y}_i - \bar{Y}_j|}{\sqrt{S_i^2/n_i + S_j^2/n_j}} > T_{\hat{\nu}_{ij}}^{(\alpha_p/2p^*)}$$

where $p^* = \binom{p}{2}$.

CHAPTER 8

Procedures for Mixed Two-Way Layouts and Designs with Random Covariates

In the present chapter we consider multiple comparison procedures (MCPs) for two distinct types of experimental designs. The first is an equireplicated two-way layout where the factor of main interest (the treatment factor) is fixed and the other factor (e.g., blocks) is random. This is the general balanced mixed model in the traditional sense. The commonly used one-way repeated measures design is a special case of this model. The model for the second experimental design considered in this chapter involves, aside from the fixed treatment factor, covariates that are random but observable. The coefficients in the linear model corresponding to these covariates are assumed to be fixed (and unknown). Thus all the effects are fixed but the observable covariates are random.

The one-way repeated measures design and the equireplicated mixed two-way layout form the setting for Section 1. Various models for these designs are introduced in Section 1.1. The corresponding procedures are described in Sections 1.2 and 1.3. A comparison of the procedures is given in Section 1.4. In section 2 we discuss MCPs for analysis of covariance (ANCOVA) designs with random covariates. The general model for such designs is described in Section 2.1 by first introducing a simple model and then building upon it. Unconditional (with respect to the values of the covariates) MCPs are discussed in Section 2.2; these MCPs are exact. Conditional (on the observed values of the covariates) MCPs are discussed in Section 2.3; these MCPs are approximate. The two types of MCPs are compared in Section 2.4.

1 PROCEDURES FOR ONE-WAY REPEATED MEASURES AND MIXED TWO-WAY DESIGNS

1.1 Models

We first discuss models for one-way repeated measures designs and then for mixed two-way designs. Both of these designs have a fixed treatment factor (say, A) and another random factor (say, B).

In a one-way repeated measures design, blocks consisting of a random sample of, say, n experimental units drawn from a large population constitute the random factor. Each unit is measured repeatedly at, say, k successive points in time or under k different conditions. The times or conditions of measurements are fixed in advance, and constitute the treatment factor. When the conditions of measurements correspond to actual treatments and when these treatments are randomly allocated within each block (possibly consisting of different but matched experimental units), then the resulting design is referred to as a randomized block design (with random blocks). The two designs are analyzed similarly; for convenience, we have presented our discussion in the context of the former.

Let $\mathbf{Y}_j = (Y_{1j}, Y_{2j}, \ldots, Y_{kj})'$ denote the vector of responses for the jth experimental unit $(1 \leqq j \leqq n)$. The following model is commonly assumed:

$$\mathbf{Y}_j = \mathbf{M}_j + \mathbf{E}_j \qquad (1 \leqq j \leqq n) \tag{1.1}$$

where all the $\mathbf{M}_j = (M_{1j}, M_{2j}, \ldots, M_{kj})'$ and $\mathbf{E}_j = (E_{1j}, E_{2j}, \ldots, E_{kj})'$ are distributed independently of each other as k-variate normal vectors, the former with mean vector $\boldsymbol{\theta} = (\theta_1, \theta_2, \ldots, \theta_k)'$ (the vector of treatment effects) and variance-covariance matrix $\boldsymbol{\Sigma}_0$ and the latter with mean vector $\mathbf{0}$ and variance-covariance matrix $\sigma^2\mathbf{I}$. Thus the $\mathbf{Y}_j$'s are independent and identically distributed (i.i.d.) $N(\boldsymbol{\theta}, \boldsymbol{\Sigma})$ random vectors where $\boldsymbol{\Sigma} = \boldsymbol{\Sigma}_0 + \sigma^2\mathbf{I}$.

We now discuss mixed two-way designs. In a mixed two-way layout the random factor may not be a blocking factor. Also, for each factor-level combination (i, j), we may have $r_{ij} \geqq 1$ observations Y_{ijl} with mutually independent replication errors E_{ijl} $(1 \leqq l \leqq r_{ij})$. Restricting to the balanced case $(r_{ij} = r \geqq 1$ for all $i, j)$, we write the following model for the observation vectors $\mathbf{Y}_{jl} = (Y_{1jl}, Y_{2jl}, \ldots, Y_{kjl})'$:

$$\mathbf{Y}_{jl} = \mathbf{M}_j + \mathbf{E}_{jl} \qquad (1 \leqq j \leqq n, 1 \leqq l \leqq r) \tag{1.2}$$

where $\mathbf{Y}_{j1}, \mathbf{Y}_{j2}, \ldots, \mathbf{Y}_{jr}$ have a joint kr-variate normal distribution with the $\mathbf{Y}_{jl}$'s having a common mean vector $\boldsymbol{\theta}$ and the following covariance structure:

$$\text{cov}(\mathbf{Y}_{jl}, \mathbf{Y}_{jl'}) = \begin{cases} \boldsymbol{\Sigma} = \boldsymbol{\Sigma}_0 + \sigma^2 \mathbf{I} & \text{if } l = l' \\ \boldsymbol{\Sigma}_0 & \text{if } l \neq l' . \end{cases} \tag{1.3}$$

When there is a single replication per cell ($r = 1$), this model reduces to that for a one-way repeated measures design.

Exact procedures for making pairwise comparisons among the θ_i's can be constructed if we impose special restrictions on the form of $\boldsymbol{\Sigma}$. (See Section 1.3.1 for an exact procedure for contrasts among the θ_i's under a general model.) The least restrictive of such models for a one-way repeated measures design was proposed by Huynh and Feldt (1970). This model assumes that $\boldsymbol{\Sigma} = \{\sigma_{ii'}\}$ is given by

$$\sigma_{ii'} = \begin{cases} 2\lambda_i + \tau^2 & \text{if } i = i' \\ \lambda_i + \lambda_{i'} & \text{if } i \neq i' \end{cases} \tag{1.4}$$

for some $\tau^2 > 0$ and real λ_i's. Note that (1.4) is equivalent to the condition that all pairwise differences of the treatment sample means $\bar{Y}_{i\cdot} = \Sigma_{j=1}^{n} Y_{ij}/n$ have the same variance given by

$$\text{var}(\bar{Y}_{i\cdot} - \bar{Y}_{i'\cdot}) = \frac{2\tau^2}{n} \qquad (1 \leqq i \neq i' \leqq k) . \tag{1.5}$$

Recall that this is the pairwise balance condition (2.9) of Chapter 3. We refer to (1.4) (or equivalently (1.5)) as the *spherical model.*

A special case of the spherical model is obtained when $\boldsymbol{\Sigma}_0$ is a "uniform" matrix having the following form:

$$\boldsymbol{\Sigma}_0 = \sigma_0^2[(1 - \rho_0)\mathbf{I} + \rho_0 \mathbf{J}] \tag{1.6}$$

where $\mathbf{I}$ is a $k \times k$ identity matrix, $\mathbf{J}$ is a $k \times k$ matrix of all 1's, and $-1/(k-1) \leqq \rho_0 \leqq 1$. This is equivalent to $\boldsymbol{\Sigma}$ also having the same form:

$$\boldsymbol{\Sigma} = (\sigma_0^2 + \sigma^2)[(1 - \rho)\mathbf{I} + \rho \mathbf{J}] \tag{1.7}$$

where

$$\rho = \frac{\rho_0 \sigma_0^2}{\sigma_0^2 + \sigma^2} \cdot \tag{1.8}$$

This model is obtained by putting $\tau^2 = \sigma_0^2(1 - \rho_0) + \sigma^2$ and $\lambda_i = \rho_0 \sigma_0^2/2$ for all i in (1.4). For balanced mixed two-way layout designs (with $r \geqq 1$),

(1.7) has been referred to as the *symmetric model* by Scheffé (1956) and Hocking (1973).

1.2 Exact Procedures for Spherical and Symmetric Models

1.2.1 One-Way Repeated Measures Designs or Mixed Two-Way Designs with a Single Replication per Cell

If the spherical model holds, then an exact Tukey-type procedure can be based on Theorem 1.1 below. The proof of this theorem follows along the lines of Huynh and Feldt (1970). These authors showed that (1.4) is a necessary and sufficient condition for the mixed two-way analysis of variance (ANOVA) F-test to be exact (see also Rouanet and Lepine 1970). In Theorem 1.1,

$$MS_{AB} = \frac{\sum_{i=1}^{k}\sum_{j=1}^{n}(Y_{ij} - \bar{Y}_{i\cdot} - \bar{Y}_{\cdot j} + \bar{Y}_{\cdot\cdot})^2}{(k-1)(n-1)}$$

denotes the mean square for interaction between A and B; here the dot notation is the usual one—a dot replacing a subscript indicates averaging over that subscript.

Theorem 1.1. The pivotal random variable (r.v.)

$$\max_{1 \le i < i' \le k} \frac{|\bar{Y}_{i\cdot} - \bar{Y}_{i'\cdot} - (\theta_i - \theta_{i'})|}{\sqrt{MS_{AB}/n}} \tag{1.9}$$

has the Studentized range distribution with parameter k and d.f. $(k - 1) \times (n - 1)$ if and only if (1.4) or equivalently (1.5) holds.

Proof. From (1.5) it follows, using the result of Hochberg (1974c), that

$$\max_{1 \le i < i' \le k} \frac{|\bar{Y}_{i\cdot} - \bar{Y}_{i'\cdot} - (\theta_i - \theta_{i'})|}{\sqrt{\tau^2/n}}$$

is distributed as the range of k i.i.d $N(0, 1)$ r.v.'s. It only remains to show that

$$MS_{AB} \sim \frac{\tau^2 \chi^2_{(k-1)(n-1)}}{(k-1)(n-1)}$$

independent of the $\bar{Y}_{i\cdot}$'s. This follows from the results of Box (1954), who showed that $SS_{AB} = (k-1)(n-1)MS_{AB}$ is distributed independently of the $\bar{Y}_{i\cdot}$'s as the weighted sum of $k-1$ independent χ^2_{n-1} r.v.'s, the weights being the positive eigenvalues of the matrix $\mathbf{\Sigma}(\mathbf{I} - \mathbf{J}/k)$. This weighted sum is distributed as $\tau^2\chi^2_{(k-1)(n-1)}$ if and only if all the eigenvalues are equal to τ^2, which is true if and only if $\mathbf{\Sigma}$ has the form (1.4). □

From this theorem and (2.4) of Chapter 3 it follows that exact $(1-\alpha)$-level simultaneous confidence intervals for all contrasts $\sum_{i=1}^k c_i\theta_i$ are given by

$$\sum_{i=1}^{k} c_i\theta_i \in \left[\sum_{i=1}^{k} c_i\bar{Y}_{i\cdot} \pm Q^{(\alpha)}_{k,(k-1)(n-1)}\sqrt{\frac{MS_{AB}}{n}}\sum_{i=1}^{k}\frac{|c_i|}{2}\right] \quad \forall\, \mathbf{c} \in \mathbb{C}^k \tag{1.10}$$

where $Q^{(\alpha)}_{k,(k-1)(n-1)}$ denotes the upper α point of the Studentized range distribution with parameter k and d.f. $(k-1)(n-1)$.

Bhargava and Srivastava (1973) independently proved Theorem 1.1 for the symmetric model using a different technique. We explain this technique since it extends readily to balanced mixed two-way layouts with multiple replications per cell.

The symmetric model in this case is equivalent to the intraclass correlation model:

$$\operatorname{corr}(Y_{ij}, Y_{i'j}) = \rho_1 = \frac{\sigma_0^2}{\sigma_0^2 + \sigma^2} \qquad (1 \leq i \neq i' \leq k,\ 1 \leq j \leq n)\,.$$

Bhargava and Srivastava transformed the Y_{ij}'s to independent normal r.v.'s Z_{ij} with common variance σ^2 by using the transformation

$$Z_{ij} = Y_{ij} - \delta\bar{Y}_{\cdot j} \qquad (1 \leq i \leq k,\ 1 \leq j \leq n) \tag{1.11}$$

where $\bar{Y}_{\cdot j} = \sum_{i=1}^k Y_{ij}/k$ and $\delta = 1 + (1-\rho_1)^{1/2}\{1 + (k-1)\rho_1\}^{-1/2}$. Note that the Z_{ij}'s are not observable because δ involves the unknown parameter ρ_1. They showed that for any contrast $\mathbf{c} \in \mathbb{C}^k$ the following relations hold:

$$\sum_{i=1}^{k} c_iY_{ij} = \sum_{i=1}^{k} c_iZ_{ij} \qquad (1 \leq j \leq n)\,,$$

$$\operatorname{var}\left(\sum_{i=1}^{k} c_iY_{ij}\right) = \operatorname{var}\left(\sum_{i=1}^{k} c_iZ_{ij}\right) = \sigma^2\sum_{i=1}^{k} c_i^2 \qquad (1 \leq j \leq n)\,, \tag{1.12}$$

and

$$SS_{AB} = (k-1)(n-1)MS_{AB} = \sum_{i=1}^{k} \sum_{j=1}^{n} (Z_{ij} - \bar{Z}_{i\cdot} - \bar{Z}_{\cdot j} + \bar{Z}_{\cdot\cdot})^2 .$$

Therefore SS_{AB} is distributed as $\sigma^2 \chi^2_{(k-1)(n-1)}$ independent of any contrast in the $\bar{Y}_{i\cdot}$'s (which can be expressed as the same contrast in the $\bar{Z}_{i\cdot}$'s). Their result then follows on realizing that (1.9) can be expressed in terms of the Z_{ij}'s as

$$\max_{1 \leq i < i' \leq k} \frac{|\bar{Z}_{i\cdot} - \bar{Z}_{i'\cdot} - (\theta_i - \theta_{i'})|}{\sqrt{MS_{AB}/n}} \tag{1.13}$$

and it has the distribution of a $Q_{k,(k-1)(n-1)}$ r.v.

We have presented here the exact Tukey-type procedure for the spherical and symmetric models. The corresponding exact Scheffé-type procedure can be constructed in an analogous manner. Stepwise procedures can also be constructed similarly.

In closing we mention that Fenech (1979) considered Kempthorne's (1952, p. 137) randomization model for the randomized block design. This model does not assume that the errors are independent, normal, and have a constant variance. He showed (under some simple conditions) that the true Type I FWE of (1.10) approaches the nominal level α as the number of blocks increases to infinity.

1.2.2 Balanced Mixed Two-Way Designs with Multiple Replications per Cell

In this case we only consider the symmetric model. For this model, using (1.3) and (1.7) we obtain the following correlation structure among the Y_{ijl}'s:

$$\text{corr}(Y_{ijl}, Y_{i'j'l'}) = \begin{cases} \rho_1 = \dfrac{\sigma_0^2}{\sigma_0^2 + \sigma^2} & \text{if } i = i',\ j = j',\ l \neq l' \\ \rho_2 = \dfrac{\rho_0 \sigma_0^2}{\sigma_0^2 + \sigma^2} & \text{if } i \neq i',\ j = j' \\ 0 & \text{if } j \neq j' . \end{cases} \tag{1.14}$$

By using two transformations similar to (1.11) in a recursive fashion, Hochberg and Tamhane (1983) showed that the r.v.

$$\max_{1 \leq i < i' \leq k} \frac{|\bar{Y}_{i\cdot\cdot} - \bar{Y}_{i'\cdot\cdot} - (\theta_i - \theta_{i'})|}{\sqrt{MS_{AB}/rn}} \tag{1.15}$$

has the Studentized range distribution with parameter k and d.f. $(k-1) \times (n-1)$ where

$$MS_{AB} = \frac{r \sum_{i=1}^{k} \sum_{j=1}^{n} (\bar{Y}_{ij\cdot} - \bar{Y}_{i\cdot\cdot} - \bar{Y}_{\cdot j\cdot} + \bar{Y}_{\cdots})^2}{(k-1)(n-1)} .$$

Thus the $100(1-\alpha)\%$ simultaneous Tukey-type intervals for all contrasts $\sum_{i=1}^{k} c_i\theta_i$ are given by

$$\sum_{i=1}^{k} c_i\theta_i \in \left[\sum_{i=1}^{k} c_i\bar{Y}_{i\cdot\cdot} \pm Q_{k,(k-1)(n-1)}^{(\alpha)} \sqrt{\frac{MS_{AB}}{rn}} \sum_{i=1}^{k} \frac{|c_i|}{2} \right] \quad \forall \mathbf{c} \in \mathbb{C}^k . \tag{1.16}$$

Note that (1.10) is a special case of (1.16) for $r = 1$. Also note that

$$MS_{\text{error}} = \frac{\sum_{i=1}^{k} \sum_{j=1}^{n} \sum_{l=1}^{r} (Y_{ijl} - \bar{Y}_{ij\cdot})^2}{kn(r-1)}$$

cannot be used in (1.16) in place of MS_{AB}.

1.3 Procedures for General Models

1.3.1 An Exact Procedure

This exact procedure was proposed by Scheffé (1959, pp. 270–274) and is based on Hotelling's T^2-test. For this procedure to be valid it is only necessary that the vectors $(\bar{E}_{ij}, \ldots, \bar{E}_{kj})$ be i.i.d. multivariate normal; here $\bar{E}_{ij} = \sum_{l=1}^{r} E_{ijl}/r$.

To apply the procedure, first compute the differences

$$D_{ij} = \bar{Y}_{ij\cdot} - \bar{Y}_{kj\cdot} \qquad (1 \leq i \leq k-1, 1 \leq j \leq n) ,$$

and the cross products

$$A_{ii'} = \sum_{j=1}^{n} (D_{ij} - \bar{D}_{i\cdot})(D_{i'j} - \bar{D}_{i'\cdot}) \qquad (1 \leq i, i' \leq k-1) . \tag{1.17}$$

Let $\mathbf{A} = \{A_{ii'}\}$ be the $(k-1) \times (k-1)$ matrix of cross products and let $\bar{\mathbf{D}} = (\bar{D}_{1\cdot}, \bar{D}_{2\cdot}, \ldots, \bar{D}_{k-1\cdot})'$ be the vector of mean differences. Using the fact that the vectors $\mathbf{D}_j = (D_{1j}, D_{2j}, \ldots, D_{k-1,j})'$, $j = 1, 2, \ldots, n$, are independent and multivariate normal, each with mean vector $\boldsymbol{\delta} =$

$(\delta_1, \delta_2, \ldots, \delta_{k-1})' = (\theta_1 - \theta_k, \theta_2 - \theta_k, \ldots, \theta_{k-1} - \theta_k)'$ and variance-covariance matrix $\boldsymbol{\Sigma}_D$, which is estimated independently by a Wishart matrix $[1/(n-1)]\mathbf{A}$, it follows that

$$T^2 = n(n-1)(\bar{\mathbf{D}} - \boldsymbol{\delta})'\mathbf{A}^{-1}(\bar{\mathbf{D}} - \boldsymbol{\delta}) \tag{1.18}$$

is distributed as a $\{(k-1)(n-1)/(n-k+1)\}F_{k-1,n-k+1}$ r.v. (where it is assumed that $n \geqq k$). Thus the T^2-statistic can be used to test the overall null hypothesis H_0: $\theta_1 = \theta_2 = \cdots = \theta_k$, which is equivalent to the hypothesis that $\boldsymbol{\delta}$ is a null vector. Scheffé (1959, p. 273) shows how his projection method can also handle random projections (necessitated by the fact that $\mathbf{A}$ is a random matrix) to obtain simultaneous confidence intervals for all linear combinations of the δ_i's. But any linear combination $\Sigma_{i=1}^{k-1} l_i\delta_i$ of the δ_i's equals the contrast $\Sigma_{i=1}^{k} c_i\theta_i$ among the θ_i's where $c_i = l_i$ $(1 \leqq i \leqq k-1)$, and $c_k = -\Sigma_{i=1}^{k-1} l_i$, and similarly $\Sigma_{i=1}^{k-1} l_i\bar{D}_{i.} = \Sigma_{i=1}^{k} c_i\bar{Y}_{i..}$. We thus obtain the following $100(1-\alpha)\%$ simultaneous confidence intervals for all contrasts $\Sigma_{i=1}^{k} c_i\theta_i$:

$$\sum_{i=1}^{k} c_i\theta_i \in \left[\sum_{i=1}^{k} c_i\bar{Y}_{i..} \pm \left\{\frac{k-1}{n(n-k+1)} F_{k-1,n-k+1}^{(\alpha)}\right\}^{1/2} (\mathbf{d}'\mathbf{A}\mathbf{d})^{1/2}\right] \quad \forall\, \mathbf{c} \in \mathbb{C}^k \tag{1.19}$$

where $\mathbf{d} = (c_1, c_2, \ldots, c_{k-1})'$.

For pairwise and some additional low order contrasts, (1.16) yields shorter intervals than (1.19). This, however, must be weighed against the fact that (1.19) is more generally valid than (1.16).

1.3.2 Approximate Procedures

An approximate procedure for pairwise comparisons can be based on the pivotal r.v.'s

$$T_{ii'} = \frac{\{\bar{Y}_{i..} - \bar{Y}_{i'..} - (\theta_i - \theta_{i'})\}\sqrt{n}}{\sqrt{S_{ii} + S_{i'i'} - 2S_{ii'}}} \qquad (1 \leqq i < i' \leqq k) \tag{1.20}$$

where

$$S_{ii'} = \frac{\sum_{j=1}^{n} (\bar{Y}_{ij.} - \bar{Y}_{i..})(\bar{Y}_{i'j.} - \bar{Y}_{i'..})}{(n-1)} \qquad (1 \leqq i, i' \leqq k)\,. \tag{1.21}$$

To obtain $100(1-\alpha)\%$ simultaneous confidence intervals for all pairwise differences $\theta_i - \theta_{i'}$, we must determine the critical constant ξ such that

$$\inf \Pr\{|T_{ii'}| \leqq \xi \ (1 \leqq i < i' \leqq k)\} = 1 - \alpha \tag{1.22}$$

where the infimum is taken over the set of all variance-covariance matrices $\boldsymbol{\Sigma}_Y$ of the independent vectors $(\bar{Y}_{1j\cdot}, \bar{Y}_{2j\cdot}, \ldots, \bar{Y}_{kj\cdot})'$.

Based on an extensive simulation study, Alberton and Hochberg (1984) conjectured that the infimum in (1.22) is attained for $\boldsymbol{\Sigma}_Y = \mathbf{I}$. This is an extension of the Tukey conjecture discussed in Chapter 3. They also found for $k = 3$ that for a large range of values of α and n, the ξ-value satisfying (1.22) is well approximated by $|M|^{(\alpha)}_{k^*, n-1}$, the upper α point of the Studentized maximum modulus distribution with parameter $k^* = k(k-1)/2$ and d.f. $n - 1$. Based on these results Alberton and Hochberg proposed the following approximate $100(1-\alpha)\%$ simultaneous confidence intervals for the pairwise differences $\theta_i - \theta_{i'}$:

$$\theta_i - \theta_{i'} \in \left[\bar{Y}_{i\cdot\cdot} - \bar{Y}_{i'\cdot\cdot} \pm |M|^{(\alpha)}_{k^*, n-1}\sqrt{\frac{S_{ii} + S_{i'i'} - 2S_{ii'}}{n}}\right] \quad (1 \leqq i < i' \leqq k) . \tag{1.23}$$

A slightly conservative approximation is obtained by employing the Bonferroni method that amounts to using $T^{(\alpha^*/2)}_{n-1}$ (where $\alpha^* = \alpha/k^*$) in place of $|M|^{(\alpha)}_{k^*, n-1}$. These intervals can be extended in the usual manner to the family of all contrasts using (2.3) of Chapter 3. Another multiple comparison technique appropriate under Scheffé's general model was given by Mudholkar and Subbaiah (1976).

We now give a comprehensive example to illustrate the various procedures.

Example 1.1. Table 1.1 shows coded data given by Scheffé (1959, p. 289) on the measurements of flow rates of a fuel through three types of nozzles (factor A) by five different operators (factor B), each of whom made three determinations on each nozzle.

We regard the nozzle as a fixed factor and the operator as a random factor. We thus have a balanced mixed two-way layout. Suppose that it is of interest to make three pairwise comparisons among the types of nozzles with Type I familywise error rate (FWE) $\alpha = 0.10$.

The cell means and the analysis of variance for the data in Table 1.1 are given in Tables 1.2 and 1.3, respectively.

Assuming the symmetric model (1.7) we can apply (1.16) to obtain the following 90% simultaneous confidence intervals for three pairwise differ-

TABLE 1.1. Measurements of Flow Rates (Coded Data)

	Operator (Factor B)				
Nozzle (Factor A)	1	2	3	4	5
1	6, 6, −15	26, 12, 5	11, 4, 4	1, 14, 7	25, 18, 25
2	13, 6, 13	4, 4, 11	17, 10, 17	−5, 2, −5	15, 8, 1
3	10, 10, −11	−35, 0, −14	11, −10, −17	12, −2, −16	−4, 10, 24

Source: Scheffé (1959).

TABLE 1.2. Cell Means $\bar{Y}_{ij\cdot}$ for Coded Data in Table 1.1

	Operator (Factor B)					
Nozzle (Factor A)	1	2	3	4	5	$\bar{Y}_{i\cdot\cdot}$
1	−1.00	14.33	6.33	14.00	22.67	11.27
2	10.67	6.33	14.67	−2.67	8.00	7.40
3	3.00	−16.33	−5.33	−2.00	10.00	−2.13
$\bar{Y}_{\cdot j\cdot}$	4.22	1.44	5.22	3.11	13.56	$\bar{Y}_{\cdots} = 5.51$

TABLE 1.3. Analysis of Variance for Coded Data in Table 1.1

Source	Sum of Squares (SS)	d.f.	Mean Square (MS)
Nozzle main effects	$SS_A = 1428.45$	2	$MS_A = 714.23$
Operator main effects	$SS_B = 799.88$	4	$MS_B = 399.94$
Nozzle × operator interaction	$SS_{AB} = 1819.51$	8	$MS_{AB} = 227.44$
Error	$SS_{\text{error}} = 3037.96$	30	$MS_{\text{error}} = 101.27$
Total	$SS_{\text{total}} = 7085.80$	44	

ences (here $Q_{3,8}^{(0.10)}\sqrt{MS_{AB}/rn} = 3.37\sqrt{227.44/15} = 13.12$):

$$\theta_1 - \theta_2 : 11.27 - 7.40 \pm 13.12 = [-9.25, 16.99]$$

$$\theta_1 - \theta_3 : 11.27 + 2.13 \pm 13.12 = [0.28, 26.52]$$

$$\theta_2 - \theta_3 : 7.40 + 2.13 \pm 13.12 = [-3.59, 22.65] .$$

Thus only the nozzle types 1 and 3 are significantly different at $\alpha = 0.10$ using the intervals (1.16). We may also note that the F-ratio for the nozzle main effects is $MS_A/MS_{AB} = 3.14$ with 2 and 8 d.f., which is just significant at $\alpha = 0.10$ ($F_{2,8}^{(0.10)} = 3.11$). Both of these procedures are, of course, valid only if the symmetric model (1.7) holds; otherwise they must be regarded as approximate.

We next consider the procedures of Section 1.3 that do not require the assumption of the symmetric model for their application. We first consider the simultaneous confidence intervals (1.19) based on Hotelling's T^2-Test. The values of the $D_{ij} = \bar{Y}_{ij\cdot} - \bar{Y}_{kj\cdot}$ computed from Table 1.2 are shown in Table 1.4. Matrix $\mathbf{A}$ defined in (1.17) is then given by

$$\mathbf{A} = \begin{bmatrix} 610.72 & 273.39 \\ 273.39 & 522.65 \end{bmatrix} .$$

TABLE 1.4. Values of the D_{ij}'s Computed From Table 1.2

	j					
i	1	2	3	4	5	
1	−4.00	30.66	11.66	16.00	12.67	$\bar{D}_{1\cdot} = 13.40$
2	7.67	22.66	20.00	−0.67	−2.00	$\bar{D}_{2\cdot} = 9.53$

The **d**-vectors corresponding to the differences $\theta_1 - \theta_2$, $\theta_1 - \theta_3$, and $\theta_2 - \theta_3$ are $(1, -1)'$, $(1, 0)'$, and $(0, 1)'$ and the corresponding values of $(\mathbf{d}'\mathbf{A}\mathbf{d})$ are 586.59, 610.72, and 522.65, respectively. The factor $\{(k-1)/n(n-k+1)\}F_{k-1,n-k+1}^{(\alpha)}$ for $\alpha = 0.10$, $k = 3$, $n = 5$ equals $(2/15) \times 5.46 = 0.728$. We thus obtain the following 90% simultaneous confidence intervals for all pairwise differences:

$$\theta_1 - \theta_2 : 3.87 \pm \{0.728 \times 586.59\}^{1/2} = [-16.79, 24.53]$$

$$\theta_1 - \theta_3 : 13.40 \pm \{0.728 \times 610.72\}^{1/2} = [-7.69, 34.49]$$

$$\theta_2 - \theta_3 : 9.53 \pm \{0.728 \times 522.65\}^{1/2} = [-9.98, 29.04] .$$

Note that these intervals are much wider than the ones obtained using the procedure (1.16) and none of the pairwise differences can be declared significant.

Finally, we apply the Alberton–Hochberg approximate procedure (1.23). For this purpose note that

$$S_{11} + S_{22} - 2S_{12} = \frac{A_{11} + A_{22} - 2A_{12}}{4} = \frac{586.59}{4} = 146.65$$

$$S_{11} + S_{33} - 2S_{13} = \frac{A_{11}}{4} = \frac{610.72}{4} = 152.68$$

$$S_{22} + S_{33} - 2S_{23} = \frac{A_{22}}{4} = \frac{522.65}{4} = 130.65 .$$

Also $|M|_{3,4}^{(0.10)} = 2.98$. From (1.23) we compute the 90% simultaneous confidence intervals for all pairwise differences as follows:

$$\theta_1 - \theta_2 : 3.87 \pm 2.98\sqrt{\frac{146.65}{5}} = [-12.27, 20.01]$$

$$\theta_1 - \theta_3 : 13.40 \pm 2.98\sqrt{\frac{152.68}{5}} = [-3.07, 29.87]$$

$$\theta_2 - \theta_3 : 9.53 \pm 2.98\sqrt{\frac{130.65}{5}} = [-5.70, 24.76] .$$

These intervals, although shorter than the ones obtained using the Scheffé-type procedure, are still too wide to enable us to claim a significant difference for any pair. □

1.4 A Comparison of Procedures

For one-way repeated measures designs it is well known that the F-test of the overall null hypothesis $H_0: \theta_1 = \theta_2 = \cdots = \theta_k$ becomes increasingly liberal (i.e., the Type I error probability exceeds the nominal level) as $\mathbf{\Sigma}$ departs from the sphericity assumption (1.4); see Rogan, Keselman, and Mendoza (1979) for a summary of simulation work on this topic. Maxwell (1980) and Mitzel and Games (1981) studied by simulation techniques the performance of the Tukey-type procedure (1.10) for pairwise comparisons when the sphericity assumption is violated. They found that it also behaves in a liberal fashion (see also Boik 1981). These authors recommend using (1.19) or (1.23) (actually the Bonferroni version of (1.23), which is slightly more conservative). For pairwise comparisons, of course, (1.19) gives overly conservative results and (1.23) is preferred. For general contrasts one may wish to use (1.19).

For one-way repeated measures designs, Stoline (1984) made a simulation study of a composite procedure for pairwise comparisons that consists of performing a preliminary test of sphericity of the $\mathbf{\Sigma}$ matrix (see Mauchly 1940 and Huynh and Feldt 1970). Depending on the outcome of this test, one uses either (1.10) (which is appropriate when the sphericity assumption holds) or the Bonferroni version of (1.23) (which is appropriate when the sphericity assumption does not hold). However, the preliminary test lacks adequate power for detecting departures from sphericity unless it is used at nominal significance levels as high as 0.50. Therefore Stoline concluded that the unconditional (without the preliminary test) use of (1.23) is generally the preferred choice for pairwise comparisons. The Tukey-type intervals (1.10) should be used unconditionally only when the sphericity assumption is known to be satisfied (which would rarely be the case in practice) or they should be used conditionally in conjunction with a preliminary test of sphericity at $\alpha = 0.50$.

2 PROCEDURES FOR ANALYSIS OF COVARIANCE DESIGNS WITH RANDOM COVARIATES

2.1 Models

The simplest ANCOVA design consists of a one-way layout with a single covariate. Thigpen and Paulson (1974) considered the following model for this design: Let X_{ij} and Y_{ij} denote the observations on the covariate and the response variable, respectively, for the jth individual in the ith treatment group $(1 \leqq i \leqq k, 1 \leqq j \leqq n_i)$. It is assumed that the pair $(X_{ij}, Y_{ij})'$ has a bivariate normal distribution with mean vector $(\xi_i, \theta_i)'$ and variance-covariance matrix

$$\mathbf{\Sigma} = \begin{bmatrix} \sigma_X^2 & \rho\sigma_X\sigma_Y \\ \rho\sigma_X\sigma_Y & \sigma_Y^2 \end{bmatrix}$$

where all the parameters are unknown and all the pairs $(X_{ij}, Y_{ij})'$ are independent $(1 \leqq i \leqq k, 1 \leqq j \leqq n_i)$. Thigpen and Paulson further assumed that $\xi_1 = \xi_2 = \cdots = \xi_k = \xi$ (say). Conditionally on X_{ij} we can write

$$Y_{ij} = \theta_i + \beta(X_{ij} - \xi) + E_{ij} \qquad (1 \leqq i \leqq k, 1 \leqq j \leqq n_i) \tag{2.1}$$

where $\beta = \rho\sigma_Y/\sigma_X$ and the E_{ij}'s are i.i.d. $N(0, \sigma^2)$ with $\sigma^2 = \sigma_Y^2(1 - \rho^2)$. Note that the slope coefficient β is fixed.

Bryant and Paulson (1976) considered a more general model allowing designs other than the one-way layout and involving more than one random covariate. However, they still retained the assumption that all random covariate vectors are identically distributed. More specifically, their model assumes that on the ith experimental unit $(1 \leqq i \leqq N)$ a vector of $q \geqq 1$ random concomitant variables $(X_{i1}, X_{i2}, \ldots, X_{iq})' = \mathbf{X}_i$ and a single response variable Y_i are measured where the vectors $(\mathbf{X}_i', Y_i)'$ are i.i.d. $(q + 1)$-variate normal with mean vector $(\boldsymbol{\xi}, \eta_i)$ and variance-covariance matrix

$$\mathbf{\Sigma} = \begin{bmatrix} \mathbf{\Sigma}_{XX} & \boldsymbol{\sigma}_{XY} \\ \boldsymbol{\sigma}'_{XY} & \sigma_Y^2 \end{bmatrix},$$

all the parameters being unknown. Here $\boldsymbol{\xi}$ is $q \times 1$, the η_i's are scalars, $\mathbf{\Sigma}_{XX}$ is $q \times q$, $\boldsymbol{\sigma}_{XY}$ is $q \times 1$, and $\sigma_Y^2 > 0$ is a scalar. The η_i's are related by a linear model

$$\boldsymbol{\eta} = (\eta_1, \ldots, \eta_N)' = \mathbf{A}\boldsymbol{\mu}$$

where $\boldsymbol{\mu}: r \times 1$ is a vector of unknown parameters, and $\mathbf{A}: N \times r$ is a known design matrix whose ith row is $\mathbf{a}_i'$ $(1 \leqq i \leqq N)$. Thus conditionally on $\mathbf{X}_i$ we can write

$$Y_i = \mathbf{a}_i'\boldsymbol{\mu} + (\mathbf{X}_i - \boldsymbol{\xi})'\boldsymbol{\beta} + E_i \qquad (1 \leqq i \leqq N) \tag{2.2}$$

where

$$\boldsymbol{\beta} = \mathbf{\Sigma}_{XX}^{-1}\boldsymbol{\sigma}_{XY}$$

and the E_i's are i.i.d. $N(0, \sigma^2)$ with $\sigma^2 = \sigma_Y^2 - \boldsymbol{\sigma}'_{XY}\boldsymbol{\Sigma}_{XX}^{-1}\boldsymbol{\sigma}_{XY}$. It is assumed that $N > q + r$ and that the rows $\mathbf{a}'_i$ of the design matrix $\mathbf{A}$ satisfy certain balance conditions (stated as (2.16), (2.17), and (2.18) in the sequel). The parameters of interest are

$$\theta_i = \mathbf{b}'_i\boldsymbol{\mu} \qquad (1 \leq i \leq k) \tag{2.3}$$

where the $\mathbf{b}_i$'s are $r \times 1$ vectors of known constants. The contrasts among the θ_i's are assumed to be estimable. It can be seen that (2.1) is a special case of (2.2).

Bryant and Bruvold (1980) further generalized the model (2.2) by relaxing the assumption of the same mean $\boldsymbol{\xi}$ for all covariate vectors. They assumed that there are m covariate groups, and if the ith observation comes from the jth covariate group, then $\mathbf{X}_i$ is q-variate normal with mean $\boldsymbol{\xi}_j$ and variance-covariance matrix $\boldsymbol{\Sigma}_{XX}$ (common for all groups), $(1 \leq i \leq N,\ 1 \leq j \leq m)$. Thus model (2.2) generalizes to

$$Y_i = \mathbf{a}'_i\boldsymbol{\mu} + (\mathbf{X}_i - \boldsymbol{\Xi}\boldsymbol{\delta}_i)'\boldsymbol{\beta} + E_i \qquad (1 \leq i \leq N) \tag{2.4}$$

where $\boldsymbol{\Xi} = [\boldsymbol{\xi}_1, \boldsymbol{\xi}_2, \ldots, \boldsymbol{\xi}_m]$ is a $q \times m$ matrix and $\boldsymbol{\delta}_i = (\delta_{i1}, \delta_{i2}, \ldots, \delta_{im})'$ where $\delta_{ij} = 1$ or 0 depending on whether the ith observation belongs to the jth covariate group or not.

Let $\mathbf{Y} = (Y_1, Y_2, \ldots, Y_N)'$, $\mathbf{X} = (\mathbf{X}_1, \mathbf{X}_2, \ldots, \mathbf{X}_N)'$, $\boldsymbol{\Delta} = (\boldsymbol{\delta}_1, \boldsymbol{\delta}_2, \ldots, \boldsymbol{\delta}_N)'$, $\mathbf{E} = (E_1, E_2, \ldots, E_N)'$, and

$$\mathbf{Z} = \mathbf{X} - \boldsymbol{\Delta}\boldsymbol{\Xi}\,. \tag{2.5}$$

Then (2.4) can be written in matrix notation as

$$\mathbf{Y} = \mathbf{A}\boldsymbol{\mu} + \mathbf{Z}\boldsymbol{\beta} + \mathbf{E}\,. \tag{2.6}$$

Example 2.2 discusses a design for which model (2.6) is appropriate.

2.2 Exact Unconditional Procedures

2.2.1 Pairwise Comparisons

In the present section we consider a simultaneous confidence procedure that exactly controls the Type I FWE unconditionally with respect to the values of the covariates for the family of all pairwise comparisons. This procedure can thus be considered a generalization of the classical T-procedure. In the following section we consider a Tukey–Kramer (TK) type procedure used conditionally on the observed values of the covariates. However, there is no mathematical proof that this latter procedure controls the FWE in all cases.

To give the formulas for the least squares (LS) estimates of the parameters of interest, we first define some notation. Let $\hat{\boldsymbol{\Xi}} = (\hat{\boldsymbol{\xi}}_1, \hat{\boldsymbol{\xi}}_2, \ldots, \hat{\boldsymbol{\xi}}_m)$ where

$$\hat{\boldsymbol{\xi}}_j = \frac{1}{\#_j} \sum_{i=1}^{N} \delta_{ij} \mathbf{X}_i \qquad (1 \leq j \leq m)\,; \tag{2.7}$$

here $\#_j = \Sigma_{i=1}^N \delta_{ij}$ is the number of experimental units in covariate group j. Thus $\hat{\boldsymbol{\xi}}_j$ is simply the vector of sample means of the covariates for observations in group j. Further let

$$\hat{\mathbf{Z}} = \mathbf{X} - \boldsymbol{\Delta}\hat{\boldsymbol{\Xi}}\,, \tag{2.8}$$

$$\mathbf{H} = \mathbf{I} - \mathbf{A}(\mathbf{A}'\mathbf{A})^{-}\mathbf{A}'\,, \tag{2.9}$$

$$\mathbf{S}_{XX} = \hat{\mathbf{Z}}'\mathbf{H}\hat{\mathbf{Z}}\,, \qquad \mathbf{S}_{XY} = \hat{\mathbf{Z}}'\mathbf{H}\mathbf{Y}\,, \qquad S_{YY} = \mathbf{Y}'\mathbf{H}\mathbf{Y}\,, \tag{2.10}$$

and

$$\nu = N - q - \text{rank}\,(\mathbf{A})\,; \tag{2.11}$$

here $(\mathbf{A}'\mathbf{A})^{-}$ denotes a generalized inverse of $\mathbf{A}'\mathbf{A}$. Then the LS estimates of $\boldsymbol{\beta}$ and $\boldsymbol{\mu}$ are given by

$$\hat{\boldsymbol{\beta}} = \mathbf{S}_{XX}^{-1}\mathbf{S}_{XY}\,, \tag{2.12}$$

$$\hat{\boldsymbol{\mu}} = (\mathbf{A}'\mathbf{A})^{-}\mathbf{A}'(\mathbf{Y} - \hat{\mathbf{Z}}\hat{\boldsymbol{\beta}})\,, \tag{2.13}$$

and the corresponding unbiased estimate of σ^2 is given by

$$S^2 = \frac{1}{\nu}\{S_{YY} - \mathbf{S}'_{XY}\mathbf{S}_{XX}^{-1}\mathbf{S}_{XY}\}\,. \tag{2.14}$$

Using (2.13) the LS estimate of θ_i can be expressed as

$$\hat{\theta}_i = \mathbf{b}'_i\hat{\boldsymbol{\mu}} = \mathbf{b}'_i(\mathbf{A}'\mathbf{A})^{-}\mathbf{A}'(\mathbf{Y} - \hat{\mathbf{Z}}\hat{\boldsymbol{\beta}}) \qquad (1 \leq i \leq k)\,. \tag{2.15}$$

The balance conditions that are necessary for the application of the exact unconditional procedure can now be given as follows:

(i) Let $\mathbf{B}(\mathbf{A}'\mathbf{A})^{-}\mathbf{B}' = \mathbf{V} = \{v_{ii'}\}$ where $\mathbf{B} = (\mathbf{b}_1, \mathbf{b}_2, \ldots, \mathbf{b}_k)'$. The first balance condition is then (cf. (2.9) of Chapter 3)

$$v_{ii} + v_{i'i'} - 2v_{ii'} = 2v_0 \qquad (1 \leqq i \neq i' \leqq k) . \qquad (2.16)$$

Bryant and Bruvold (1980) postulate the condition that the v_{ii}'s are all equal and the $v_{ii'}$'s for $i \neq i'$ are all equal, which is less general than (2.16). For model (2.1), this condition reduces to $n_1 = \cdots = n_k = n$ (say) and we get $v_0 = 1/n$.

(ii) The second condition is

$$\mathbf{b}_i'(\mathbf{A}'\mathbf{A})^{-}\mathbf{A}'\mathbf{\Delta} = \mathbf{d}' \qquad (1 \leqq i \leqq k) \qquad (2.17)$$

where $\mathbf{d} = (d_1, d_2, \ldots, d_m)'$ is a vector of constants.

(iii) The third condition is that there exists an $r \times m$ matrix of constants $\mathbf{\Gamma} = \{\gamma_{ij}\}$ such that

$$\mathbf{\Delta} = \mathbf{A\Gamma} . \qquad (2.18)$$

In other words, the columns of $\mathbf{\Delta}$ are in the column space of $\mathbf{A}$.

Under these three conditions, Bryant and Bruvold (1980) derived the unconditional (with respect to the $\mathbf{X}_i$'s) distribution of

$$\bar{Q}_{q,k,\nu} = \max_{1 \leqq i < i' \leqq k} \frac{|(\hat{\theta}_i - \hat{\theta}_{i'}) - (\theta_i - \theta_{i'})|}{(v_0 S^2)^{1/2}} . \qquad (2.19)$$

They showed that this distribution depends only on q, k, and ν (and does not depend on m and any unknown nuisance parameters). This distribution is the same as that derived by Bryant and Paulson (1976) under the assumption of a common mean vector for all covariate vectors, and it can be obtained as a special case of Theorem 2.1; see the discussion following that theorem. For $q = 0$ this distribution is identical to the Studentized range distribution. If $\bar{Q}^{(\alpha)}_{q,k,\nu}$ denotes the upper α point of the distribution of $\bar{Q}_{q,k,\nu}$, then $100(1-\alpha)\%$ unconditional simultaneous confidence intervals for all pairwise differences among the θ_i's are given by

$$\theta_i - \theta_{i'} \in [\hat{\theta}_i - \hat{\theta}_{i'} \pm \bar{Q}^{(\alpha)}_{q,k,\nu} S\sqrt{v_0}] \qquad (1 \leqq i < i' \leqq k) . \qquad (2.20)$$

These intervals can be extended to the family of all contrasts by applying (2.4) of Chapter 3. Bryant and Paulson (1976) have given tables of $\bar{Q}^{(\alpha)}_{q,k,\nu}$ for selected values of α, k, q, and ν. A simple approximation to $\bar{Q}^{(\alpha)}_{q,k,\nu}$ in terms of $Q^{(\alpha)}_{k,\nu} = \bar{Q}^{(\alpha)}_{0,k,\nu}$ is given by (2.26) in the sequel.

One can also develop unconditional step-down test procedures based on the ranges of the $\hat{\theta}_i$'s. Bryant and Bruvold (1980) have given a Duncan-type step-down procedure with associated tables of $\bar{Q}^{(\alpha_p)}_{q,p,\nu}$ for the nominal significance levels $\alpha_p = 1 - (1 - \alpha)^{p-1}$ $(2 \leqq p \leqq k)$. We omit the details regarding the development of such procedures since they are quite similar to those described in Chapter 4.

2.2.2 General Contrasts

Bryant and Fox (1985) showed that the extended T-procedure just discussed is a special case of the following general result: Any simultaneous confidence procedure for a given family of contrasts among the treatment means in an ANOVA design can be extended to the same family of contrasts in an ANCOVA design with random covariates. This extension is based on a general distributional result that we now discuss.

Consider model (2.6) and let $\boldsymbol{\theta} = (\theta_1, \theta_2, \ldots, \theta_k)'$ be the parameter vector of interest where the θ_i's are given by (2.3). The LS estimate $\hat{\boldsymbol{\theta}}$ of $\boldsymbol{\theta}$ is given by (2.15) and the corresponding unbiased estimate S^2 of σ^2 is given by (2.14). Under the assumption that $\boldsymbol{\beta} = \mathbf{0}$ (i.e., there are no covariates in the model) denote the LS estimate of $\boldsymbol{\theta}$ by $\hat{\boldsymbol{\theta}}_0$ (obtained by putting $\hat{\boldsymbol{\beta}} = \mathbf{0}$ in (2.15)) and the corresponding unbiased estimate of σ^2 by S_0^2, which has $\nu_0 = N - \text{rank}\,(\mathbf{A})$ d.f.; here $S_0^2 = S_{YY}/\nu_0$. We then have the following theorem.

Theorem 2.1 (Bryant and Fox 1985). Let $T(\mathbf{x})$, $\mathbf{x} \in \mathbb{R}^k$ be any real valued function satisfying the following conditions:

(i) $T(u\mathbf{x}) = u^{\gamma}T(\mathbf{x})$ for any $u \geqq 0$ and for some real constant γ.
(ii) $T(\mathbf{x} + r\mathbf{1}) = T(\mathbf{x})$ for any real r (where $\mathbf{1}$ is a vector of all 1's).

Let $F_q(t; k, \nu)$ denote the distribution function of $T((\hat{\boldsymbol{\theta}} - \boldsymbol{\theta})/S)$ under model (2.6) and let $F_0(t; k, \nu)$ denote the distribution function of $T((\hat{\boldsymbol{\theta}}_0 - \boldsymbol{\theta})/S_0)$ under the additional hypothesis that $\boldsymbol{\beta} = \mathbf{0}$. Then

$$F_q(t; k, \nu) = \int_0^1 F_0(\sqrt{u^{\gamma}}t; k, \nu)g(u)\, du \tag{2.21}$$

where $g(u)$ is the beta density with parameters $(\nu + 1)/2$ and $q/2$, that is,

$$g(u) = \frac{\Gamma([\nu + q + 1]/2)}{\Gamma([\nu + 1]/2) \cdot \Gamma(q/2)}\, u^{(\nu-1)/2}(1 - u)^{(q/2)-1}\,, \qquad 0 \leqq u \leqq 1\,.$$

□

The choice of the function $T(\cdot)$ depends on the particular family of contrasts of interest. Assuming the balance conditions (2.16)–(2.18), the choice

$$T(\mathbf{x}) = \max_{1 \leq i < i' \leq k} \frac{|x_i - x_{i'}|}{\sqrt{v_0}}, \qquad (2.22)$$

corresponds to the family of pairwise comparisons. For the family of comparisons with a control, the balance condition (2.16) modifies to the following two conditions:

$$v_{ii} + v_{kk} - 2v_{ik} = 2v_0 \qquad (1 \leq i \leq k-1), \qquad (2.23)$$

and

$$v_{ii'} - v_{ik} - v_{i'k} + v_{kk} = 2\rho v_0 \qquad (1 \leq i \neq i' \leq k-1) \qquad (2.24)$$

where $-1/(k-1) \leq \rho \leq 1$. In this case, for two-sided comparisons we have

$$T(\mathbf{x}) = \max_{1 \leq i \leq k-1} \frac{|x_i - x_k|}{\sqrt{2v_0}}. \qquad (2.25)$$

For model (2.1), conditions (2.23) and (2.24) reduce to $n_1 = \cdots = n_{k-1} = n$ (say), and we get $2v_0 = 1/n + 1/n_k$ and $\rho = n/(n + n_k)$.

Both (2.22) and (2.25) satisfy conditions (i) and (ii) of Theorem 2.1 for $\gamma = 1$. In the case of (2.22), the distribution $F_q(\cdot\,; k, \nu)$ is that of the $\bar{Q}_{q,k,\nu}$ r.v. defined in (2.19) and it is obtained from (2.21) by substituting for $F_0(\cdot\,; k, \nu)$ the distribution function of the $Q_{k,\nu}$ r.v. In the case of (2.25), the distribution $F_q(\cdot\,; k, \nu)$ is obtained from (2.21) by substituting for $F_0(\cdot\,; k, \nu)$ the distribution function of $\max_{1 \leq i \leq k-1} |T_i|$ where the T_i's have a $(k-1)$-variate equicorrelated t-distribution with ν d.f. and common correlation ρ. Thus this theorem provides a general method for obtaining the distribution function of the pivotal r.v. associated with any unconditional procedure for the ANCOVA design with random covariates (satisfying appropriate balance conditions) given the distribution function of the corresponding pivotal r.v. for the ANOVA design.

Although relation (2.21) can be utilized to evaluate the exact critical points associated with any unconditional procedure, in practice it would be desirable to have simple approximations to these critical points in terms of the critical points of the corresponding procedure for the ANOVA design. Let $D_{q,k,\nu}^{(\alpha)}$ denote the upper α point of the r.v. $D_{q,k,\nu}$, which has the distribution $F_q(\cdot\,; k, \nu)$. Bryant and Fox (1985) proposed the approximation

$$D_{q,k,\nu}^{(\alpha)} \cong \sqrt{\eta} D_{0,k,\phi}^{(\alpha)} \qquad (2.26)$$

where the constants ϕ and η are chosen so as to match the first two moments of the r.v.'s $D^2_{q,k,\nu}$ and $\eta D^2_{0,k,\phi}$. For $\gamma = 1$, these constants are given by

$$\phi = \frac{4t_1 - 2t_2}{t_1 - t_2}, \qquad \eta = \frac{(\phi - 2)\nu(\nu + q - 1)}{\phi(\nu - 2)(\nu - 1)} \tag{2.27}$$

where

$$t_1 = (\nu - 1)(\nu - 2)(\nu + q - 3), \qquad t_2 = (\nu - 3)(\nu - 4)(\nu + q - 1). \tag{2.28}$$

Bryant and Fox made an extensive numerical investigation of this approximation in selected cases and found it to be accurate to within ± 0.02 whenever $\nu \geqq 10$ and $q \leqq 3$. Thus they concluded that it is not necessary to compile separate tables of $D^{(\alpha)}_{q,k,\nu}$ if the tables of $D^{(\alpha)}_{0,k,\nu}$ applicable to the ANOVA design are available. The following example provides an illustration of this approximation.

Example 2.1. Let us approximate $\bar{Q}^{(0.10)}_{1,6,19}$ using (2.26). This critical point is needed in Example 2.3.

First calculate, using (2.28), $t_1 = (19 - 1)(19 - 2)(19 + 1 - 3) = 5202$ and $t_2 = (19 - 3)(19 - 4)(19 + 1 - 1) = 4560$. Next calculate, using (2.27), $\phi = 18.206$ and $\eta = 1.050$. Thus the desired approximation is $\bar{Q}^{(0.10)}_{1,6,19} \cong \sqrt{1.050}\,Q^{(0.10)}_{6,18.206}$. Interpolating linearly in $1/\nu$ between $Q^{(0.10)}_{6,18} = 3.98$ and $Q^{(0.10)}_{6,19} = 3.97$ we get $Q^{(0.10)}_{6,18.206} \cong 3.978$. Substituting this value in the above we get $\bar{Q}^{(0.10)}_{1,6,19} \cong 4.076$. □

2.3 An Approximate Conditional Procedure

In Chapter 3 we discussed the TK-procedure for pairwise comparisons in general fixed-effects linear models. In fact, one of the early applications of the TK-procedure was to the one-way layout design with a single fixed covariate. Hochberg and Varon-Salomon (1984) proposed to use the same procedure (generalized here to handle any number of covariates) in the ANCOVA design conditionally on the observed values of the random covariates. This procedure is based on the following distributional result.

Theorem 2.2. For model (2.6), conditionally on $\mathbf{X} = (\mathbf{X}_1, \mathbf{X}_2, \ldots, \mathbf{X}_N)'$ the estimator $\hat{\boldsymbol{\mu}}$ given by (2.13) is multivariate normal with mean vector

$$E(\hat{\boldsymbol{\mu}}|\mathbf{X}) = (\mathbf{A}'\mathbf{A})^{-}\mathbf{A}'\mathbf{A}\boldsymbol{\mu} + (\mathbf{A}'\mathbf{A})^{-}\mathbf{A}'\boldsymbol{\Delta}(\hat{\boldsymbol{\Xi}} - \boldsymbol{\Xi})\boldsymbol{\beta} \tag{2.29}$$

and variance-covariance matrix

$$\begin{aligned}\text{cov}(\hat{\boldsymbol{\mu}}|\mathbf{X}) &= \sigma^2\{(\mathbf{A}'\mathbf{A})^- + (\mathbf{A}'\mathbf{A})^-\mathbf{A}'\hat{\mathbf{Z}}(\hat{\mathbf{Z}}'\mathbf{H}\hat{\mathbf{Z}})^{-1}\hat{\mathbf{Z}}'\mathbf{A}(\mathbf{A}'\mathbf{A})^-\}\\ &= \sigma^2\mathbf{R} \quad \text{(say)}\end{aligned} \tag{2.30}$$

where $\hat{\mathbf{Z}}$ and $\mathbf{H}$ are defined in (2.8) and (2.9), respectively. Furthermore, $\nu S^2/\sigma^2$ (where ν is given by (2.11) and S^2 is given by (2.14)) is distributed as a χ^2 r.v. with ν d.f. independently of $\hat{\boldsymbol{\mu}}$.

Proof. Conditionally on $\mathbf{X}$, $\hat{\mathbf{Z}}$ is a matrix of constants and $\mathbf{Y}$ is a vector of independent normal r.v.'s; each Y_i has variance σ^2. Thus from the standard results about fixed-effects linear models (see, e.g., Scheffé 1959, Chapter 2) we can show that $\hat{\boldsymbol{\beta}}$ given by (2.12) is normal with $E(\hat{\boldsymbol{\beta}}|\mathbf{X}) = \boldsymbol{\beta}$ and $\text{cov}(\hat{\boldsymbol{\beta}}|\mathbf{X}) = \sigma^2(\hat{\mathbf{Z}}'\mathbf{H}\hat{\mathbf{Z}})^{-1}$. Also $\nu S^2/\sigma^2$ is distributed as χ^2_ν. Furthermore $(\mathbf{A}'\mathbf{A})^-\mathbf{A}'\mathbf{Y}$, $\hat{\boldsymbol{\beta}}$, and S^2 are mutually independent. Therefore $\hat{\boldsymbol{\mu}} = (\mathbf{A}'\mathbf{A})^-\mathbf{A}'(\mathbf{Y} - \hat{\mathbf{Z}}\hat{\boldsymbol{\beta}})$ is distributed independently of S^2 as a multivariate normal vector with $E(\hat{\boldsymbol{\mu}}|\mathbf{X})$ and $\text{cov}(\hat{\boldsymbol{\mu}}|\mathbf{X})$ given by (2.29) and (2.30), respectively. □

Using (2.15) and (2.17) together with Theorem 2.2, we see that the $\hat{\theta}_i$'s are jointly normally distributed with

$$\begin{aligned}E(\hat{\theta}_i|\mathbf{X}) &= \mathbf{b}_i'(\mathbf{A}'\mathbf{A})^-\mathbf{A}'\mathbf{A}\boldsymbol{\mu} + \mathbf{b}_i'(\mathbf{A}'\mathbf{A})^-\mathbf{A}'\boldsymbol{\Delta}(\hat{\boldsymbol{\Xi}} - \boldsymbol{\Xi})\boldsymbol{\beta}\\ &= \mathbf{b}_i'(\mathbf{A}'\mathbf{A})^-\mathbf{A}'\mathbf{A}\boldsymbol{\mu} + \mathbf{d}'\boldsymbol{\Delta}(\hat{\boldsymbol{\Xi}} - \boldsymbol{\Xi})\boldsymbol{\beta} \qquad (1 \le i \le k),\end{aligned} \tag{2.31}$$

and

$$\text{cov}(\hat{\boldsymbol{\theta}}|\mathbf{X}) = \sigma^2\mathbf{BRB}' = \sigma^2\mathbf{W} \quad \text{(say)} \tag{2.32}$$

where $\mathbf{B} = (\mathbf{b}_1, \mathbf{b}_2, \ldots, \mathbf{b}_k)'$ and $\mathbf{R}$ is defined in (2.30). From (2.31) we note that contrasts among the $\hat{\boldsymbol{\theta}}_i$'s are conditionally unbiased. This follows because if $\mathbf{c} \in \mathbb{C}^k$ then

$$\begin{aligned}E(\mathbf{c}'\hat{\boldsymbol{\theta}}|\mathbf{X}) &= \mathbf{c}'\mathbf{B}(\mathbf{A}'\mathbf{A})^-\mathbf{A}'\mathbf{A}\boldsymbol{\mu} + (\mathbf{d}'\boldsymbol{\Delta}(\hat{\boldsymbol{\Xi}} - \boldsymbol{\Xi})\boldsymbol{\beta})\mathbf{c}'\mathbf{1}\\ &= \mathbf{c}'\mathbf{B}(\mathbf{A}'\mathbf{A})^-\mathbf{A}'\mathbf{A}\boldsymbol{\mu}\\ &= \mathbf{c}'\boldsymbol{\theta}\end{aligned}$$

where the last step follows from the estimability assumption made following (2.3). Thus conditionally on $\mathbf{X}$, each

$$T_{ii'} = \frac{\hat{\theta}_i - \hat{\theta}_{i'} - (\theta_i - \theta_{i'})}{S\sqrt{w_{ii} + w_{i'i'} - 2w_{ii'}}} \qquad (1 \le i \ne i' \le k)$$

has Student's t-distribution with ν d.f. Here the $w_{ii'}$'s are elements of the matrix $\mathbf{W}$ defined in (2.32). The extended Tukey conjecture (see Chapter 3, Section 3.2.1) states that $\max_{1\leqq i<i'\leqq k}|T_{ii'}|$ is stochastically smaller than a $Q_{k,\nu}/\sqrt{2}$ r.v. If this conjecture is correct, then 100 $(1-\alpha)\%$ simultaneous (conditional) confidence intervals for pairwise differences $\theta_i - \theta_{i'}$ are given by

$$\theta_i - \theta_{i'} \in \left[\hat{\theta}_i - \hat{\theta}_{i'} \pm Q_{k,\nu}^{(\alpha)} S\sqrt{\frac{w_{ii} + w_{i'i'} - 2w_{ii'}}{2}}\right] \quad (1 \leqq i < i' \leqq k)\,. \tag{2.33}$$

If the conditional joint confidence level of intervals (2.33) is at least $1-\alpha$ then, of course, the unconditional joint confidence level will also be at least $1-\alpha$.

Although we have discussed only the case of pairwise comparisons here, it should be clear that conditional procedures can also be developed for other families of comparisons, for example, comparisons with a control. Such conditional procedures are not restricted to designs satisfying certain balance conditions as are the corresponding unconditional procedures.

The following example demonstrates, for a randomized complete block design with fixed block effects, the computations required for implementing the unconditional and conditional procedures for pairwise comparisons. A numerical illustration of the procedures is given in Example 2.3.

Example 2.2. Consider a randomized complete block design with k treatments, n blocks, and $N = kn$ experimental units where both the treatments and the blocking factor are fixed. Let $(X_{ij}, Y_{ij})'$ be a bivariate normal observation on the ith treatment in the jth block. Conditional on X_{ij} we assume that

$$Y_{ij} = \theta_i + \psi_j + \beta(X_{ij} - \xi_j) + E_{ij} \qquad (1 \leqq i \leqq k, 1 \leqq j \leqq n) \tag{2.34}$$

where θ_i is the ith treatment effect, ψ_j is the jth block effect, ξ_j is the common mean of the covariates X_{ij} from block j, β is the common slope coefficient, and the E_{ij}'s are i.i.d. $N(0, \sigma^2)$ random errors; here all the unknown parameters are regarded as fixed and the ψ_j's satisfy the condition $\Sigma_{j=1}^{n} \psi_j = 0$.

We establish a correspondence between (2.34) and (2.6) by noting that $\boldsymbol{\mu} = (\theta_1, \ldots, \theta_k; \psi_1, \ldots, \psi_n)'$, $\boldsymbol{\Xi} = (\xi_1, \xi_2, \ldots, \xi_n)$, $q = 1$, $r = k + n$,

$$\mathbf{A}=\begin{array}{r} \\ n\left\{\begin{array}{c}\\ \\ \\ \end{array}\right. \\ n\left\{\begin{array}{c}\\ \\ \\ \end{array}\right. \\ \vdots \\ n\left\{\begin{array}{c}\\ \\ \\ \end{array}\right. \end{array}\overset{\qquad\overbrace{\qquad\qquad}^{k}\qquad\overbrace{\qquad\qquad}^{n}}{\begin{bmatrix} 1 & & & & 1 & & \\ \vdots & & & & & \ddots & \\ 1 & & & & & & 1 \\ & 1 & & & 1 & & \\ & \vdots & & & & \ddots & \\ & 1 & & & & & 1 \\ & & \ddots & & & & \\ & & & 1 & 1 & & \\ & & & \vdots & & \ddots & \\ & & & 1 & & & 1 \end{bmatrix}}_{kn\times(k+n)}, \tag{2.35}$$

which is the incidence matrix of the design, and

$$\boldsymbol{\Delta}=\begin{array}{r} n\left\{\begin{array}{c}\\ \\ \\ \end{array}\right. \\ n\left\{\begin{array}{c}\\ \\ \\ \end{array}\right. \\ \vdots \\ n\left\{\begin{array}{c}\\ \\ \\ \end{array}\right. \end{array}\begin{bmatrix} 1 & & \\ & \ddots & \\ & & 1 \\ 1 & & \\ & \ddots & \\ & & 1 \\ \vdots & & \vdots \\ 1 & & \\ & \ddots & \\ & & 1 \end{bmatrix}_{kn\times n}. \tag{2.36}$$

The parameters of interest here are $\theta_i = \mathbf{b}_i'\boldsymbol{\mu}$ $(1 \leqq i \leqq k)$ where $\mathbf{b}_i$ is a $(k+n)$-vector with 1 in the ith place and 0 everywhere else. For this design the balance conditions (2.16)–(2.18) can be verified as follows:

(i) First check that $\mathbf{B}(\mathbf{A}'\mathbf{A})^{-}\mathbf{B}' = \mathbf{V} = (1/n)\mathbf{I}$ and thus condition (2.16) is satisfied with $v_0 = 1/n$.

(ii) Next check that $\mathbf{b}_i'(\mathbf{A}'\mathbf{A})^{-}\mathbf{A}'$ is a $1 \times kn$ row vector consisting of k blocks of n components each, with the ith block consisting of all entries $= 1/n$ and all other blocks consisting of zero entries only. Thus $\mathbf{b}_i'(\mathbf{A}'\mathbf{A})^{-}\mathbf{A}'\mathbf{\Delta}$ is a $1 \times n$ row vector $\mathbf{d}' = (1/n, \ldots, 1/n)$ for all $i = 1, 2, \ldots, k$, and hence condition (2.17) is satisfied.

(iii) Condition (2.18) is obviously satisfied since the columns of $\mathbf{\Delta}$ are the same as the last n columns of $\mathbf{A}$ as is seen by inspecting (2.35) and (2.36).

Hence the exact unconditional procedure (2.20) can be used for making pairwise comparisons among the θ_i's.

The estimates of interest are derived as follows. Let

$$S_{XY} = \sum_{i=1}^{k} \sum_{j=1}^{n} (X_{ij} - \bar{X}_{i\cdot} - \bar{X}_{\cdot j} + \bar{X}_{\cdot\cdot})(Y_{ij} - \bar{Y}_{i\cdot} - \bar{Y}_{\cdot j} + \bar{Y}_{\cdot\cdot}) ,$$

$$S_{XX} = \sum_{i=1}^{k} \sum_{j=1}^{n} (X_{ij} - \bar{X}_{i\cdot} - \bar{X}_{\cdot j} + \bar{X}_{\cdot\cdot})^2 , \tag{2.37}$$

$$S_{YY} = \sum_{i=1}^{k} \sum_{j=1}^{n} (Y_{ij} - \bar{Y}_{i\cdot} - \bar{Y}_{\cdot j} + \bar{Y}_{\cdot\cdot})^2 .$$

The LS estimates of the θ_i's are

$$\hat{\theta}_i = \bar{Y}_{i\cdot} - \hat{\beta}(\bar{X}_{i\cdot} - \bar{X}_{\cdot\cdot}) \qquad (1 \leqq i \leqq k) \tag{2.38}$$

where

$$\hat{\beta} = \frac{S_{XY}}{S_{XX}} . \tag{2.39}$$

Also

$$S^2 = \frac{1}{\nu}\left[S_{YY} - \frac{S_{XY}^2}{S_{XX}}\right] \tag{2.40}$$

where

$$\nu = kn - 1 - (k + n - 1) = kn - k - n . \tag{2.41}$$

Thus for a randomized complete block design with a single random covariate, all the quantities needed for applying the exact unconditional procedure (2.20) are available through the formulas (2.37)–(2.41).

For applying the approximate conditional procedure (2.33) we need, in addition, the matrix $\mathbf{W} = \mathbf{BRB}'$ where $\mathbf{B} = [\mathbf{I}_{k\times k}|\mathbf{0}_{k\times n}]$ and $\mathbf{R}$ is given by (2.30). It can be checked that

$$\sigma^2\{w_{ii} + w_{i'i'} - 2w_{ii'}\} = \sigma^2\left\{\frac{2}{n} + \frac{(\bar{X}_{i\cdot} - \bar{X}_{i'\cdot})^2}{S_{XX}}\right\} \qquad (1 \leqq i < i' \leqq k) \tag{2.42}$$

which, of course, is the formula for the variance of $\hat{\theta}_i - \hat{\theta}_{i'}$ in the corresponding fixed covariate model. □

Example 2.3 ***(Randomized Block Design with a Random Covariate).*** We use the data of Example 1.2 of Chapter 3 to illustrate the conditional and unconditional procedures. Recall that in that example (and in Example 3.2 of Chapter 3) the covariate (percentage of dry matter) was regarded as fixed for illustration purposes. Here we regard it as random.

Suppose that we wish to construct simultaneous 90% unconditional confidence intervals (2.20) for all pairwise differences $\theta_i - \theta_{i'}$ $(1 \leqq i < i' \leqq 6)$. In Example 1.2 of Chapter 3 we calculated the LS-estimates $\hat{\theta}_i$ that are given in Table 1.4 of that chapter. In that example we also calculated $S = 7.106$ with $\nu = 19$ d.f. In Example 2.1 of this chapter we obtained the value of the critical constant $\bar{Q}^{(0.10)}_{1,6,19} \cong 4.076$. Thus the common allowance for the desired intervals is $\bar{Q}^{(0.10)}_{1,6,19}S/\sqrt{n} \cong 4.076 \times 7.106/\sqrt{5} = 12.95$.

The intervals given by (2.33) corresponding to the approximate conditional procedure are the same as the TK-intervals, the computational details of which are given in Example 3.2 of Chapter 3.

For these data, the unconditional intervals work out to be shorter than the conditional intervals for all pairs except (1, 3). The allowance for (1, 3) using the latter interval is 12.69 (see Example 3.2 of Chapter 3); using the former interval it is 12.95. As we see in Section 2.4, however, generally the conditional procedure yields shorter intervals than the unconditional procedure. □

2.4 A Comparison of the Conditional and Unconditional Procedures

As noted before, the unconditional procedures are applicable only in certain balanced designs, while the conditional procedures are not so restricted. Thus, for example, the latter can be used for making pairwise or treatments versus control comparisons in ANCOVA designs that do not satisfy the appropriate balance conditions given earlier.

It should be reemphasized, however, that for the family of pairwise comparisons there is as yet no analytical proof for $k > 3$ that the conditional TK-type procedure (2.33) controls the unconditional FWE. This result would follow if the Tukey conjecture were true for $k > 3$ for general nondiagonal convariance matrices; see the discussion in Section 3.2.1 of Chapter 3. Brown (1984) proved this conjecture for $k = 3$. For $k > 3$ Hochberg and Varon-Salomon (1984) studied the unconditional Type I FWE of this procedure for the one-way ANCOVA design with a single random covariate and equal number of replications per treatment. They showed that this FWE depends on k, α, and n, but it does not depend on the parameters of the joint normal distribution of (X, Y). They also conducted a simulation study and found that the unconditional FWE $\leqq \alpha$ in all the cases that they studied. This lends further support to the conjecture that the conditional procedure (2.33) is always conservative.

Of course, instead of using the Tukey–Kramer version of the conditional procedure, one could use the GT2 or the Dunn–Šidák versions, which are known to be conservative. Alternatively one could use Scheffé's S-procedure conditionally, which would provide exact control of the unconditional FWE for the family of all contrasts but which would be overly conservative for the family of pairwise comparisons. In general, if we have a procedure that is known to control the FWE for a given family of contrasts in ANCOVA designs with fixed covariates, then that procedure when used in a conditional manner would also control the FWE unconditionally in designs with random covariates. Whether to use a conditional or an unconditional procedure in a given situation when both can be applied depends on the following considerations among others: (i) guaranteed control of the FWE, (ii) convenience of use, and (iii) lengths of the confidence intervals.

With regard to (i), we already have commented in detail as far as the conditional procedures are concerned. The unconditional procedures, of course, control the FWE exactly. With regard to (ii), the conditional procedures are more convenient to use because their critical points are more widely tabulated. Nonetheless, (iii) is likely to be the major consideration in the choice of a procedure and hence we discuss it in detail next.

Hochberg and Varon-Salomon (1984) compared the lengths of the pairwise confidence intervals yielded by the two procedures, (2.20) and (2.33), for the balanced one-way layout design with a single random covariate. Note that the unconditional procedure (2.20) yields the same length for each pairwise interval, namely $2\bar{Q}^{(\alpha)}_{1,k,\nu}S/\sqrt{n}$, while the conditional procedure (2.33) yields different lengths. The ratio of the lengths of the two intervals for the (i, i')th comparison is given by

$$R_{\alpha;1,k,\nu} = \frac{Q^{(\alpha)}_{k,\nu}S\sqrt{1/n + (\bar{X}_{i\cdot} - \bar{X}_{i'\cdot})^2/2S_{XX}}}{\bar{Q}^{(\alpha)}_{1,k,\nu}S/\sqrt{n}},$$

where $S_{XX} = \Sigma_{i=1}^{k}\Sigma_{j=1}^{n}(X_{ij} - \bar{X}_{i\cdot})^2$. For the general case of $q \geqq 1$ random covariates this ratio can be expressed as

$$R_{\alpha;q,k,\nu} = \frac{Q^{(\alpha)}_{k,\nu}}{\bar{Q}^{(\alpha)}_{q,k,\nu}}\sqrt{1 + \frac{q}{\nu+1}F_{q,\nu+1}}$$

where $F_{m,n}$ is an F r.v. with m and n d.f. Hochberg and Varon-Salomon (1984) evaluated $E\{R_{\alpha;1,k,\nu}\}$ for various combinations of α, k, and ν, and found it to be generally less than unity, thus favoring the conditional procedure over the unconditional one. They also evaluated $\Pr\{R_{\alpha;1,k,\nu} > 1\}$ and found that it generally lies between 0.2 and 0.3.

Bryant and Fox (1985) investigated the case $q = 3$. Their results are in general agreement with those of Hochberg and Varon-Salomon for the case $q = 1$. They found that the distribution of $R_{\alpha;3,k,\nu}$ is tightly concentrated near 1 particularly for $\nu \geqq 15$. Thus although $\Pr\{R_{\alpha;3,k,\nu} > 1\}$ is small (0.2 to 0.4), $E\{R_{\alpha;3,k,\nu}\}$ is only slightly less than unity. The superiority of the conditional procedure over the unconditional one is substantial only for very small values of ν. The same conclusion was arrived at when they considered alternative measures such as the expected squared lengths and the root mean square lengths of the $\binom{k}{2}$ pairwise intervals associated with the two procedures.

To summarize, for the family of pairwise comparisons the conditional procedure (2.33) is preferred over the unconditional procedure (2.20) for small d.f. ν. For moderate to large ν the two procedures give about the same results and hence the choice would depend on other considerations. Analogous comparisons of the conditional and unconditional procedures for other families of contrasts have not been carried out, but one would expect the final conclusions to be not too different from the above.

CHAPTER 9

Distribution-Free and Robust Procedures

This chapter is primarily devoted to multiple comparison procedures (MCPs) that are distribution-free (also commonly referred to as nonparametric) in the sense that their Type I error rates do not depend (under relatively mild assumptions) on the distributions that generate the sample data. In the context of testing of hypotheses, this means that the marginal or also the joint null distributions of the test statistics do not depend on the underlying distributions of the observations. This corresponds to the notion of a testing family and of a joint testing family (discussed in Section 1.2 of Appendix 1), respectively. Some of the procedures that are surveyed here are distribution-free in a more restrictive sense, as is seen in the sequel. We also discuss in this chapter some robust MCPs that are not exactly distribution-free but that are resistant to substantial departures from the usual distributional assumptions (in particular, the normality assumption) and to gross outlier observations.

It is well known (see, e.g., Scheffé 1959, Chapter 10) that single inference procedures based on t- and F-statistics are fairly robust to nonnormality. However, as noted by Ringland (1983), the problem of robustness becomes more serious in the case of multiple inferences. To illustrate this point Ringland used the example of the normal theory Bonferroni procedure for a finite family of inferences on scalar parametric functions. This procedure uses the critical points from the extreme tail portion of the t-distribution, and this is the portion that is most sensitive to nonnormality. Furthermore, even if the discrepancy due to nonnormality in the Type I error probability for a single inference is not large, for multiple inferences this discrepancy gets magnified by a factor equal to the number of inferences. Since many finite union-intersection (UI) procedures are approximated by the Bonferroni procedure, they also suffer from the same lack of robustness. Hence the MCPs of this chapter should be used when the normality assumption is suspect.

To keep the presentation simple, the main body of the discussion in this chapter is restricted to simple statistics such as sign, rank, and signed rank statistics. Extensions to general linear rank statistics are briefly summarized in Section 3.3.

The one-way layout design is the setting for Section 1. In Sections 1.1 and 1.2 we consider single-step test procedures for two families—comparisons with a control and all pairwise comparisons, respectively. We also explain how these procedures yield the associated simultaneous confidence intervals for appropriate contrasts among location parameters when a location model is assumed. Step-down procedures for the family of subset hypotheses are discussed in Section 1.3. A comparison among the various procedures is given in Section 1.4.

The randomized complete block design is the setting for Section 2. The underlying models and the associated hypotheses of interest are explained in Section 2.1. In Section 2.2 we consider the family of comparisons with a control and discuss Steel's procedure based on sign statistics and Nemenyi's procedure based on signed rank statistics. Section 2.3 is devoted to single-step test procedures for the family of all pairwise comparisons. Analogs of the Steel and Nemenyi procedures are discussed as well as another procedure proposed by Nemenyi that is based on Friedman rank statistics. Section 2.4 describes some step-down procedures. A comparison among the various procedures is made in Section 2.5.

Some other nonparametric procedures and problems are discussed in Section 3. They include permutation procedures, median tests, and general linear rank statistics. Finally in Section 4 a discussion of MCPs based on some robust estimators is given.

1 PROCEDURES FOR ONE-WAY LAYOUTS

1.1 Comparisons with a Control

Consider a control treatment labeled k and test treatments labeled $1, 2, \ldots, k-1$ where $k \geqq 3$. Let $\{Y_{ij}\ (1 \leqq j \leqq n_i)\}$ be a random sample of size n_i from treatment i $(1 \leqq i \leqq k)$. We assume that the Y_{ij}'s come from a continuous distribution F_i $(1 \leqq i \leqq k)$. We also assume that $n_1 = n_2 = \cdots = n_{k-1} = n$ (say), which may be different from n_k. Under this assumption the computation of the exact critical points necessary to implement the MCPs described below is greatly simplified.

Steel (1959b) proposed a single-step test procedure for the family of hypotheses $H_{0i}: F_i \equiv F_k$ $(1 \leqq i \leqq k-1)$, which we now describe. In this procedure the n observations from F_i and the n_k observations from F_k are pooled and rank ordered from the smallest to the largest. Because the observations from only the treatments being compared are ranked, this is referred to as the method of *separate rankings*. Let R_{ij} be the rank of Y_{ij} in this ranking $(1 \leqq j \leqq n)$ and let RS_{ik}^{+} be the Wilcoxon rank sum statistic:

$$RS_{ik}^{+} = \sum_{j=1}^{n} R_{ij} \qquad (1 \leqq i \leqq k-1).$$

Suppose that the alternatives to the H_{0i}'s are the one-sided hypotheses $H_{1i}^{(+)}: F_i < F_k$ $(1 \leqq i \leqq k-1)$ (where $F < G$ means that $F(x) \leqq G(x)$ for all $x \in \mathbb{R}$ with a strict inequality for at least some x, that is, a random observation from F is stochastically larger than an independent random observation from G). In this case Steel's procedure rejects H_{0i} if

$$RS_{ik}^{+} > RS_{k-1}^{+(\alpha)} \qquad (1 \leqq i \leqq k-1) \tag{1.1}$$

where $RS_{k-1}^{+(\alpha)}$ is the upper α point of the distribution of

$$RS_{k-1}^{+} = \max_{1 \leqq i \leqq k-1} RS_{ik}^{+}$$

under the overall null hypothesis $H_0: F_1 \equiv F_2 \equiv \cdots \equiv F_k$.

If the alternatives to the H_{0i}'s are the one-sided hypotheses $H_{1i}^{(-)}: F_i > F_k$ $(1 \leqq i \leqq k-1)$, then Steel's procedure rejects H_{0i} if

$$RS_{ik}^{-} = n(n_k + n + 1) - RS_{ik}^{+} > RS_{k-1}^{+(\alpha)} \qquad (1 \leqq i \leqq k-1). \tag{1.2}$$

(Note that RS_{ik}^{-} is the rank sum for sample i if all $n + n_k$ observations from treatments i and k are assigned ranks in the reverse order. The same critical point $RS_{k-1}^{+(\alpha)}$ is used in both (1.1) and (1.2) because the joint distribution of the RS_{ik}^{+}'s is the same as that of the RS_{ik}^{-}'s under H_0.)

For the two-sided alternative $H_{1i} = H_{1i}^{(+)} \cup H_{1i}^{(-)}$, Steel's procedure rejects H_{0i} if

$$RS_{ik} = \max(RS_{ik}^{+}, RS_{ik}^{-}) > RS_{k-1}^{(\alpha)} \qquad (1 \leqq i \leqq k-1) \tag{1.3}$$

where $RS_{k-1}^{(\alpha)}$ is the upper α point of the null distribution (under H_0) of

$$RS_{k-1} = \max_{1 \leqq i \leqq k-1} RS_{ik}. \tag{1.4}$$

Note that the critical points $RS_{k-1}^{+(\alpha)}$ and $RS_{k-1}^{(\alpha)}$ are functions of k, n, n_k, and α only, and do not depend on the specific common distribution function under H_0. Therefore these procedures control the Type I familywise error rate (FWE) under H_0 at level α. A natural question to ask is whether the Type I FWE is controlled under configurations other than H_0. The following theorem answers this question.

Theorem 1.1. Each one of the single-step test procedures (1.1)–(1.3) strongly (that is, under all configurations) controls the Type I FWE at the designated level α.

Proof. We prove the result only for (1.1); the proofs for the other two procedures are similar. Let P be any nonempty subset of $\{1, 2, \ldots, k-1\}$ of cardinality p. Then under $\bigcap_{i \in P} H_{0i}$, the joint distribution of the RS_{ik}^{+}'s for $i \in P$ does not depend on the common distribution $F_i = F_k$, $i \in P$. This joint distribution is simply the permutational distribution obtained from the equally likely (under $\bigcap_{i \in P} H_{0i}$) orderings of the $n_k + pn$ observations where in computing RS_{ik}^{+} we only consider the relative ranks of the observations from the treatments $i \in P$ and k. Therefore $\{(H_{0i}, RS_{ik}^{+}),\ i = 1, \ldots, k-1\}$ forms a joint testing family. From Theorem 1.2 of Appendix 1 it follows that the Type I FWE of (1.1) is maximized under the overall null hypothesis $H_0 = \bigcap_{i=1}^{k-1} H_{0i} : F_1 = F_2 = \cdots = F_k$. Since $RS_{k-1}^{+(\alpha)}$ is the upper α point of the distribution of $\max_{1 \leq i \leq k-1} RS_{ik}^{+}$ under H_0, the theorem follows. □

Steel (1959b) computed exact upper tail probabilities of the null distribution of RS_{k-1} for $k = 3, 4$; $n = 3, 4, 5$ where $n = n_i$ $(1 \leq i \leq k)$. For large n_i's the joint null distribution of the RS_{ik}^{+}'s can be approximated by a multivariate normal distribution with the following parameters:

$$E(RS_{ik}^{+}) = \frac{n_i(n_i + n_k + 1)}{2} \qquad (1 \leq i \leq k-1),$$

$$\operatorname{var}(RS_{ik}^{+}) = \frac{n_i n_k(n_i + n_k + 1)}{12} \qquad (1 \leq i \leq k-1), \tag{1.5}$$

$$\operatorname{corr}(RS_{ik}^{+}, RS_{jk}^{+}) = \sqrt{\frac{n_i n_j}{(n_i + n_k + 1)(n_j + n_k + 1)}} \qquad (1 \leq i < j \leq k-1).$$

Thus for $n_1 = n_2 = \cdots = n_{k-1} = n$ (say), a large sample approximation to $RS_{k-1}^{+(\alpha)}$ is given by

$$RS_{k-1}^{+(\alpha)} \cong \left\lfloor \frac{n(n + n_k + 1)}{2} + \frac{1}{2} + Z_{k-1,\rho}^{(\alpha)} \sqrt{\frac{n n_k(n + n_k + 1)}{12}} \right\rfloor \tag{1.6}$$

where $\lfloor x \rfloor$ denotes the smallest integer greater than or equal to x and $Z_{k-1,\rho}^{(\alpha)}$ is the one-sided upper α equicoordinate point of $k-1$ equicorrelated standard normal variables with common correlation $\rho = n/(n + n_k + 1)$. Similarly a large sample approximation to $RS_{k-1}^{(\alpha)}$ is given by

$$RS_{k-1}^{(\alpha)} \cong \left\lfloor \frac{n(n+n_k+1)}{2} + \frac{1}{2} + |Z|_{k-1,\rho}^{(\alpha)} \sqrt{\frac{nn_k(n+n_k+1)}{12}} \right\rfloor \quad (1.7)$$

where $|Z|_{k-1,\rho}^{(\alpha)}$ is the corresponding two-sided upper α equicoordinate point. The critical points $Z_{k-1,\rho}^{(\alpha)}$ and $|Z|_{k-1,\rho}^{(\alpha)}$ are given in Tables 2 and 3 respectively, of Appendix 3, for selected values of k, α, and ρ. The constant $\frac{1}{2}$ in (1.6) and (1.7) is the standard continuity correction.

Suppose now that the F_i's are members of a location parameter family, that is,

$$F_i(y) = F(y - \theta_i) \qquad (1 \leqq i \leqq k) \quad (1.8)$$

where $F(\cdot)$ is an unknown continuous distribution function and the θ_i's are unknown location parameters. We now explain how to construct simultaneous confidence intervals (one-sided or two-sided) for $\theta_i - \theta_k$ $(1 \leqq i \leqq k-1)$. The basic theory of nonparametric confidence intervals based on rank statistics for location parameters is given in Lehmann (1963) and Hodges and Lehmann (1963).

First consider one-sided intervals. Let $RS_{ik}^{+}(\delta)$ be the rank sum for sample i when a constant δ is subtracted from each observation from sample i, which is then ranked together with sample k $(1 \leqq i \leqq k-1)$. The set of values of δ that satisfy the inequality $RS_{ik}^{+}(\delta) \leqq RS_{k-1}^{+(\alpha)}$ is an interval of the form $[\delta_i, \infty)$. The collection of these intervals forms simultaneous lower one-sided confidence intervals for the differences $\theta_i - \theta_k$ $(1 \leqq i \leqq k-1)$ with a joint confidence level of $1-\alpha$. Simultaneous upper one-sided intervals are obtained similarly by considering the inequalities $RS_{ik}^{-}(\delta) \leqq RS_{k-1}^{+(\alpha)}$. Simultaneous two-sided intervals are obtained by considering the inequalities $RS_{ik}(\delta) = \max\{RS_{ik}^{+}(\delta), RS_{ik}^{-}(\delta)\} \leqq RS_{k-1}^{(\alpha)}$.

Explicit formulas for these intervals can be given as follows. Let $D_{ik,(1)} < D_{ik,(2)} < \cdots < D_{ik,(nn_k)}$ be the ordered values of the differences $D_{ik,jl} = Y_{ij} - Y_{kl}$ $(1 \leqq j \leqq n,\ 1 \leqq l \leqq n_k)$. Then simultaneous $(1-\alpha)$-level lower one-sided confidence intervals on the $(\theta_i - \theta_k)$'s are given by $[D_{ik,(N_{k-1}^{+(\alpha)}+1)}, \infty)$, $i = 1, \ldots, k-1$, where

$$N_{k-1}^{+(\alpha)} = nn_k + \frac{n(n+1)}{2} - RS_{k-1}^{+(\alpha)}\,. \quad (1.9)$$

The corresponding upper one-sided intervals are given by $(-\infty, D_{ik,(nn_k - N_{k-1}^{+(\alpha)})}]$, $i = 1, \ldots, k-1$, and the two-sided intervals are given by

$$[D_{ik,(N_{k-1}^{(\alpha)}+1)}, D_{ik,(nn_k - N_{k-1}^{(\alpha)})}], \qquad i = 1, \ldots, k-1,$$

where

$$N_{k-1}^{(\alpha)} = nn_k + \frac{n(n+1)}{2} - RS_{k-1}^{(\alpha)}. \qquad (1.10)$$

These confidence intervals can be obtained graphically as follows (Miller 1981 has credited this graphical method to L.E. Moses). To obtain the lower one-sided intervals plot the nn_k pairs of values (Y_{kl}, Y_{ij}) $(1 \leqq j \leqq n,\ 1 \leqq l \leqq n_k)$ as points on a graph paper (with the Y_{kl}'s as abscissa and the Y_{ij}'s as ordinate values). Draw a 45° line through one of these points such that exactly $N_{k-1}^{+(\alpha)}$ points fall below the line. Then $D_{ik,(N_{k-1}^{+(\alpha)}+1)}$ is the intercept of this line on the ordinate axis $(1 \leqq i \leqq k-1)$. The upper one-sided intervals are obtained by following the same procedure but by drawing a 45° line through one of the points such that $N_{k-1}^{+(\alpha)}$ points lie above it. The two-sided confidence intervals are also obtained similarly but now two 45° lines are drawn through two of these points such that the lower line has $N_{k-1}^{(\alpha)}$ points below it and the upper line has $N_{k-1}^{(\alpha)}$ points above it. The interval intercepted by these two lines on the vertical axis is then $[D_{ik,(N_{k-1}^{(\alpha)}+1)}, D_{ik,(nn_k - N_{k-1}^{(\alpha)})}]$, $i = 1, \ldots, k-1$.

Example 1.1. The following data in Table 1.1 on the rate of dust removal from three groups of subjects to assess mucociliary efficiency are taken from Hollander and Wolfe (1973, p. 116).

We wish to make two-sided comparisons of the two disease groups of subjects with the normal (control) group. We use Steel's procedure (1.3)

TABLE 1.1. Half-Time of Mucociliary Clearance (in Hours)

Subjects with Obstructive Airway Disease (1)	Subjects with Asbestosis (2)	Normal Subjects (3)
3.8	2.8	2.9
2.7	3.4	3.0
4.0	3.7	2.5
2.4	2.2	2.6
	2.0	3.2

Source: Hollander and Wolfe (1973).

for this purpose. Separate pairwise rankings result in $RS_{13}^+ = 22$, $RS_{13}^- = 18$, and $RS_{23}^+ = 27$, $RS_{23}^- = 28$. Exact critical points of the null distribution of $RS_{k-1} = \max_{1 \le i \le k-1}\{\max(RS_{ik}^+, RS_{ik}^-)\}$ are not available when the sample sizes are unequal. We explore the applicability of the large sample approximation (although the sample sizes here are quite small) based on the asymptotic joint normality of the rank sums with parameters given by (1.5). To control the FWE asymptotically at level α, we can compare the test statistic $RS_{i3} = \max(RS_{i3}^+, RS_{i3}^-)$ with the critical constant (obtained using (1.7))

$$RS_{i3}^{(\alpha)} \cong \left\lfloor \frac{n_i(n_i + n_3 + 1)}{2} + \frac{1}{2} + |Z|_{2,\rho}^{(\alpha)} \sqrt{\frac{n_i n_3 (n_i + n_3 + 1)}{12}} \right\rfloor \qquad (i = 1, 2) \quad (1.11)$$

where

$$\rho = \operatorname{corr}(RS_{13}^+, RS_{23}^+) = \sqrt{\frac{n_1 n_2}{(n_1 + n_3 + 1)(n_2 + n_3 + 1)}} . \qquad (1.12)$$

(Note a slight change in notation between (1.7) and (1.11). In (1.7) the subscript on $RS^{(\alpha)}$ denotes the *number* of treatment-control comparisons while that in (1.11) denotes the *particular* treatment-control comparison.) For the equal sample sizes case, (1.11) gives a reasonably good approximation even for small sample sizes. For example, if $n_1 = n_2 = n_3 = 5$, then for $\alpha = 0.25$ we have $|Z|_{2,0.5}^{(0.25)} = 1.4538$ from Odeh's (1982) tables, and thus we obtain $RS_{13}^{(0.25)} = RS_{23}^{(0.25)} \cong \lfloor 34.96 \rfloor = 35$. If we use Steel's (1959b) exact tables, then we get the two-sided critical point corresponding to $\alpha = 0.2486$ to be 36.

Note that in (1.6) and (1.7) we assumed that $n_1 = n_2 = \cdots = n_{k-1}$ so as to obtain a common correlation ρ between all the RS_{ik}^+'s since tables of the maximum of correlated standard normal variables are available only for this special case. However, for $k - 1 = 2$ there is only a single correlation to consider, which makes this assumption unnecessary.

From (1.12) we calculate $\rho = 0.4264$. A conservative approximation to $|Z|_{2,0.4264}^{(\alpha)}$ is $|Z|_{2,0.4}^{(\alpha)}$, which for $\alpha = 0.25$ equals 1.4706 as seen from Odeh's (1982) tables. From (1.11) we then calculate $RS_{13}^{(0.25)} \cong \lfloor 26.50 \rfloor = 27$ and $RS_{23}^{(0.25)} \cong \lfloor 35.04 \rfloor = 36$. Comparing these critical points with $\max\{RS_{13}^+, RS_{13}^-\} = 22$ and $\max\{RS_{23}^+, RS_{23}^-\} = 28$, respectively, we conclude that neither comparison is significant at $\alpha = 0.25$.

If we assume the location model (1.8) for the present data, then simultaneous confidence intervals on $\theta_i - \theta_3$ $(i = 1, 2)$ can be obtained using the graphical procedure described earlier. Consider $\theta_1 - \theta_3$ first.

Plot the pairs (Y_{3l}, Y_{1j}) $(1 \leqq j \leqq 4, 1 \leqq l \leqq 5)$ as shown in Figure 1.1. To obtain a 75% simultaneous two-sided confidence interval on $\theta_1 - \theta_3$ draw two 45° lines through two of these points such that $N_{13}^{(0.25)}$ points fall above the upper line and below the lower line, where (from (1.10))

$$N_{13}^{(0.25)} = 4 \times 5 + \frac{4 \times 5}{2} - 27 = 3\,;$$

here we have used the approximation $RS_{13}^{(0.25)} \cong 27$. (Note again a slight change in notation compared to (1.10), where the subscript on $N^{(\alpha)}$ denotes the *number* of treatment-control comparisons; here it denotes the *particular* treatment-control comparison.) The desired interval is given by the interval intercepted on the vertical axis by the two 45° lines. From Figure 1.1 this interval is seen to be $[-0.50, 1.20]$.

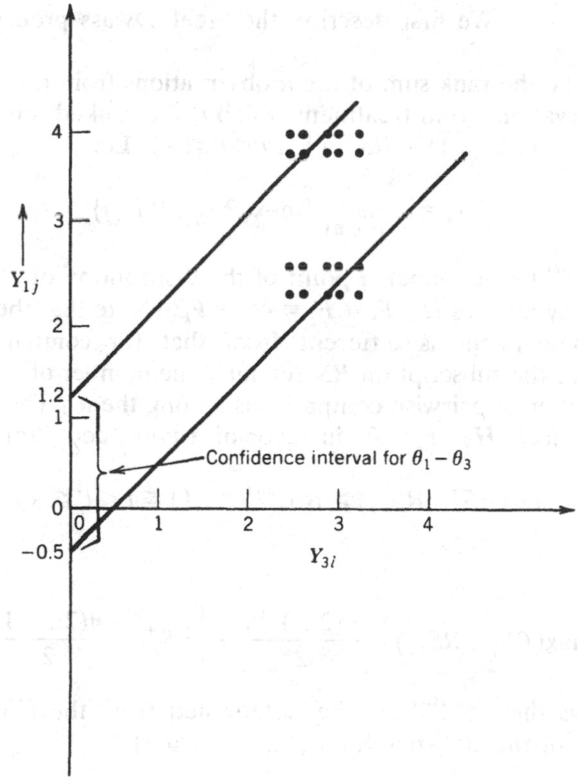

Figure 1.1. Graphical construction of a simultaneous 75% two-sided confidence interval for $\theta_1 - \theta_3$.

A similar graphical construction gives $[-0.80, 0.80]$ as a simultaneous 75% two-sided confidence interval for $\theta_2 - \theta_3$. Here the number of points falling above and below the two 45° lines is (using $RS_{23}^{(0.25)} \cong 36$)

$$N_{23}^{(0.25)} = 5 \times 5 + \frac{5 \times 6}{2} - 36 = 4 . \qquad \square$$

1.2 Pairwise Comparisons

1.2.1 A Procedure Based on Rank Sum Statistics Using Separate Rankings

Consider the setup of Section 1.1 but now consider the family of all pairwise hypotheses $H_{ii'}: F_i \equiv F_{i'}$ $(1 \leqq i < i' \leqq k)$ against two-sided alternatives. Steel (1960) and Dwass (1960) independently proposed a single-step test procedure for this family that is based on separate pairwise rankings of observations. They considered only the balanced case $n_1 = n_2 = \cdots = n_k = n$. We first describe the Steel–Dwass procedure for this balanced case.

Let $RS_{ii'}^+$ be the rank sum of the n observations from treatment i when the $2n$ observations from treatments i and i' are ranked together, and let $RS_{ii'}^- = RS_{i'i}^+ = n(2n+1) - RS_{ii'}^+$ $(1 \leqq i < i' \leqq k)$. Let

$$RS_k^* = \max_{1 \leqq i < i' \leqq k} \{\max(RS_{ii'}^+, RS_{i'i}^+)\} ,$$

and let $RS_k^{*(\alpha)}$ be the upper α point of the distribution of RS_k^* under the overall null hypothesis $H_0: F_1 \equiv F_2 \equiv \cdots \equiv F_k$. (Note that the notation for pairwise comparisons is different from that for comparisons with a control. Here the subscript on RS^* denotes the number of treatments and not the number of pairwise comparisons among them.) The Steel–Dwass procedure rejects $H_{ii'}: F_i \equiv F_{i'}$ in favor of a two-sided alternative if

$$\max(RS_{ii'}^+, RS_{i'i}^+) > RS_k^{*(\alpha)} \qquad (1 \leqq i < i' \leqq k) . \tag{1.13}$$

Note that

$$\max(RS_{ii'}^+, RS_{i'i}^+) = \frac{n(2n+1)}{2} + \left| RS_{ii'}^+ - \frac{n(2n+1)}{2} \right| , \tag{1.14}$$

which shows that $RS_k^{*(\alpha)}$ can be determined from the $\binom{k}{2}$-variate joint distribution of the statistics $RS_{ii'}^+$ $(1 \leqq i < i' \leqq k)$.

Theorem 1.2. The single-step test procedure (1.13) strongly controls the Type I FWE at the designated level α.

Proof. The proof is similar to that of Theorem 1.1. It is based on verifying that $\{(H_{ii'}, RS^+_{ii'}), 1 \leqq i < i' \leqq k\}$ is a joint testing family and then applying Theorem 1.2 of Appendix 1. To do this, note that the intersection of any subset of pairwise hypotheses $H_{ii'}$ is equivalent to a multiple subset hypothesis: $H_{P_1} \cap \cdots \cap H_{P_r}$ where the P_j's are disjoint subsets each of size at least two and H_{P_j} postulates that all treatments belonging to subset P_j have the same distribution $(1 \leqq j \leqq r)$. The $RS^+_{ii'}$'s in disjoint subsets are independent. Hence it is enough to verify that the joint distribution of these statistics for $i, i' \in P_j$ is distribution-free. This is readily verified using an argument similar to that used in the proof of Theorem 1.1. □

The exact null distribution of RS^*_k is obtained by considering the $(kn)!$ different orderings (which are equally likely under H_0) of all the observations and computing

$$\Pr\{RS^*_k = r\} = \frac{\text{number of orderings giving } RS^*_k = r}{(kn)!} .$$

Steel (1960) computed this exact distribution for $k = 3$ and $n = 2, 3, 4$. For large n, an approximation due to Dwass (1960) may be used. Dwass showed that the joint asymptotic $(n \to \infty)$ distribution of the statistics $RS^+_{ii'}$ $(1 \leqq i < i' \leqq k)$ under H_0 is multivariate normal with the following parameters:

$$\begin{aligned} E(RS^+_{ii'}) &= \frac{n(2n+1)}{2} \qquad (1 \leqq i < i' \leqq k), \\ \text{var}(RS^+_{ii'}) &= \frac{n^2(2n+1)}{12} \qquad (1 \leqq i < i' \leqq k), \\ \text{corr}(RS^+_{ii'}, RS^+_{jj'}) &= \begin{cases} 0 & i \neq j,\ i' \neq j' \\ \dfrac{n}{2n+1} & i \neq j,\ i' = j' \text{ or } i = j,\ i' \neq j' \\ \dfrac{-n}{2n+1} & i \neq j',\ i' = j \text{ or } i = j',\ i' \neq j . \end{cases} \end{aligned} \tag{1.15}$$

The correlation structure (1.15) is asymptotically $(n \to \infty)$ identical to the correlation structure of $\binom{k}{2}$ pairwise differences among k independent and homoscedastic random variables (r.v.'s). Therefore, the asymptotic distribution of

$$\max_{1 \leqq i < i' \leqq k} \frac{|RS^+_{ii'} - E(RS^+_{ii'})|}{\sqrt{\text{var}(RS^+_{ii'})}}$$

can be approximated by that of a $Q_{k,\infty}/\sqrt{2}$ r.v. where $Q_{k,\infty}$ is the range of k independent and identically distributed (i.i.d.) $N(0, 1)$ r.v.'s. This yields the approximation

$$RS_k^{*(\alpha)} \cong \left\lfloor \frac{n(2n+1)}{2} + \frac{1}{2} + Q_{k,\infty}^{(\alpha)} \sqrt{\frac{n^2(2n+1)}{24}} \right\rfloor \tag{1.16}$$

where $\frac{1}{2}$ is the standard continuity correction and $Q_{k,\infty}^{(\alpha)}$ is the upper α point of the $Q_{k,\infty}$ r.v.

For unequal but large sample sizes we can use a Tukey–Kramer (TK) type approximation; see Chapter 3, Section 3.2.1. The rank sums $RS_{ii'}^+$ and $RS_{i'i}^+$ are computed from the pairwise rankings of the $n_i + n_{i'}$ observations from F_i and $F_{i'}$. Under $H_{ii'}: F_i \equiv F_{i'}$, the mean and variance of $RS_{ii'}^+$ are given by

$$E(RS_{ii'}^+) = \frac{n_i(n_i + n_{i'} + 1)}{2} ,$$
$$\mathrm{var}(RS_{ii'}^+) = \frac{n_i n_{i'}(n_i + n_{i'} + 1)}{12} . \tag{1.17}$$

The asymptotic joint normality of the $RS_{ii'}^+$'s can be deduced, for example, from the general results of Koziol and Reid (1977). The large sample α-level TK-type approximate procedure would then reject $H_{ii'}$ in favor of a two-sided alternative if $\max\{RS_{ii'}^+, n_i(n_{i'}+1) - RS_{ii'}^+\} > RS_{ii'}^{*(\alpha)}$ where

$$RS_{ii'}^{*(\alpha)} \cong \left\lfloor \frac{n_i(n_i + n_{i'} + 1)}{2} + \frac{1}{2} + Q_{k,\infty}^{(\alpha)} \sqrt{\frac{n_i n_{i'}(n_i + n_{i'} + 1)}{24}} \right\rfloor$$
$$(1 \leqq i < i' \leqq k) . \tag{1.18}$$

The single-step test procedure described above can be inverted to obtain simultaneous confidence intervals for pairwise differences $\theta_i - \theta_{i'}$ $(1 \leqq i < i' \leqq k)$ if we assume the location model (1.1). For this purpose we can use a graphical method similar to the one described in Section 1.1. Here we plot $(Y_{ij}, Y_{i'j'})$. The critical number of points that must fall above and below the two 45° lines is now

$$N_{ii'}^{*(\alpha)} = n_i n_{i'} + \frac{n_i(n_i + 1)}{2} - RS_{ii'}^{*(\alpha)} . \tag{1.19}$$

1.2.2 Procedures Based on Rank Sum Statistics Using Joint Rankings

Dunn (1964) proposed a single-step test procedure that is based on joint rankings of observations from all the treatments. Let R_{ij} be the rank of Y_{ij}

in the pooled sample of $N = \Sigma_{i=1}^{k} n_i$ observations and let

$$\bar{R}_{i\cdot} = \frac{1}{n_i} R_{i\cdot} = \frac{1}{n_i} \sum_{j=1}^{n_i} R_{ij} \qquad (1 \leq i \leq k).$$

The Dunn procedure rejects $H_{ii'}: F_i \equiv F_{i'}$ in favor of a two-sided alternative if

$$|\bar{R}_{i\cdot} - \bar{R}_{i'\cdot}| > Z^{(\alpha^*)} \left\{ \frac{N(N+1)}{12} \left(\frac{1}{n_i} + \frac{1}{n_{i'}} \right) \right\}^{1/2} \qquad (1 \leq i < i' \leq k) \tag{1.20}$$

where $Z^{(\alpha^*)}$ is the upper α^* point of the standard normal distribution, $\alpha^* = \frac{1}{2}\{1 - (1-\alpha)^{1/k^*}\}$, and $k^* = \binom{k}{2}$. (Actually, Dunn noted that it would be desirable to use $Z^{(\alpha^*)}$, but she instead used a slightly larger critical value $Z^{(\alpha/2k^*)}$.)

Dunn's procedure is based on the asymptotic ($n_i \to \infty$, $n_i/N \to \lambda_i$, $0 < \lambda_i < 1$ for all i) multivariate normality of $(\bar{R}_{1\cdot}, \bar{R}_{2\cdot}, \ldots, \bar{R}_{k\cdot})$ with the following parameters under the overall null hypothesis $H_0: F_1 \equiv F_2 \equiv \cdots \equiv F_k$:

$$E(\bar{R}_{i\cdot}) = \frac{(N+1)}{2} \qquad (1 \leq i \leq k),$$

$$\text{var}(\bar{R}_{i\cdot}) = \frac{(N+1)(N-n_i)}{12 n_i} \qquad (1 \leq i \leq k), \tag{1.21}$$

$$\text{cov}(\bar{R}_{i\cdot}, \bar{R}_{i'\cdot}) = -\frac{(N+1)}{12} \qquad (1 \leq i < i' \leq k).$$

This asymptotic distribution result was derived by Kruskal and Wallis (1952) in the context of their analysis of variance test; see (1.23) below. From (1.21) it follows that, under H_0, the differences $\bar{R}_{i\cdot} - \bar{R}_{i'\cdot}$ have marginal normal distributions with zero means and variances equal to $N(N+1)(n_i^{-1} + n_{i'}^{-1})/12$. Application of the Dunn–Šidák inequality then shows that the single-step test procedure (1.20) asymptotically controls the FWE at level α under the overall null hypothesis H_0 (but not necessarily under other configurations).

If the sample sizes n_i are all equal, then the $\bar{R}_{i\cdot}$'s have equal variances and equal covariances under H_0. Thus in large samples we can use Tukey's T-procedure for pairwise comparisons, which involves replacing $Z^{(\alpha^*)}$ in (1.20) by a sharper critical point $Q_{k,\infty}^{(\alpha)}/\sqrt{2}$. If we assume that the Tukey conjecture stated in Section 3.2.1 of Chapter 3 holds for general nondiagonal covariance matrices, then we can use this critical point even in the case of unequal sample sizes. This leads to a TK-type modification of the Dunn procedure that rejects $H_{ii'}: F_i \equiv F_{i'}$ if

$$|\bar{R}_{i\cdot} - \bar{R}_{i'\cdot}| > \left(\frac{Q_{k,\infty}^{(\alpha)}}{\sqrt{2}}\right)\left\{\frac{N(N+1)}{12}\left(\frac{1}{n_i} + \frac{1}{n_{i'}}\right)\right\}^{1/2} \qquad (1 \leqq i < i' \leqq k) . \tag{1.22}$$

For the case of equal sample sizes, McDonald and Thompson (1967) have tabled the critical points of $\max_{1 \leqq i < i' \leqq k} |R_{i\cdot} - R_{i'\cdot}|$. Their computations are based on the exact (not asymptotic) null distribution, but they employ certain probability inequalities; thus the tabled critical points are slightly on the conservative side. See also the tables of Tobach et al. (1967).

Nemenyi (1963) proposed a procedure based on an application of Scheffé's projection method (see Section 2.2 of Chapter 2) to the Kruskal–Wallis test. The large sample α-level Kruskal–Wallis test rejects the overall null hypothesis H_0 if

$$KW = \frac{12}{N(N+1)} \sum_{i=1}^{k} n_i \left(\bar{R}_{i\cdot} - \frac{N+1}{2}\right)^2 > \chi^2_{k-1}(\alpha) \tag{1.23}$$

where $\chi^2_{k-1}(\alpha)$ is the upper α point of the chi-square distribution with $k-1$ d.f. Application of the projection method to (1.23) leads to a single-step test procedure of level α that rejects $H_{ii'}$: $F_i \equiv F_{i'}$ if

$$|\bar{R}_{i\cdot} - \bar{R}_{i'\cdot}| > \{\chi^2_{k-1}(\alpha)\}^{1/2}\left\{\frac{N(N+1)}{12}\left(\frac{1}{n_i} + \frac{1}{n_{i'}}\right)\right\}^{1/2} \qquad (1 \leqq i < i' \leqq k) . \tag{1.24}$$

This procedure is clearly more conservative than (1.22) and hence is not recommended.

The single-step test procedures of this section cannot be used for constructing simultaneous confidence intervals since they are not based on testing families. Also they do not strongly control the FWE under all configurations although they do control it (at least in large samples) under H_0; see Section 1.4 for a fuller discussion.

Example 1.2 **(*Continuation of Example 1.1*).** If all three pairwise comparisons are of interest for the data in Table 1.1, then the Steel–Dwass procedure based on the TK-type approximation to the critical points (1.18) can be used. Clearly, the comparisons (1, 3) and (2, 3) will still be nonsignificant because we now have a larger family. For the comparison (1, 2) we calculate $RS^+_{12} = 24$, $RS^-_{12} = n_1(n_1 + n_2 + 1) - RS^+_{12} = 16$, and thus $\max\{RS_{12}^{+}, RS_{12}^{-}\} = 24$. This is to be compared with the critical point

$$RS_{12}^{*(\alpha)} \cong \left\lfloor \frac{n_1(n_1+n_2+1)}{2} + \frac{1}{2} + Q_{3,\infty}^{(\alpha)}\sqrt{\frac{n_1 n_2(n_1+n_2+1)}{24}} \right\rfloor .$$

For $\alpha = 0.20$ we have $Q_{3,\infty}^{(\alpha)} = 2.42$ from Table 8 in Appendix 3, and hence $RS_{12}^{*(0.20)} \cong \lfloor 32.50 \rfloor = 33$. Thus this comparison is also not significant at $\alpha = 0.20$. Simultaneous confidence intervals for $\theta_i - \theta_{i'}$ can be constructed by using a graphical method similar to that in Example 1.1. □

1.3 Step-Down Procedures

Nonparametric step-down procedures have not been developed to the same extent as their normal theory counterparts. The possibility and desirability of using nonparametric test statistics in step-down procedures have been noted by several authors (e.g., Ryan 1960, Steel 1961, Miller 1981, Campbell 1980). Lehmann and Shaffer (1979) have investigated some properties of such procedures, for example, separability, and control of the Type I FWE. Recently Campbell and Skillings (1985) studied these procedures further.

Consider the class of step-down test procedures discussed in Section 4.3.3 of Chapter 2 for the family of subset homogeneity hypotheses. Campbell and Skillings (1985) proposed the following three choices of test statistics in that class for the balanced one-way layout case:

(i) Separate ranks Steel–Dwass statistics:

$$RS_P^* = \max_{i,i' \in P} \{\max(RS_{ii'}^{+}, RS_{i'i}^{+})\} .$$

(ii) Joint ranks range statistics:

$$R_P = \max_{i,i' \in P} |\bar{R}_{i\cdot}^{(P)} - \bar{R}_{i'\cdot}^{(P)}|$$

where $\bar{R}_{i\cdot}^{(P)}$ is the average rank of the observations from treatment i when the observations from all treatments in set P are jointly ranked.

(iii) Kruskal–Wallis statistics:

$$KW_P = \frac{12}{p(np+1)} \sum_{i \in P} \left(\bar{R}_{i\cdot}^{(P)} - \frac{np+1}{12}\right)^2 .$$

It is clear that for any "partition" $(P_1, P_2, \ldots, P_r)$ of $\{1, 2, \ldots, k\}$

(where the P_i's are disjoint subsets of cardinality $p_i \geqq 2$ and $\Sigma_{i=1}^{r} p_i \leqq k$) the statistics $RS_{P_1}^*, RS_{P_2}^*, \ldots, RS_{P_r}^*$ are mutually independent. The same holds for the other two choices of test statistics because of the reranking for each subset. Also the step-down procedures based on these statistics possess the separability property; this was shown for the Kruskal–Wallis statistics by Lehmann and Shaffer (1979), and similar proofs can be given for the other two statistics. Thus for all three choices of test statistics we can apply Theorem 3.2 of Appendix 1 to obtain the upper bound (3.15) stated there on the Type I FWE in terms of the nominal levels α_p. To control the Type I FWE at a designated level α we can therefore use one of the standard choices for the α_p's, for example, the Tukey–Welsch (TW) choice:

$$\alpha_p = 1 - (1 - \alpha)^{p/k} \ (2 \leqq p \leqq k - 2), \qquad \alpha_{k-1} = \alpha_k = \alpha .$$

The discreteness of the distribution of nonparametric statistics can present problems in achieving the desired α_p levels, however.

Since this procedure involves testing too many subsets, Campbell and Skillings considered an ad hoc shortcut version of this procedure as follows. At the initial step the treatments are ordered and labeled $1, 2, \ldots, k$ according to their rank sums computed from the joint rankings and this same labeling is used in the subsequent steps. At the $(k - p + 1)$th step ($p = 2, \ldots, k$), subsets of the form $\{j, j+1, \ldots, j+p-1\}$ are tested if and only if they have not been retained as homogeneous by implication at a previous step. When testing the homogeneity of subset $\{j, j+1, \ldots, j+p-1\}$, the observations from those treatments are reranked and the difference in the rank sums of the "extreme" treatments, j and $j+p-1$, is tested for significance at nominal level α_p. If this difference is found significant, then treatments j and $j+p-1$ are concluded to be different and one proceeds in the next step to test the homogeneity of subsets $\{j, j+1, \ldots, j+p-2\}$ and $\{j+1, j+2, \ldots, j+p-1\}$. If the rank sum difference is found to be not significant at nominal level α_p, then the subset $\{j, j+1, \ldots, j+p-1\}$ and all of its subsets are retained as homogeneous.

Campbell and Skillings showed that under the location model this shortcut procedure asymptotically ($n \to \infty$) has the same upper bound on the Type I FWE as given by (3.15) of Appendix 1. Thus by choosing the α_p's appropriately (e.g., the TW-choice) the Type I FWE can be strongly controlled. Other statistics and ordering rules can also be used.

One would expect this shortcut step-down procedure to be generally more powerful than its nonparametric single-step competitors as well as the corresponding all-subsets step-down procedure. This was confirmed in the simulation study made by Campbell and Skillings. It should, however, be noted that the shortcut procedure can in some instances suffer from lack of power and inconsistent decisions because the ordering of the rank sums after reranking may not agree with the initial ordering of the rank sums based on the joint ranking of all the observations.

The subject of nonparametric stepwise procedures is still far from fully developed. Further research is needed in the areas of choice of test statistics, directional decisions, and step-up procedures. So the present section should be viewed only as preliminary in nature.

1.4 A Comparison of Procedures

We first restrict attention to single-step test procedures and discuss the choice between separate rankings and joint rankings. The first point to note is that the test statistics computed from joint rankings do not yield testing families, while those computed from separate rankings do. For example, the distribution of the test statistics $|\bar{R}_{i\cdot} - \bar{R}_{i'\cdot}|$ used for testing $H_{ii'}$ in the procedures of Section 1.2.2 is not determined under $H_{ii'}$ since it also depends on the distributions of the observations from other treatments that are not postulated to be homogeneous under $H_{ii'}$. For this reason these procedures do not strongly control the Type I FWE. Oude Voshaar (1980) showed that under the slippage configuration (i.e., $k-1$ treatments have the same distribution and one treatment has a different distribution) the Type I FWE of the (improved) Dunn procedure (1.22) based on joint rankings can exceed the nominal α level for highly skewed distributions even under the assumption of the location model. The lack of the testing family property also implies that single-step test procedures based on joint rankings do not have corresponding confidence analogs. The Steel–Dwass procedure (1.13) based on separate rankings does not suffer from these drawbacks. Fligner (1984) has shown similar drawbacks for procedures based on joint rankings in treatments versus control problems.

Although procedures based on separate rankings would generally be preferred in view of the advantages mentioned above, one should not overlook power considerations when making a choice of the method of ranking to use. Koziol and Reid (1977) have shown that asymptotically ($n_i \rightarrow \infty$, $n_i/N \rightarrow \lambda_i$, $0 < \lambda_i < 1 \; \forall \; i$), the Steel–Dwass procedure (1.13) based on separate rankings and the (improved) Dunn procedure (1.22) based on joint rankings have the same power properties under local alternatives for location or scale shifts. This extends the result of Sherman (1965), who showed that the two procedures have the same Pitman efficiencies. However, these equivalence results are asymptotic and for local alternatives; the powers of the two procedures can be different in small samples and under nonlocal alternatives. Skillings (1983) provided some useful guidelines for these latter situations based on a simulation study. He found that neither procedure is uniformly superior in terms of power for all nonnull configurations. The (improved) Dunn procedure (1.22) is more powerful for detecting differences between extreme treatments when there are intermediate treatments present. On the other hand, the Steel–Dwass procedure (1.13) is not affected by the presence of other treatments and hence has higher power for detecting differences between adjacent treatments. Fairly and Pearl (1984) arrived at the same conclusion analytically by comparing the Bahadur efficiencies of the two procedures.

Lehmann (1975, pp. 244–245) has noted that joint rankings may provide more information than separate rankings in location problems. He has also noted a lack of transitivity that can arise with separate rankings in which, say, treatment 1 can be declared better than treatment 2, and treatment 2 can be declared better than treatment 3, but treatment 1 cannot be declared better than treatment 3. Such intransitivities cannot arise with joint rankings. Despite these limitations separate rankings are still generally preferred in practice.

Comparisons between single-step and step-down procedures have been made by Lin and Haseman (1978) and Campbell and Skillings (1985) using Monte Carlo methods. The conclusions here are generally similar to those obtained in the normal case (see Section 5.3 of Chapter 4), namely that step-down procedures are more powerful than single-step procedures particularly for certain definitions of power.

2 PROCEDURES FOR RANDOMIZED COMPLETE BLOCK DESIGNS

2.1 Models and Hypotheses

We consider n complete blocks of common size k. The k treatments are assigned randomly within each block. Let Y_{ij} be the observation on the ith treatment in the jth block ($1 \leqq i \leqq k$, $1 \leqq j \leqq n$). As in Section 1 we assume that the Y_{ij}'s are continuous r.v.'s so that ties occur among them with zero probability.

The most general nonparametric model for this setting postulates only that the random vectors $(Y_{1j}, Y_{2j}, \ldots, Y_{kj})$ are mutually independent for $j = 1, 2, \ldots, n$; the distributions of these vectors can be different in different blocks. For this general model there are at least three common ways of specifying the null hypothesis of equality of any pair of treatments i and i'. They are as follows:

$$H_{ii'}^{(1)}: \text{median}(Y_{ij} - Y_{i'j}) = 0 \qquad (1 \leqq j \leqq n); \tag{2.1a}$$

$$H_{ii'}^{(2)}: \text{distribution of } Y_{ij} - Y_{i'j} \text{ is symmetric about zero} \qquad (1 \leqq j \leqq n); \tag{2.1b}$$

$$H_{ii'}^{(3)}: Y_{ij} \text{ and } Y_{i'j} \text{ are exchangeable within each block } j \qquad (1 \leqq j \leqq n). \tag{2.1c}$$

Note that $H_{ii'}^{(3)} \Rightarrow H_{ii'}^{(2)} \Rightarrow H_{ii'}^{(1)}$; thus $H_{ii'}^{(1)}$ is the most general hypothesis of the three while $H_{ii'}^{(3)}$ is the least general.

Many nonparametric statistics can be used to test the pairwise hypotheses (2.1a)–(2.1c), the simplest among them being the sign statistic $S_{ii'}^{+}$ = the number of blocks with positive differences $Y_{ij} - Y_{i'j}$. Now $S_{ii'}^{+}$ is marginally distribution-free under each one of the three hypotheses, but the collection of the $S_{ii'}^{+}$'s for $i, i' \in P$ (where P is any subset of $K = \{1, 2, \ldots, k\}$ of size $p \geqq 3$) is not jointly distribution-free under the corresponding intersection hypotheses $\bigcap_{i,i' \in P} H_{ii'}^{(l)}$ for $l = 1, 2, 3$. In other words, $\{(H_{ii'}^{(l)}, S_{ii'}^{+}), 1 \leqq i < i' \leqq k\}$ forms a testing family but not a joint testing family $(1 \leqq l \leqq 3)$. As explained in Section 1.4 of Appendix 1, the Type I FWE can be controlled (exactly) for a simultaneous test procedure if it is based on a joint testing family. With just a testing family, (exact) control of the Type I FWE is not possible but the Type I per-family error rate (PFE) can be readily controlled (by using the Bonferroni inequality), which thus provides conservative control of the FWE. The same difficulty arises with other nonparametric test statistics.

Joint testing families are obtained only if further assumptions are added to our general nonparametric model. One such assumption is that of within block independence. This assumption makes $\{(H_{ii'}^{(3)}, S_{ii'}^{+}), 1 \leqq i < i' \leqq k\}$ a joint testing family but not the other two. Under this assumption $H_{ii'}^{(3)}$ is simply the hypothesis that Y_{ij} and $Y_{i'j}$ have the same distribution for each j. All three collections $\{(H_{ii'}^{(l)}, S_{ii'}^{+}), 1 \leqq i < i' \leqq k\}$, for $l = 1, 2, 3$ become joint testing families under the model

$$Y_{ij} = \mu_{ij} + E_{ij} \qquad (1 \leqq i \leqq k,\ 1 \leqq j \leqq n) \tag{2.2}$$

where the μ_{ij}'s are unknown mean effects (fixed) and the error vectors $(E_{1j}, \ldots, E_{kj})$ are independently distributed with zero mean vectors but with possibly different (continuous) distributions, each of which is symmetric in its k arguments. All three hypotheses $H_{ii'}^{(l)}$ are given by $\mu_{ij} = \mu_{i'j}$ $(1 \leqq j \leqq n)$ under (2.2).

In some problems we assume that the treatment and block effects are additive, that is, $\mu_{ij} = \theta_i + \beta_j$ where the θ_i's and β_j's are unknown fixed treatment and block effects, respectively. The marginal distribution of Y_{ij} is then of the form

$$F_{ij}(y) = F(y - \theta_i - \beta_j) \qquad (1 \leqq i \leqq k, 1 \leqq j \leqq n) \tag{2.3}$$

for some unknown continuous distribution function $F(\cdot)$. (Actually the assumption of fixed block effects can be relaxed.) We refer to (2.2) as the *location model* and to (2.3) as the *additive location model*. Under the additive location model the hypotheses $H_{ii'}^{(l)}$ $(1 \leqq l \leqq 3)$ are given by $\theta_i = \theta_{i'}$.

2.2 Comparisons with a Control

2.2.1 A Procedure Based on Sign Statistics

Let us suppose, as in Section 1.1, that the kth treatment is a control and the first $k - 1$ treatments are to be compared with it. Under model (2.2) consider the hypotheses

$$H_{0i}: \mu_{ij} = \mu_{kj} \qquad \forall\, j \; (1 \leqq i \leqq k - 1)\,. \tag{2.4}$$

Steel (1959a) proposed a single-step test procedure based on sign statistics for testing the hypotheses (2.4) simultaneously. Let S_{ik}^+ be the number of blocks in which $Y_{ij} - Y_{kj} > 0$, $S_{ik}^- = n - S_{ik}^+$, and $S_{ik} = \max(S_{ik}^+, S_{ik}^-)$ $(1 \leqq i \leqq k - 1)$. Next let

$$S_{k-1}^+ = \max_{1 \leqq i \leqq k-1} S_{ik}^+ \qquad \text{and} \qquad S_{k-1} = \max_{1 \leqq i \leqq k-1} S_{ik}\,.$$

Further let $S_{k-1}^{+(\alpha)}$ and $S_{k-1}^{(\alpha)}$ be the upper α points of the distributions of S_{k-1}^+ and S_{k-1}, respectively, under the overall null hypothesis $H_0 = \bigcap_{i=1}^{k-1} H_{0i}$. The Steel procedure rejects H_{0i} in favor of the one-sided alternative that $\mu_{ij} > \mu_{kj}$ for some j if

$$S_{ik}^+ > S_{k-1}^{+(\alpha)} \qquad (1 \leqq i \leqq k - 1) \tag{2.5}$$

and in favor of the two-sided alternative that $\mu_{ij} \neq \mu_{kj}$ for some j if

$$S_{ik} > S_{k-1}^{(\alpha)} \qquad (1 \leqq i \leqq k - 1)\,. \tag{2.6}$$

The null distributions of S_{k-1}^+ and S_{k-1} can be obtained by utilizing the fact that the $k!$ permutations of the observations in each block are equally likely under H_0. The distribution-free nature of S_{k-1}^+ and S_{k-1} is a consequence of this fact. This also implies that under model (2.2), the procedures (2.5) and (2.6) are based on joint testing families and they strongly control the Type I FWE at the designated level α.

Steel (1959a) gave a few exact values of the critical points $S_{k-1}^{+(\alpha)}$ and $S_{k-1}^{(\alpha)}$ for $k = 2$ and 3, and for $n = 3, 4, 5$. Additional tables were given by Rhyne and Steel (1965). For large n, approximations to the critical points can be obtained by using the asymptotic multivariate normality of the vector $(S_{1k}^+, \ldots, S_{k-1,k}^+)$ with the following parameters under H_0:

$$E(S_{ik}^+) = \frac{n}{2}, \qquad \text{var}(S_{ik}^+) = \frac{n}{4}, \qquad \text{corr}(S_{ik}^+, S_{i'k}^+) = \frac{1}{3}.$$

Thus the desired large sample approximations are given by

$$S_{k-1}^{+(\alpha)} \cong \left\lfloor \frac{n}{2} + \frac{1}{2} + \frac{\sqrt{n}}{2} Z_{k-1,1/3}^{(\alpha)} \right\rfloor \tag{2.7}$$

and

$$S_{k-1}^{(\alpha)} \cong \left\lfloor \frac{n}{2} + \frac{1}{2} + \frac{\sqrt{n}}{2} |Z|_{k-1,1/3}^{(\alpha)} \right\rfloor. \tag{2.8}$$

Under models (2.1a)–(2.1c) the correlations among the S_{ik}^+'s depend on the unknown distribution functions of the Y_{ij}'s (even under the appropriate null hypotheses) and thus we do not obtain joint testing families. However, one can use the Bonferroni inequality to conservatively control the FWE in these cases. This implies that $Z_{k-1,1/3}^{(\alpha)}$ and $|Z|_{k-1,1/3}^{(\alpha)}$ in (2.7) and (2.8) must be replaced by $Z^{(\alpha/(k-1))}$ and $Z^{(\alpha/2(k-1))}$, respectively. The Bonferroni approximation can also be employed with small sample sizes based on the exact binomial distributions of the S_{ik}^+'s and S_{ik}^-'s.

We now discuss simultaneous confidence intervals derived from the test procedures (2.5) and (2.6) under the additive-location model (2.3). Lower one-sided confidence intervals on the differences $\theta_i - \theta_k$ ($1 \leq i \leq k-1$) are obtained as follows. Let $S_{ik}^+(\delta)$ be the number of blocks for which $Y_{ij} - Y_{kj} > \delta$. It can be shown that the set of values of δ that satisfy the inequality $S_{ik}^+(\delta) \leq S_{k-1}^{+(\alpha)}$ is an interval of the form $[\delta_i, \infty)$, and the collection of such intervals for $i = 1, 2, \ldots, k-1$ forms simultaneous $(1-\alpha)$-level lower one-sided confidence intervals on the $k-1$ differences $\theta_i - \theta_k$. Simultaneous upper one-sided intervals are obtained similarly by considering the inequalities $S_{ik}^-(\delta) \leqq S_{k-1}^{-(\alpha)} = S_{k-1}^{+(\alpha)}$, and simultaneous two-sided intervals are obtained by considering the inequalities $S_{ik}(\delta) \leqq S_{k-1}^{(\alpha)}$.

These confidence intervals can be conveniently computed as follows (Lehmann 1963). Order the n differences $D_{ij} = Y_{ij} - Y_{kj}$ $(1 \leqq j \leqq n)$ and let $D_{i(1)} < D_{i(2)} < \cdots < D_{i(n)}$ be the ordered values $(1 \leqq i \leqq k-1)$. Simultaneous $(1-\alpha)$-level lower one-sided intervals are given by

$$[D_{i(l_\alpha^+)}, \infty) \quad (1 \leqq i \leqq k-1) \tag{2.9}$$

where $l_\alpha^+ = n - S_{k-1}^{+(\alpha)} + 1$. Simultaneous two-sided intervals are given by

$$[D_{i(l_\alpha)}, D_{i(u_\alpha)}] \qquad (1 \leqq i \leqq k-1)\,. \tag{2.10}$$

where $l_\alpha = n - S_{k-1}^{(\alpha)} + 1$ and $u_\alpha = S_{k-1}^{(\alpha)}$.

2.2.2 *A Procedure Based on Signed Rank Statistics*

Based on what we know about the relative performances of the sign and signed rank tests for the paired samples problem, it is natural to expect that if an MCP for comparisons with a control can be based on signed rank statistics, then it would generally be more powerful than the Steel procedure based on sign statistics. Nemenyi (1963) proposed such a procedure. However, his procedure is not distribution-free even under the restrictive model (2.3).

Let $D_{ij} = Y_{ij} - Y_{kj}$ $(1 \leqq j \leqq n)$ and let $|D|_{i(1)} < |D|_{i(2)} < \cdots < |D|_{i(n)}$ be the ordered absolute values of the D_{ij}'s. Let R_{ij} be the rank of D_{ij} in this ranking. Then the signed rank statistic for testing the hypothesis $H_{0i}: \theta_i = \theta_k$ (under model (2.3)) against the upper one-sided alternative that $\theta_i > \theta_k$ is given by

$$SR_{ik}^+ = \sum_{j=1}^{n} R_{ij} I(D_{ij} > 0) \qquad (1 \leqq i \leqq k-1) \tag{2.11}$$

where $I(A)$ is an indicator function of event A. The signed rank statistic for testing H_{0i} against the two-sided alternative that $\theta_i \neq \theta_k$ is given by

$$SR_{ik} = \max\left\{SR_{ik}^+, \frac{n(n+1)}{2} - SR_{ik}^+\right\} \qquad (1 \leqq i \leqq k-1)\,. \tag{2.12}$$

Let $SR_{k-1}^+ = \max_{1 \leqq i \leqq k-1} SR_{ik}^+$ and $SR_{k-1} = \max_{1 \leqq i \leqq k-1} SR_{ik}$. Further let $SR_{k-1}^{+(\alpha)}$ and $SR_{k-1}^{(\alpha)}$ be the upper α points of the distributions of SR_{k-1}^+ and SR_{k-1}, respectively, under the overall null hypothesis $H_0 = \bigcap_{i=1}^{k-1} H_{0i}$. (As noted in the sequel, SR_{k-1}^+ and SR_{k-1} are not truly distribution-free under

H_0, and therefore their critical points depend on the distribution F under model (2.3).) Nemenyi's (1963) procedure rejects H_{0i} in favor of the upper one-sided alternative if

$$SR_{ik}^{+} > SR_{k-1}^{+(\alpha)} \qquad (1 \leq i \leq k-1) \tag{2.13}$$

and in favor of the two-sided alternative if

$$SR_{ik} > SR_{k-1}^{(\alpha)} \qquad (1 \leq i \leq k-1)\,. \tag{2.14}$$

Now although the SR_{ik}^{+}'s are marginally distribution-free, they are not jointly distribution-free under H_0. (The latter statement is true, more generally, for any subset of at least two of the SR_{ik}^{+}'s under the intersection of the corresponding H_{0i}'s.) In other words, $\{(H_{0i}, SR_{ik}^{+}),\ 1 \leq i \leq k-1\}$ is a testing family but not a joint testing family. The dependence of the joint distribution of $SR_{1k}^{+}, \ldots, SR_{k-1,k}^{+}$ under H_0 on the common distribution of the Y_{ij}'s in each block persists even when the Y_{ij}'s are assumed to be independent within blocks and even asymptotically as $n \rightarrow \infty$. These facts were proved by Hollander (1966). Of course, the same statements also apply to the statistics SR_{ik}.

Hollander (1966) showed that the asymptotic distribution of $(SR_{1k}^{+}, \ldots, SR_{k-1,k}^{+})$ is multivariate normal with the following parameters under H_0:

$$E(SR_{ik}^{+}) = \frac{n(n+1)}{4}\,, \qquad \mathrm{var}(SR_{ik}^{+}) = \frac{n(n+1)(2n+1)}{24}\,,$$
$$\mathrm{corr}(SR_{ik}^{+}, SR_{i'k}^{+}) = \rho(F) = 12\lambda(F) - 3\,. \tag{2.15}$$

Here F is the unknown distribution function in the model (2.3) and

$$\lambda(F) = \Pr\{X_1 + X_2 < X_3 + X_4,\ X_1 + X_5 < X_6 + X_7\}\,,$$

where all the X_i's are independent with common distribution F. Thus the correlations among the SR_{ik}^{+}'s depend on the unknown F. Therefore procedures (2.13) and (2.14) are not even asymptotically distribution-free.

It can be shown using the results of Lehmann (1964) that $0 < \rho(F) \leq \frac{1}{2}$ and for many of the common distributions, $\rho(F)$ is between 0.45 and 0.50 (see Miller 1981, p. 162). Nemenyi (1963) used Monte Carlo methods to estimate $\rho(F)$ for the uniform distribution and found it to be close to $\frac{1}{2}$. Based on this result he proposed the following large sample approximation to the critical point $SR_{k-1}^{+(\alpha)}$:

$$SR_{k-1}^{+(\alpha)} \cong \left\lfloor \frac{n(n+1)}{4} + \frac{1}{2} + Z_{k-1,1/2}^{(\alpha)} \sqrt{\frac{n(n+1)(2n+1)}{24}} \right\rfloor . \quad (2.16)$$

A corresponding approximation to $SR_{k-1}^{(\alpha)}$ is obtained by replacing $Z_{k-1,1/2}^{(\alpha)}$ by $|Z|_{k-1,1/2}^{(\alpha)}$ in (2.16). Both of these approximations are slightly on the liberal side because the upper bound on $\rho(F)$, namely $\frac{1}{2}$, is used in their derivation. A conservative approximation to $SR_{k-1}^{+(\alpha)}$ (respectively, $SR_{k-1}^{(\alpha)}$) can be obtained by acting as if $\rho(F) = 0$, which amounts to replacing $Z_{k-1,1/2}^{(\alpha)}$ in (2.16) by $Z^{(\alpha^*)}$ (respectively, $Z^{(\alpha^*/2)}$) where $\alpha^* = 1 - (1-\alpha)^{1/(k-1)}$. Alternatively, a consistent estimate of $\rho(F)$ can be used (see Miller 1981, p. 162) to obtain asymptotically exact critical points.

We now discuss how simultaneous confidence intervals can be obtained under model (2.3) for $\theta_i - \theta_k$ $(1 \leqq i \leqq k-1)$ based on signed rank statistics. These intervals, being based on single-step test procedures (2.13) and (2.14), are also not truly distribution-free.

Let $SR_{ik}^{+}(\delta)$ and $SR_{ik}(\delta)$ be the signed rank statistics as in (2.11) and (2.12), respectively, but after subtracting δ from each Y_{ij} $(1 \leqq i \leqq k-1, 1 \leqq j \leqq n)$. It can be shown that the set of δ-values that satisfy the inequality $SR_{ik}^{+}(\delta) \leqq SR_{k-1}^{+(\alpha)}$ is an interval of the form $[\delta_i, \infty)$ and for $i = 1, 2, \ldots, k-1$ they give simultaneous $(1-\alpha)$-level lower one-sided confidence intervals on the $(\theta_i - \theta_k)$'s. Similarly simultaneous $(1-\alpha)$-level upper one-sided intervals can be obtained from the inequalities $SR_{ik}^{-}(\delta) = n(n+1)/2 - SR_{ik}^{+}(\delta) \leqq SR_{k-1}^{+(\alpha)}$, and two-sided intervals can be obtained from the inequalities $SR_{ik}(\delta) \leqq SR_{k-1}^{(\alpha)}$ $(1 \leqq i \leqq k-1)$.

Tukey proposed the following graphical method for constructing these intervals. First, for fixed i, plot the n differences $Y_{ij} - Y_{kj}$ $(1 \leqq j \leqq n)$ on the vertical axis. Next draw two lines, one with slope $+1$ and the other with slope -1, from each one of these n points extending in the positive right half of the plane. These lines intersect in $\binom{n}{2}$ points, which together with the n points on the vertical axis give $n(n+1)/2$ points. Now draw a horizontal line through one of these points so that $n(n+1)/2 - SR_{k-1}^{+(\alpha)}$ points are below it. Let this horizontal line intersect the vertical axis at δ_i. Repeat this procedure for $i = 1, 2, \ldots, k-1$. Then the desired simultaneous lower one-sided intervals for the $(\theta_i - \theta_k)$'s are given by $[\delta_i, \infty)$ $(1 \leqq i \leqq k-1)$. Miller (1981, p. 164) explains why this graphical procedure works.

For two-sided intervals two horizontal lines must be drawn through two points such that $n(n+1)/2 - SR_{k-1}^{(\alpha)}$ points fall below the lower line and above the upper line. Let $[\delta_{i1}, \delta_{i2}]$ be the interval intercepted by the two lines on the vertical axis. Repeat this procedure for $i = 1, 2, \ldots, k-1$. The desired simultaneous $(1-\alpha)$-level two-sided intervals for the $(\theta_i - \theta_k)$'s are then $[\delta_{i1}, \delta_{i2}]$ $(1 \leqq i \leqq k-1)$.

The procedures of this section are similar to those of the following section for pairwise comparisons and hence they are not illustrated here separately; see Example 2.1 for pairwise comparisons instead.

2.3 Pairwise Comparisons

2.3.1 *A Procedure Based on Sign Statistics*

Assume model (2.2) and consider the hypotheses

$$H_{ii'}: \mu_{ij} = \mu_{i'j} \qquad \forall\, j\ (1 \leqq i < i' \leqq k)\,.$$

Nemenyi (1963) and Rhyne and Steel (1967) independently proposed a single-step test procedure based on sign statistics $S_{ii'} = \max(S^+_{ii'}, S^-_{ii'})$ that are defined in analogy with the S_{ik} statistics of Section 2.2.1. Let $S^*_k = \max_{1 \leqq i < i' \leqq k} S_{ii'}$ and let $S^{*(\alpha)}_k$ be the upper point of the distribution of S^*_k under the overall null hypothesis $H_0 = \bigcap_{1 \leqq i < i' \leqq k} H_{ii'}$. The Nemenyi procedure rejects $H_{ii'}$ in favor of the two-sided alternative if

$$S_{ii'} > S^{*(\alpha)}_k \qquad (1 \leqq i < i' \leqq k)\,. \tag{2.17}$$

It can be shown that under the location model (2.2), the collection $\{(H_{ii'}, S_{ii'}),\ 1 \leqq i < i' \leqq k\}$ forms a joint testing family. As a result, the single-step test procedure (2.17) strongly controls the Type I FWE at level α.

Selected values of the critical points $S^{*(\alpha)}_k$ were calculated by Nemenyi (1963) for $k = 3$ and for small values of n. More detailed tables of cumulative probabilities and critical points for $k = 3$ have been given by Rhyne and Steel (1967). To obtain a large sample $(n \to \infty)$ approximation to $S^{*(\alpha)}_k$, note that the asymptotic joint distribution of the $S^+_{ii'}$'s is a $\binom{k}{2}$-variate singular normal distribution with the following parameters under H_0:

$$E(S^+_{ii'}) = \frac{n}{2}\,, \qquad \operatorname{var}(S^+_{ii'}) = \frac{n}{4}\,,$$

$$\operatorname{corr}(S^+_{ii'}, S^+_{jj'}) = \begin{cases} 0 & i \neq j,\ i' \neq j' \\ \frac{1}{3} & i \neq j,\ i' = j' \text{ or } i = j,\ i' \neq j' \\ -\frac{1}{3} & i \neq j',\ i' = j \text{ or } i = j',\ i' \neq j\,. \end{cases} \tag{2.18}$$

Using the fact that $S_{ii'} = n/2 + |S^+_{ii'} - n/2|$ $(1 \leqq i < i' \leqq k)$, we can approximate $S_k^{*(\alpha)}$ by

$$S_k^{*(\alpha)} \cong \left\lfloor \frac{n}{2} + \frac{1}{2} + \frac{\sqrt{n}}{2} |Z|_{k^*,\mathbf{R}}^{(\alpha)} \right\rfloor \tag{2.19}$$

where $|Z|_{k^*,\mathbf{R}}^{(\alpha)}$ is the upper α point of the maximum of the absolute values of $k^* = \binom{k}{2}$ standard normal r.v.'s having the correlation matrix $\mathbf{R}$ whose entries are given by (2.18). The values of $|Z|_{k^*,\mathbf{R}}^{(\alpha)}$ have not been tabulated for the special correlation structure $\mathbf{R}$. But by using the Dunn–Šidák inequality we can obtain the following bounds on $|Z|_{k^*,\mathbf{R}}^{(\alpha)}$:

$$\frac{Q_{k,\infty}^{(\alpha)}}{\sqrt{2}} < |Z|_{k^*,\mathbf{R}}^{(\alpha)} < Z^{(\alpha^*/2)}$$

where $\alpha^* = 1 - (1-\alpha)^{1/k^*}$. Here the left hand side inequality is obtained by replacing all $\pm\frac{1}{3}$'s by $\pm\frac{1}{2}$'s, and the right hand side inequality is obtained by replacing all $\pm\frac{1}{3}$'s by 0's. Nemenyi (1963) proposed approximating $|Z|_{k^*,\mathbf{R}}^{(\alpha)}$ by

$$|Z|_{k^*,\mathbf{R}}^{(\alpha)} \cong \frac{2}{3}\frac{Q_{k,\infty}^{(\alpha)}}{\sqrt{2}} + \frac{1}{3}Z^{(\alpha^*/2)} \tag{2.20}$$

where the weights $\frac{2}{3}$ and $\frac{1}{3}$ are chosen because they give the weighted average of the correlation coefficients corresponding to the upper and lower bounds on $|Z|_{k^*,\mathbf{R}}^{(\alpha)}$, namely, $\frac{1}{2}$ and 0, to be the desired value $\frac{1}{3}$. A comparison of the large sample approximation to $S_k^{*(\alpha)}$ given by (2.19) (where $|Z|_{k^*,\mathbf{R}}^{(\alpha)}$ is approximated by (2.20)) with some exact values shows a good agreement and therefore use of this approximation is recommended.

The method of constructing simultaneous two-sided confidence intervals for $\theta_i - \theta_{i'}$ $(1 \leqq i < i' \leqq k)$ is similar to that described in Section 2.2. To obtain the interval for $\theta_i - \theta_{i'}$ we order the differences $D_{ii',j} = Y_{ij} - Y_{i'j}$ $(1 \leqq j \leqq n)$ as $D_{ii',(1)} < D_{ii',(2)} < \cdots < D_{ii',(n)}$. The desired interval for $\theta_i - \theta_{i'}$ is then given by (in analogy with (2.10)):

$$[D_{ii',(l_\alpha)}, D_{ii',(u_\alpha)}] \qquad (1 \leqq i < i' \leqq k) \tag{2.21}$$

where $l_\alpha = n - S_k^{*(\alpha)} + 1$ and $u_\alpha = S_k^{*(\alpha)}$.

2.3.2 *A Procedure Based on Signed Rank Statistics*

The Nemenyi procedure of Section 2.2.2 for comparisons with a control can be extended in an obvious manner to the problem of all pairwise comparisons (for which Nemenyi 1963 proposed it originally). Formally, this procedure rejects $H_{ii'}: F_i \equiv F_{i'}$ in favor of a two-sided alternative if

$$SR_{ii'} > SR_k^{*(\alpha)} \qquad (1 \leqq i < i' \leqq k) \tag{2.22}$$

where the $SR_{ii'}$'s are defined analogously to (2.12) and $SR_k^{*(\alpha)}$ is the upper α point of the r.v.

$$SR_k^* = \max_{1 \leqq i < i' \leqq k} SR_{ii'}$$

under the overall null hypothesis $H_0 = \bigcap_{1 \leqq i < i' \leqq k} H_{ii'}$. However, this procedure inherits all the difficulties that are present in the case of comparisons with a control. In particular, the collection $\{(H_{ii'}, SR_{ii'}),\ 1 \leqq i < i' \leqq k\}$ is not a joint testing family and the distribution of SR_k^* under H_0 depends on the common distribution F under model (2.3). In large samples ($n \to \infty$) the joint distribution of the $SR_{ii'}$'s can be approximated by a $\binom{k}{2}$-variate normal distribution with means and variances given by (2.15) and with nonnegative correlations that depend on F. By setting these correlations equal to zero we obtain the following conservative approximation to $SR_k^{*(\alpha)}$:

$$SR_k^{*(\alpha)} \cong \left[\frac{n(n+1)}{4} + \frac{1}{2} + Z^{(\alpha^*/2)} \sqrt{\frac{n(n+1)(2n+1)}{24}} \right] \tag{2.23}$$

where $\alpha^* = 1 - (1-\alpha)^{1/k^*}$, $k^* = \binom{k}{2}$, and $Z^{(\alpha^*/2)}$ is the upper $\alpha^*/2$ point of the standard normal distribution. The resulting procedure (2.22) conservatively controls the Type I FWE in large samples.

The simultaneous confidence intervals of Section 2.2.2 can also be extended in an obvious manner to all pairwise differences $\theta_i - \theta_{i'}$ ($1 \leqq i < i' \leqq k$); see Example 2.1.

2.3.3 *A Procedure Based on Within-Block Rank Statistics*

Nemenyi (1963) proposed a single-step procedure for testing all treatment differences in a randomized block design based on an application of Scheffé's projection method (see Section 2.2 of Chapter 2) to the Friedman (1937) test. The large sample ($n \to \infty$) Friedman 2-way analysis of variance (ANOVA) test rejects the null hypothesis of exchangeability of the Y_{ij}'s ($1 \leqq i \leqq k$) in each block j ($1 \leqq j \leqq n$) if

$$F = \frac{12n}{k(k+1)} \sum_{i=1}^{k} \left(\bar{R}_{i.} - \frac{k+1}{2} \right)^2 > \chi^2_{k-1}(\alpha) \tag{2.24}$$

where

$$\bar{R}_{i\cdot} = \frac{1}{n} \sum_{j=1}^{n} R_{ij} \qquad (1 \leqq i \leqq k)$$

and R_{ij} is the rank of Y_{ij} among the observations in the jth block $(1 \leqq i \leqq k,\ 1 \leqq j \leqq n)$. By applying the projection method to (2.24) Nemenyi obtained a single-step test procedure that rejects the pairwise null hypothesis $H_{ii'}$ of the exchangeability of observations from treatments i and i' within each block if

$$|\bar{R}_{i\cdot} - \bar{R}_{i'\cdot}| > \{\chi^2_{k-1}(\alpha)\}^{1/2} \left\{ \frac{k(k+1)}{6n} \right\}^{1/2} \qquad (1 \leqq i < i' \leqq k) . \tag{2.25}$$

In large samples this procedure controls the Type I FWE conservatively at level α under the overall null hypothesis H_0. At other configurations it may fail to control the Type I FWE because of the lack of joint testing family property.

A more powerful procedure for all pairwise comparisons (again with guaranteed control of the Type I FWE in large samples only under H_0) is obtained by noting that asymptotically under H_0 the $\bar{R}_{i\cdot}$'s are jointly normally distributed with $E(\bar{R}_{i\cdot}) = (k+1)/2$, $\text{var}(\bar{R}_{i\cdot}) = (k^2 - 1)/12n$, and $\text{corr}(\bar{R}_{i\cdot}, \bar{R}_{i'\cdot}) = -1/(k-1)$ for $i \neq i'$. Thus the pairwise balance condition (2.9) of Chapter 3 is satisfied and hence the results of Section 2.2 of Chapter 3 can be applied. The resulting improved procedure rejects $H_{ii'}$ if

$$|\bar{R}_{i\cdot} - \bar{R}_{i'\cdot}| > (Q^{(\alpha)}_{k,\infty}/\sqrt{2}) \left\{ \frac{k(k+1)}{6n} \right\}^{1/2} \qquad (1 \leqq i < i' \leqq k) . \tag{2.26}$$

McDonald and Thompson (1967) have tabulated selected critical points of the exact small sample null distribution of $\max_{1 \leqq i < i' \leqq k} |R_{i\cdot} - R_{i'\cdot}|$ where $R_{i\cdot} = \Sigma_{j=1}^{n} R_{ij}$. These critical points are slightly on the conservative side because of certain inequalities used in their computation.

2.3.4 Procedures Based on Aligned Rank Statistics

Procedures based on sign statistics and within-block rank statistics utilize only intra-block comparisons among the treatments. Inter-block information cannot be utilized under the location model (2.2) because of the difficulty in adjusting for block effects. However, under the additive-location model (2.3) it is possible to exploit the inter-block information. We saw an example of this in Section 2.3.2 where we discussed a procedure based on signed rank statistics. This is a special case of "aligned rank statistics" that use the inter-block information. Single-step procedures based on general "aligned rank statistics" were proposed by Sen (1969a).

Aligning the Y_{ij}'s amounts to transforming them into new variables whose distributions do not depend on the block effects. For example, for $k = 2$, if we subtract the average of each block, $(Y_{1j} + Y_{2j})/2$, from each observation Y_{ij}, then under model (2.3) we get pairs of aligned variables, $(Y_{1j} - Y_{2j})/2$ and $(Y_{2j} - Y_{1j})/2$, $j = 1, \ldots, n$. Ranking after alignment means jointly ranking all the transformed variables across blocks and treatments. A test of $\theta_1 = \theta_2$ can then be based on the rank sum differences between the treatments. For $k = 2$, use of the signed rank statistic is readily seen to be equivalent to the rank sum test applied to the joint ranking of aligned variables $(Y_{1j} - Y_{2j})$'s and $(Y_{2j} - Y_{1j})$'s.

More generally for $k > 2$, the alignment of the Y_{ij}'s can be achieved by subtracting some estimate of the location of the block (e.g., the block mean or median) from each observation in a given block. All aligned observations are then jointly ranked. The ranks assigned to the ordered aligned observations (referred to as the aligned ranks) are then summed across blocks for each treatment. A conditionally distribution-free test of the overall null hypothesis $H_0: \theta_1 = \cdots = \theta_k$ based on aligned ranks utilizes the fact that under this hypothesis, all $(k!)^n$ within-block permutations are equally likely. Puri and Sen (1971, Chapter 7) show that general rank score statistics obtained from aligned ranks are unconditionally distribution-free in large samples.

The theory of aligned rank statistics developed in Sen (1968) was applied to the problem of all pairwise comparisons by Sen (1969a). He used aligned rank statistics obtained from the joint rankings of observations from all treatments. Alternatively, one may adopt the following separate rankings procedure to obtain a (conditional and asymptotically unconditional) testing family. When a test for the homogeneity of a subset P of treatments is desired, align observations only from those treatments within each block (by subtracting some translation invariant and symmetric function of the Y_{ij}'s for $i \in P$). Next rank order these aligned observations across treatments in subset P and across blocks, and let the test statistic of the corresponding hypothesis be a suitable function of these aligned ranks.

Example 2.1. Table 2.1 gives some data on the effectiveness of hypnosis (higher skin potential indicating higher effectiveness) under four requested emotions (treatments) on eight subjects (blocks). These data are taken from Lehmann (1975, p. 264).

TABLE 2.1. Skin Potential Adjusted for Initial Level (in Millivolts) under Four Hypnotically Requested Emotions

	Subjects							
Emotions	1	2	3	4	5	6	7	8
Fear (1)	23.1	57.6	10.5	23.6	11.9	54.6	21.0	20.3
Happiness (2)	22.7	53.2	9.7	19.6	13.8	47.1	13.6	23.6
Depression (3)	22.5	53.7	10.8	21.1	13.7	39.2	13.7	16.3
Calmness (4)	22.6	53.1	8.3	21.6	13.3	37.0	14.8	14.8

Source: Lehmann (1975).

Suppose that we wish to make all pairwise comparisons between the four emotions. We assume the additive-location model (2.3) and use the procedures of Section 2.3. To apply these procedures we calculate the sign and signed rank statistics as shown in Table 2.2. The ties occurring in the calculation of signed rank statistics are handled by assigning average ranks. They affect the variance of the $SR_{ii'}$'s but this effect is ignored here.

From Table 2.2 we see that the comparison (1, 4) between the emotions of fear and calmness produces both the largest sign statistic and the largest signed rank statistic. We compare the sign statistic $S_{14} = 7$ with the large sample critical point (2.19) where $|Z|_{k^*,\mathbf{R}}^{(\alpha)}$ is approximated by (2.20). For $\alpha = 0.20$ we obtain

$$|Z|_{6,\mathbf{R}}^{(0.20)} \cong \frac{2}{3}\frac{Q_{4,\infty}^{(0.20)}}{\sqrt{2}} + \frac{1}{3}Z^{(0.0183)}$$

$$= \frac{2}{3} \times \frac{2.78}{\sqrt{2}} + \frac{1}{3} \times 2.09 = 2.01 .$$

Then from (2.19) we obtain

$$S_4^{*(0.20)} \cong \left\lfloor 4 + \frac{1}{2} + \frac{\sqrt{8}}{2} \times 2.01 \right\rfloor = \lfloor 7.34 \rfloor = 8 .$$

Thus we see that even at $\alpha = 0.20$ we cannot find any significant differences between the four emotions. This is, of course, due to the fact that with just eight blocks, the MCP based on sign statistics lacks sufficient power to detect any differences.

Let us next apply the signed rank procedure of (2.22). The $SR_{ii'}$'s are not jointly distribution-free (both in finite samples and asymptotically) even under the restrictive model (2.3). Hence we use the conservative (in

TABLE 2.2. Calculation of the Sign and Signed Rank Statistics for Data in Table 2.1[a]

Emotion Pair (i, i')	Subjects								$S_{ii'}$	$SR_{ii'}$
	1	2	3	4	5	6	7	8		
(1, 2)	0.4 1	4.4 6	0.8 2	4.0 5	−1.9 3	7.5 8	7.4 7	−3.3 4	6	29
(1, 3)	0.6 2	3.9 5	−0.3 1	2.5 4	−1.8 3	15.4 8	7.3 7	4.0 6	6	32
(1, 4)	0.5 1	4.5 5	2.2 4	2.0 3	−1.4 2	17.6 8	6.2 7	5.5 6	7	34
(2, 3)	0.2 3	−0.5 4	−1.1 5	−1.5 6	0.1 1.5	7.9 8	−0.1 1.5	7.3 7	4	19.5
(2, 4)	0.1 1.5	0.1 1.5	1.4 5	−2.0 6	0.5 3	10.1 8	−1.2 4	8.8 7	6	26
(3, 4)	−0.1 1	0.6 4	2.5 8	−0.5 3	0.4 2	2.2 7	−1.1 5	1.5 6	5	27

[a] The upper entry in each cell is $D_{ii'.j} = Y_{ij} - Y_{i'j}$ and the lower entry is the rank of $|D_{ii'.j}|$.

large samples) upper bound on $SR_k^{*(\alpha)}$ given by (2.23) to obtain

$$SR_4^{*(0.20)} \cong \left\lfloor \frac{8 \times 9}{4} + \frac{1}{2} + 2.09\sqrt{\frac{8 \times 9 \times 17}{24}} \right\rfloor = \lfloor 33.42 \rfloor = 34 .$$

Thus we find that the comparison (1, 4) with $SR_{14} = 34$ is barely significant at $\alpha = 0.20$.

We next demonstrate how to construct simultaneous pairwise confidence intervals based on sign and signed rank statistics. We again use the pair (1, 4) for illustration purposes. The confidence intervals for the pairwise differences $\theta_i - \theta_{i'}$ based on sign statistics are given by (2.21) where for $\alpha = 0.20$ we have $l_\alpha = n - S_k^{*(\alpha)} + 1 = 8 - 8 + 1 = 1$ and $u_\alpha = S_k^{*(\alpha)} = 8$. The desired simultaneous two-sided 80% confidence interval for $\theta_1 - \theta_4$ is therefore given by $[D_{14,(1)}, D_{14,(8)}]$, which equals [0.5, 17.6] as seen from Table 2.2.

To calculate the interval based on signed rank statistics we employ the graphical method given in Section 2.2.2 (modified here for pairwise differences). First, as shown in Figure 2.1, plot the differences $Y_{1j} - Y_{4j}$ $(1 \leqq j \leqq 8)$ on the vertical axis and draw lines with slopes ± 1 through these points. This gives a total of $n(n+1)/2 = 36$ points as explained in Section 2.2.2. Next draw two horizontal lines through two of these points such that $n(n+1)/2 - SR_k^{*(\alpha)} = 36 - 34 = 2$ points fall above and below each line. The desired 80% simultaneous two-sided confidence interval for $\theta_1 - \theta_4$ is then given by the interval intercepted on the vertical axis by the two horizontal lines, which as seen from Figure 2.1 equals [0.30, 11.55].

To apply procedure (2.26) we assign ranks 1–4 to the observations under four different conditions for each subject. The resulting rank sums are $R_{1.} = 27$, $R_{2.} = 20$, $R_{3.} = 19$, and $R_{4.} = 14$. The maximum rank sum difference is $|R_{1.} - R_{4.}| = 13$. This difference just fails to be significant at $\alpha = 0.05$ as can be seen by comparing it with the critical value

$$\left(\frac{Q_{k,\infty}^{(\alpha)}}{\sqrt{2}}\right)\left\{\frac{nk(k+1)}{6}\right\}^{1/2} = \left(\frac{3.63}{\sqrt{2}}\right)\left\{\frac{8 \times 4 \times 5}{6}\right\}^{1/2} = 13.25 .$$

The corresponding exact critical value can be found from McDonald and Thompson's (1967) tables. This value is 14 with an exact upper tail probability of 0.034, which agrees with the large sample approximation.

It is of some interest to note that in this example a much more significant result is obtained with procedure (2.26) than with procedures (2.17) and (2.22). It is not known whether this is generally true, that is, whether procedure (2.26) is generally more powerful.

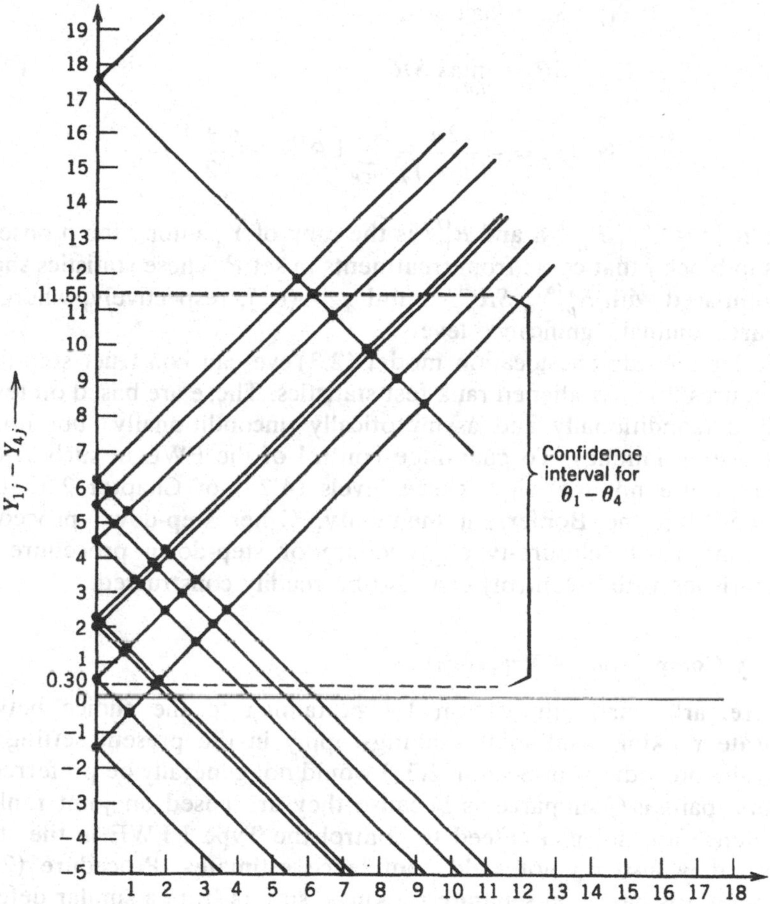

Figure 2.1. Graphical construction of a simultaneous 80% two-sided confidence interval for $\theta_1 - \theta_4$ based on signed rank statistics.

□

2.4 Step-Down Procedures

Here we discuss only the choice of distribution-free test statistics that can be used in different step-down procedures. Other aspects of these procedures are essentially as discussed in Section 1.3. Consider the hypothesis H_P of homogeneity of treatments in subset P of cardinality $p \geqq 2$. For testing this hypothesis we can use a variety of nonparametric test statistics; the following three are simple extensions of the statistics presented earlier for pairwise comparisons:

$$\text{(i)}\quad S_P = \max_{i,i' \in P} S_{ii'}\,,$$
$$\text{(ii)}\quad SR_P = \max_{i,i' \in P} SR_{ii'}\,, \tag{2.27}$$
$$\text{(iii)}\quad F_P = \frac{12n}{p(p+1)} \sum_{i \in P} \left(\bar{R}_{i\cdot}^{(P)} - \frac{p+1}{2} \right)^2 .$$

Here $\bar{R}_{i\cdot}^{(P)} = \Sigma_{j=1}^{n} R_{ij}^{(P)}/n$ and $R_{ij}^{(P)}$ is the rank of Y_{ij} among the p observations in block j that come from treatments in set P. These statistics should be compared with $S_p^{*(\alpha_p)}$, $SR_p^{*(\alpha_p)}$, and $\chi^2_{p-1}(\alpha_p)$, respectively, where the α_p's are nominal significance levels.

Under the additive-location model (2.3) we can construct step-down procedures that use aligned rank test statistics. These are based on testing families (conditionally and asymptotically unconditionally) but not on joint testing families. To guarantee control of the FWE in such settings one may use nominal significance levels (4.20) of Chapter 2 that are derived from the Bonferroni inequality. Other step-down procedures (e.g., the Peritz closure-type procedure or step-down procedures for comparisons with a control) can also be readily constructed.

2.5 A Comparison of Procedures

The remarks made in Section 1.4 pertaining to the choice between separate rankings and joint rankings apply in the present setting too. Thus the procedures of Section 2.3.3 would not generally be preferred for making pairwise comparisons because they are based on joint rankings and hence are not guaranteed to control the Type I FWE in the strong sense; they also do not yield confidence estimates. Procedure (2.22), although it is based on separate rankings, suffers from a similar defect in that it is not truly distribution-free. However, this can be remedied (at least in large samples) by using an easily determined conservative upper bound on its critical point. Procedure (2.17) is distribution-free but may suffer from lack of power in some situations. Thus the signed rank procedure (2.22) with a suitable correction of its critical point may offer the best choice in practice.

If one is interested only in tests and not in confidence estimation, then step-down procedures should be used because of their higher power. No studies have been conducted to assess the relative powers of step-down procedures based on different test statistics given in (2.27). Therefore no clear recommendation can be given. It should be noted that these statistics are based on separate rankings for different subsets. Thus the step-down procedures based on them do not have the drawbacks of the single-step procedure (2.25), which was based on joint rankings.

3 PROCEDURES BASED ON OTHER APPROACHES

3.1 Procedures Based on Permutation Tests

To bring out the main ideas behind permutation MCPs (also referred to as randomization or rerandomization MCPs) we consider the problem of pairwise comparisons among all treatments in the simplest setting of a balanced one-way layout. Miller (1981, p. 179) proposed a procedure for this problem based on the randomization distribution of the range of the sample means. This distribution is obtained by regarding all $(kn)!/(n!)^k$ permutations of the observations among the k treatments as equally likely. The upper α point of this distribution is used as the common critical point for testing the homogeneity of any subset of treatments by using the range of the sample means for that subset as the test statistic. Shuster and Boyett (1979) extended this procedure to randomized block designs.

Petrondas and Gabriel (1983) showed that these procedures are not valid because, for example, a pairwise null hypothesis implies equal probabilities only for the permutations of the observations from those two treatments, and not for the permutations of the observations from all the treatments. As a result, the use of a common critical point obtained from the randomization distribution of the range of all the sample means does not control the Type I FWE of the procedure. To obtain valid permutation tests, one must consider separately for each hypothesis only the subreference set of permutations that become equally likely under that hypothesis. Thus, for example, for testing the equality of any two treatments, one must consider the randomization distribution induced by the $(2n)!/(n!)^2$ equally likely permutations of the $2n$ observations from those two treatments with n observations allocated to each. The test is conducted by referring the observed value of the test statistic to the upper α point of the randomization distribution of that test statistic. Such a test is clearly distribution-free.

In the same manner, a distribution-free test can be obtained for testing the homogeneity of any subset of treatments. A family of such subset hypotheses, together with the corresponding test statistics (with separate randomization distributions), forms a testing family but not a joint testing family. However, one can use the Bonferroni method to construct a procedure that controls the FWE conservatively. For example, to control the FWE for the family of all pairwise comparisons, each pairwise test can be carried out at level $\alpha/\binom{k}{2}$. To achieve this, a separate critical point is used for each test, which is the upper $\alpha/\binom{k}{2}$ point of the randomization distribution (appropriate for that test) of the test statistic. Recently Gabriel and Robinson (1986) have shown how simultaneous confidence intervals can be derived from rerandomization tests.

Petrondas and Gabriel (1983) also discuss a Peritz-type closed step-down procedure that uses tests based on separate randomization distributions. All of these procedures can be extended to the randomized block design setting. Here the randomization distributions are obtained by regarding within block permutations as equally likely.

3.2 A Procedure Based on Median Statistics

Nemenyi (1963) proposed a single-step procedure for the problem of pairwise comparisons in the one-way layout setting based on Mood's two-sample median test. This procedure was rediscovered by Levy (1979). As we point out below, this procedure is plagued with serious problems and must be modified for use in practice.

Let $n_1, n_2, \ldots, n_k$ be the sizes of independent samples corresponding to k treatments. Let F_i be the common continuous distribution of the n_i independent observations on treatment i and put $N = \sum_{i=1}^{k} n_i$. Nemenyi proposed the following procedure for testing all pairwise hypotheses $H_{ii'}: F_i \equiv F_{i'}$ $(1 \leqq i < i' \leqq k)$: Find the grand median of the pooled sample of all N observations and calculate M_i, the number of observations from the ith treatment group that exceed the grand median $(1 \leqq i \leqq k)$. Reject $H_{ii'}$ if $|M_i/n_i - M_{i'}/n_{i'}|$ is large $(1 \leqq i < i' \leqq k)$. The common critical value is obtained from the joint distribution of $M_1, M_2, \ldots, M_k$ under the overall null hypothesis $H_0: F_1 \equiv F_2 \equiv \cdots \equiv F_k$. This distribution is given by

$$\Pr\{M_1 = m_1, \ldots, M_k = m_k\} = \frac{\binom{n_1}{m_1}\binom{n_2}{m_2}\cdots\binom{n_k}{m_k}}{\binom{N}{M}} \tag{3.1}$$

where

$$M = \sum_{i=1}^{k} M_i = \begin{cases} \dfrac{N}{2} & \text{if } N \text{ is even} \\ \dfrac{N-1}{2} & \text{if } N \text{ is odd}\,. \end{cases} \tag{3.2}$$

For the case $n_1 = \cdots = n_k = n$, Nemenyi obtained exact critical points of the distribution of $\max_{1 \leq i < i' \leq k} |M_i - M_{i'}|$ using (3.1) for a few selected values of k and n. In other cases one can use the asymptotic $(n_i \to \infty \;\forall\, i)$ multivariate normal approximation to the distribution of the M_i's with the following parameters under H_0:

$$E(M_i) = \frac{n_i}{N} M \qquad (1 \leqq i \leqq k),$$

$$\text{var}(M_i) = \frac{n_i(N-n_i)(N-M)M}{N^2(N-1)} \qquad (1 \leqq i \leqq k), \tag{3.3}$$

$$\text{cov}(M_i, M_{i'}) = -\frac{n_i n_{i'}(N-M)M}{N^2(N-1)} \qquad (1 \leqq i < i' \leqq k).$$

If the n_i's are equal to a common value, say, n, then the M_i's have equal variances and covariances, and

$$\text{var}(M_i - M_{i'}) = \frac{2n(N-M)MN}{N^2(N-1)} \cong \frac{n}{2}. \tag{3.4}$$

Therefore (as noted in Section 2.2 of Chapter 3), the r.v.

$$\max_{1 \leqq i < i' \leqq k} \frac{|M_i - M_{i'}|}{\sqrt{n/4}} \tag{3.5}$$

is asymptotically distributed as a $Q_{k,\infty}$ r.v. under H_0. Thus for the large equal sample sizes case, $\sqrt{(n/4)}Q_{k,\infty}^{(\alpha)}$ serves as the common critical point for testing $H_{ii'}: F_i \equiv F_{i'}$ using the test statistic $|M_i - M_{i'}|$ $(1 \leqq i < i' \leqq k)$.

One basic difficulty with this procedure is that it is not based on a testing family since for any treatment pair (i, i'), the distribution of $|M_i - M_{i'}|$ is not determined under $H_{ii'}: F_i \equiv F_{i'}$. This difficulty is, of course, common to all nonparametric MCPs that use the method of joint rankings. Ryan and Ryan (1980) have pointed out another difficulty with this procedure. They give an example where four treatment groups are equispaced in their locations and the interval supports of their distributions are nonoverlapping. Therefore we have $M_1 = 0$, $M_2 = 0$, $M_3 = n$, and $M_4 = n$ with probability one. Using the median procedure, the treatment pairs (1, 2) and (3, 4) will be declared nonsignificant and all other pairs will be declared significant (if n and/or α are large enough). Notice that, although the location separations between 1 and 2 and between 2 and 3 are the same, this procedure declares the former pair as nonsignificant but the latter pair as significant.

A way to overcome both of these difficulties is to base the procedure on separate median statistics for each of the $\binom{k}{2}$ pairwise comparisons. This gives a joint testing family. Therefore the necessary critical point can be found from the distribution of the maximum of these median statistics under the overall null hypothesis.

3.3 Procedures Based on General Linear Rank Statistics

Sen (1966) proposed the following nonparametric generalization of Tukey's T-procedure. This generalization uses general linear rank statistics, and is applicable to balanced one-way layouts.

Denote by n the common sample size from each treatment group. Let $J_n(u)$ be a monotone nondecreasing function defined on the [0, 1] interval satisfying the regularity conditions stated in Chernoff and Savage (1958, p. 972). Let $E_{m,n} = J_n(m/2n)$ $(1 \leqq m \leqq 2n)$, and let

$$\bar{E}_n = \frac{1}{2n} \sum_{m=1}^{2n} E_{m,n}, \qquad V_n = \frac{1}{2n-1} \sum_{m=1}^{2n} (E_{m,n} - \bar{E}_n)^2 . \tag{3.6}$$

For comparing treatments i and i', Sen proposed the two-sample Chernoff–Savage statistic:

$$T_{ii'} = \sum_{m=1}^{2n} E_{m,n} Z_m^{(ii')} \qquad (1 \leqq i < i' \leqq k) \tag{3.7}$$

where $Z_m^{(ii')} = 1$ if the mth smallest observation among the pooled sample of $2n$ observations from treatments i and i' comes from treatment i, and $Z_m^{(ii')} = 0$ otherwise. The Wilcoxon statistic $RS^+_{ii'}$ considered in Section 1.2 is a special case of (3.7).

It can be shown that $\{(H_{ii'}, T_{ii'}), 1 \leqq i < i' \leqq k\}$ forms a joint testing family. Therefore the common critical point for the single-step test procedure that controls the Type I FWE at level α can be obtained as the upper α point of the distribution of the maximum of the (suitably standardized) statistics $T_{ii'}$. Under the overall null hypothesis Sen (1966) showed that

$$\max_{1 \leqq i < i' \leqq k} \left\{ 2\sqrt{\frac{n}{V_n}}\, |T_{ii'} - \bar{E}_n| \right\} \tag{3.8}$$

is distributed asymptotically $(n \to \infty)$ as the range of k i.i.d. $N(0, 1)$ r.v.'s. Thus Sen's procedure concludes that treatments i and i' are significantly different if

$$2\sqrt{\frac{n}{V_n}}\, |T_{ii'} - \bar{E}_n| > Q_{k,\infty}^{(\alpha)} \qquad (1 \leqq i < i' \leqq k) . \tag{3.9}$$

Scheffé's S-procedure can also be generalized to the nonparametric setting in a similar manner as above. Both of these procedures can be extended to more complex designs. The review article by Sen (1980) is a good reference for these developments and their multivariate extensions.

We now turn to the problem of simultaneous confidence estimation. Sen (1966) showed how the Hodges–Lehmann (1963) method can be used to obtain simultaneous confidence intervals for $\theta_i - \theta_{i'}$ $(1 \leqq i < i' \leqq k)$ under the location model for the one-way layout design. The method involves a somewhat complicated inversion scheme. Sen (1969a) indicated that the same method can be extended to the randomized block design (under model (2.3)) using general aligned rank scores. Wei (1982) proposed a simpler method that does not require numerical inversion. Wei's intervals are based on Mann–Whitney statistics (applied to aligned observations), and they can be given in an explicit form as follows: Let $\tilde{Y}_{ij} = Y_{ij} - \bar{Y}_{\cdot j}$ be the aligned observations where $\bar{Y}_{\cdot j}$ is the jth block mean $(1 \leqq i \leqq k,\ 1 \leqq j \leqq n)$. Let

$$D_{(1)}^{(ii')} < D_{(2)}^{(ii')} < \cdots < D_{(n^2)}^{(ii')}$$

be the ordered values of the differences $D_{jj'}^{(ii')} = \tilde{Y}_{ij} - \tilde{Y}_{i'j'}$ $(1 \leqq j, j' \leqq n)$. Then simultaneous confidence intervals for $\theta_i - \theta_{i'}$ with a large sample confidence level of $1 - \alpha$ are given by

$$D_{(l_\alpha)}^{(ii')} < \theta_i - \theta_{i'} < D_{(u_\alpha)}^{(ii')} \qquad (1 \leqq i < i' \leqq k) \tag{3.10}$$

where $l_\alpha = \lfloor n^2/2 - (Q_{k,\infty}^{(\alpha)} n^{3/2}/6) \rfloor$, $u_\alpha = \lfloor n^2/2 + (Q_{k,\infty}^{(\alpha)} n^{3/2}/6) \rfloor + 2$ and $\lfloor x \rfloor$ denotes the greatest integer $\leqq x$.

4 ROBUST PROCEDURES

In recent years considerable research effort has been directed toward the development of robust estimates of location and their standard errors. Although the problem of robust estimation has received much attention, very little effort has been made to develop MCPs that employ these robust estimates. Two recent contributions in this direction have been made by Dunnett (1982) and Ringland (1983).

Dunnett conducted an extensive simulation study to compare simultaneous confidence procedures based on different estimators of location in terms of two criteria: (i) stability of their joint confidence levels under different nonnormal but symmetric and unimodal distributions, and (ii) their average confidence interval widths. For the family of all pairwise comparisons among the location parameters θ_i $(1 \leqq i \leqq k)$ in a one-way layout setting, he considered the following class of simultaneous confidence intervals:

$$\theta_i - \theta_j \in \left[\hat{\theta}_i - \hat{\theta}_j \pm \xi_{ij,\alpha} \sqrt{\frac{\hat{\sigma}_i^2}{\hat{n}_i} + \frac{\hat{\sigma}_j^2}{\hat{n}_j}} \right] \qquad (1 \leqq i < j \leqq k). \tag{4.1}$$

In (4.1), $\hat{\theta}_i$ is a robust estimate of θ_i, $\hat{\sigma}_i$ is an associated robust estimate of the scale parameter σ_i of the distribution of the observations from treatment i, $\hat{n}_i$ is an "effective" sample size for $\hat{\theta}_i$, and $\xi_{ij,\alpha}$ is a critical constant that is chosen so as to guarantee the desired joint confidence coefficient of $1-\alpha$. Dunnett approximated $\xi_{ij,\alpha}$ by $T_{\hat{\nu}_{ij}}^{(\alpha^*)}$, the upper α^* point of Student's t-distribution with $\hat{\nu}_{ij}$ d.f. where

$$\hat{\nu}_{ij} = \frac{(\hat{\sigma}_i^2/\hat{n}_i + \hat{\sigma}_j^2/\hat{n}_j)^2}{[\hat{\sigma}_i^4/\hat{n}_i^2\hat{\nu}_i + \hat{\sigma}_j^4/\hat{n}_j^2\hat{\nu}_j]}, \tag{4.2}$$

$\hat{\nu}_i$ is an "effective" d.f. associated with $\hat{\sigma}_i^2$, and α^* is chosen so that the intervals in (4.1) have a joint confidence coefficient of $1-\alpha$ when the underlying distributions are normal. A conservative choice for α^* is $\alpha^* = \frac{1}{2}\{1-(1-\alpha)^{1/k^*}\}$ where $k^* = \binom{k}{2}$. As an example, if $\hat{\theta}_i$ is the β-trimmed mean, then $\hat{n}_i = n_i(1-2\beta)$ and $\hat{\nu}_i = \hat{n}_i - 1$. For some other robust estimators such as adaptive estimators (the particular adaptive estimators studied were wave, bi-square, and Hampel) the choice of $\hat{n}_i$ and $\hat{\nu}_i$ is not so clear, and Dunnett took them to be n_i and $n_i - 1$, respectively.

Dunnett's simulation study was restricted to the balanced one-way layout setting. His overall conclusions were as follows: Tukey's T-procedure becomes increasingly conservative and yields very wide confidence intervals for long-tailed and outlier prone distributions. As expected, procedures based on robust estimators roughly maintain the designated joint confidence level. They generally have narrower confidence intervals for nonnormal distributions at the expense of somewhat wider confidence intervals relative to the T-procedure for the normal distribution. Procedures based on trimmed mean type estimators perform better for short-tailed distributions such as the uniform, while procedures based on adaptive estimators (wave, bi-square, and Hampel) perform better for long-tailed distributions such as the Cauchy.

An important outcome of Dunnett's work is the finding that the distribution-free procedure of Steel and Dwass (which controls the FWE for all distributions) is the best choice for near-normal distributions. The Steel–Dwass procedure is dominated by other procedures based on robust estimators only for extremely long-tailed distributions. Thus to make a proper choice among the various rival procedures, it is necessary to have a rough idea about the nature of the underlying distributions.

Ringland (1983) considered the problem of simultaneous confidence estimation of the location parameters θ_i $(1 \leqq i \leqq k)$ in a one-way layout under the assumption of a common scale parameter σ for all treatment distributions. Let $Z_i = \sqrt{n_i}(\hat{\theta}_i - \theta_i)/\hat{\tau}_i$ where $\hat{\theta}_i$ is an M-estimate of θ_i and

$\hat{\tau}_i$ is an associated Studentizing factor $(1 \leqq i \leqq k)$; the formulas for evaluating the $\hat{\theta}_i$'s and $\hat{\tau}_i$'s are given in Ringland's article. He considered the following three methods of constructing simultaneous confidence regions for the θ_i's:

(i) The Bonferroni method:

$$\{\boldsymbol{\theta} : |Z_i| \leqq T_{\nu}^{(\alpha/2k)} \ (1 \leqq i \leqq k)\} .$$

(ii) The Studentized maximum modulus method:

$$\{\boldsymbol{\theta} : |Z_i| \leqq |M|_{k,\nu}^{(\alpha)} \ (1 \leqq i \leqq k)\} .$$

(iii) The Scheffé projection method:

$$\left\{\boldsymbol{\theta} : \sum_{i=1}^{k} Z_i^2 \leqq kF_{k,\nu}^{(\alpha)}\right\} .$$

Here ν may be taken to be the usual error d.f., $\nu = \Sigma_{i=1}^{k} n_i - k$.

To study the error probabilities associated with these three confidence regions, Ringland derived the first order Edgeworth expansion for the joint distribution of the Z_i's from which he obtained approximations to the desired probabilities. He evaluated these approximations numerically (and estimated the corresponding exact probabilities by simulation) for different choices of M-estimators of the θ_i's and σ and for different underlying distributions. His main conclusions were as follows: The Bonferroni and Studentized maximum modulus methods are relatively nonrobust and exceed the nominal error probability α when k is large and the distributions are heavier tailed than the normal. On the other hand, the Scheffé projection method controls the FWE quite accurately under different situations and is thus robust. These conclusions are fairly independent of different choices of estimators. In summary, this study points to the lack of robustness of the classical (based on the least squares estimates) Tukey-type (finite UI) MCPs and the need for accurate approximations to the critical points of the distributions of the maximums of Studentized M-estimators.

CHAPTER 10

Some Miscellaneous Multiple Comparison Problems

In this chapter we study procedures for some miscellaneous multiple comparison problems. Some of these problems are related to the ones considered in the earlier chapters. But for the most part these problems and associated procedures do not fit neatly into one of the earlier chapters and hence are discussed separately here.

In Section 1 we discuss some multiple comparison procedures (MCPs) for categorical data. Section 2 considers the problem of multiple comparisons of variances; the problem of simultaneous confidence regions for variance components in random-effects models is also discussed here. Section 3 describes some graphical procedures for multiple comparisons of means. These procedures are for the most part informal in nature (i.e., they are not designed to control a specified error rate) as opposed to those discussed in Chapter 3. Section 4 is concerned with multiple comparisons of means under order restrictions. Section 5 deals with the problem of simultaneous inferences on interactions in two-way layouts. Finally Section 6 considers the problem of partitioning (clustering) of treatment means in a one-way layout.

1 MULTIPLE COMPARISON PROCEDURES FOR CATEGORICAL DATA

We review here procedures for the following problems involving categorical data:

(i) Multiple comparisons among k treatments with Bernoulli responses.

(ii) Multiple comparisons among the cell probabilities of a single multinomial distribution.

(iii) Multiple comparisons among the cross product ratios in two and higher dimensional contingency tables.

(iv) Multiple comparisons among k logistic response curves.

Even for making a single inference a very few exact small sample procedures are available for categorical data. Thus it is not surprising that almost all MCPs are based on large sample normal approximations.

1.1 Treatments with Bernoulli Responses

Suppose that we wish to compare k independent treatments with Bernoulli responses in terms of their success probabilities $\pi_1, \pi_2, \ldots, \pi_k$. The data consist of n_i independent observations on the ith treatment of which Y_i are successes, the sample proportion of successes being $\hat{\pi}_i = Y_i/n_i$ $(1 \leq i \leq k)$. For pairwise comparisons among the π_i's, $(1-\alpha)$-level large sample $(n_i \to \infty\ \forall i)$ simultaneous confidence intervals based on the Tukey–Kramer (TK) procedure (see Chapter 3, Section 3.2.1) are given by

$$\pi_i - \pi_j \in \left[\hat{\pi}_i - \hat{\pi}_j \pm Q_{k,\infty}^{(\alpha)} \sqrt{\frac{1}{2}\left\{\frac{\hat{\pi}_i(1-\hat{\pi}_i)}{n_i} + \frac{\hat{\pi}_j(1-\hat{\pi}_j)}{n_j}\right\}}\right] \qquad (1 \leq i < j \leq k) \qquad (1.1)$$

where $Q_{k,\infty}^{(\alpha)}$ is the upper α point of the range of k independent and identically distributed (i.i.d.) standard normal variables. Simultaneous confidence intervals for m prespecified contrasts among the π_i's can be obtained by applying the Bonferroni or the somewhat sharper Dunn–Šidák inequality (see Appendix 2).

Knoke (1976) considered a so-called maximal contrast test that is essentially the Scheffé procedure applied to dichotomous data. This procedure can be used to data-snoop contrasts among the π_i's. Using Monte Carlo simulations he studied the significance levels and powers of several competing procedures to test the overall null hypotheses $H_0: \pi_1 = \pi_2 = \cdots = \pi_k$ and/or to make multiple comparisons among the π_i's.

Bhapkar and Somes (1976) considered a matched samples design for comparing k treatments with Bernoulli responses. In this design there are n matched sets of observations or blocks where each block consists of responses of the same subject to k treatments or responses of k different subjects who are matched according to relevant criteria and then assigned randomly to k treatments. Let $X_{ij} = 1$ or 0 depending on whether the outcome for the ith treatment is a success or failure in the jth block $(1 \leqq i \leqq k, 1 \leqq j \leqq n)$. For this experimental setup Bhapkar and Somes assumed a 2^k-cell multinomial model where the cells are indexed by $\boldsymbol{\delta} = (\delta_1, \delta_2, \ldots, \delta_k)'$ with $\delta_i = 1$ or 0 indicating a success or failure, respectively, on the ith treatment $(1 \leqq i \leqq k)$. Let $\pi_{\boldsymbol{\delta}}$ be the cell probability for cell $\boldsymbol{\delta}$, $\Sigma_{\boldsymbol{\delta}} \pi_{\boldsymbol{\delta}} = 1$. The marginal success probability π_i for the ith treatment is given by the sum of the $\pi_{\boldsymbol{\delta}}$'s over all $\boldsymbol{\delta}$'s with $\delta_i = 1$ $(1 \leqq i \leqq k)$ and the joint success probability π_{ij} for the ith and jth treatment is given by the sum of the $\pi_{\boldsymbol{\delta}}$'s over all $\boldsymbol{\delta}$'s with $\delta_i = \delta_j = 1$ $(1 \leqq i \neq j \leqq k)$. It is desired to make comparisons among the π_i's.

Let $Y_i = \Sigma_{j=1}^n X_{ij}$ and $\hat{\pi}_i = Y_i/n$ be the number of successes and the proportion of successes on the ith treatment, respectively $(1 \leqq i \leqq k)$. Also let $Y_{ij} = \Sigma_{l=1}^n X_{il}X_{jl}$ be the number of blocks that result in successes on both the ith and jth treatment and let $\hat{\pi}_{ij} = Y_{ij}/n$ $(1 \leqq i \neq j \leqq n)$. It is well known that the vector $\sqrt{n}(\hat{\pi}_1 - \pi_1, \hat{\pi}_2 - \pi_2, \ldots, \hat{\pi}_k - \pi_k)$ has an asymptotically $(n \rightarrow \infty)$ multivariate normal distribution with zero mean vector and variance-covariance matrix $\mathbf{V} = \{v_{ij}\}$ where $v_{ii} = \pi_i(1 - \pi_i)$ and $v_{ij} = \pi_{ij} - \pi_i\pi_j$ $(1 \leqq i \neq j \leqq k)$. Furthermore, the matrix $\mathbf{V}$ can be consistently estimated by $\hat{\mathbf{V}} = \{\hat{v}_{ij}\}$ where $\hat{v}_{ii} = \hat{\pi}_i(1 - \hat{\pi}_i)$ and $\hat{v}_{ij} = \hat{\pi}_{ij} - \hat{\pi}_i\hat{\pi}_j$ $(1 \leqq i \neq j \leqq k)$. Using these facts Bhapkar and Somes (1976) showed that asymptotically $(1 - \alpha)$-level simultaneous confidence intervals for all contrasts $\Sigma_{i=1}^k c_i\pi_i$ are given by

$$\sum_{i=1}^k c_i\pi_i \in \left[\sum_{i=1}^k c_i\hat{\pi}_i \pm \sqrt{\chi^2_{k-1}(\alpha) \cdot \frac{(\mathbf{c}'\hat{\mathbf{V}}\mathbf{c})}{n}}\right] \quad \forall\, \mathbf{c} \in \mathbb{C}^k \qquad (1.2)$$

where $\mathbb{C}^k$ is the k-dimensional contrast space and $\chi^2_{k-1}(\alpha)$ is the upper α point of the chi-square distribution with $k - 1$ degrees of freedom (d.f.). These intervals are derived by the Scheffé projection method (see Section 2.2 of Chapter 2). If one is interested in a prespecified finite set of contrasts among the π_i's, then generally narrower intervals can be derived by using the Bonferroni inequality or the slightly sharper Dunn–Šidák inequality. The large sample $(1 - \alpha)$-level Dunn–Šidák simultaneous intervals for all pairwise comparisons are given by

$$\pi_i - \pi_j \in \left[\hat{\pi}_i - \hat{\pi}_j \pm Z^{(\alpha^*)}\sqrt{\frac{\hat{\pi}_i + \hat{\pi}_j - 2\hat{\pi}_{ij} - (\hat{\pi}_i - \hat{\pi}_j)^2}{n}}\right]$$

$$(1 \leqq i < j \leqq k)\,; \qquad (1.3)$$

here $Z^{(\alpha^*)}$ is the upper $\alpha^* = \frac{1}{2}\{1 - (1 - \alpha)^{1/k^*}\}$ point of the standard normal distribution and $k^* = \binom{k}{2}$. Note that because the $\hat{\pi}_i$'s are not independent it is not known whether the TK-procedure can be applied here to yield guaranteed (in large samples) simultaneous $(1-\alpha)$-level pairwise confidence intervals except when $k=3$; see the discussion in Section 3.2.1 of Chapter 3.

1.2 Multinomial Cell Probabilities

Consider a single k-cell multinomial distribution with cell probabilities $\pi_1, \pi_2, \ldots, \pi_k$, $\Sigma_{i=1}^{k} \pi_i = 1$. Let $Y_1, Y_2, \ldots, Y_k$ be the observed cell frequencies with $\Sigma_{i=1}^{k} Y_i = n$ and let $\hat{\pi}_i = Y_i/n$ $(1 \leq i \leq k)$. Also let $\boldsymbol{\pi} = (\pi_1, \pi_2, \ldots, \pi_{k-1})'$ and $\hat{\boldsymbol{\pi}} = (\hat{\pi}_1, \hat{\pi}_2, \ldots, \hat{\pi}_{k-1})'$. Asymptotically $(n \to \infty)$, $\sqrt{n}(\hat{\boldsymbol{\pi}} - \boldsymbol{\pi})$ has a $(k-1)$-variate normal distribution with zero mean vector and variance-covariance matrix $\mathbf{V} = \{v_{ij}\}$ where $v_{ii} = \pi_i(1 - \pi_i)$, $v_{ij} = -\pi_i \pi_j$ $(1 \leq i \neq j \leq k-1)$. Hence

$$n(\hat{\boldsymbol{\pi}} - \boldsymbol{\pi})'\mathbf{V}^{-1}(\hat{\boldsymbol{\pi}} - \boldsymbol{\pi}) = n \sum_{i=1}^{k} \frac{(\hat{\pi}_i - \pi_i)^2}{\pi_i}$$

is distributed as χ^2_{k-1}. By solving the inequality $n\,\Sigma_{i=1}^{k} \{(\hat{\pi}_i - \pi_i)^2/\pi_i\} \leq \chi^2_{k-1}(\alpha)$, Quesenberry and Hurst (1964) derived the following asymptotic $(1-\alpha)$-level simultaneous confidence intervals for the π_i's:

$$\pi_i \in \left[\frac{\xi + 2Y_i \pm \sqrt{\xi[\xi + 4Y_i(n - Y_i)/n]}}{2(N + \xi)}\right] \qquad (1 \leq i \leq k) \qquad (1.4)$$

where $\xi = \chi^2_{k-1}(\alpha)$. These intervals are shorter in small samples than the following intervals proposed by Gold (1963) but are asymptotically equivalent to them:

$$\pi_i \in \left[\hat{\pi}_i \pm \xi\sqrt{\frac{\hat{\pi}_i(1 - \hat{\pi}_i)}{n}}\right] \qquad (1 \leq i \leq k)\,. \qquad (1.5)$$

Note that both (1.4) and (1.5) are conservative, being based on the Scheffé projection method.

Goodman (1965) proposed Bonferroni intervals for the π_i's that are shorter than both (1.4) and (1.5) for the usual values of k and α. (See Alt and Spruill 1978 and Savin 1980 for more detailed comparisons between the Bonferroni and Scheffé intervals.) These intervals use $\xi = \chi^2_1(\alpha/k) = (Z^{(\alpha/2k)})^2$ in (1.4). Goodman (1965) has also given Bonferroni intervals for pairwise differences $\pi_i - \pi_j$ and for pairwise ratios π_i/π_j.

Bailey (1980) has studied the use of the arcsine transformation of the multinomial proportions $\hat{\pi}_i = Y_i/n$ to obtain simultaneous confidence intervals for the π_i's. He found that these intervals are preferable to (1.4) in small samples.

For multiple comparisons among the cell probabilities of several independent k-cell multinomial distributions, see Gold (1963) and Goodman (1964a).

1.3 Contingency Tables

Goodman (1964b) considered the problem of constructing simultaneous confidence intervals for cross product ratios

$$\psi_{ij} = \frac{\pi_{11}\pi_{ij}}{\pi_{i1}\pi_{1j}} \qquad (2 \leqq i \leqq a, 2 \leqq j \leqq b) \tag{1.6}$$

in an $a \times b$ table. For convenience, assume that sampling is multinomial with a total sample of size n and cell probabilities π_{ij} $(1 \leqq i \leqq a, 1 \leqq j \leqq b)$. Let Y_{ij} denote the observed cell frequencies with $\Sigma_{i=1}^{a} \Sigma_{j=1}^{b} Y_{ij} = n$. The natural estimate of ψ_{ij} is (assuming that all the Y_{ij}'s > 0)

$$\hat{\psi}_{ij} = \frac{Y_{11}Y_{ij}}{Y_{i1}Y_{1j}} \qquad (2 \leqq i \leqq a, 2 \leqq j \leqq b)\,. \tag{1.7}$$

Let $\theta_{ij} = \log_e \psi_{ij}$ and $\hat{\theta}_{ij} = \log_e \hat{\psi}_{ij}$. It can be shown (Plackett 1962) that for studying the joint distribution of any set of contrasts among the $\log_e Y_{ij}$'s, the latter can be regarded as asymptotically $(n \rightarrow \infty)$ uncorrelated with variances $1/n\pi_{ij}$. From this result one can calculate the asymptotic covariance matrix $\mathbf{V}$ of the θ_{ij}'s. The diagonal entries of $\mathbf{V}$ are $v_{ij} = \text{var}(\hat{\theta}_{ij}) = n^{-1}(\pi_{ij}^{-1} + \pi_{i1}^{-1} + \pi_{1j}^{-1} + \pi_{11}^{-1})$ and the off-diagonal entries are $v_{ij,i'j'} = \text{cov}(\hat{\theta}_{ij}, \hat{\theta}_{i'j'}) = n^{-1}(\pi_{11}^{-1} + \delta_{ii'}\pi_{i1}^{-1} + \delta_{jj'}\pi_{j1}^{-1})$ where $\delta_{ii'}$ and $\delta_{jj'}$ are Kronecker δ's. Moreover, asymptotically the $\hat{\theta}_{ij}$'s are jointly normal with means θ_{ij}. Let $\boldsymbol{\theta}$ denote the vector of the θ_{ij}'s and $\hat{\boldsymbol{\theta}}$ the corresponding vector of the $\hat{\theta}_{ij}$'s. Let $\hat{\mathbf{V}}$ be a consistent estimate of $\mathbf{V}$ obtained by estimating the $n\pi_{ij}$'s with Y_{ij}'s. Then the asymptotic distribution of $(\hat{\boldsymbol{\theta}} - \boldsymbol{\theta})'\hat{\mathbf{V}}^{-1}(\hat{\boldsymbol{\theta}} - \boldsymbol{\theta})$ is χ^2_ν where $\nu = (a-1) \times (b-1)$. By Scheffé's projection method it then follows that asymptotic $(1-\alpha)$-level simultaneous confidence intervals for all contrasts among the θ_{ij}'s are given by

$$\sum_{i=2}^{a}\sum_{j=2}^{b} c_{ij}\theta_{ij} \in \left[\sum_{i=2}^{a}\sum_{j=2}^{b} c_{ij}\hat{\theta}_{ij} \pm \sqrt{\chi^2_\nu(\alpha)(\mathbf{c}'\hat{\mathbf{V}}\mathbf{c})}\right] \qquad \forall\, \mathbf{c} \in \mathbb{C}^\nu \tag{1.8}$$

where $\mathbf{c}$ denotes the contrast vector formed from the c_{ij}'s satisfying $\Sigma_{i=2}^{a}\Sigma_{j=2}^{b} c_{ij} = 0$ and $\mathbb{C}^\nu$ denotes the corresponding ν-dimensional ($\nu = (a-1)(b-1)$) contrast space.

If only pairwise differences between the θ_{ij}'s (which correspond to pairwise ratios between the ψ_{ij}'s) are of interest, then shorter intervals can be obtained by using the TK-procedure. These intervals are given by

$$\theta_{ij} - \theta_{i'j'} \in [\hat{\theta}_{ij} - \hat{\theta}_{i'j'} \pm Q^{(\alpha)}_{(a-1)(b-1),\infty} \sqrt{\tfrac{1}{2}(\hat{v}_{ij} + \hat{v}_{i'j'} - 2\hat{v}_{ij,i'j'})}]$$

$$(2 \leqq i, i' \leqq a, 2 \leqq j, j' \leqq b) . \qquad (1.9)$$

Gabriel (1966) proposed a simultaneous test procedure for testing independence hypotheses for all subtables of an $a \times b$ table. If the sampling is product multinomial, that is, if the rows can be viewed as independent multinomial populations, then any independence hypothesis for a subtable can be interpreted as a homogeneity hypothesis concerning the given subset of cell probabilities of the corresponding row populations.

Let A and B denote the sets of all rows and columns, respectively, and let $P \subseteq A$ and $Q \subseteq B$ denote specified subsets of rows and columns with cardinalities $p \leqq a$ and $q \leqq b$, respectively. For testing the hypothesis of independence H_{PQ} for the subtable formed by the rows in P and columns in Q, Gabriel proposed the log-likelihood ratio statistic

$$Z_{PQ} = 2\Big\{\sum_{i \in P}\sum_{j \in Q} Y_{ij} \log_e Y_{ij} - \sum_{i \in P} Y_{iQ} \log_e Y_{iQ}$$

$$- \sum_{j \in Q} Y_{Pj} \log_e Y_{Pj} + Y_{PQ} \log_e Y_{PQ}\Big\} \qquad (1.10)$$

where $Y_{iQ} = \Sigma_{j \in Q} Y_{ij}$, $Y_{Pj} = \Sigma_{i \in P} Y_{ij}$, and $Y_{PQ} = \Sigma_{i \in P} \Sigma_{j \in Q} Y_{ij}$. The hypothesis H_{PQ} is rejected if $Z_{PQ} > \chi^2_\nu(\alpha)$ where $\nu = (a-1)(b-1)$. Note that the common critical point $\chi^2_\nu(\alpha)$ is based on the asymptotic distribution of $Z_{AB} = \max_{P,Q} Z_{PQ}$ under the overall independence hypothesis $H_{AB} = \bigcap_{P,Q} H_{PQ}$; this distribution is chi-square with $\nu = (a-1)(b-1)$ d.f. It follows from Theorem 1.2 of Appendix 1 that this procedure controls the Type I FWE at level α in large samples for the family of hypotheses H_{PQ}. Furthermore, this simultaneous test procedure is coherent because the statistics Z_{PQ} are monotone (i.e, if $P \supseteq P'$, $Q \supseteq Q'$ then $Z_{PQ} \geqq Z_{P'Q'}$). Gabriel noted that the family of hypotheses tested by this procedure can be expanded (without exceeding the upper limit α on the FWE) to include hypotheses of the type $\bigcap_{i=1}^{r} H_{P_iQ_i}$ where (P_1, Q_1), $(P_2, Q_2), \ldots, (P_r, Q_r)$ are disjoint combinations. It is concluded that at least one $H_{P_iQ_i}$ is false if $\Sigma_{i=1}^{r} Z_{P_iQ_i} > \chi^2_\nu(\alpha)$.

Hirotsu (1983) considered the problem of simultaneous testing of all null hypotheses concerning rowwise interactions (and similarly but separately columnwise interactions) in an $a \times b$ table. Assume multinomial sampling with π_{ij} and Y_{ij} denoting the probability and observed frequency for the (i, j)th cell, respectively $(1 \leqq i \leqq a, 1 \leqq j \leqq b)$. The rowwise interaction null hypothesis for the row pair (i, i') is given by

$$H_0(i, i'): \frac{\pi_{ij}}{\pi_{i\cdot}} = \frac{\pi_{i'j}}{\pi_{i'\cdot}} \qquad (1 \leqq j \leqq b) \tag{1.11}$$

where $\pi_{i\cdot} = \Sigma_{j=1}^{b} \pi_{ij}$ and $\pi_{\cdot j} = \Sigma_{i=1}^{a} \pi_{ij}$ $(1 \leqq i \leqq a, 1 \leqq j \leqq b)$. If the hypotheses $H_0(i, i')$ hold simultaneously for all row pairs (i, i'), then the overall independence hypothesis

$$H_0: \pi_{ij} = \pi_{i\cdot}\pi_{\cdot j} \qquad (1 \leqq i \leqq a, 1 \leqq j \leqq b) \tag{1.12}$$

holds. Note that the rowwise interaction null hypotheses are special cases of the subtable independence hypotheses considered by Gabriel (1966). But Hirotsu (1983) also includes in his family the null hypotheses on the interactions between all pairs (P, P') of exclusive subsets of the rows of an $a \times b$ table. These hypotheses, namely,

$$H_0(P, P'): \frac{\sum_{i \in P} \pi_{ij}}{\sum_{i \in P} \pi_{i\cdot}} = \frac{\sum_{i \in P'} \pi_{ij}}{\sum_{i \in P'} \pi_{i\cdot}}, \qquad P \cap P' = \phi ,$$

are not addressed by Gabriel's (1966) procedure.

Let $Y_{iB} = \Sigma_{j=1}^{b} Y_{ij}$ and $Y_{Aj} = \Sigma_{i=1}^{a} Y_{ij}$ denote the ith row sum and the jth column sum, respectively $(1 \leqq i \leqq a, 1 \leqq j \leqq b)$. Further let $Y_{PB} = \Sigma_{i \in P} Y_{iB}$ and $Y_{Pj} = \Sigma_{i \in P} Y_{ij}$. Then Hirotsu's statistic for testing $H_0(P, P')$ is given by

$$Z(P, P') = n\left(\frac{1}{Y_{PB}} + \frac{1}{Y_{P'B}}\right)^{-1} \sum_{j=1}^{b} \frac{1}{Y_{Aj}} \left(\frac{Y_{Pj}}{Y_{PB}} - \frac{Y_{P'j}}{Y_{P'B}}\right)^2 \tag{1.13}$$

where n is the total sample size. Hirotsu showed that the asymptotic null distribution of the maximum of all statistics $Z(P, P')$ is that of the largest root of a standard Wishart matrix of dimension $\min(a-1, b-1)$ and d.f. $= \max(a-1, b-1)$. Thus the upper α point of this distribution (which has been tabulated in selected cases by Hanumara and Thompson 1968, for example) can be used as a common critical point. This choice improves upon the larger critical point $\chi^2_{\nu}(\alpha)$ (where $\nu = (a-1)(b-1)$) used in Hirotsu (1978). An exactly parallel development can be given for testing columnwise interactions.

Relatively little attention has been paid to multiple comparison problems arising in higher dimensional tables. Bjørnstad (1982) considered the problem of ordering several dependent two-way tables according to some measure of association. Such tables arise as slices of larger multidimensional tables. One of the examples he offered deals with the association between voters' participation in an election and their occupations, measured over different elections. He focused attention on the probability of making an error in ordering the tables. For each one of the selected measures of association, he evaluated this probability using the large sample joint distribution of the estimates of the corresponding measure of association for the different tables.

Aitkin (1979) proposed an MCP for fitting a parsimonious log-linear model to data in a multidimensional table. The problem here is in principle similar to that of choosing the best subset of variables in regression for which Aitkin (1974) proposed an analogous procedure. (For an alternative multiple comparison approach to selecting the best subset of variables in regression, see Spjøtvoll 1972c.)

1.4 Logistic Response Curves

Consider a one-way layout experiment in which the k treatments are quantitative in nature, that is, each one of them can be applied at a level that can be varied over a continuous numerical scale. Further suppose that the response variable is binary, that is, each response is a "success" or "failure." As an example, suppose that the treatments are different drugs. The level at which a given drug is administered to a subject is referred to as its dose-level. The probability of success is a function of the dose-level of the drug. We assume that for the ith drug this is a logisitic function given by

$$\pi_i(x) = \frac{1}{1 + \exp\{-\alpha_i - \beta_i x\}} \qquad (1 \leqq i \leqq k) \qquad (1.14)$$

where x denotes the dose-level.

Reiersøl (1961) considered the problem of making all pairwise comparisons among the k drugs in terms of their response functions (1.14). In particular, he considered the following problems:

(i) Pairwise comparisons between the β_i's $(1 \leqq i \leqq k)$.

(ii) Pairwise comparisons between the quantities $\alpha_i + \beta_i x$ $(1 \leqq i \leqq k)$ for some specified dose-level x.

(iii) Pairwise comparisons between the qth quantiles:

$$\theta_i^{(q)} = \frac{\log_e\{q/(1-q)\} - \alpha_i}{\beta_i} \qquad (1 \leqq i \leqq k)$$

for some specified $q \in (0, 1)$.

Reiersøl assumed a fixed-sample setup in which for the ith drug, n_{ij} independent observations are taken at dose-level x_{ij} ($j = 1, 2, \ldots, m_i$, $i = 1, 2, \ldots, k$). Based on the minimum logit chi-square estimates of α_i and β_i computed separately for each i, Reiersøl derived simultaneous tests of pairwise null hypotheses concerning the above-listed parameters. He used Scheffé's projection method to derive these tests, but if only pairwise comparisons are of interest, then the TK-procedure can be used to obtain less conservative tests. The reader is referred to Reiersøl's (1961) article for additional details regarding his tests.

2 MULTIPLE COMPARISONS OF VARIANCES

The problem of comparisons of variances usually arises in one of the following settings:

(i) In fixed-effects models with unequal error variances sometimes the variances rather than the mean effects are the parameters of main interest. For example, it may be desired to compare several measuring instruments in terms of their lack of precision or reliability. For each instrument this can be assessed by the variability of repeat measurements made with it.

(ii) In random-effects models with several random factors the ratios of the variance components of different factors or the ratios of these components to the experimental error variance are usually of interest.

(iii) A preliminary test of the assumption of equal error variances for difference treatments is often desirable in an analysis of variance (ANOVA). The usual tests for the equality of variances (e.g., the Box–Bartlett test or the Levene test) are not designed to provide protection against a Type II error (concluding that the variances are equal when they are not), which is more pertinent in such applications. A possible approach here is to test the hypothesis that for at least one pair of treatments the ratio of the larger to the smaller variance is more than some critical threshold against the alternative that the variances are equal. This is an intersection-union test. From Berger (1982) it follows that each pair of variances should be tested separately at level α, and one may proceed with the assumption of equal variances if and only if all such tests are rejected.

We discuss the multiple comparison problems arising in settings (i) and (ii) in Sections 2.1 and 2.2, respectively.

2.1 Error Variances in Fixed-Effects Models

Hartley (1950) proposed a procedure for testing the equality of variances in a one-way layout that, being based on the union-intersection (UI) principle, can be used to make all pairwise comparisons among the variances.

Consider a balanced one-way layout with observations Y_{ij} on the ith treatment being i.i.d. normal with mean θ_i and variance σ_i^2 $(1 \leq i \leq k, 1 \leq j \leq n)$. Let $S_i^2 = \Sigma_{j=1}^{n} (Y_{ij} - \bar{Y}_i)^2/(n-1)$ be the usual unbiased estimate of σ_i^2 $(1 \leq i \leq k)$. Then the $U_i = S_i^2/\sigma_i^2$ are distributed independently as $\chi^2_{n-1}/(n-1)$ random variables (r.v.'s) and

$$F_{\max} = \max_{1 \leq i, j \leq k} \left(\frac{U_i}{U_j} \right) = \frac{U_{\max}}{U_{\min}} \tag{2.1}$$

has the distribution of the maximum of $\binom{k}{2}$ correlated F r.v.'s each with numerator and denominator d.f. equal to $n-1$. Denoting by $F_{\max}^{(\alpha)}$ the upper α point of the distribution of $F_{\max}$ (which also depends on k and n), it is straightforward to show that $(1-\alpha)$-level simultaneous confidence intervals for all variance ratios σ_i^2/σ_j^2 are given by

$$\frac{1}{F_{\max}^{(\alpha)}} \left(\frac{S_i^2}{S_j^2} \right) \leq \frac{\sigma_i^2}{\sigma_j^2} \leq F_{\max}^{(\alpha)} \left(\frac{S_i^2}{S_j^2} \right) \qquad (1 \leq i < j \leq k) . \tag{2.2}$$

Hartley (1950) tabulated $F_{\max}^{(\alpha)}$ for $\alpha = 0.05$ and for selected values of k and n. David (1952) gave more accurate tables of $F_{\max}^{(\alpha)}$ for $\alpha = 0.01$ and 0.05, which are included in Pearson and Hartley (1962).

If comparisons of treatments with respect to a control are desired in terms of their variances, then the distributions of the ratios

$$G_{\min} = \frac{\min_{1 \leq i \leq k-1} U_i}{U_k} \qquad \text{and} \qquad G_{\max} = \frac{\max_{1 \leq i \leq k-1} U_i}{U_k}$$

are required (assuming that the kth treatment is a control). The distribution of $G_{\min}$ has been studied by Gupta and Sobel (1962) and that of $G_{\max}$ has been studied by Gupta (1963b). These authors have tabulated the lower α points of $G_{\min}$ (denoted by $G_{\min}^{(1-\alpha)}$) and the upper α points of

$G_{\max}$ (denoted by $G_{\max}^{(\alpha)}$), respectively, for selected values of α, k and n. Using these critical points we can obtain $(1-\alpha)$-level simultaneous lower one-sided confidence intervals for the variance ratios σ_i^2/σ_k^2 as

$$\frac{\sigma_i^2}{\sigma_k^2} \geqq G_{\min}^{(1-\alpha)}\left(\frac{S_i^2}{S_k^2}\right) \qquad (1 \leqq i \leqq k-1) \tag{2.3}$$

and upper one-sided intervals as

$$\frac{\sigma_i^2}{\sigma_k^2} \leqq G_{\max}^{(\alpha)}\left(\frac{S_i^2}{S_k^2}\right) \qquad (1 \leqq i \leqq k-1)\,. \tag{2.4}$$

The solution to the problem of constructing two-sided simultaneous intervals for σ_i^2/σ_k^2 $(1 \leqq i \leqq k-1)$ is also straightforward but the corresponding critical points have not been tabulated.

Bechhofer (1968) considered the problem of comparing treatments with a control in terms of their variances in a two-way layout design with fixed factors A and B. For an $a \times b$ table with n independent, normally distributed observations per cell, he assumed the following multiplicative model for the cell variances:

$$\sigma_{ij}^2 = \alpha_i \beta_j \qquad (1 \leqq i \leqq a, 1 \leqq j \leqq b)\,.$$

Regarding the ath row as a control, he derived simultaneous one-sided confidence intervals for α_i/α_a $(1 \leqq i \leqq a-1)$. We refer the reader to Bechhofer's article for additional details.

Han (1969) extended Hartley's (1950) $F_{\max}$-test for the equality of treatment variances to randomized complete block designs. He assumed the model

$$Y_{ij} = \mu + \alpha_i + \beta_j + E_{ij} \qquad (1 \leqq i \leqq k, 1 \leqq j \leqq n) \tag{2.5}$$

with fixed parameters μ, α_i, and β_j $(\Sigma_{i=1}^k \alpha_i = \Sigma_{j=1}^n \beta_j = 0)$. He assumed that the error variances E_{ij} are distributed as $N(0, \sigma_i^2)$ r.v.'s and that they follow the intraclass correlation model:

$$\operatorname{corr}(E_{ij}, E_{i'j'}) = \begin{cases} 0 & \text{if } j \neq j' \\ \rho & \text{if } j = j',\ i \neq i'\,. \end{cases} \tag{2.6}$$

Thus the observations in each block are equicorrelated while those in different blocks are independent.

The unbiased estimator of σ_i^2 is

$$S_i^2 = \frac{\sum_{j=1}^{n} (Y_{ij} - \bar{Y}_{i\cdot} - \bar{Y}_{\cdot j})^2}{n-1},$$

which is distributed as $\sigma_i^2 \chi_{n-1}^2/(n-1)$ for $i = 1, 2, \ldots, k$. Note that the S_i^2's are dependent r.v.'s (because of assumption (2.6)) and hence the distribution of $F_{\max}$ (defined as in (2.1)) is rather intractable. Han obtained the following large sample ($n \to \infty$) approximation to $F_{\max}^{(\alpha)}$ in this case:

$$F_{\max}^{(\alpha)} \cong \exp\left\{Q_{k,\infty}^{(\alpha)}\sqrt{\frac{2}{n-1}\left[1 - \frac{1}{(k-1)^2}\right]}\right\}. \tag{2.7}$$

Simultaneous $(1-\alpha)$-level confidence intervals for σ_i^2/σ_j^2 $(1 \leqq i < j \leqq k)$ are then given by (2.2) where $F_{\max}^{(\alpha)}$ is approximated by (2.7).

Levy (1975) proposed a Newman–Keuls (NK) type step-down procedure for detailed comparisons among the σ_i^2's for the model given by (2.5) and (2.6). He used the $F_{\max}$-statistics for testing the homogeneity of the σ_i^2's for any subset of treatments and the approximate critical points given by (2.7) (with k replaced by the size of the subset under test).

Some other references on multiple comparisons of variances in fixed-effects models include Gnanadesikan (1959), Krishnaiah (1965), and Ramachandran (1956a,b). A general note of caution regarding the procedures of this section is that they are not particularly robust to the violation of the normality assumption.

2.2 Variance Components in Random-Effects Models

Let $\sigma_1^2, \sigma_2^2, \ldots, \sigma_k^2$ be the variance components corresponding to $k \geqq 2$ independent random factors in a balanced k-factor random-effects design and let σ^2 be the variance of the independent experimental error. Broemeling (1969) considered the problem of constructing a simultaneous confidence region for the ratios σ_i^2/σ^2 $(1 \leqq i \leqq k)$.

Let S_i^2 be the mean square associated with the ith random factor $(1 \leqq i \leqq k)$ in a standard ANOVA and let S^2 be the mean square error (MSE). It can be shown that, in a balanced design, the S_i^2's and S^2 are mutually independent, $S_i^2/\tau_i^2 \sim \chi_{\nu_i}^2/\nu_i$ and $S^2/\sigma^2 \sim \chi_\nu^2/\nu$ where $\tau_i^2 = E(S_i^2)$ is a known linear combination of $\sigma_1^2, \sigma_2^2, \ldots, \sigma_k^2$ $(1 \leqq i \leqq k)$. The d.f. ν_i, ν, and the linear combinations τ_i^2 depend on the specific design employed.

Broemeling derived a simultaneous confidence region for the ratios

τ_i^2/σ^2 by starting with the probability statement

$$\Pr\left\{\left(\frac{\tau_i^2}{\sigma^2}\right)\left(\frac{S^2}{S_i^2}\right) \leqq F_{\nu,\nu_i}^{(\alpha_i)}\ (i=1,2,\ldots,k)\right\} \geqq 1-\alpha \tag{2.8}$$

where $F_{\nu,\nu_i}^{(\alpha_i)}$ is the upper α_i point of the F-distribution with d.f. ν and ν_i and where the α_i's are chosen so that $\Pi_{i=1}^k (1-\alpha_i) = 1-\alpha$. The inequality in (2.8) follows from Kimball's inequality (see Appendix 2). By expressing each τ_i^2 in terms of the known linear combination of the σ_i^2's a conservative $(1-\alpha)$-level simultaneous confidence region for the ratios σ_i^2/σ^2 can be derived from (2.8). In general, it is very difficult to obtain an exact confidence region even for a single linear combination of variance components; in the case of multiple linear combinations, approximations are almost always necessary. For the special case $k=2$, however, Sahai and Anderson (1973) derived an exact expression for the left hand side of (2.8) in terms of the inverted Dirichlet distribution and found that Kimball's inequality is quite sharp in this case; see also Sahai (1974).

Broemeling and Bee (1976) used a similar technique to obtain simultaneous confidence intervals for the parameters of a balanced incomplete block (BIB) design in which both treatments and blocks are regarded as random. Broemeling (1978) attempted to extend (2.8) to two-sided intervals using a two-sided version of Kimball's multiplicative inequality, but Tong (1979) showed that such an extension is invalid.

Khuri (1981) gave an overview of the work on simultaneous confidence estimation of functions of variance components. He also proposed a conservative procedure for the simultaneous confidence estimation of all continuous functions of the variance components in a general balanced random-effects model.

3 GRAPHICAL PROCEDURES

Graphical procedures are receiving increasing attention in all areas of statistics and the subject of multiple comparisons is no exception. In this section we review some of these procedures. While the graphical procedures discussed in Chapter 3 are formal in nature in that they attempt to control the Type I FWE, the ones discussed here are somewhat informal and descriptive.

3.1 Normal Plots

Nelder, in his discussion of O'Neill and Wetherill (1971), stated

> When the data consist of an unstructured single set of means . . . there is no prior pattern except the null one, the treatments are all the same. Possible posterior patterns would include . . . (configurations in which) the means divide into two or more groups within which they look like samples from normal distributions.

Plackett, in the same discussion, suggested that such a posterior pattern can be revealed by plotting the treatment means on a normal probability paper. We now describe Plackett's suggestion in more detail.

Let $\bar{Y}_{(1)} \leqq \bar{Y}_{(2)} \leqq \cdots \leqq \bar{Y}_{(k)}$ be the ordered sample means obtained from a balanced one-way layout with independent, normal homoscedastic errors. The procedure consists of plotting the $\bar{Y}_{(i)}$'s on the normal probability paper and checking, "Whether by suitable shifts parallel to the axis of means, we can arrange that the means all lie close to a single line," with slope $1/S$ where S is the sample estimate of σ. Patterns of the type mentioned above are indicated when the plotted points fall along a few parallel lines (with slope $1/S$).

Example 3.1. Applying this procedure to Duncan's (1955) barley data (see Example 1.1 of Chapter 3) Plackett obtained the plot shown in Figure 3.1. This plot suggests that the varieties separate into three groups: $\{A\}$, $\{F, G, D\}$, and $\{C, B, E\}$. □

It should be noted that this graphical procedure does not control any particular error rate and the decisions as to what constitutes a homogeneous subset and what constitutes a significant separation between two homogeneous subsets are quite ad hoc. For example, the difference between the fourth and fifth ordered means (which fall into separate groups according to the normal plot) is 6.10 while at 5% level the least significant difference (LSD) using separate pairwise t-tests is 10.52.

Of course, one may argue that for purely exploratory applications, where such a plot is likely to be used, a formal control of any error rate is irrelevant. But even Nelder granted that such a plot should be accompanied by a consideration of the significance levels of the Studentized gaps and of Studentized residuals of the sample means from a regression line on the normal scores. We now review some literature on the distributions of the spacings (gaps) in an ordered sample from a normal distribution.

Tukey (1949) was the first to consider the use of spacings in his gap-straggler test to group the means (discussed in Section 6.2). The joint distribution of the spacings was studied by Pyke (1965). His following result is particularly relevant here: Let $Z_{(1)} \leqq Z_{(2)} \leqq \cdots \leqq Z_{(k)}$ be an

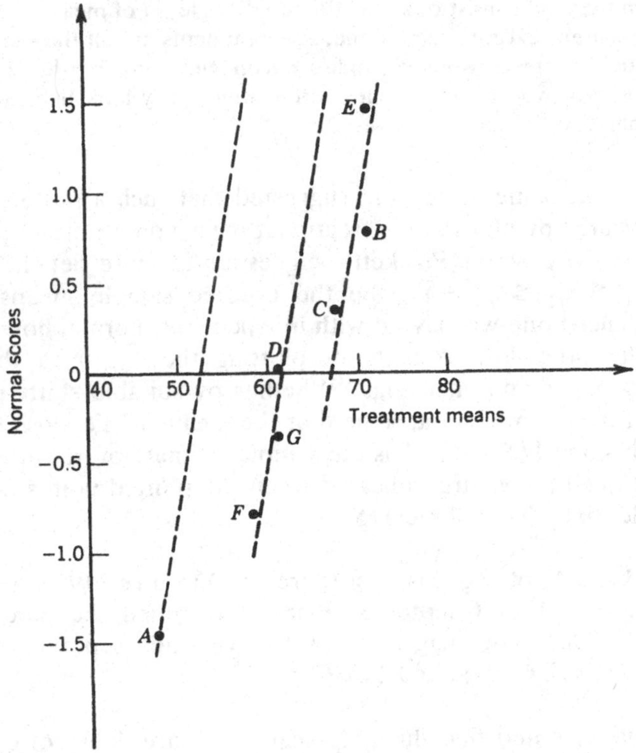

Figure 3.1. Normal plot of the varietal sample means for the barley data. Reproduced with slight modifications with the kind permission of the Royal Statistical Society.

ordered random sample from a $N(0, 1)$ distribution and let $\zeta_i = E(Z_{(i)})$ $(1 \leqq i \leqq k)$. Then as $k \rightarrow \infty$, the standardized spacings

$$W_i = \frac{Z_{(i)} - Z_{(i-1)}}{\zeta_i - \zeta_{i-1}} \qquad (2 \leqq i \leqq k)$$

are distributed as independent unit exponential r.v.'s. This result can be used to perform significance tests on the spacings if σ is known and k is large. If σ is not known, then one can obtain Studentized spacings by replacing σ by its estimate S. Andrews et al. (1980) have argued that in this case too, even when k is moderate, the asymptotic approximation based on the unit exponential distribution is valid. Finally we note that

Feder (1974) has derived an approximation to the distribution of the W_i's using a Taylor series expansion.

3.2 Miscellaneous Techniques

Use of side-by-side box plots for making pairwise comparisons between treatment means is becoming increasingly common. Some other miscellaneous graphical techniques are as follows.

Sampson (1980) proposed a graphical display technique wherein the treatments are denoted by equispaced points on a circle. Significant pairwise differences are indicated by drawing chords between the corresponding points. This technique is useful when pairwise comparisons are made on several occasions over time. By inspecting such circular displays for successive time periods, the researcher can assess how and when the treatment differences change over time.

Renner (1969) discussed the use of a graphical version of Duncan's step-down procedure for the comparison of binomial proportions. This method uses a binomial graph paper and arcsine transformations of the proportions.

Schweder and Spjøtvoll (1982) proposed a graphical procedure for examining multiple test results. Suppose that $m \geqq 2$ hypotheses are tested and the corresponding P-values are calculated. It is well known that the P-value is uniformly distributed over $[0, 1]$ when the hypothesis under test is true. Let N_p be the number of hypotheses with P-values $\geqq p$ $(0 < p < 1)$. If m_0 denotes the unknown number of true hypotheses, then for p not too small it is clear that $E(N_p) \cong m_0(1-p)$. For p small, $E(N_p)$ will be greater than $m_0(1-p)$ since false hypotheses will generally have small P-values and as a result they will also be included in N_p. Based on these considerations Schweder and Spjøtvoll proposed plotting N_p versus $1-p$. On this plot the points corresponding to the true hypotheses should roughly fall along a straight line for large values of p and the points corresponding to the false hypotheses should deviate upward from this straight line for small values of p. Furthermore, the slope of the straight line fitted to the points with large values of p should give an estimate of m_0, say, $\hat{m}_0$. The usual Bonferroni procedure with Type I familywise error rate (FWE) $\leqq \alpha$ (Type I per-family error rate (PFE) $= \alpha$) compares each P-value with α/m. Schweder and Spjøtvoll suggested that a sharper Bonferroni procedure can be obtained by using $\alpha/\hat{m}_0$ in place of α/m. The exact probabilistic properties of this procedure are difficult to determine.

4 MULTIPLE COMPARISONS OF MEANS UNDER ORDER RESTRICTIONS

In many practical problems of interest the levels of the treatment factor are ordinal, and they often imply a corresponding ordering of the mean responses. Thus if $\theta_1, \theta_2, \ldots, \theta_k$ denote the k treatment means, it may be known *a priori* that (say) $\theta_1 \leqq \theta_2 \leqq \cdots \leqq \theta_k$. An example of this arises in dose-response studies where k treatments consist of monotonically increasing levels of the dose of a certain drug, and it is postulated that the mean response is a nondecreasing function of the dose-level. Inferences about the θ_i's under such order restrictions are known as *isotonic inferences*. A comprehensive reference on this topic is the book by Barlow et al. (1972).

If an underlying interval scale variable (e.g., the dose-level) is available and if a natural class of regression models exists relating the mean response to the underlying variable, then a proper approach is to obtain a well-fitting model in that class and draw inferences from that model. But sometimes such a class of models is not available. In such situations one may fit orthogonal polynomials. However, this latter approach ignores specified order restrictions on the θ_i's. Some authors (e.g., Chew 1976b and Dawkins 1983) have categorically advised against the use of MCPs in settings such as dose-response studies. But the present discussion shows that MCPs under order restrictions may have a role to play in such settings and definitely in other settings where treatments are purely ordinal (i.e., no underlying interval scale variable is available).

4.1 All Monotone Contrasts

We assume the usual one-way layout model given in Chapter 2, Example 1.1. Let us first review the likelihood ratio (LR) test of the null hypothesis $H_0: \theta_1 = \theta_2 = \cdots = \theta_k$ against the alternative $H_1: \theta_1 \leqq \theta_2 \leqq \cdots \leqq \theta_k$ (with at least one strict inequality), which is described in Barlow et al. (1972, Chapter 3). Let $\hat{\theta}_i$ be the maximum likelihood estimate of θ_i $(1 \leqq i \leqq k)$ under the order restriction $\theta_1 \leqq \theta_2 \leqq \cdots \leqq \theta_k$. The $\hat{\theta}_i$'s are obtained by the well-known amalgamation procedure (Barlow et al. 1972, Chapter 2), and they can be expressed as

$$\hat{\theta}_i = \max_{1 \leqq u \leqq i} \min_{i \leqq v \leqq k} \frac{\sum_{j=u}^{v} n_j \bar{Y}_j}{\sum_{j=u}^{v} n_j} \qquad (1 \leqq i \leqq k) . \tag{4.1}$$

If the common error variance σ^2 is known, then the LR test statistic is given by

$$\frac{\sum_{i=1}^{k} n_i(\hat{\theta}_i - \bar{\bar{Y}})^2}{\sigma^2} \equiv \bar{\chi}_k^2 \tag{4.2}$$

where $\bar{\bar{Y}} = \Sigma_{i=1}^k n_i \bar{Y}_i / \Sigma_{i=1}^k n_i$. The null distribution of $\bar{\chi}_k^2$ is that of a weighted sum of independent χ_ν^2's ($\nu = 1, 2, \ldots, k-1$). If σ^2 is unknown, then the LR test statistic is given by (Barlow et al. 1972, Chapter 3)

$$\frac{\sum_{i=1}^{k} n_i(\hat{\theta}_i - \bar{\bar{Y}})^2}{\sum_{i=1}^{k}\sum_{j=1}^{n_i} (Y_{ij} - \hat{\theta}_i)^2} \equiv \bar{E}_k^2 . \tag{4.3}$$

The null distribution of $\bar{E}_k^2$ is given by a weighted sum of ratios of incomplete beta functions.

We next discuss the application of the above to multiple comparisons under order restrictions. Let

$$\mathbb{C}'^k = \left\{ \mathbf{c} = (c_1, c_2, \ldots, c_k)' \in \mathbb{C}^k : c_1 \leqq c_2 \leqq \cdots \leqq c_k, \sum_{i=1}^{k} n_i c_i = 0 \right\}$$

be the space of all k-dimensional "monotone" contrasts. Marcus and Peritz (1976) showed that the $\bar{\chi}_k^2$ statistic given by (4.2) is the supremum of the statistics

$$Z^2(\mathbf{c}) = \frac{\left(\sum_{i=1}^{k} n_i c_i \bar{Y}_i \right)^2}{\sigma^2 \sum_{i=1}^{k} n_i c_i^2}$$

over all $\mathbf{c} \in \mathbb{C}'^k$. Using this fact, Marcus and Peritz derived the following $(1-\alpha)$-level simultaneous one-sided confidence intervals for all monotone contrasts among the θ_i's:

$$\sum_{i=1}^{k} n_i c_i \theta_i \geqq \sum_{i=1}^{k} n_i c_i \bar{Y}_i - \bar{\chi}_k^{(\alpha)} \sigma \left(\sum_{i=1}^{k} n_i c_i^2 \right)^{1/2} \quad \forall \mathbf{c} \in \mathbb{C}'^k \tag{4.4}$$

where $\bar{\chi}_k^{(\alpha)}$ is the upper α point of the distribution of $\sqrt{\bar{\chi}_k^2}$.

When σ^2 is unknown, one must modify (4.4) by replacing σ^2 with its unbiased estimate S^2 (where it is assumed that S^2 is distributed independently of the $\bar{Y}_i$'s as $\sigma^2\chi_\nu^2/\nu$) and $\bar{\chi}_k^{(\alpha)}$ by $\bar{E}_k^{(\alpha)}$, the upper α point of the distribution of $\sqrt{\bar{E}_k^2}$.

Marcus (1982) showed that a two-sided version of (4.4) is obtained by replacing $\bar{\chi}_k^{(\alpha)}$ with $\bar{\chi}_k^{(\alpha/2)}$. If σ^2 is unknown then, of course, one must use the critical point $\bar{E}_k^{(\alpha/2)}$. Note that these two-sided intervals can be regarded as analogs in the ordered case (i.e., order restrictions on the θ_i's) of Scheffé's S-procedure in the unordered case.

Williams (1977) derived two-sided simultaneous confidence intervals for all monotone contrasts $\sum_{i=1}^k c_i\theta_i$ in the case of a balanced one-way layout ($n_i = n \ \forall i$). When σ^2 is known, these intervals are given by

$$\sum_{i=1}^k c_i\theta_i \in \left[\sum_{i=1}^k c_i\bar{Y}_i \pm R_k^{(\alpha/2)} \frac{\sigma}{\sqrt{n}} \sum_{i=1}^k \frac{|c_i|}{2}\right] \quad \forall \mathbf{c} \in \mathbb{C}'^k . \tag{4.5}$$

Here $R_k^{(\alpha/2)}$ is the upper $\alpha/2$ point of

$$R_k = \max_{1 \le r \le k} \frac{1}{r} \sum_{i=1}^r Z_i - \min_{1 \le r \le k} \frac{1}{k-r+1} \sum_{i=r}^k Z_i$$

where $Z_1, Z_2, \ldots, Z_k$ are i.i.d. $N(0, 1)$ r.v.'s. If σ^2 is unknown, then a Studentized version of R_k must be used in (4.5). Its distribution depends also on ν, the d.f. on which the estimate S^2 of σ^2 is based.

Marcus (1976, 1982) has tabulated selected values of $\bar{\chi}_k^{(\alpha)}$, $R_k^{(\alpha)}$, and $\bar{E}_k^{(\alpha)}$. The upper α points of the Studentized version of R_k have not yet been tabulated, but Williams has suggested that in that case the limiting values of $R_k^{(\alpha)}$ for $k \to \infty$ can be used as good approximations, which he has tabulated for selected α.

For the equal sample size case and for known σ^2, Marcus (1982) compared the widths of her Scheffé-type two-sided intervals with the widths of the corresponding Williams' Tukey-type intervals (4.5) for selected contrasts $\sum_{i=1}^k c_i\theta_i$. She concluded that these two procedures compare very much like their unordered case counterparts, that is, the Tukey-type intervals are superior for pairwise and other lower order contrasts while the Scheffé-type intervals are superior for higher order contrasts.

A closed step-down procedure (see Chapter 4, Section 3) for the family of all partition hypotheses under order restrictions was proposed by Marcus, Peritz, and Gabriel (1976). An NK-type step-down procedure for making pairwise comparisons among the θ_i's under the order restriction was proposed by Spjøtvoll (1977).

4.2 Comparisons with a Control

Williams (1971, 1972) proposed a step-down procedure for comparing the mean responses θ_i corresponding to increasing dose-levels indexed by $i = 1, 2, \ldots, k$, with the mean response θ_0 corresponding to the zero dose-level (control). As in the previous section, the usual one-way layout model with independent, homoscedastic, normally distributed errors is assumed and the order restriction $\theta_0 \leqq \theta_1 \leqq \cdots \leqq \theta_k$ is imposed. The problem of interest is to detect the lowest dose-level for which the mean response is greater than θ_0.

In his 1971 paper Williams assumed that the sample sizes for all $k+1$ dose-levels are the same. In this case formula (4.1) for isotonic estimates $\hat{\theta}_i$ simplifies to

$$\hat{\theta}_i = \max_{0 \leqq u \leqq i} \min_{i \leqq v \leqq k} \frac{\sum_{j=u}^{v} \bar{Y}_j}{(v - u + 1)} \qquad (0 \leqq i \leqq k) .$$

Let

$$\bar{T}_i = \frac{\hat{\theta}_i - \bar{Y}_0}{\sqrt{2S^2/n}} \qquad (1 \leqq i \leqq k)$$

where S^2 is the usual unbiased estimate of σ^2 based on $\nu = (k+1)(n-1)$ d.f. Note that $\bar{T}_1 \leqq \bar{T}_2 \leqq \cdots \leqq \bar{T}_k$. Let $\bar{T}_r^{(\alpha)}$ denote the upper α point of the distribution of $\bar{T}_r$ under the null hypothesis $\theta_0 = \theta_1 = \cdots = \theta_r$ $(1 \leqq r \leqq k)$. The $\bar{T}_r^{(\alpha)}$'s form an increasing sequence for $r = 1, 2, \ldots, k$. The values of $\bar{T}_r^{(\alpha)}$ for $\alpha = 0.01, 0.05$, and $r = 1(1)10$ are given in Williams (1971).

Williams' step-down procedure starts by comparing $\bar{T}_k$ with $\bar{T}_k^{(\alpha)}$. If $\bar{T}_k > \bar{T}_k^{(\alpha)}$, then the overall null hypothesis $\theta_0 = \theta_1 = \cdots = \theta_k$ is rejected and the test proceeds to compare $\bar{T}_{k-1}$ with $\bar{T}_{k-1}^{(\alpha)}$; otherwise testing stops with the conclusion that there are no significant differences between the zero dose-level and any of the higher dose-levels. In general, for $r = 1, 2, \ldots, k$ the hypothesis $\theta_0 = \theta_1 = \cdots = \theta_r$ is rejected if and only if

$$\bar{T}_j > \bar{T}_j^{(\alpha)} \qquad \text{for } j = r, r+1, \ldots, k . \tag{4.6}$$

If r is the last dose-level for which (4.6) is satisfied, then we conclude that it is the lowest dose-level for which the mean response is greater than θ_0.

Williams (1972) extended this procedure to the case of unequal sample sizes, especially the case where $n_0 > n_1 = n_2 = \cdots = n_k$. Chase (1974) also considered this same case and derived an LR test. Shirley (1977) gave a nonparametric version of Williams' (1972) procedure, which was improved by Williams (1986). House (1986) extended this improved procedure to randomized block designs.

Shirley (1979) made a Monte Carlo based comparison between Bartholomew's LR test (as modified by Marcus 1976 to operate in a step-down manner) based on the $\bar{E}_k^2$ statistic (4.3), Williams' test, and a new test proposed by her for the problem considered by Williams (1971, 1972). The main finding of this simulation study was that Bartholomew's modified test is generally more powerful than its competitors although it is more cumbersome to apply and has tables available only for the equal sample size case.

5 INTERACTIONS IN TWO-WAY LAYOUTS

Most of the multiple comparison problems considered thus far have dealt with a single treatment factor that may be embedded in a complex design. When two or more treatment factors are of simultaneous interest, the inferences become more involved, particularly because of the possibility of interactions. If the interactions can be safely ignored, then we get an additive model in which each factor can be analyzed (e.g., pairwise comparisons may be performed between the levels of each factor) essentially independently of the others. Thus it is not surprising that when analyzing a higher way factorial design, usually attention is first focused on assessing the significance of interactions. Of course, in some problems interactions are the parameters of primary interest and not some "nuisances" to be wished away; see Neyman (1977) for some examples of situations where synergistic effects are important and are to be detected.

Gabriel, Putter, and Wax (1973) noted that in order to arrive at a parsimonious nonadditive model, one may employ a systematic search procedure involving a finite set of interaction contrasts of a given type. To make valid simultaneous inferences on a given set of interaction contrasts appropriate MCPs are needed. In the present section we discuss such MCPs for selected families of interaction contrasts in balanced two-way layouts.

It is worthwhile to note here that all of these MCPs are designed to give protection against Type I errors of rejecting null interactions. In practice, often the primary goal is to identify null interactions that lead to a simplified model. Therefore MCPs giving protection against Type II errors (erroneously eliminating nonnull interactions) would be more appropriate (Neyman 1977). But such MCPs have not been developed yet.

5.1 Types of Interaction Contrasts

Consider an equireplicated two-way factorial layout for which the usual model is

$$Y_{ijk} = \theta_{ij} + E_{ijk} = \mu + \alpha_i + \beta_j + \gamma_{ij} + E_{ijk}$$
$$(1 \leqq i \leqq I, 1 \leqq j \leqq J, 1 \leqq k \leqq n) \qquad (5.1)$$

where the E_{ijk}'s are i.i.d. $N(0, \sigma^2)$ r.v.'s, and $\Sigma_{i=1}^{I} \alpha_i = \Sigma_{j=1}^{J} \beta_j = 0$, $\Sigma_{i=1}^{I} \gamma_{ij} = 0$ for all j, and $\Sigma_{j=1}^{J} \gamma_{ij} = 0$ for all i. The following definitions of different types of interaction contrasts were given by Bradu and Gabriel (1974). Let $\mathbf{\Theta}$ and $\mathbf{\Gamma}$ be $I \times J$ matrices of θ_{ij}'s and γ_{ij}'s, respectively, and let $\mathbf{C} = \{c_{ij}\}$ be an $I \times J$ matrix such that $\Sigma_{i=1}^{I} c_{ij} = \Sigma_{j=1}^{J} c_{ij} = 0$ for all i, j, that is, each row and column of $\mathbf{C}$ is a contrast vector. Then a *generalized interaction contrast* is defined as

$$\mathrm{tr}(\mathbf{C}'\mathbf{\Theta}) = \sum_{i=1}^{I} \sum_{j=1}^{J} c_{ij}\theta_{ij} = \sum_{i=1}^{I} \sum_{j=1}^{J} c_{ij}\gamma_{ij} = \mathrm{tr}(\mathbf{C}'\mathbf{\Gamma}) . \qquad (5.2)$$

If $\mathbf{C}$ is a matrix of rank one, that is, if $\mathbf{C} = \mathbf{ab}'$ where $\mathbf{a} = (a_1, a_2, \ldots, a_I)'$ and $\mathbf{b} = (b_1, b_2, \ldots, b_J)'$ are such that $\Sigma_{i=1}^{I} a_i = \Sigma_{j=1}^{J} b_j = 0$, then the resulting contrast is referred to as a *product-type interaction contrast*, which can be written as

$$\mathbf{a}'\mathbf{\Theta}\mathbf{b} = \mathbf{a}'\mathbf{\Gamma}\mathbf{b} . \qquad (5.3)$$

These types of contrasts were considered earlier by Gabriel, Putter, and Wax (1973). Note that each γ_{ij} in (5.1), referred to as an *interaction residual*, is a product-type interaction contrast that is obtained by setting $a_i = (I-1)/I$, $a_{i'} = -1/I$ for $i' \neq i$, and $b_j = (J-1)/J$, $b_{j'} = -1/J$ for $j' \neq j$. Another special case of a product-type interaction contrast is referred to as a *tetrad difference*:

$$\psi_{ij,i'j'} = \theta_{ij} - \theta_{i'j} - \theta_{ij'} + \theta_{i'j'} , \qquad (5.4)$$

which is obtained by setting all elements of $\mathbf{a}$ equal to zero except $a_i = +1$, $a_{i'} = -1$ and all elements of $\mathbf{b}$ equal to zero exact $b_j = +1$, $b_{j'} = -1$ $(1 \leqq i \neq i' \leqq I, 1 \leqq j \neq j' \leqq J)$. For some other special cases of product-type interaction contrasts, see Gabriel, Putter, and Wax (1973).

A least squares (LS) estimate of any contrast is obtained by replacing the θ_{ij}'s by their LS estimates $\hat{\theta}_{ij} = \bar{Y}_{ij.}$ or equivalently by replacing the

γ_{ij}'s by their LS estimates $\hat{\gamma}_{ij} = \bar{Y}_{ij\cdot} - \bar{Y}_{i\cdot\cdot} - \bar{Y}_{\cdot j\cdot} + \bar{Y}_{\cdots}$. In the sequel we denote by $\hat{\mathbf{\Theta}}$ and $\hat{\mathbf{\Gamma}}$ the matrices $\{\hat{\theta}_{ij}\}$ and $\{\hat{\gamma}_{ij}\}$, respectively.

5.2 Procedures for Selected Families

5.2.1 Generalized Interaction Contrasts

We first consider the family of all generalized interaction constrasts: $\text{tr}(\mathbf{C}'\mathbf{\Gamma}) = \Sigma_{i=1}^{I} \Sigma_{j=1}^{J} c_{ij}\gamma_{ij} = \Sigma_{i=1}^{I} \Sigma_{j=1}^{J} c_{ij}\theta_{ij}$ where the c_{ij}'s satisfy $\Sigma_{i=1}^{I} c_{ij} = \Sigma_{j=1}^{J} c_{ij} = 0$ for all i, j. Let $\mathbb{C}^{I,J}$ be the set of all such $I \times J$ matrices $\mathbf{C} = \{c_{ij}\}$. Bradu and Gabriel (1974) proposed two single-step procedures for this family, both based on the UI method. The first procedure uses the quantity $\{\text{tr}(\mathbf{C}'\mathbf{C})\}^{1/2} = (\Sigma\,\Sigma\, c_{ij}^2)^{1/2}$ as the norm for the pivotal r.v. for $\text{tr}(\mathbf{C}'\mathbf{\Gamma})$, which leads to the following r.v.:

$$T = \max_{\mathbf{C} \in \mathbb{C}^{I,J}} \left\{ \frac{\sqrt{n}\,|\text{tr}[\mathbf{C}'(\hat{\mathbf{\Gamma}} - \mathbf{\Gamma})]|}{\{\text{tr}(\mathbf{C}'\mathbf{C})\}^{1/2}} \right\}. \tag{5.5}$$

The second procedure uses the norm $\Sigma\,\Sigma\,|c_{ij}|/2$ in place of $\{\text{tr}(\mathbf{C}'\mathbf{C})\}^{1/2}$.

It can be shown that $T^2 = n \times \text{tr}[(\hat{\mathbf{\Gamma}} - \mathbf{\Gamma})(\hat{\mathbf{\Gamma}} - \mathbf{\Gamma})'] =$ the sum of the characteristic roots of $n(\hat{\mathbf{\Gamma}} - \mathbf{\Gamma})(\hat{\mathbf{\Gamma}} - \mathbf{\Gamma})'$. If S^2 denotes the usual MSE estimate of σ^2 with $\nu = IJ(n-1)$ d.f., then it can be shown that $T^2/\{(I-1)(J-1)S^2\}$ has an F-distribution with $(I-1)(J-1)$ and ν d.f. In fact, if all the γ_{ij}'s are zero, then T^2 is simply the interaction sum of squares. Thus we obtain the following $(1-\alpha)$-level simultaneous confidence intervals for all contrasts among the interactions:

$$\Sigma\,\Sigma\, c_{ij}\gamma_{ij} \in \left[\Sigma\,\Sigma\, c_{ij}\hat{\gamma}_{ij} \pm \left\{ (I-1)(J-1) F_{(I-1)(J-1),\nu}^{(\alpha)} \right\}^{1/2} \right.$$
$$\left. \times \left\{ \frac{S^2}{n} \Sigma\,\Sigma\, c_{ij}^2 \right\}^{1/2} \right] \quad \forall\, \mathbf{C} \in \mathbb{C}^{I,J}. \tag{5.6}$$

Bradu and Gabriel referred to this procedure as a Scheffé-type procedure.

For the second procedure Bradu and Gabriel noted that instead of $\mathbb{C}^{I,J}$ consisting of the set of interaction contrast matrices if we consider the set of *all* contrast matrices $\mathbf{C}$ satisfying $\Sigma_{i=1}^{I} \Sigma_{j=1}^{J} c_{ij} = 0$, then we obtain Tukey's T-procedure. Therefore by restricting the inferences to $\mathbb{C}^{I,J}$ the following conservative $(1-\alpha)$-level simultaneous confidence intervals are obtained:

$$\sum\sum c_{ij}\gamma_{ij} \in \left[\sum\sum c_{ij}\hat{\gamma}_{ij} \pm Q^{(\alpha)}_{IJ,\nu} \frac{S}{\sqrt{n}} \cdot \frac{1}{2}\sum\sum |c_{ij}|\right] \quad \forall \mathbf{c} \in \mathbb{C}^{I,J} . \tag{5.7}$$

The same intervals were derived earlier by Sen (1969b) using an alternative method (which is applicable in some other problems as well). Sen's method is based on the following lemma.

Lemma 5.1 **(Sen 1969b).** Let r.v.'s Z_{ij} $(1 \leqq i \leqq I, 1 \leqq j \leqq J)$ have a joint normal distribution with zero means, common variance σ^2, and correlation matrix given by

$$\operatorname{corr}(Z_{ij}, Z_{i'j'}) = \begin{cases} \rho_1 & i \neq i', j = j' \\ \rho_2 & i = i', j \neq j' \\ \rho_3 & i \neq i', j \neq j' \end{cases} \tag{5.8}$$

where $-1/(I-1) \leqq \rho_1 < 1$, $-1/(J-1) \leqq \rho_2 < 1$, and $1 - \rho_1 - \rho_2 + \rho_3 > 0$. Let $\nu S^2/\sigma^2$ be distributed as a χ^2_ν r.v. independent of the Z_{ij}'s. Let $\mathbb{C}^{I,J}$ be as defined before. Then

$$\Pr\left\{\left|\sum\sum c_{ij}Z_{ij}\right| \leqq Q^{(\alpha)}_{IJ,\nu} S(1-\rho_1-\rho_2+\rho_3)^{1/2}\tfrac{1}{2}\sum\sum |c_{ij}| \ \forall \mathbf{c} \in \mathbb{C}^{I,J}\right\}$$

$$\geqq 1 - \alpha . \tag{5.9}$$

□

Now let $Z_{ij} = [IJn/(I-1)(J-1)]^{1/2}(\hat{\gamma}_{ij} - \gamma_{ij})$ for $1 \leqq i \leqq I$, $1 \leqq j \leqq J$. The Z_{ij}'s have a joint normal distribution with zero means, common variance σ^2, and correlation matrix given by

$$\operatorname{corr}(Z_{ij}, Z_{i'j'}) = \begin{cases} \rho_1 = -\dfrac{1}{I-1} & i \neq i', j = j' \\ \rho_2 = -\dfrac{1}{J-1} & i = i', j \neq j' \\ \rho_3 = \dfrac{1}{(I-1)(J-1)} & i \neq i', j \neq j' . \end{cases} \tag{5.10}$$

Then by applying (5.9) we obtain the same intervals as those in (5.7).

In analogy with the performance comparison between the S- and T-procedures for contrasts among means in a one-way layout, it may be conjectured that the intervals (5.7) will perform better than the intervals

(5.6) whenever $\mathbf{C}$ is a sparse matrix (i.e., the rows and columns of $\mathbf{C}$ are low order contrasts). This fact was confirmed in a numerical comparison by Johnson (1976).

5.2.2 Product-Type Interaction Contrasts

We next consider the family of all product-type interaction contrasts. To derive a single-step test procedure for this family Bradu and Gabriel (1974) applied the UI method in the same way as in (5.5). This leads to the r.v.

$$U = \max_{\substack{\mathbf{a}\in\mathbb{C}^I \\ \mathbf{b}\in\mathbb{C}^J}} \left\{ \frac{\sqrt{n}|\mathbf{a}'(\hat{\mathbf{\Gamma}} - \mathbf{\Gamma})\mathbf{b}|}{(\mathbf{a}'\mathbf{a})^{1/2}(\mathbf{b}'\mathbf{b})^{1/2}} \right\} \tag{5.11}$$

obtained by taking the maximum of the pivotal r.v.'s for $\mathbf{a}'\mathbf{\Gamma}\mathbf{b}$ over all $\mathbf{a}\in\mathbb{C}^I$ and $\mathbf{b}\in\mathbb{C}^J$ where $\mathbb{C}^I$ and $\mathbb{C}^J$ are sets of all I- and J-dimensional contrast vectors, respectively. They showed that U^2/σ^2 = the largest characteristic root of a standard Wishart matrix $n(\hat{\mathbf{\Gamma}} - \mathbf{\Gamma})(\hat{\mathbf{\Gamma}} - \mathbf{\Gamma})'/\sigma^2$ with parameters $p = \min(I-1, J-1)$ and $q = \max(I-1, J-1)$. The same result was earlier obtained by Johnson and Graybill (1972). Thus if σ^2 is known, then exact $(1-\alpha)$-level simultaneous confidence intervals for all product-type interaction contrasts are given by

$$\sum\sum a_i b_j \gamma_{ij} \in \left[\sum\sum a_i b_j \hat{\gamma}_{ij} \pm (W_{p,q}^{(\alpha)})^{1/2} \left\{ \frac{\sigma^2}{n} \sum a_i^2 \sum b_j^2 \right\}^{1/2} \right]$$

$$\forall\, \mathbf{a}\in\mathbb{C}^I, \mathbf{b}\in\mathbb{C}^J \tag{5.12}$$

where $W_{p,q}^{(\alpha)}$ is the upper α point of the largest characteristic root of the Wishart matrix referred to above. If σ^2 is not known and is estimated by S^2, then the critical points of the distribution of U^2/S^2 are needed. Unfortunately, these have not been tabulated. In this case Johnson (1976) proposed to approximate the distribution of U^2/S^2 by that of a scaled F r.v. obtained by matching the first two moments of U^2/σ^2 with those of a scaled χ^2 r.v. The resulting approximation to the upper α critical point of U^2/S^2 is given by $\mu_1 F_{m,\nu}^{(\alpha)}$ where $m = 2\mu_1^2/\mu_2'$, and μ_1 and μ_2' are the mean and variance of the largest characteristic root of a standard Wishart matrix with parameters p and q, respectively. The moments μ_1 and μ_2' have been tabled by Mandel (1971). Bradu and Gabriel (1974) refer to (5.12) as a Roy-type procedure.

Clearly, since the family of generalized interaction contrasts contains the family of product-type interaction contrasts, it follows that $T^2 \geq U^2$ and therefore (5.12) yields shorter intervals than (5.6) for all product-type interaction contrasts; (5.12) also yields shorter intervals than (5.7) for most product-type interaction contrasts.

5.2.3 *Tetrad Differences*

We next consider the family of all tetrad differences $\psi_{ij,i'j'}$ defined in (5.4). By applying the UI method to this finite family, Bradu and Gabriel (1974) arrived at the r.v.:

$$V = \max_{i \neq i', j \neq j'} \left\{ \frac{\sqrt{n}|\hat{\psi}_{ij,i'j'} - \psi_{ij,i'j'}|}{2} \right\}. \tag{5.13}$$

If $\xi^{(\alpha)}$ denotes the upper α point of V/S, then $(1-\alpha)$-level simultaneous confidence intervals for all tetrad differences are given by

$$\psi_{ij,i'j'} \in [\hat{\psi}_{ij,i'j'} \pm \xi^{(\alpha)} S\sqrt{4/n}] \qquad (1 \leq i \neq i' \leq I, 1 \leq j \neq j' \leq J). \tag{5.14}$$

If $I = 2$ (respectively, $J = 2$), then it can be shown that $\xi^{(\alpha)}$ is given by $Q^{(\alpha)}_{J,\nu}/\sqrt{2}$ (respectively, $Q^{(\alpha)}_{I,\nu}/\sqrt{2}$). But for other values of I and J the exact distribution of V/S is not known and therefore one must approximate $\xi^{(\alpha)}$. The first order Bonferroni upper bound on $\xi^{(\alpha)}$ is $T_{\nu}^{(\alpha/2I^*J^*)}$ where $I^* = \binom{I}{2}$ and $J^* = \binom{J}{2}$. Bradu and Gabriel (1974) compared the first and second order Bonferroni bounds in this case and found them to be quite close. A slightly sharper upper bound than the first order Bonferroni upper bound is $|M|^{(\alpha)}_{I^*J^*,\nu}$; this critical point from the Studentized maximum modulus distribution is based on the Kimball (1951) inequality. Hochberg (1974a) has studied some sharper upper bounds whose analytical validity depends on the truth of the Khatri conjecture (see the discussion following Theorem 2.3 in Appendix 2).

Gabriel, Putter, and Wax (1973) pointed out that the intervals (5.14) can be extended to the family of all product-type interaction contrasts to yield the following $(1-\alpha)$-level simultaneous confidence intervals:

$$\sum\sum a_i b_j \gamma_{ij} \in \left[\sum\sum a_i b_j \hat{\gamma}_{ij} \pm \xi^{(\alpha)} \frac{S}{\sqrt{n}} \left\{ \frac{\sum |a_i| \sum |b_j|}{2} \right\}\right]$$

$$\forall\, \mathbf{a} \in \mathbb{C}^I, \mathbf{b} \in \mathbb{C}^J. \tag{5.15}$$

5.2.4 *Interaction Residuals*

In some applications the family of IJ interaction residuals γ_{ij} might be of interest, particularly when hunting for cells with localized large interactions. In this case instead of using the procedure for all product-type interaction contrasts, one could base the procedure on the distribution of the r.v. $\max_{i,j} |T_{ij}|$ where

$$T_{ij} = \sqrt{\frac{IJn}{(I-1)(J-1)}} \cdot \left(\frac{\hat{\gamma}_{ij} - \gamma_{ij}}{S}\right) \qquad (1 \leqq i \leqq I, 1 \leqq j \leqq J). \quad (5.16)$$

The T_{ij}'s have a joint IJ-variate t-distribution with an associated correlation matrix given by (5.10). Exact upper α critical points of $\max_{i,j} |T_{ij}|$ are not available, but the $|M|_{IJ,\nu}^{(\alpha)}$'s can be used as upper bounds. Thus conservative $(1-\alpha)$-level simultaneous confidence intervals for all γ_{ij}'s are given by

$$\gamma_{ij} \in \left[\hat{\gamma}_{ij} \pm |M|_{IJ,\nu}^{(\alpha)} S \sqrt{\frac{(I-1)(J-1)}{IJn}}\right] \qquad (1 \leqq i \leqq I, 1 \leqq j \leqq J). \quad (5.17)$$

5.2.5 *Interaction Elements*

Sometimes it is desired to make comparisons among the levels of one factor separately for each level of the other factor. Harter (1970) and Cicchetti (1972) have proposed step-down procedures for making pairwise comparisons between the rows separately for each column, and between the columns separately for each row. Here the parameters of interest are $\theta_{ij} - \theta_{i'j}$ $(1 \leqq i \neq i' \leqq I, 1 \leqq j \leqq J)$ and $\theta_{ij} - \theta_{ij'}$ $(1 \leqq i \leqq I, 1 \leqq j \neq j' \leqq J)$, respectively. Harter (1970) referred to these parameters as *interaction elements*. Note that the interaction elements by themselves shed no light on the nonadditivity present in a two-way table. Also note that any tetrad difference (5.4) is obtained by taking the difference of two appropriate interaction elements.

It can be readily shown using Kimball's (1951) inequality that conservative $(1-\alpha)$-level simultaneous confidence intervals for all $\theta_{ij} - \theta_{i'j}$ are given by

$$\theta_{ij} - \theta_{i'j} \in \left[\bar{Y}_{ij\cdot} - \bar{Y}_{i'j\cdot} \pm \frac{Q_{I,\nu}^{(\alpha_J)} S}{\sqrt{n}}\right] \qquad (1 \leqq i \neq i' \leqq I, 1 \leqq j \leqq J) \quad (5.18)$$

where $\alpha_J = 1 - (1-\alpha)^{1/J}$. Similarly conservative $(1-\alpha)$-level simultaneous confidence intervals for all $\theta_{ij} - \theta_{ij'}$ are given by

$$\theta_{ij} - \theta_{ij'} \in \left[\bar{Y}_{ij\cdot} - \bar{Y}_{ij'\cdot} \pm \frac{Q_{J,\nu}^{(\alpha_I)} S}{\sqrt{n}}\right] \qquad (1 \leqq i \leqq I, 1 \leqq j \neq j' \leqq J) \quad (5.19)$$

where $\alpha_I = 1 - (1-\alpha)^{1/I}$.

The intervals (5.18) and (5.19) can be extended to the family of all contrasts among the levels of one factor at each one of the levels of the other factor by using Lemma 2.1 of Chapter 3. For the same family Johnson (1976) proposed a modification of the procedure given in Section 5.2.2. This modification uses the critical point $W_{p,q}^{(\alpha)}$ (in case of known σ) where $p = \min(I, J-1)$, $q = \max(I-1, J)$ when making comparisons among the rows in each column, and $p = \min(I-1, J)$, $q = \max(I, J-1)$ when making comparisons among the columns in each row.

We now illustrate all of the procedures described above by means of a comprehensive example.

Example 5.1. Bradu and Gabriel (1974) report data from an experiment designed to study the effects of soil type and phosphatic fertilizers on the weight of lettuce crop. The experimental design consists of a balanced two-way layout (3 soil types × 4 fertilizer types) with $n = 4$ replication per cell. The cell averages are given in Table 5.1. The MSE computed from the raw data is $S^2 = 0.16$ with $\nu = 36$ d.f.

The interaction residuals $\hat{\gamma}_{ij} = \bar{Y}_{ij\cdot} - \bar{Y}_{i\cdot\cdot} - \bar{Y}_{\cdot j\cdot} + \bar{Y}_{\cdot\cdot\cdot}$ constitute the entries of matrix $\hat{\Gamma}$ given below:

$$\hat{\Gamma} = \begin{bmatrix} -1.12 & 0.10 & 1.02 \\ 0.34 & 0.23 & -0.57 \\ 0.08 & -0.16 & 0.08 \\ 0.70 & -0.17 & -0.53 \end{bmatrix} .$$

The interaction sum of squares (SS_{Int}) is given by

$$SS_{\text{Int}} = n \sum \sum \hat{\gamma}_{ij}^2 = n \times \text{tr}[\hat{\Gamma}\hat{\Gamma}'] = 14.546 ,$$

TABLE 5.1. Cell Averages (Grams of Dry Matter per Plot)

	Soil Type		
Fertilizer[a]	A	B	C
F_0	0.53	0.47	2.32
F_1	3.63	2.23	2.36
F_2	3.52	2.00	3.16
F_3	4.06	1.90	2.47

Source: Bradu and Gabriel (1974).
[a] F_0 = no fertilizer, F_1 = monocalcium phosphate, F_2 = potassium metaphosphate, F_3 = magnesium phosphate.

and has 6 d.f. Comparing this with the critical value $6S^2 \times F_{6,36}^{(0.05)} = 6 \times 0.16 \times 2.37 = 2.275$, it is seen that the interaction effect is highly significant. We now proceed to identify specific interaction contrasts contributing to this significance.

First, by using the Cauchy–Schwarz inequality it can be shown that the maximum of T (see (5.5)) under the hypothesis of no interactions is achieved when $\mathbf{C} = \mathbf{C}_0$ is any matrix proportional to $\hat{\mathbf{\Gamma}}$, and this maximum equals SS_{Int} just computed. By taking $\mathbf{C}_0 = \hat{\mathbf{\Gamma}}$, the most significant generalized interaction contrast, $\Sigma\,\Sigma\, c_{0ij}\gamma_{ij}$, is estimated to be 3.636. Since this contrast is obtained by data-snooping, (5.6) or (5.7) should be used to construct a confidence interval for it.

It is difficult to give a practical interpretation to this generalized interaction contrast. To look for a more meaningful significant interaction contrast, we turn to the class of product-type contrasts. The most significant contrast in this class is $\mathbf{a}_0'\mathbf{\Gamma}\mathbf{b}_0$ where $\mathbf{a}_0$ and $\mathbf{b}_0$ maximize $(\mathbf{a}'\hat{\mathbf{\Gamma}}\mathbf{b})^2/(\mathbf{a}'\mathbf{a})(\mathbf{b}'\mathbf{b})$ among all contrast vectors $\mathbf{a} \in \mathbb{C}^I$ and $\mathbf{b} \in \mathbb{C}^J$. Again, using the Cauchy–Schwarz inequality it follows that $\mathbf{a}_0$ is the characteristic vector corresponding to the largest characteristic root of $\hat{\mathbf{\Gamma}}\hat{\mathbf{\Gamma}}'$ and $\mathbf{b}_0$ is any vector proportional to $\mathbf{a}_0'\hat{\mathbf{\Gamma}}$. Bradu and Gabriel computed $\mathbf{a}_0 = (-0.815, 0.336, 0.007, 0.472)'$ and $\mathbf{b}_0 = (0.730, -0.047, -0.682)'$, which they approximated by $\mathbf{a} = (-1, \frac{1}{2}, 0, \frac{1}{2})'$ and $\mathbf{b} = (1, 0, -1)'$, respectively. The product-type contrast resulting from this $\mathbf{a}$ and $\mathbf{b}$ can be interpreted as the difference in comparisons of average of fertilizers F_1 and F_3 with F_0 for soil types A and C. This contrast is estimated to be $\Sigma\,\Sigma\, a_i b_j \hat{\gamma}_{ij} = 3.22$ and the associated U^2/S^2 statistic equals

$$\frac{n\left(\sum\sum a_i b_j \hat{\gamma}_{ij}\right)^2}{S^2\left(\sum a_i^2\right)\left(\sum b_j^2\right)} = \frac{4(3.22)^2}{0.16(1.5 \times 2)} = 86.40\,.$$

If we ignore the variability in S^2, then for a 0.05 level test this can be compared with $W_{2,3}^{(0.05)} = 10.74$ (from Hanumara and Thompson 1968). If we take into account the variability in S^2, then the approximation to the corresponding critical point suggested by Johnson (1976) can be computed as follows: First find from Mandel (1971) the two moments $\mu_1 = 5.03$ and $\mu_2' = (3.08)^2$ of the largest characteristic root of a standard Wishart matrix with parameters $p = 2$ and $q = 3$. This gives $m = 2\mu_1^2/\mu_2' = 5.33$. Then compute the desired approximation: $\mu_1 F_{m,\nu}^{(0.05)} = 5.03 \times 2.436 = 12.25$. In both cases we see that the selected contrast is highly significant. Using the latter critical point in place of $W_{p,q}^{(\alpha)}$ and with σ^2

estimated by S^2, we obtain 3.22 ± 1.212 as a 95% confidence interval for the selected contrast.

Simpler significant interaction contrasts can be found among the tetrad differences. For this family an approximate upper 5% point of V/S is given by $|M|_{18,36}^{(0.05)} = 3.19$. Thus any tetrad difference $\psi_{ij,i'j'}$ is significant at the 5% level if $|\hat{\psi}_{ij,i'j'}| > |M|_{18,36}^{(0.05)} S\sqrt{4/n} = 3.19 \times 0.16 = 1.276$. This test is passed by the following tetrad differences: $\hat{\psi}_{12,12}$, $\hat{\psi}_{12,13}$, $\hat{\psi}_{12,23}$, $\hat{\psi}_{13,12}$, $\hat{\psi}_{13,13}$, $\hat{\psi}_{14,12}$, $\hat{\psi}_{14,13}$, and $\hat{\psi}_{14,23}$. These results suggest that the first row (corresponding to F_0) contributes significantly to the nonadditivity. Any 2×2 subtable that excludes the first row is roughly additive.

If we confine attention to the family of interaction residuals, then from (5.17) it follows that any $|\hat{\gamma}_{ij}|$ exceeding the critical value

$$|M|_{12,36}^{(0.05)} S\sqrt{\frac{(I-1)(J-1)}{IJn}} = 3.04 \times 0.4 \times \sqrt{\frac{3 \times 2}{4 \times 3 \times 4}} = 0.43$$

is significant at the 5% level. This test is passed by $\hat{\gamma}_{12}$, $\hat{\gamma}_{13}$, $\hat{\gamma}_{23}$, $\hat{\gamma}_{41}$, and $\hat{\gamma}_{43}$. These results are in accord with those obtained for tetrad differences.

Finally we consider the family of interaction elements $\theta_{ij} - \theta_{ij'}$ $(1 \leq i \leq 4,\ 1 \leq j \neq j' \leq 3)$, which corresponds to pairwise comparisons between the soil types for each type of fertilizer. The common critical value for each pairwise comparison can be computed using (5.19). Here $\alpha_I = 1 - (0.95)^{1/4} = 0.013$ for $\alpha = 0.05$, and

$$\frac{Q_{3,36}^{(0.013)} S}{\sqrt{n}} = \frac{4.258 \times 0.4}{\sqrt{4}} = 0.852$$

where $Q_{3,36}^{(0.013)}$ is found from Table 8 in Appendix 3 by linearly interpolating in $\log_e(1/\alpha)$ and $1/\nu$. Using this critical value we find the following significant pairwise differences among the soil types for each type of fertilizer: (A, C) and (B, C) for F_0, (A, B) and (A, C) for F_1, (A, B) and (B, C) for F_2, and (A, B) and (A, C) for F_3. Pairwise comparisons between the fertilizers for each soil type can be carried out in the same manner using (5.18). Recall that these pairwise differences by themselves are not indicative of the nonadditivity in the table. □

6 PARTITIONING TREATMENT MEANS INTO GROUPS

Partitioning (or clustering) treatments based on their means into separate groups (clusters) such that each group is reasonably homogeneous and the number of groups is relatively small is an important practical problem. For example, Tukey (1949) stated that "At a low and practical level . . . we wish to separate the varieties into distinguishable groups as often as we can without too frequently separating varieties which should stay together." Also see the discussion of O'Neill and Wetherill (1971) by Plackett and Nelder.

MCPs for testing pairwise null hypotheses (or more generally subset homogeneity hypotheses) are not well suited for partitioning means because it can happen that overlapping subsets of treatments are declared homogeneous but their union is not. For example, consider three sample means $\bar{Y}_1 \leqq \bar{Y}_2 \leqq \bar{Y}_3$. Suppose that an MCP finds the differences $|\bar{Y}_1 - \bar{Y}_2|$ and $|\bar{Y}_2 - \bar{Y}_3|$ not significant but the difference $|\bar{Y}_1 - \bar{Y}_3|$ significant. Then both (1, 2) and (2, 3) are "acceptable" homogeneous groups. Because of the overlap between the two groups, it is not clear how to partition the three means. Shaffer (1981) suggested the so-called complexity criterion for studying this phenomenon.

Many of the procedures for partitioning of treatments are based on the methods of cluster analysis. These procedures are the topic of Section 6.1. Two procedures that are not based on cluster analysis methods are the topic of Section 6.2. Finally Section 6.3 gives a comparison of different cluster analysis based procedures.

6.1 Procedures Based on Cluster Analysis

The possibility of using cluster analysis in order to partition means was suggested by Plackett in his discussion of O'Neill and Wetherill (1971). Scott and Knott (1974) pursued this idea and proposed the procedure described below.

Consider a balanced one-way layout with k treatments and n replications on each, and assume the usual normal theory model; for notation see Example 1.1 of Chapter 2. Scott and Knott's procedure starts out by determining the best possible partition of the k means into two groups and testing whether this partition is acceptable. This is posed as a problem of testing

$$H_0: \theta_1 = \theta_2 = \cdots = \theta_k \text{ versus } H_1: \begin{array}{l} \theta_i\text{'s equal for } i \in P_1 \\ \theta_j\text{'s equal for } j \in P_2 \end{array}$$

where $\{P_1, P_2\}$ is any partition of $\{1, 2, \ldots, k\}$ into two mutually exclusive and collectively exhaustive nonempty subsets. The LR test for this problem works out to be the following: Let B_0 be the maximum value, taken over all partitions $\{P_1, P_2\}$ of the between groups sum of squares, that is,

$$B_0 = \max_{P_1,P_2} \left\{ n \sum_{j=1}^{2} p_j(\bar{\bar{Y}}_{P_j} - \bar{\bar{Y}})^2 \right\} \tag{6.1}$$

where p_j is the cardinality of group P_j, $\bar{\bar{Y}}_{P_j} = \Sigma_{i\in P_j} \bar{Y}_i/p_j$ is the pooled sample mean for treatments in group P_j ($j = 1, 2$), and $\bar{\bar{Y}} = \Sigma_{i=1}^{k} \bar{Y}_i/k$ is the grand sample mean. Let

$$\hat{\sigma}^2 = \frac{\sum_{i=1}^{k}\sum_{j=1}^{n} (Y_{ij} - \bar{\bar{Y}})^2}{kn} \tag{6.2}$$

be the maximum likelihood estimate of σ^2 under H_0. Then the LR test rejects H_0 if $B_0/\hat{\sigma}^2$ is large.

It is readily seen that $B_0/\hat{\sigma}^2$ is the maximum of correlated F-type statistics. Scott and Knott (1974) gave some analytical and numerical evidence to show that the distribution of the statistic

$$\lambda = \frac{\pi}{2(\pi - 2)} \cdot \frac{B_0}{\hat{\sigma}^2}$$

can be well approximated by that of a χ^2 r.v. with $\nu_0 = k/(\pi - 2)$ d.f.

The calculation of B_0 involves finding the partition of k treatments into two groups that maximizes the between groups sum of squares (referred to as the "best" partition) or equivalently minimizes the within groups sum of squares. This is a univariate version of the cluster analysis method proposed by Edwards and Cavalli-Sforza (1965). There are $2^{k-1} - 1$ possible partitions of k means into two groups. However, to calculate B_0 it is only necessary to consider $k - 1$ partitions formed by using $k - 1$ gaps between the ordered $\bar{Y}_i$'s as separation points.

If the LR test is significant at some chosen level α (i.e., if $\lambda > \chi^2_{\nu_0}(\alpha)$), then the same test is applied to each group separately; if the LR test is not significant, then it is concluded that all k means fall in the same cluster. This procedure is continued, splitting each group into two "best" subgroups if the LR test for that group is significant, otherwise concluding that the group is homogeneous, until no groups remain to be split further. Worsley (1977) has given a nonparametric version of Scott and Knott's procedure.

Clearly, the choice of α is extremely important in this procedure. A small α will lead to an early termination of the splitting process, while a large α will take the process too far. If there are exactly $r \geqq 1$ homogeneous groups and if this true partition is tested by this procedure, then the Type I error probability that at least one homogeneous group will be split further can be shown to be bounded above by $1-(1-\alpha)^r$. This probability can become rather large for large r.

Calinski (1969) proposed cluster analysis methods based on the distance (or dissimilarity) measure $D_{ij} = |\bar{Y}_i - \bar{Y}_j|$ between the ith and jth treatments. Jolliffe (1975) proposed the P-value of the difference between $\bar{Y}_i$ and $\bar{Y}_j$ (taking into account the number of other sample means lying between the two) as an alternative dissimilarity measure. The Studentized range distribution (with appropriate number of means) is used in calculating the P-value. If the difference of the two sample means is large, then the P-value will be small, indicating dissimilarity between the corresponding treatments. For the same magnitude of difference between two sample means, if they are separated by more treatments whose sample means fall between the two, then the P-value will be correspondingly larger, that is, their difference will be less significant. Jolliffe used single linkage cluster analysis and showed the connection of the resulting procedure with the NK-procedure.

Calinski and Corsten (1985) proposed two procedures (for balanced one-way layouts) that, in contrast to the previous procedures, are designed to formally control the Type I error rate of erroneous grouping. Error rate control is achieved by embedding the simultaneous test procedures based on F-statistics (see Section 1.2 of Chapter 3) and Studentized range statistics (see Section 2.3 of Chapter 3) as stopping criteria in two different clustering algorithms. We now describe these procedures.

The first procedure uses a hierarchical, agglomerative clustering algorithm. (Here "hierarchical" means that a treatment that is included in a cluster at a certain step is not subsequently deleted, and "agglomerative" means that at each step two adjacent clusters are fused to form a new cluster.) To test the homogeneity of two or more disjoint subsets simultaneously (i.e., to test a multiple subset hypothesis) they proposed using the simultaneous test procedure (2.13) of Chapter 3. The algorithm begins with k treatments forming k separate clusters. At the first step two closest treatments (in terms of $|\bar{Y}_i - \bar{Y}_j|$) are fused together and the resulting range, $\min_{1 \leqq i < j \leqq k} |\bar{Y}_i - \bar{Y}_j|$, is compared with the critical value $Q_{k,\nu}^{(\alpha)} S/\sqrt{n}$. In general, at any step a new cluster is formed by fusing two adjacent clusters having the smallest overall range. The procedure terminates when this range exceeds the common critical value $Q_{k,\nu}^{(\alpha)} S/\sqrt{n}$. At that point the grouping arrived at the preceding step is admitted as acceptable.

The second procedure begins with all k treatments forming a single group. At the rth step ($r = 1, 2, \ldots$), the "best" partition into r groups, $(P_1, P_2, \ldots, P_r)$, is found by minimizing the within groups sum of squares:

$$n \sum_{j=1}^{r} \sum_{i \in P_j} (\bar{Y}_i - \bar{\bar{Y}}_{P_j})^2 .$$

If this minimum exceeds the common critical value $(k-1)F_{k-1,\nu}^{(\alpha)}S^2$, then the grouping is regarded as acceptable. The procedure is terminated as soon as an acceptable grouping is found (i.e., the first time the corresponding within groups sum of squares does not exceed $(k-1)F_{k-1,\nu}^{(\alpha)}S^2$). This procedure is not hierarchical because the groups at any step are not necessarily nested inside the groups at the previous step.

To conclude this section we remark that all of the procedures discussed above can be extended to unbalanced one-way layouts (and to more general designs). To extend the procedures based on the Studentized ranges, one would need to use the Tukey–Kramer or some other approximation.

6.2 Other Procedures

Tukey's (1949) gap-straggler test was the first procedure proposed in the literature for partitioning of treatment means in a one-way layout under the usual normal theory assumptions. In this procedure first the ordered sample means are divided into disjoint groups by using significantly large gaps between consecutive means as separation points. The appropriate critical constant is based on the distribution of the Studentized maximum gap statistic (under the overall null hypothesis). Tables of this distribution are not available, but an asymptotic approximation can be based on Pyke's (1965) result given in Section 3.1. Tukey suggested approximating the upper α point of the Studentized maximum gap statistic by the upper $\alpha/2$ point of the t-distribution with d.f. $\nu = \Sigma_{i=1}^{k} n_i - k$. All subsequent tests are applied to each group of sample means separately.

After partitioning the treatments into groups, the stragglers (outliers) in each group (consisting of at least three means) are identified by applying the Studentized extreme deviate (from the group mean) test (see Section 4.2 of Chapter 5). If any outlier means are found, then the test is reapplied to the remaining means in that group. Finally, the F-test is applied to check that each group (obtained from the first two steps) is homogeneous.

The overall properties of this procedure are unknown. Thus it is not clear what probability statements, if any, can be made about the disjoint groups that are obtained.

Another procedure for partitioning of means was proposed by Cox and Spjøtvoll (1982). This procedure is different from the others in that, instead of a single "best" partition, it gives all simple partitions that are consistent with the data. It is similar in operation to the second procedure of Calinski and Corsten (1985). Thus at the first step the overall null hypothesis of all k treatments forming a single homogeneous cluster is tested using the usual F-test at some chosen level of significance α. If this overall null hypothesis is rejected, then the next step is to check if any two-group partitions are consistent with the data. These are usually the ones that involve groups separated by large gaps. The acceptability of any partition is tested using the F-test, that is, a partition $(P_1, P_2, \ldots, P_r)$ is regarded as acceptable if

$$F = \frac{n \sum_{j=1}^{r} \sum_{i \in P_j} (\bar{Y}_i - \bar{\bar{Y}}_{P_j})^2}{(k-r)S^2} \leqq F_{k-r,\nu}^{(\alpha)} .$$

If some partition is found acceptable, then any further finer splitting will generally give a better fit to the data but only at the expense of a more complicated model. Therefore the objective is to find the simplest acceptable partitions.

Let r^* be the minimum number of groups so obtained (i.e., for any $r < r^*$, all the partitions are inconsistent with the data according to the F-test). In general, there may be several consistent partitions of the k treatments into r^* groups. For example, with $k = 6$ the consistent partitions for $r^* = 2$ might be (ABC, DEF), $(AB, CDEF)$, (ABD, CEF), and $(ABCD, EF)$. These partitions can be summarized by simply stating that the treatment pairs (A, B) and (E, F) always fall in different groups. Thus the grouping is conclusive with respect to these four treatments but not with respect to C and D. Cox and Spjøtvoll argue that the consistent partitions given by their procedure provide an approximate $(1-\alpha)$-level confidence set for the true partition.

6.3 A Comparison of Procedures

Willavize, Carmer, and Walker (1980) studied using Monte Carlo simulations the error rates of the Scott–Knott procedure and some other partitioning procedures based on the cluster analysis; the Calinski–Corsten procedures were not available at the time of their study. They found that Type I error rates (classifying two equal treatment means into different groups) and Type III error rates (concluding that the mean of one treatment is greater than the mean of another when the reverse is true) of all the clustering procedures are substantially higher than those of Fisher's LSD procedure (which itself is quite liberal as noted in Example 4.6 of Chapter 2) when both use the same nominal α for each stage of testing; the Type II error rates for the former are lower. Moreover, while the Type I error rate for the LSD procedure is determined by the choice of α alone, that for a clustering procedure depends additionally on the unknown σ^2. The clustering procedures have such excessive Type I error rates because they force nonoverlapping grouping on the treatment means. Thus one should be cautious in using these procedures unless high Type I error rates can be tolerated.

The Calinski–Corsten (1985) procedures correct the problem of lack of control of Type I and Type III error rates by incorporating simultaneous test procedures in clustering algorithms. In this respect they are superior to the clustering procedures of Scott and Knott, and Jolliffe. Calinski and Corsten gave five numerical examples in all of which their two procedures gave identical results, provided that α was not too small. The same results were also obtained using the Scott–Knott and Jolliffe procedures in those examples where comparisons with the latter were possible. But these examples were perhaps atypical in that, according to Calinski and Corsten, they exhibited a clear grouping of treatments that would be discovered by other procedures, too. For further comparisons between different cluster analysis based procedures we refer the reader to Calinski and Corsten's article.

CHAPTER 11

Optimal Procedures Based on Decision-Theoretic, Bayesian, and Other Approaches

Most of the present book is devoted to multiple comparison procedures (MCPs) that are intuitive in nature and are based on the classical approach of controlling the Type I familywise error rate (FWE). In this chapter we review selected approaches for deriving *optimal* MCPs in some specified sense. Much of this work is decision-theoretic in nature. MCPs are viewed as multiple decision procedures and the losses that result from various decisions are explicitly considered. In a Bayesian approach there is the added feature of a prior distribution over the space of unknown parameters.

Lehmann (1957a, b) was the first author to consider the problem of multiple comparisons from a decision-theoretic viewpoint. His results are discussed in Section 1. Spjøtvoll's (1972a) work on optimal MCPs is discussed in Section 2. Spjøtvoll does not use an explicit decision-theoretic formulation; rather his approach is in the classical Neyman-Pearson spirit of maximizing the power subject to a given constraint on a Type I error rate. Section 3 discusses a long line of work over the last twenty five years by Duncan and his associates who developed a Bayesian approach to the problem of pairwise comparisons of means in a one-way layout. Duncan's work makes rather stringent assumptions about loss functions and prior distributions but the resulting MCP has some attractive properties. Section 4 discusses a combination of a Bayesian approach and the classical Neyman–Pearson approach to the problem of making pairwise comparisons with directional decisions in a one-way layout. Finally Section 5 describes a so-called Γ-minimax approach to the multiple comparison problem, which involves specifying a class of priors instead of a unique prior assumed in the Bayesian approach.

A feature common to all of the formulations considered in the present chapter is the assumption (implicit in Spjøtvoll's work) of an additive loss function. Because of this assumption, the problem of minimizing the risk (expected loss) or the Bayes risk, can be solved by minimizing the separate contributions due to component problems (comparisons). This leads to a procedure that is optimal for each component comparison regardless of other comparisons. The Bayesian approach brings in the information available on other treatments through the assumption of the common prior distribution for each treatment mean.

1 A DECISION-THEORETIC APPROACH

Lehmann (1957a, b) adopted a decision-theoretic approach to study multiple comparison problems. He first considered the problem of testing multiple hypotheses:

$$H_{0i}: \boldsymbol{\theta} \in \Theta_{0i} \text{ versus } H_{1i}: \boldsymbol{\theta} \in \Theta_{0i}^{-1} = \Theta - \Theta_{0i}, \qquad i \in I. \tag{1.1}$$

Here $\boldsymbol{\theta}$ is a (possibly vector) parameter of interest that belongs to the parameter space Θ, Θ_{0i}'s are proper subsets of Θ, and I is an arbitrary (possibly infinite) index set. The hypotheses (1.1) induce a partition of the parameter space Θ into disjoint subsets

$$\Theta_j = \bigcap_{i \in I} \Theta_{0i}^{\varepsilon_{ij}}, \qquad j \in J \tag{1.2}$$

where $\varepsilon_{ij} = 1$ (respectively, -1) if H_{0i} is true (respectively, false) and J is the set indexing all vectors $(\varepsilon_{ij}, i \in I)$ with each $\varepsilon_{ij} = \pm 1$. In the following we restrict attention to only nonempty Θ_j's and to decision procedures δ that classify $\boldsymbol{\theta}$ into exactly one of the nonempty Θ_j's.

Without loss of generality we may restrict to nonrandomized decision procedures δ. A nonrandomized decision procedure can be specified in terms of the partition $\{\mathscr{X}_j\}$ of the sample space $\mathscr{X}$ such that if the sample outcome $\mathbf{X} \in \mathscr{X}_j$ then $\delta(\mathbf{X}) = d_j$, which is the decision that $\boldsymbol{\theta}$ lies in Θ_j. If a family of tests for (1.1) is available with A_i being the acceptance region and A_i^{-1} (the complement of A_i) the rejection region for H_{0i} $(i \in I)$, then it is natural to consider the associated decision procedure having

$$\mathscr{X}_j = \bigcap_{i \in I} A_i^{\varepsilon_{ij}}, \qquad j \in J \tag{1.3}$$

where $e_{ij} = 1$ (respectively, -1) if H_{0i} is accepted (respectively, rejected) and, as before, J is the index set of all vectors $(e_{ij}, i \in I)$ with each $e_{ij} = \pm 1$. The restriction on the decision procedures stated above implies that

$$\Pr_{\boldsymbol{\theta}}\left\{\bigcup_{j\in J} (\delta(\mathbf{X}) = d_j)\right\} = \Pr_{\boldsymbol{\theta}}\left\{\bigcup_{j\in J} \mathscr{X}_j\right\} = 1 \qquad \forall\, \boldsymbol{\theta} \in \Theta\,. \tag{1.4}$$

Lehmann refers to a family of tests defined by $\{A_i\}$ that leads to the classification regions $\mathscr{X}_j$ (given by (1.3)) satisfying (1.4) as *compatible*. For a compatible family of tests Lehmann showed that there is a one-to-one correspondence between the $\mathscr{X}_j$'s and the A_i's (except possibly on a set of measure zero).

In the above we can think of the classification procedure δ as being obtained by forming a Cartesian product of the decision procedures for component testing problems. Because of the restriction on δ that it must not classify $\boldsymbol{\theta}$ to an empty set Θ_j, Lehmann refers to it as a *restricted product procedure*. We now consider in detail Lehmann's formulation for the case of two decisions.

Consider two different decision spaces D_1 and D_2 with the associated loss functions $L_1(\boldsymbol{\theta}, d_1)$ for $d_1 \in D_1$ and $L_2(\boldsymbol{\theta}, d_2)$ for $d_2 \in D_2$ where $\boldsymbol{\theta}$ indexes some fixed family of probability distributions for data $\mathbf{X}$ where $\boldsymbol{\theta}$ and $\mathbf{X}$ are common to both the decision problems. Simultaneous consideration of the two component problems leads to a new problem in which the decision space is $D = D_1 \times D_2$. Assume that the loss function for decision $\mathbf{d} = (d_1, d_2) \in D$ is

$$L(\boldsymbol{\theta}, \mathbf{d}) = L_1(\boldsymbol{\theta}, d_1) + L_2(\boldsymbol{\theta}, d_2)\,.$$

As before, some decisions (d_1, d_2) correspond to an empty set and hence must be eliminated from $D_1 \times D_2$. We assume below that D is a subset of $D_1 \times D_2$ obtained by deleting such (d_1, d_2). Then given procedures δ_1 and δ_2 for the two component problems, a *restricted product procedure* $\boldsymbol{\delta} = (\delta_1, \delta_2)$ makes decision $\boldsymbol{\delta}(\mathbf{X}) = \mathbf{d} = (d_1, d_2) \in D$ when $\delta_1(\mathbf{X}) = d_1$ and $\delta_2(\mathbf{X}) = d_2$. The pair (δ_1, δ_2) is called *compatible* if

$$\Pr_{\boldsymbol{\theta}}\{(\delta_1(\mathbf{X}), \delta_2(\mathbf{X})) \in D\} = 1 \qquad \forall\, \boldsymbol{\theta} \in \Theta\,.$$

For this general setting also, a one-to-one correspondence holds between compatible pairs (δ_1, δ_2) and restricted product procedures $\boldsymbol{\delta}$. This and many of the following results concerning restricted product procedures extend to infinite products.

Let the risk function of δ_i be $R_i(\boldsymbol{\theta}, \delta_i) = E_{\boldsymbol{\theta}}\{L_i(\theta, \delta_i(\mathbf{X}))\}$ for $i = 1, 2$, and that of $\boldsymbol{\delta} = (\delta_1, \delta_2)$ be $R(\boldsymbol{\theta}, \boldsymbol{\delta}) = R_1(\boldsymbol{\theta}, \delta_1) + R_2(\boldsymbol{\theta}, \delta_2)$. Let $\mathcal{D}_1$ and $\mathcal{D}_2$ be two classes of procedures for the two decision problems and let $\mathcal{D}$ be the subset of $\mathcal{D}_1 \times \mathcal{D}_2$ corresponding to compatible pairs (δ_1, δ_2). If δ_i^0 is a uniformly (in $\boldsymbol{\theta}$) minimum risk procedure in class $\mathcal{D}_i$ (i.e., if $R_i(\boldsymbol{\theta}, \delta_i^0) \leqq R_i(\boldsymbol{\theta}, \delta_i)$ for all $\delta_i \in \mathcal{D}_i$ and for all $\boldsymbol{\theta} \in \Theta$) for $i = 1, 2$, it follows that if $\boldsymbol{\delta}^0 = (\delta_1^0, \delta_2^0) \in \mathcal{D}$, then it is a uniformly minimum risk procedure in $\mathcal{D}$. Lehmann also showed a similar result concerning the unbiasedness of $\boldsymbol{\delta}$ where the unbiasedness is defined by the following condition:

$$E_{\boldsymbol{\theta}}\{L(\boldsymbol{\theta}', \boldsymbol{\delta}(\mathbf{X}))\} \geqq E_{\boldsymbol{\theta}}\{L(\boldsymbol{\theta}, \boldsymbol{\delta}(\mathbf{X}))\} \qquad \forall\, \boldsymbol{\theta} \text{ and } \boldsymbol{\theta}' \in \Theta .$$

In particular, Lehmann showed that the restricted product procedure defined by (1.3) for the multiple hypothesis testing problem (1.1) is unbiased if H_{0i} is tested by an unbiased test procedure at level $\alpha_i = b_i/(a_i + b_i)$ where a_i and b_i are the losses for falsely rejecting and falsely accepting H_{0i}, respectively ($i \in I$).

Finally, for the case of a finite family, Lehmann showed that if the test of each H_{0i} uniformly minimizes the risk among all tests that are similar on the boundary of Θ_{0i} and Θ_{0i}^{-1} (for all values of nuisance parameters if any are present) at level $\alpha_i = b_i/(a_i + b_i)$ and if the collection of such tests for all $i \in I$ is compatible, then under certain regularity conditions the restricted product procedure is unbiased and uniformly minimizes the risk among all unbiased procedures.

In a later paper Lehmann (1957b) relaxed the compatibility requirement and extended his main theorem to more general settings. He noted that if the acceptance of a hypothesis is interpreted as making no statement about $\boldsymbol{\theta}$, then the compatibility requirement can be relaxed. Thus rejection of H_{0i} implies that $\boldsymbol{\theta} \in \Theta_{0i}^{-1}$ but acceptance implies nothing more than $\boldsymbol{\theta} \in \Theta$. This is often the way acceptance of a hypothesis is interpreted in practice in any case. Analogous to (1.2), here we have

$$\Theta_j = \bigcap_{\substack{i \in I \\ \varepsilon_{ij} = -1}} \Theta_{0i}^{\varepsilon_{ij}}, \qquad j \in J$$

and analogous to (1.3), here we have

$$\mathscr{X}_j = \bigcap_{\substack{i \in I \\ e_{ij} = -1}} A_i^{e_{ij}}, \qquad j \in J .$$

Using the same additive loss function assumed earlier with losses a_i and b_i for falsely rejecting and falsely accepting H_{0i}, respectively ($i \in I$), it follows that the optimality property stated in the preceding paragraph holds for the (unrestricted) product procedure.

Example 1.1 ***(Pairwise Comparisons of Means).*** Consider the usual one-way layout model and the family of all pairwise hypothesis testing problems

$$H_{0,ij}: \theta_i = \theta_j \text{ versus } H_{1,ij}: \theta_i \neq \theta_j \qquad (1 \leqq i < j \leqq k). \tag{1.5}$$

It is well known (Lehmann 1986) that for testing $H_{0,ij}$ the t-test which rejects when

$$|\bar{Y}_i - \bar{Y}_j| > T_{\nu}^{(\alpha/2)} S \sqrt{\frac{1}{n_i} + \frac{1}{n_j}} \qquad (1 \leqq i < j \leqq k) \tag{1.6}$$

is a uniformly minimum risk procedure in the class of all similar procedures for a suitable choice of the losses a, b that give $\alpha = b/(a+b)$; here we are assuming the same loss function for all component problems (1.5). Note, however, that this family of tests is not compatible since with positive probability $H_{0,ij}$ and $H_{0,jl}$ may be accepted but $H_{0,il}$ may be rejected. However, by interpreting acceptance as making no statement about $\theta_i - \theta_j$ at all, it is seen that all pairwise t-tests, when considered simultaneously, possess the optimality property stated above. □

Lehmann's (1957a,b) basic work has been generalized and extended by many authors; see, e.g., Schaafsma (1969). Gupta and Huang (1981) have given a comprehensive summary of this research.

2 A NEYMAN–PEARSON TYPE APPROACH

Spjøtvoll (1972a) adopted the classical Neyman–Pearson approach for deriving the "most powerful" (in a sense to be made precise below) MCPs for a finite family of hypotheses testing problems. As in the Neyman–Pearson formulation, a constraint on a suitable Type I error rate is needed; Spjøtvoll chose the per-family error rate (PFE) rather than the familywise error rate (FWE) for this purpose because of the following reasons:

(i) The PFE is technically easier to work with than the FWE because it corresponds to an additive loss function.

(ii) It is "More instructive to think in terms of the expected number of false rejections than in terms of the probability of at least one false rejection."

(iii) Since FWE $\leqq$ PFE, controlling the PFE at some level γ also provides an upper bound on the FWE, whereas the converse is not true.

As in the preceding section let $\mathbf{X}$ be a random variable (r.v.) whose probability distribution is indexed by a parameter $\boldsymbol{\theta} \in \Theta$ and consider a finite family of hypothesis testing problems:

$$H_{0i}: \boldsymbol{\theta} \in \Theta_{0i} \text{ versus } H_{1i}: \boldsymbol{\theta} \in \Theta_{1i} \qquad (1 \leqq i \leqq m). \tag{2.1}$$

Let $\boldsymbol{\delta} = (\delta_1, \delta_2, \ldots, \delta_m)$ be a vector of randomized test procedures such that if $\mathbf{X} \in \mathscr{X}$ is observed, then $\delta_i(\mathbf{X})$ is the probability of rejecting H_{0i} $(1 \leqq i \leqq m)$. Thus $\boldsymbol{\delta}$ is a product procedure in the Lehmann sense. (Spjøtvoll implicitly assumed that $\bigcap_{i=1}^{m} \Theta_{1i}$ is not empty or that $\Pr\{\delta_i(\mathbf{X}) = 1,\ i = 1, 2, \ldots, m\} = 0$, which implies that $\boldsymbol{\delta}$ is compatible.) The power function of $\boldsymbol{\delta} = (\delta_1, \delta_2, \ldots, \delta_m)$ is defined to be the vector $\boldsymbol{\beta}(\boldsymbol{\theta}) = (\beta_1(\boldsymbol{\theta}), \beta_2(\boldsymbol{\theta}), \ldots, \beta_m(\boldsymbol{\theta}))$, where $\beta_i(\boldsymbol{\theta}) = E_{\boldsymbol{\theta}}\delta_i(\mathbf{X})$ (i.e., $\beta_i(\boldsymbol{\theta})$ is the power function of the ith test), $1 \leqq i \leqq m$. For specified γ $(0 < \gamma < m)$ let $S(\gamma)$ be the set of all test procedures $\boldsymbol{\delta} = (\delta_1, \delta_2, \ldots, \delta_m)$ such that

$$\sum_{i=1}^{m} E_{\boldsymbol{\theta}}(\delta_i(\mathbf{X})) \leqq \gamma \qquad \forall\, \boldsymbol{\theta} \in \Theta_0 = \bigcap_{i=1}^{m} \Theta_{0i}. \tag{2.2}$$

Note that the left hand side of (2.2) is the PFE when all the null hypotheses H_{0i} are true.

Spjøtvoll considered two optimality criteria:

(i) A test $\boldsymbol{\delta}^0 = (\delta_1^0, \delta_2^0, \ldots, \delta_m^0) \in S(\gamma)$ is said to *maximize the minimum average power* over specified subsets $\Theta'_{1i} \subseteq \Theta_{1i}$ $(1 \leqq i \leqq m)$ if it maximizes

$$\sum_{i=1}^{m} \inf_{\Theta'_{1i}} E_{\boldsymbol{\theta}}(\delta_i(\mathbf{X})) \tag{2.3}$$

among all tests $\boldsymbol{\delta} \in S(\gamma)$.

(ii) A test $\boldsymbol{\delta}^0 = (\delta_1^0, \delta_2^0, \ldots, \delta_m^0) \in S(\gamma)$ is said to *maximize the minimum power* over specified subsets $\Theta'_{1i} \subseteq \Theta_{1i}$ $(1 \leqq i \leqq m)$ if it maximizes

$$\min_{1 \leqq i \leqq m} \inf_{\Theta'_{1i}} E_{\boldsymbol{\theta}}(\delta_i(\mathbf{X})) \tag{2.4}$$

among all tests $\boldsymbol{\delta} \in S(\gamma)$.

According to Spjøtvoll these criteria are directed toward improving the powers of individual tests. This is in response to the common complaint against MCPs that the powers of individual tests are very small.

The optimal tests $\boldsymbol{\delta}^0$ for cases (i) and (ii) above are characterized in the following two theorems by Spjøtvoll. In these theorems $f_{01}, \ldots, f_{0m}$ and $f_{11}, \ldots, f_{1m}$ are integrable functions (which can be interpreted as probability densities of $\mathbf{X}$ under certain "least favorable" configurations $\boldsymbol{\theta} \in \Theta_{0i}$ and $\boldsymbol{\theta} \in \Theta'_{1i}$, $i = 1, \ldots, m$, respectively) with respect to a σ-finite measure μ defined on $\mathscr{X}$. Let $S'(\gamma) \subseteq S(\gamma)$ be the set of all tests $\boldsymbol{\delta} = (\delta_1, \delta_2, \ldots, \delta_m)$ satisfying

$$\sum_{i=1}^{m} \int \delta_i(\mathbf{x}) f_{0i}(\mathbf{x})\, d\mu(\mathbf{x}) = \gamma \,.$$

Theorem 2.1 **(Spjøtvoll 1972a).** Suppose that there exists a test $\boldsymbol{\delta}^0 \in S'(\gamma)$ defined by

$$\delta_i^0(\mathbf{X}) = \begin{cases} 1 & \text{if } f_{1i}(\mathbf{X}) > c f_{0i}(\mathbf{X}) \\ a_i & \text{if } f_{1i}(\mathbf{X}) = c f_{0i}(\mathbf{X}) \\ 0 & \text{if } f_{1i}(\mathbf{X}) < c f_{0i}(\mathbf{X}) \end{cases} \qquad (1 \leqq i \leqq m)$$

where $a_1, \ldots, a_m$ and $c > 0$ are constants. Then $\boldsymbol{\delta}^0$ maximizes

$$\sum_{i=1}^{m} \int \delta_i(\mathbf{x}) f_{1i}(\mathbf{x})\, d\mu(\mathbf{x})$$

among all tests $\boldsymbol{\delta} \in S'(\gamma)$. □

Theorem 2.2 **(Spjøtvoll 1972a).** Suppose that there exists a test $\boldsymbol{\delta}^0 \in S'(\gamma)$ defined by

$$\delta_i^0(\mathbf{X}) = \begin{cases} 1 & \text{if } f_{1i}(\mathbf{X}) > c_i f_{0i}(\mathbf{X}) \\ a_i & \text{if } f_{1i}(\mathbf{X}) = c_i f_{0i}(\mathbf{X}) \\ 0 & \text{if } f_{1i}(\mathbf{X}) < c_i f_{0i}(\mathbf{X}) \end{cases} \quad (1 \leqq i \leqq m)$$

where the constants $a_1, \ldots, a_m, c_1, \ldots, c_m$ are such that the integrals (which correspond to individual powers of $\boldsymbol{\delta}^0$ at specific alternatives)

$$\int \delta_i^0(\mathbf{x}) f_{1i}(\mathbf{x})\, d\mu(\mathbf{x}) = \int \delta_j^0(\mathbf{x}) f_{1j}(\mathbf{x})\, d\mu(\mathbf{x}) \qquad (1 \leqq i \neq j \leqq m)\,, \quad (2.5)$$

and $c_i \geqq 0$, $\Sigma_{i=1}^m c_i > 0$. Then $\boldsymbol{\delta}^0$ maximizes

$$\min_{1 \leqq i \leqq m} \int \delta_i(\mathbf{x}) f_{1i}(\mathbf{x})\, d\mu(\mathbf{x})$$

among all tests $\boldsymbol{\delta} \in S'(\gamma)$. □

Example 2.1 ***(Spjøtvoll 1972a).*** Assume the usual one-way layout model with independent normal errors and constant known variance, which may be taken to be unity. Consider the following multiple testing problem concerning linear functions of the treatment means θ_i:

$$H_{0i}: \sum_{j=1}^{k} a_{ij}\theta_j = b_i \text{ versus } H_{1i}: \sum_{j=1}^{k} a_{ij}\theta_j > b_i \qquad (1 \leqq i \leqq m)$$

where the a_{ij}'s and b_i's are specified constants. Let

$$\Theta'_{1i} = \left\{\boldsymbol{\theta} = (\theta_1, \ldots, \theta_k)' : \sum_{j=1}^{k} a_{ij}\theta_j - b_i \geqq \Delta_i\right\} \qquad (1 \leqq i \leqq m)$$

where the $\Delta_i > 0$ are specified constants. Assume that $\Theta_0 = \bigcap_{i=1}^m \Theta_{0i}$ is not empty.

For this setup, letting $f_0 = f_{0i}$ $(1 \leqq i \leqq m)$ be the density function of the observations under $\boldsymbol{\theta}^0$ where $\boldsymbol{\theta}^0 = (\theta_1^0, \theta_2^0, \ldots, \theta_k^0)$ is any point in Θ_0, and letting f_{1i} be the density function under a suitably chosen point on the boundary of Θ'_{1i} (i.e., when $\Sigma_{j=1}^k a_{ij}\theta_j - b_i = \Delta_i$), $i = 1, 2, \ldots, m$, Spjøtvoll derived the optimal tests $\boldsymbol{\delta}^0 = (\delta_1^0, \delta_2^0, \ldots, \delta_m^0)$ for maximizing the criteria (2.3) and (2.4), respectively. First, he showed that the condition $f_{1i} > c_i f_{0i}$ for rejection according to Theorems 2.1 and 2.2 is equivalent to

$$\left(\sum_{j=1}^{k} \frac{a_{ij}^2}{n_j}\right)^{-1/2} \left(\sum_{j=1}^{k} a_{ij}\bar{Y}_j - b_i\right) > \tfrac{1}{2}\left(\sum_{j=1}^{k} \frac{a_{ij}^2}{n_j}\right)^{-1/2} \Delta_i + \left(\sum_{j=1}^{k} \frac{a_{ij}^2}{n_j}\right)^{1/2} \Delta_i^{-1} \log_e c_i \tag{2.6}$$

which is simply the usual z-test $(1 \leqq i \leqq m)$. For the optimality criterion (2.4), the c_i's are chosen to satisfy

$$\sum_{i=1}^{m} \Phi\left[-\tfrac{1}{2}\Delta_i\left(\sum_{j=1}^{k} \frac{a_{ij}^2}{n_j}\right)^{-1/2} - \left(\sum_{j=1}^{k} \frac{a_{ij}^2}{n_j}\right)^{1/2} \Delta_i^{-1} \log_e c_i\right] = \gamma\,, \tag{2.7}$$

which is condition (2.2), and

$$\beta_i(\boldsymbol{\theta})\Big|_{\sum_{j=1}^{k} a_{ij}\theta_j - b_i = \Delta_i} = \Phi\left[\left(\sum_{j=1}^{k} \frac{a_{ij}^2}{n_j}\right)^{-1/2} \frac{\Delta_i}{2} - \left(\sum_{j=1}^{k} \frac{a_{ij}^2}{n_j}\right)^{1/2} \Delta_i^{-1} \log_e c_i\right]$$

$$= \text{const.} \qquad (1 \leqq i \leqq m),$$

which is condition (2.5). For the optimality criterion (2.3), the c_i's are equal and this common value is obtained by solving (2.7).

The tests simplify considerably if $\Sigma_{j=1}^{k} a_{ij}^2/n_j = A$ and $\Delta_i = \Delta$, in which case the optimal tests for the two optimality criteria coincide and are given by choosing the right hand side of (2.6) equal to $Z^{(\gamma/m)}$, the upper γ/m point of the standard normal distribution. This result holds uniformly in Δ.

The problems of all pairwise comparisons and comparisons with a control are special cases of the above. If the common variance is unknown, then the restriction of unbiasedness is placed on the δ_i's and the optimal tests are found to be t-tests that (as seen in the preceding section) also have uniformly minimum risk among all unbiased procedures. □

Spjøtvoll also showed how to use the optimal tests derived using Theorems 2.1 and 2.2 to construct optimal simultaneous confidence regions.

Before closing this section we comment on the relation between Lehmann's and Spjøtvoll's works. Essentially both arrived at the result that if an optimal procedure is used for each component problem (e.g., the t-test for testing a hypothesis about a linear parametric function of the means in a one-way layout), then the resulting product procedure is optimal in the simultaneous setting. This result holds because the loss function is additive. If the losses from the component problems are not additive (e.g., if the FWE is to be controlled instead of the PFE), then the nature of the optimal procedures may change. Such loss functions have not been studied in the context of multiple comparison problems.

3 A BAYESIAN APPROACH

3.1 A Two-Decision Student-*t* Problem

Duncan (1961, 1965) developed a Bayesian MCP for all pairwise comparisons of means in a balanced one-way layout (assuming the usual normal theory model) by first considering the component problem of comparing two means. Let θ_1 and θ_2 be two treatment means and let $\bar{Y}_1$ and $\bar{Y}_2$ be the corresponding sample means each based on n observations. Let S^2 be an estimate of the error variance σ^2, which distributed as $\sigma^2\chi^2_\nu/\nu$ independently of the $\bar{Y}_i$'s. Then it follows that the Student t-statistic

$$T = \frac{(\bar{Y}_1 - \bar{Y}_2)}{S}\sqrt{\frac{n}{2}}$$

has a noncentral t-distribution with ν degrees of freedom (d.f.) and noncentrality parameter

$$\tau = \frac{(\theta_1 - \theta_2)}{\sigma}\sqrt{\frac{n}{2}}\,.$$

Consider the problem of choosing between the following two decisions:

$$d^0_+ : \tau \leqq 0 \qquad \text{and} \qquad d^1_+ : \tau > 0\,. \tag{3.1}$$

For a decision $d_+ (= d^0_+$ or $d^1_+)$ consider a simple linear loss function

$$L_+(\tau, d_+) = \begin{cases} 0 & \tau \leqq 0,\ d_+ = d^0_+ \\ k_0\tau & \tau > 0,\ d_+ = d^0_+ \\ k_1|\tau| & \tau \leqq 0,\ d_+ = d^1_+ \\ 0 & \tau > 0,\ d_+ = d^1_+ \end{cases} \tag{3.2}$$

where $k_1 > k_0 > 0$. The ratio $K = k_1/k_0$ is a measure of the relative seriousness of a Type I error versus a Type II error.

Duncan assumed a normal prior density for τ,

$$p(\tau) = \frac{1}{\gamma\sqrt{2\pi}} \exp\left(\frac{-\tau^2}{2\gamma^2}\right), \qquad -\infty < \tau < \infty \tag{3.3}$$

and showed that the Bayes rule has the form

$$\delta_+(T) = \begin{cases} d^0_+ & \text{if } T \leqq t_* \\ d^1_+ & \text{if } T > t_* \end{cases} \tag{3.4}$$

where $t_* = t_*(K, \nu, \gamma^2)$ is a critical constant that can be obtained by solving a certain integral equation; this equation is the result of minimizing the Bayes risk. Duncan (1961) has provided a short table of t_*-values.

To use this table it is necessary to specify K and γ^2. For $\nu = \infty$ (σ^2 known) the analysis is simpler since, in that case, T is normal with a nonzero mean. For this case Duncan (1965) showed that

$$t_*(K, \infty, \gamma^2) = (1 + \gamma^{-2})^{1/2} t_*(K, \infty, \infty) \tag{3.5}$$

where $t_*(K, \infty, \infty)$ is the critical constant for the Bayes rule when $\gamma^2 = \infty$ (diffuse prior) and is obtained by solving the equation

$$\frac{\phi(t) + t\Phi(t)}{\phi(-t) - t\Phi(-t)} = K\,. \tag{3.6}$$

Here $\Phi(\cdot)$ and $\phi(\cdot)$ are the standard normal distribution and density functions, respectively. We discuss below how to specify K and γ^2.

It is clear from symmetry considerations that for the problem of choosing between the decisions

$$d_-^0 : \tau \geqq 0 \qquad \text{and} \qquad d_-^1 : \tau < 0 \tag{3.7}$$

under the prior density (3.3) and loss function

$$L_-(\tau, d_-) = \begin{cases} 0 & \tau \geqq 0,\ \ d_- = d_-^0 \\ k_0|\tau| & \tau < 0,\ \ d_- = d_-^0 \\ k_1\tau & \tau \geqq 0,\ \ d_- = d_-^1 \\ 0 & \tau < 0,\ \ d_- = d_-^1 \end{cases}, \tag{3.8}$$

the Bayes rule is

$$\delta_-(T) = \begin{cases} d_-^0 & \text{if } T \geqq -t_* \\ d_-^1 & \text{if } T < -t_*\,. \end{cases} \tag{3.9}$$

Here $t_* = t_*(K, \nu, \gamma^2)$ is the same critical constant that is used in (3.4).

If the prior density $p(\tau)$ has a nonzero mean τ_0, say, then for $\nu = \infty$ it can be shown that the critical constant to be used in Bayes rules (3.4) and (3.9) is $t_* - \tau_0$. For finite ν the adjustment for a nonzero mean is not so straightforward.

3.2 A Three-Decision Student-*t* Problem

Duncan (1961) next considered the problem of choosing between the following three decisions:

$$d^0 : \text{decide } \tau = 0\,, \qquad d^1 : \text{decide } \tau > 0\,, \qquad d^2 : \text{decide } \tau < 0\,. \tag{3.10}$$

(More realistically, the decisions are $d^0: |\tau| \leqq \varepsilon$, $d^1: \tau > \varepsilon$, and $d^2: \tau < -\varepsilon$, interpreted as τ is not substantially different from zero, τ is substantially greater than zero, and τ is substantially less than zero, respectively, where $\varepsilon > 0$ is some unspecified threshold value; (3.10) is a mathematical simplification of this more realistic decision problem. The same comment applies to (3.1) and (3.7). The consequences of using the simpler decisions (3.1), (3.7), and (3.10) are trivial here, but in the multiple comparison problem discussed in the next section, the more realistic interpretation of the decisions becomes essential to enhance compatibility; see (3.13) and the parenthetical remark following it.)

We now describe Duncan's (1961) approach to the three-decision problem (3.10). A simple linear loss function (analogous to (3.2)) for this problem is

$$L(\tau, d) = \begin{cases} 0 & \tau = 0, \quad d = d^0 \\ c_0|\tau| & \tau \neq 0, \quad d = d^0 \\ c_1|\tau| & \tau \leqq 0, \quad d = d^1 \\ 0 & \tau > 0, \quad d = d^1 \\ 0 & \tau < 0, \quad d = d^2 \\ c_1\tau & \tau \geqq 0, \quad d = d^2 \end{cases} \tag{3.11}$$

where $c_1 - c_0 > c_0 > 0$. Assume the prior density (3.3) for τ.

To see the connection between this three-decision problem and the two-decision problems considered previously, note that d^0, d^1, and d^2 can be expressed as products of component decisions d^i_+ and d^i_- $(i = 0, 1)$ as follows:

$$d^0 = (d^0_+, d^0_-), \qquad d^1 = (d^1_+, d^0_-), \qquad d_2 = (d^0_+, d^1_-).$$

Thus in general, we have $d = (d_+, d_-)$ ruling out the incompatible pair (d^1_+, d^1_-). With this notation it is also seen that

$$L(\tau, d) = L_+(\tau, d_+) + L_-(\tau, d_-)$$

by putting $k_0 = c_0$ and $k_1 = c_1 - c_0$ in (3.2) and (3.8). (Note, e.g., that for $\tau \leqq 0$, the decision $d = d^1 = (d^1_+, d^0_-)$ leads to a Type I error loss of $k_1|\tau|$ due to d^1_+ and a Type II error loss of $k_0|\tau|$ due to d^0_-, which sum to $(k_0 + k_1)|\tau| = c_1|\tau|$.) Thus we have the setting of Lehmann's multiple decision problem discussed in Section 1. Because of the additive loss structure (which results in the Bayes risk being additive) it turns out that the Bayes rule $\delta(T)$ for the three-decision problem is given by the (restricted) product of Bayes rules $\delta_+(T)$ and $\delta_-(T)$ for the component problems. Thus

$$\delta(T) = \begin{cases} d^0 & \text{if } |T| \leqq t_* \\ d^1 & \text{if } T > t_* \\ d^2 & \text{if } T < -t_* \end{cases}$$

where $t_* = t_*(K, \nu, \gamma^2)$ is the same critical constant as before with $K = k_1/k_0 = (c_1/c_0) - 1$.

3.3 Pairwise Comparisons

Now consider the problem of making all pairwise comparisons between $k \geqq 2$ treatments in a balanced one-way layout. The marginal distribution of each pairwise t-statistic

$$T_{ij} = \frac{(\bar{Y}_i - \bar{Y}_j)}{S} \sqrt{\frac{n}{2}} \qquad (1 \leqq i < j \leqq k) \tag{3.12}$$

is noncentral t with ν d.f. and noncentrality parameter

$$\tau_{ij} = \frac{(\theta_i - \theta_j)}{\sigma} \sqrt{\frac{n}{2}} \qquad (1 \leqq i < j \leqq k)\,.$$

A common multiple comparison problem is that of choosing between the three decisions

$$d_{ij}^0 : |\tau_{ij}| \leqq \varepsilon\,, \qquad d_{ij}^1 : \tau_{ij} > \varepsilon\,, \qquad d_{ij}^2 : \tau_{ij} < -\varepsilon \tag{3.13}$$

for each pair (i, j) $(1 \leqq i < j \leqq k)$ where $\varepsilon > 0$ is some unspecified threshold constant. (Note that in a pairwise comparisons problem, this interpretation of decisions is not only more realistic as noted earlier, but it also leads to a greater number of compatible decisions. For example, for $k = 3$ the decision $(d_{12}^0, d_{23}^0, d_{13}^1)$ is compatible if $\varepsilon > 0$ but not if $\varepsilon = 0$. This point was first noted by Lehmann 1957a.)

Consider the product decision

$$d_l = \bigcap_{1 \leqq i < j \leqq k} d_{ij}^{\alpha}$$

where l indexes all compatible decisions from the set of $\binom{k}{2}^3$ decisions corresponding to the choices $\alpha = 0, 1,$ or 2 for each d_{ij}. Duncan assumed the additive loss function

$$L(\boldsymbol{\tau}, d_l) = \sum_{1 \leqq i < j \leqq k} L_{ij}(\tau_{ij}, d_{ij}^{\alpha}) \qquad (3.14)$$

where $\boldsymbol{\tau}$ is the vector of the τ_{ij}'s and each $L_{ij}(\cdot,\cdot)$ is given by (3.11).

To find the Bayes rule for this problem, Duncan (1965) restricted attention, based on invariance considerations, to $k-1$ normalized orthogonal contrasts among the $\bar{Y}_i$'s. More specifically, let $\mathbf{A}$ be a $(k-1) \times k$ matrix whose rows form normalized orthogonal contrasts and let $\mathbf{Z} = \mathbf{A}\bar{\mathbf{Y}}$ where $\mathbf{Z} = (Z_1, Z_2, \ldots, Z_{k-1})'$ and $\bar{\mathbf{Y}} = (\bar{Y}_1, \bar{Y}_2, \ldots, \bar{Y}_k)'$. Now the $(Z_j\sqrt{n}/S)$'s have a joint $(k-1)$-variate noncentral t-distribution with ν d.f. and noncentrality parameter vector $\boldsymbol{\lambda} = (\sqrt{n}/\sigma)\mathbf{A}\boldsymbol{\theta}$ where $\boldsymbol{\theta}$ is the vector of treatment means. Duncan postulated identical independent priors of the form (3.3) on the $k-1$ elements of $\boldsymbol{\lambda}$, that is, the prior for $\boldsymbol{\lambda}$ is

$$p(\boldsymbol{\lambda}) = (2\pi\gamma^2)^{-(k-1)/2} \times \exp(-\boldsymbol{\lambda}'\boldsymbol{\lambda}/2\gamma^2). \qquad (3.15)$$

Then using the additive structure of the loss (and hence the risk) function (3.14) he derived the Bayes rule for all pairwise comparisons, which is the product of the following component rules:

$$\delta_{ij}(\bar{\mathbf{Y}}, S^2) = \begin{cases} d_{ij}^0 & \text{if } |T_{ij}| \leqq t_* R_{ij}^{-1} \\ d_{ij}^1 & \text{if } T_{ij} > t_* R_{ij}^{-1} \\ d_{ij}^2 & \text{if } T_{ij} < -t_* R_{ij}^{-1} \end{cases} \qquad (1 \leqq i < j \leqq k) \qquad (3.16)$$

where $t_* = t_*(K, \nu', \gamma^2)$ is the same critical constant as before, $\nu' = \nu + k - 2$, and

$$R_{ij}^2 = \frac{\nu'}{\left[\nu + \dfrac{1}{(1+\gamma^2)}\left\{\sum_{l=1}^{k-1} Z_l^2 - T_{ij}^2\right\}\right]} \qquad (1 \leqq i < j \leqq k).$$

Note that $\Sigma_{l=1}^{k-1} Z_l^2$ equals $(k-1)$ times the F-statistic for the treatment effects. Thus an alternative formula for R_{ij}^2 is

$$R_{ij}^2 = \frac{\nu'}{[\nu + (1/(1+\gamma^2)S^2)\{SS_{\text{treat}} - SS_{ij}\}]^{1/2}} \qquad (1 \leqq i < j \leqq k) \qquad (3.17)$$

where SS_{treat} is the treatment sum of squares in the analysis of variance (ANOVA) and $SS_{ij} = (n/2)(\bar{Y}_i - \bar{Y}_j)^2$. The quantity inside the braces in (3.17) is referred to as the residual sum of squares between treatments (after subtracting the contribution due to the (i, j)th comparison). Note that $R_{ij}^2 = 1$ for $k = 2$. Thus in the following discussion we assume that $k > 2$.

Some insight into (3.17) and the modified d.f. $\nu' = \nu + k - 2$ is obtained if we rewrite (3.16) in terms of the modified t-statistics

$$T'_{ij} = T_{ij}R_{ij} = \frac{\bar{Y}_i - \bar{Y}_j}{S_{ij}} \sqrt{\frac{n}{2}} \qquad (1 \leqq i < j \leqq k),$$

which are compared with the critical constant t_*; here

$$S_{ij}^2 = \frac{\nu S^2 + (1 + \gamma^2)^{-1}(SS_{\text{treat}} - SS_{ij})}{\nu'} \qquad (1 \leqq i < j \leqq k). \quad (3.18)$$

It can be shown that the prior (3.15) on $\boldsymbol{\lambda}$ results in $SS_{\text{treat}} - SS_{ij}$ having the prior distribution of $(1 + \gamma^2)\sigma^2\chi_{k-2}^2$ and thus (3.18) is the pooled estimate of σ^2 with pooled d.f. $\nu' = \nu + k - 2$.

In the absence of any knowledge of γ^2, by noting that $E(MS_{\text{treat}}) = (1 + \gamma^2)\sigma^2$ under the given prior (3.15), Duncan suggested that γ^2 be estimated by $\hat{\gamma}^2 = F - 1$. Then the following approximation to the exact Bayes rule (3.16) can be used:

$$\delta_{ij}(\bar{\mathbf{Y}}, S^2) = \begin{cases} d_{ij}^0 & \text{if } |T_{ij}| \leqq t_* \\ d_{ij}^1 & T_{ij} > t_* \\ d_{ij}^2 & T_{ij} < -t_* \end{cases} \qquad (1 \leqq i < j \leqq k) \qquad (3.19)$$

where now $t_* \cong t_*(K, \nu, F - 1)$. Moreover, if ν is large, then using (3.5) this t_* can be further approximated by

$$t_*(K, \nu, F - 1) \cong \left(\frac{F}{F - 1}\right)^{1/2} t_*(K, \infty, \infty) \qquad (3.20)$$

where $t_*(K, \infty, \infty)$ is obtained by solving (3.6). An exact empirical Bayes procedure for unknown γ^2 was later developed by Waller and Duncan (1969) and is discussed in the following section. But, for now, further insight may be obtained by examining the approximate Bayes procedure given by (3.19) and (3.20).

The most important point is that the critical constant t_* does not depend on k, the number of treatments. In this sense, the Bayes procedure has the nature of a per-comparison error rate (PCE) controlling MCP. However, t_* does depend on the extent of heterogeneity among the k treatment means (except when $k=2$). To see this relation consider the plot of (3.20) as a function of F shown in Figure 3.1. From this plot it is seen that if F is large (indicating heterogeneity between treatments means), then a small critical value t_* is used and the resulting procedure resembles Fisher's least significant difference (LSD) procedure, which, for each pairwise comparison, uses a critical value appropriate for a single comparison if the preliminary F-test is significant. If F is small (indicating homogeneity among treatments) then a large critical value t_* is used. In fact, the Bayes procedure can be thought of as a continuous extension of the LSD. While at the second step the LSD uses either a small critical value ($=T_\nu^{(\alpha/2)}$) if $F > F_{k-1,\nu}^{(\alpha)}$ or an infinite critical value (i.e., accepts all pairwise null hypotheses automatically) if $F \leq F_{k-1,\nu}^{(\alpha)}$, the Bayes procedure uses a critical value that varies continuously with F.

Finally, a comment on the choice of K is in order. The choice of K is more basic than that of the level of significance α in a conventional hypothesis test, but there is a one-to-one relationship between the two. The more serious a Type I error is relative to a Type II error (i.e., the larger the K), the smaller the value of α that should be used. By matching the critical values for $k=2$ (in which case t_* does not depend on the F-statistic), Waller and Duncan (1969) calculated that the values $K=50$, 100, and 500 roughly correspond to $\alpha=0.10$, 0.05, and 0.01, respectively. These values may be used as benchmarks in selecting an appropriate value of K. We emphasize that in choosing K one must

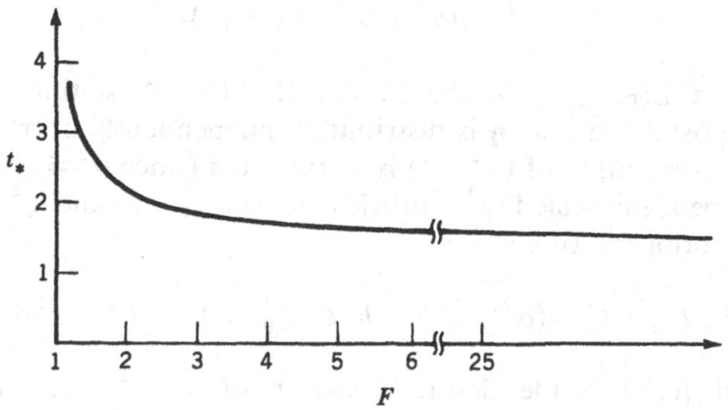

Figure 3.1. Critical constant t_* as a function of F-ratio for $K=50$, $f=120$, and $q=\infty$.

consider the errors resulting in a *two* treatment comparison to the exclusion of the other treatments. Because of the critical role played by K in these Bayes procedures, Duncan refers to them as *K-ratio t-tests*.

3.4 Pairwise Comparisons: A Bayesian Approach Using Hyper Priors, and Other Extensions

Waller and Duncan (1969) considered the same setup as in the preceding section, but instead of assuming that γ^2 is known or using a point estimate of it, they assumed a prior distribution (the so-called hyper prior) for a related parameter as well as for the error variance σ^2 and worked out a Bayes rule for the resulting problem. We now describe their approach.

Suppose that the prior distributions for the θ_i's $(1 \leqq i \leqq k)$ are independent normal each with mean θ_0 and variance σ_θ^2. (This is like the Model II ANOVA assumption.) Then it follows that the components of $\boldsymbol{\eta} = (\sigma/\sqrt{n})\boldsymbol{\lambda} = \mathbf{A}\boldsymbol{\theta}$ are independent normal each with mean zero and variance σ_θ^2, that is, the prior for $\boldsymbol{\eta}$ is

$$p_1(\boldsymbol{\eta}) = (2\pi\sigma_\theta^2)^{-(k-1)/2} \times \exp\left(\frac{-\boldsymbol{\eta}'\boldsymbol{\eta}}{2\sigma_\theta^2}\right). \tag{3.21}$$

Note from (3.15) that

$$\gamma^2 = \frac{\sigma_\theta^2}{(\sigma^2/n)} . \tag{3.22}$$

Let

$$\zeta^2 = n\sigma_\theta^2 + \sigma^2 = (1+\gamma^2)\sigma^2 \tag{3.23}$$

denote the $E(MS_{\text{treat}})$ in the Model II ANOVA setting. Waller and Duncan postulated that $\boldsymbol{\eta}$ is distributed independently of σ^2 and ζ^2 and the joint distribution of (σ^2, ζ^2) is a truncated (since $\sigma^2 \leqq \zeta^2$) product of two independent scaled χ^2 densities: $\sigma^2 \sim \sigma_0^2\chi_{\nu_0}^2/\nu_0$ and $\zeta^2 \sim \zeta_0^2\chi_{q_0}^2/q_0$. Thus the prior for (σ^2, ζ^2) is

$$p_2(\sigma^2, \zeta^2) = C \cdot h(\sigma^2|\sigma_0^2, \nu_0) \cdot h(\zeta^2|\zeta_0^2, q_0) \qquad (0 \leqq \sigma^2 \leqq \zeta^2 < \infty)$$

where $h(\cdot|c, \nu)$ is the density function of a $c\chi_\nu^2/\nu$ r.v. and C is a normalizing constant. (In Duncan 1965 a chi-square prior density was postulated only for σ_θ^2 and not for σ^2.) The full prior density is then

$$p(\eta, \sigma^2, \zeta^2) = p_1(\eta)p_2(\sigma^2, \zeta^2) \tag{3.24}$$

where $p_1(\eta)$ is given by (3.21) with $\sigma_\theta^2 = (\zeta^2 - \sigma^2)/n$.

Starting with this prior and the loss function given in (3.14) and (3.11), Waller and Duncan derived the Bayes rule, which is the product of the following component Bayes rules:

$$\delta_{ij}(\bar{\mathbf{Y}}, S^2) = \begin{cases} d_{ij}^0 & \text{if } |\bar{Y}_i - \bar{Y}_j| \leq t^* S\sqrt{\dfrac{2}{n}} \\ d_{ij}^1 & \text{if } \bar{Y}_i - \bar{Y}_j > t^* S\sqrt{\dfrac{2}{n}} \\ d_{ij}^2 & \text{if } \bar{Y}_i - \bar{Y}_j < -t^* S\sqrt{\dfrac{2}{n}} \end{cases} \quad (1 \leq i < j \leq k) \tag{3.25}$$

where $t^* = t^*(K, F, q, f)$ is a critical constant that depends on K, $f = \nu_0 + \nu = \nu_0 + k(n-1)$, $q = k - 1 + q_0$,

$$F = \frac{\{(k-1)MS_{\text{treat}} + q_0\zeta_0^2\}/q}{\{\nu MS_{\text{error}} + \nu_0\sigma_0^2\}/f}, \tag{3.26}$$

and

$$S^2 = \left(\frac{\nu MS_{\text{error}} + \nu_0\sigma_0^2}{f}\right). \tag{3.27}$$

Here MS_{treat} and MS_{error} are the usual mean squares in the ANOVA table.

The values of $t^*(K, F, q, f)$ have been tabulated by Waller and Duncan (1972). We give these values in Table 9 of Appendix 3. The derivations for the computations of these tables are given in Waller and Duncan (1974) and a computer program is given in Waller and Kemp (1975). We note the following relation between the $t^*(K, F, q, f)$-values of this section and $t_*(K, \nu, \gamma^2)$-values of the preceding section:

$$\begin{aligned} t^*(K, F, \infty, \infty) &= t_*(K, \infty, F-1) \\ &= \sqrt{\frac{F}{F-1}}\, t_*(K, \infty, \infty). \end{aligned} \tag{3.28}$$

For using the table of the t^*-values one needs to specify K, q_0 (with $q = k - 1 + q_0$) and ν_0 (with $f = \nu + \nu_0$); in addition the value of the F-ratio is needed, which can be calculated from (3.26) once ζ_0^2 and σ_0^2 are specified. As noted before, $K = 50, 100$, and 500 correspond roughly to $\alpha = 0.10, 0.05$, and 0.01, respectively. The values of q_0 and ν_0 are chosen to indicate the relative weights that one wishes to attach to the prior estimates ζ_0^2 and σ_0^2 with reference to $k - 1$ and ν, which are the weights (d.f.) associated with MS_{treat} and MS_{error}, respectively. If prior estimates ζ_0^2 and σ_0^2 of ζ^2 and σ^2 are not available, then one can set $q_0 = \nu_0 = 0$. In this case $p_2(\sigma^2, \zeta^2)$ does not depend on σ_0^2 and ζ_0^2 and becomes a uniform density in $\log_e \sigma^2$ and $\log_e \zeta^2$. In this case (3.26) reduces to the usual ANOVA F-statistic.

We now illustrate the use of the Waller–Duncan procedure by reanalyzing Duncan's (1955) barley data given in Example 1.1 of Chapter 3.

Example 3.1. Recall that we have $MS_{\text{varieties}} = 366.97$ with 6 d.f. and $MS_{\text{error}} = S^2 = 79.64$ with 30 d.f. Let us suppose that prior estimates ζ_0^2 and σ_0^2 are not available and thus $q_0 = \nu_0 = 0$, $q = 6$, $f = 30$, and $F = 4.61$. Next we must choose a value for K. Corresponding to a PCE of $\alpha = 0.05$ we choose $K = 100$. Interpolating linearly in Table 9 of Appendix 3 with respect to $a = 1/\sqrt{F} = 0.466$ as suggested by Waller and Duncan, we find that the critical $t^* = 2.108$. The critical value for the mean difference is thus

$$t^* S \sqrt{\frac{2}{n}} = 2.108 \times 8.924 \sqrt{\frac{2}{6}} = 10.86\,.$$

This value is slightly larger than the value $T_{30}^{(0.05)} S\sqrt{2/n} = 10.52$ used in the second step of testing by Fisher's LSD at $\alpha = 0.05$, but is much smaller than the value 16.26 used by Tukey's T-procedure, which controls the FWE at level 0.05. Thus here the Waller–Duncan procedure is slightly more conservative than Fisher's LSD, but is much less conservative than Tukey's procedure. In fact, it leads to the same significant pairwise differences as found by Fisher's LSD. □

Duncan (1975) has extended this approach to the family of all contrasts (still restricting attention to the balanced one-way layout setting), but the validity of this extension is not clear (at least to us) because of the technical difficulties of defining an additive loss function for the infinite family of inferences on all contrasts or for a random finite subset selected from this family. He assumed noninformative priors for σ^2 and ζ^2, that is, $\nu_0 = q_0 = 0$ (which is not essential to the final result), and showed that the same critical value $t^* = t^*(K, F, q, f)$ encountered in (3.25) can be used to test any contrast $\Sigma_{i=1}^k c_i\theta_i$, either specified *a priori* or selected *a posteriori* by data-snooping. The decision rule for each contrast is similar to (3.25) except that $\bar{Y}_i - \bar{Y}_j$ is replaced by the estimated contrast $\Sigma_{i=1}^k c_i\bar{Y}_i$, and $S\sqrt{2/n}$ is replaced by the estimated standard error of the contrast, namely, $S(\Sigma_{i=1}^k c_i^2/n)^{1/2}$.

Duncan and Godbold (1979) further extended this Bayesian approach (under the usual additive linear losses, a normal symmetric prior on the θ_i's, and a noninformative prior on σ^2 and ζ^2) to pairwise comparisons in an unbalanced one-way layout. The procedure given by Duncan and Godbold is only an approximation to the exact Bayes rule.

The main difficulty that arises in the unbalanced case is that the quantities

$$E(MS_{\text{treat}}) = \tilde{n}\sigma_\theta^2 + \sigma^2 = \zeta^2 \tag{3.29}$$

and

$$E(MS_{ij}) = E(SS_{ij}) = \tilde{n}_{ij}\sigma_\theta^2 + \sigma^2 = \zeta_{ij}^2 \tag{3.30}$$

are not equal for all $i \neq j$; here

$$\tilde{n} = \frac{\left(\sum_{i=1}^k n_i\right)^2 - \left(\sum_{i=1}^k n_i^2\right)}{(k-1)\left(\sum_{i=1}^k n_i\right)},$$

$$\tilde{n}_{ij} = \frac{2}{1/n_i + 1/n_j},$$

and

$$SS_{ij} = \frac{(\bar{Y}_i - \bar{Y}_j)^2}{1/n_i + 1/n_j}.$$

To overcome this difficulty, Duncan and Godbold suggested using separate critical constants $t^*(K, F_{ij}, q_{ij}, f)$ for separate pairwise comparisons. The "F"-statistic, F_{ij}, for the (i, j)th pairwise comparison (with q_{ij} being its numerator d.f. given by (3.32) below) is arrived at by first obtaining an unbiased estimate $\hat{\sigma}_\theta^2 = (MS_{\text{treat}} - MS_{\text{error}})/\tilde{n}$ of the variance component

σ_θ^2 and then computing F_{ij} from

$$F_{ij} = \frac{\tilde{n}_{ij}\hat{\sigma}_\theta^2 + \hat{\sigma}^2}{\hat{\sigma}^2} = r_{ij}\frac{MS_{\text{treat}}}{MS_{\text{error}}} + (1 - r_{ij})$$

$$= r_{ij}F + (1 - r_{ij}) \qquad (1 \leqq i < j \leqq k) \tag{3.31}$$

where $r_{ij} = \tilde{n}_{ij}/n$ and $F = MS_{\text{treat}}/MS_{\text{error}}$ is the usual F-statistic. Note that $F_{ij} - 1 = r_{ij}(F - 1)$ can be thought of as an estimate of γ_{ij}^2 where, in analogy with (3.22),

$$\gamma_{ij}^2 = \frac{\sigma_\theta^2}{(\sigma^2/\tilde{n}_{ij})} = r_{ij}\left(\frac{\zeta^2}{\sigma^2}\right) \qquad (1 \leqq i < j \leqq k);$$

here ζ^2 is given by (3.29). Now F_{ij} does not have an F-distribution because its numerator is not distributed as a (scaled) χ^2 r.v. But the distribution of the numerator can be approximated by that of a $\zeta_{ij}^2 \chi_{q_{ij}}^2 / q_{ij}$ r.v. where, using the Satterthwaite (1946) method of matching moments,

$$q_{ij} = \frac{r_{ij}^2 F^2}{(k-1)F_{ij}^2} + \frac{(1 - r_{ij}^2)^2}{fF_{ij}^2} \qquad (1 \leqq i < j \leqq k). \tag{3.32}$$

The Duncan–Godbold procedure is now clear: First choose the error weight ratio K using the guidelines given earlier. Next let $f =$ the d.f. available for estimating σ^2, which in a one-way layout equals $\nu = \Sigma_{i=1}^k n_i - k$. Then for each pairwise comparison compute F_{ij} from (3.31), q_{ij} from (3.32), and find $t^*(K, F_{ij}, q_{ij}, f)$ from Table 9 in Appendix 3 using interpolation if necessary. Compare the t-statistic

$$T_{ij} = \frac{(\bar{Y}_i - \bar{Y}_j)}{S\sqrt{1/n_i + 1/n_j}}$$

for the (i, j)th treatment difference against the critical value $t^*(K, F_{ij}, q_{ij}, f)$ and use the usual three-decision rule. Duncan and Godbold have also extended this procedure to the case of correlated estimates $\hat{\theta}_i$ of the treatment effects θ_i (which arise, e.g., in a one-way layout with a fixed covariate); we refer the reader to their paper for details.

Stevenson and Bland (1982), using an approach similar to that of Waller and Duncan (1969), have derived a Bayesian multiple comparison procedure for the θ_i's when the treatment variances σ_i^2 are unknown and unequal.

3.5 Bayesian Simultaneous Confidence Intervals

Dixon and Duncan (1975) derived minimum Bayes risk simultaneous confidence intervals for all pairwise differences $\Delta_{ij} = \theta_i - \theta_j \ (1 \leqq i < j \leqq k)$ in a balanced one-way layout. They assumed the following additive loss structure for the total loss resulting from simultaneously estimating the Δ_{ij}'s by intervals $[a_{ij}, b_{ij}]$:

$$\sum_{1 \leqq i < j \leqq k} L(\Delta_{ij}, [a_{ij}, b_{ij}]) \tag{3.33}$$

where

$$L(\Delta, [a, b]) = \begin{cases} k_1(a - \Delta)^2 + k_0(b - \Delta)^2 , & \Delta < a \\ k_0(a - \Delta)^2 + k_0(b - \Delta)^2 , & a \leqq \Delta \leqq b \\ k_0(a - \Delta)^2 + k_1(b - \Delta)^2 , & \Delta > b \end{cases} \tag{3.34}$$

and $k_1 \geqq k_0 > 0$. The loss function (3.34) is derived as follows: Consider the problem of estimating Δ by a lower one-sided interval $[a, \infty)$ and let the corresponding loss function be

$$L_l(\Delta, a) = \begin{cases} k_1(a - \Delta)^2 , & \Delta < a \\ k_0(a - \Delta)^2 , & \Delta \geqq a . \end{cases}$$

Here $k_1 \geqq k_0$ implies that the loss incurred if Δ is overestimated by an amount $|a - \Delta|$ is $K = k_1/k_0 \geqq 1$ times the loss incurred if Δ is underestimated by the same amount. Similarly for the problem of estimating Δ by an upper one-sided interval $(-\infty, b]$, let the loss function be

$$L_u(\Delta, b) = \begin{cases} k_1(b - \Delta)^2 , & \Delta > b \\ k_0(b - \Delta)^2 , & \Delta \leqq b . \end{cases}$$

Then

$$L(\Delta, [a, b]) = L_l(\Delta, a) + L_u(\Delta, b) .$$

For the loss function given by (3.33)-(3.34) and under the same prior density model (3.24) assumed for the testing problem, the Bayes intervals $[a_{ij}^*, b_{ij}^*]$ are obtained by separately minimizing the individual components of the Bayes risk. This leads to a pair of integral equations for each (i, j) whose solutions give a_{ij}^* and b_{ij}^* $(1 \leqq i < j \leqq k)$. Dixon and Duncan showed that each interval $[a_{ij}^*, b_{ij}^*]$ corresponds to the set of values of Δ_0 for which decision d_{ij}^0 is made in the three-decision problem of choosing between

$$d_{ij}^0 : \text{decide } \Delta_{ij} = \Delta_0 , \qquad d_{ij}^1 : \text{decide } \Delta_{ij} > \Delta_0 , \qquad d_{ij}^2 : \text{decide } \Delta_{ij} < \Delta_0$$

using the Bayes test of Waller and Duncan (1969) given in the preceding section but modified to take into account the fact that Δ_{ij} is now tested against a nonzero value Δ_0. This test has the form (3.25) except that the critical constant t^* now depends also on $\tau_0 = \Delta_0/S\sqrt{2/n}$ (in addition to K, F, q, and f); the critical constant t^* used in (3.25) corresponds to $\tau_0 = 0$. It may be noted that the equivalence between the Bayes tests and Bayes confidence intervals depends on the assumption of additive linear losses for the former and additive squared error losses for the latter. The squared error losses can be thought of as integrals of the linear losses. This equivalence result is shown to hold for more general loss functions by Dixon (1976).

The exact evaluation of the intervals $[a_{ij}^*, b_{ij}^*]$ poses a formidable computational problem. However, the intervals turn out to have a particularly simple form when σ_θ^2 and σ^2 are assumed to be known (i.e., $q = f = \infty$). In this case the simultaneous confidence intervals are given by

$$\theta_i - \theta_j \in \psi\left[\bar{Y}_i - \bar{Y}_j \pm t_*(K, \infty, \infty)\psi^{-1/2}\sigma\sqrt{\frac{2}{n}}\right] \qquad (1 \leqq i < j \leqq k) \quad (3.35)$$

where the multiplying constant

$$\psi = \frac{\gamma^2}{1+\gamma^2} = \frac{\sigma_\theta^2}{\sigma_\theta^2 + \sigma^2/n}$$

is called a shrinkage factor and $t_*(K, \infty, \infty) = t^*(K, \infty, \infty, \infty)$ is the solution to (3.6). Note from (3.5) that $t_*(K, \infty, \infty)\psi^{-1/2} = t_*(K, \infty, \gamma^2)$. Also note that for $K = 1$, $t_*(K, \infty, \infty) = 0$ and thus (3.35) reduces to $\psi(\bar{Y}_i - \bar{Y}_j)$, which is the well-known Bayes point estimate of $\theta_i - \theta_j$ with respect to a symmetric (since $k_0 = k_1$; see (3.34)) squared error loss function.

In practice, γ^2 and σ^2 are not known but may be consistently estimated by $\hat{\gamma}^2 = F - 1$ (=0 if $F \leqq 1$) and S^2, respectively (where F is calculated using (3.26) and S^2 is calculated using (3.27)). Substitution of these estimates in (3.35) and use of (3.28) results in the following confidence intervals:

$$\theta_i - \theta_j \in \left(1 - \frac{1}{F}\right)\left[\bar{Y}_i - \bar{Y}_j \pm t^*(K, F, \infty, \infty)S\sqrt{\frac{2}{n}}\right] \qquad (1 \leqq i < j \leqq k)\,. \quad (3.36)$$

These confidence intervals can be readily extended to all contrasts; see Duncan (1975) for details.

It should be emphasized that (3.35) and (3.36) are not derived so as to have a given joint confidence coefficient. A confidence interval with a posterior probability content of $1-\alpha$ can be derived for any pairwise difference $\theta_i - \theta_j$ by noting that the posterior distribution of $\theta_i - \theta_j$ is normal with mean $= \psi(\bar{Y}_i - \bar{Y}_j)$ and variance $= 2\psi\sigma^2/n$. Estimating ψ and σ^2 consistently by $(1-1/F)$ and S^2, respectively, yields the following $(1-\alpha)$-level (per-comparison) large sample confidence intervals for $\theta_i - \theta_j$:

$$\theta_i - \theta_j \in \left(1-\frac{1}{F}\right)\left[\bar{Y}_i - \bar{Y}_j \pm Z^{(\alpha/2)}\left(1-\frac{1}{F}\right)^{-1/2} S\sqrt{\frac{2}{n}}\right] \quad (1 \leqq i < j \leqq k) \tag{3.37}$$

where $Z^{(\alpha/2)}$ is the upper $\alpha/2$ point of the standard normal distribution. A comparison of (3.36) with (3.37) shows that these intervals are identical if $t_*(K, \infty, \infty) = Z^{(\alpha/2)}$, which can be achieved by an appropriate choice of K and α.

We finally note that if three-decision tests of hypotheses concerning $\theta_i - \theta_j$ ($\theta_i - \theta_j = 0$ or >0 or <0) are carried out using the confidence intervals (3.36) in the usual manner, then the resulting decision rule has exactly the same form as (3.25) except that t^* now equals $t^*(K, F, \infty, \infty)$, which is the result of the large sample ($q = f = \infty$) assumption.

3.6 Concluding Comments

An attractive feature of the Bayesian procedures discussed in this section (also summarized in Duncan and Dixon 1983) is the adaptive, post-hoc manner in which the critical t^*-values depend on the extent of heterogeneity between treatments. A large F-ratio implying large heterogeneity between treatments suggests less cautionary (small) t^*-values. A small F-ratio implying small heterogeneity between treatments suggests more cautionary (large) t^*-values. Thus these procedures adaptively adjust the critical values without sacrificing power when the treatment means are *a posteriori* indicated to be heterogeneous.

Another attractive feature is that the choice of α in conventional statistical procedures is replaced by that of a more basic and understandable quantity K.

A crucial assumption made throughout Duncan's work (as well as that of Lehmann's and Spjøtvoll's works) is that the total loss for the simultaneous problem is the sum of the losses for the component problems. It is because of this assumption that the optimal procedure for the simultaneous problem turns out to be the product of the optimal procedures for the corresponding component problems. Clearly, in many cases the assumption of an additive loss function is not appropriate; for example, if the simultaneous correctness requirement (see Chapter 1, Section 2) is to be satisfied, then the total loss is a multiplicative function of the component (zero-one) losses. The additive loss assumption gives the Bayes procedure the nature of a PCE controlling procedure in the sense that the critical t^* does not depend on the number of comparisons. The assumption of linear losses present in Duncan's work may serve as a good local approximation for small differences $\theta_i - \theta_j$, but for large differences the validity of this assumption may be in doubt.

There is a major difference between the unprotected LSD (i.e., multiple t-tests each at level α) and Duncan's Bayesian test, although both are PCE controlling type procedures. While in the unprotected LSD the comparison between any pair of treatments is independent of the sample means of the other treatments present in the experiment, in the Bayesian procedure it is dependent on those results through the F-ratio, which is a measure of heterogeneity among all treatments. For example, suppose that we have five treatments under study, labeled A, B, C, D, and E, which are ordered according to their sample means. The unprotected LSD may find the difference $E - A$ significant but the Bayesian procedure at a comparable α-level (K-ratio) may find that difference not significant because the F-ratio is lowered (and hence the critical t^* is increased) due to the presence of B, C, and D, whose sample means fall between those of A and E. Duncan and Brant (1983) used this example to refute O'Brien's (1983) suggestion that the comparison between a given pair of treatments must not be affected by the presence of other treatments in the same experiment.

4 A COMBINED BAYESIAN AND NEYMAN–PEARSON TYPE APPROACH

In Section 3.4 we studied Waller and Duncan's (1969) procedure for pairwise comparisons of means in a balanced one-way layout that minimizes the posterior expectation of a loss function (given by (3.14) and (3.11)) assuming certain prior distributions for the unknown parameters. As we saw, their procedure turns out to be a "continuous" version of Fisher's LSD procedure. Lewis (1984) considered the problem of making directional decisions on all pairwise differences using a mathematical framework similar to that in Waller and Duncan (1969) but without assuming an explicit loss structure. We now describe his approach.

Lewis (1984) considered the following modification of Fisher's LSD procedure: Compare the usual ANOVA F-statistic with a suitably chosen critical constant F^*. If $F \leqq F^*$, then stop testing and conclude that no significant differences exist among treatments. If $F > F^*$, then proceed to make directional decisions as follows: Decide that $\theta_i > \theta_j$ (respectively, $\theta_i < \theta_j$) if the T_{ij}-statistic given by (3.12) is greater than t^* (respectively, less than $-t^*$) where t^* is another suitably chosen critical constant; make no directional decision on pair (i, j) if $|T_{ij}| \leqq t^*$ $(1 \leqq i < j \leqq k)$. Note that here t^* does not depend on the value of the F-statistic as in the Waller–Duncan procedure.

To determine the critical constants F^* and t^*, Lewis (1984) proposed the following Bayesian modification of the classical error rate control approach. Let PFE($\boldsymbol{\theta}$, σ^2) denote the Type III per-family error rate, that is, PFE is the expected number of misclassifications of signs. In the classical approach one would require that PFE($\boldsymbol{\theta}$, σ^2) $\leqq \gamma$ for all $\boldsymbol{\theta}$ and σ^2 where γ $(0 < \gamma < \binom{k}{2})$ is a prespecified constant. Lewis proposed to control a weighted average of PFE($\boldsymbol{\theta}$, σ^2)-values, the average being taken with respect to a weighting function defined over the space of $\boldsymbol{\theta}$. This weighting function is formally similar to the prior distribution for $\boldsymbol{\theta}$. As in Waller and Duncan (1969), Lewis assumed that the θ_i's are independent $N(\theta_0, \sigma_\theta^2)$ r.v.'s. He referred to this assumption as the *single-cluster* model for the θ_i's with θ_0 and σ^2 being, respectively, the center and spread of the cluster. Because of location invariance, we can take $\theta_0 = 0$ without loss of generality. Denote the weighted average of PFE($\boldsymbol{\theta}$, σ^2) with respect to this weighting function by $\overline{\text{PFE}}(\sigma_\theta^2, \sigma^2)$. Because of the symmetry in the procedure and in the weighting function, we get

$$\overline{\text{PFE}}(\sigma_\theta^2, \sigma^2) = k(k-1)\Pr\{F > F^*, T_{12} > t^*, \theta_1 - \theta_2 < 0 | \sigma_\theta^2, \sigma^2\} . \quad (4.1)$$

The corresponding expression for the expected number of correct classifications is obtained by reversing the inequality on $\theta_1 - \theta_2$. Lewis proposed to find F^* and t^* subject to

$$\overline{\text{PFE}}(\sigma_\theta^2, \sigma^2) \leqq \gamma \qquad \forall\, \sigma_\theta^2, \sigma^2$$

where γ is prespecified.

By making an orthonormal transformation similar to that in Waller and Duncan (1969), Lewis was able to express (4.1) as

$$k(k-1)\Pr\{(k-2)V_1 + V_2 > (1-\rho^2)(k-1)F^*,$$
$$V_2 > (1-\rho^2)^{1/2}t^*, V_3 < 0\} \quad (4.2)$$

where V_2 and V_3 have a bivariate t-distribution with d.f. ν and an associated correlation coefficient

$$\rho = \left(\frac{n\sigma_\theta^2}{n\sigma_\theta^2 + \sigma^2} \right)^{1/2} ;$$

furthermore, conditional on (V_2, V_3),

$$\frac{\nu + 2}{\nu(1 + Q/\nu)} V_1 \sim F_{k-2,\nu+2}$$

where

$$Q = \frac{V_2^2 - 2\rho V_2 V_3 + V_3^2}{1 - \rho^2} .$$

Thus (4.2) can be evaluated by two-dimensional numerical integration of a cumulative F-probability with respect to the bivariate t-density of (V_2, V_3) over the quadrant: $V_2 > (1 - \rho^2)^{1/2} t^*$ and $V_3 < 0$. The corresponding expression for the expected number of correct classifications is obtained by reversing the inequality on V_3 to $V_3 > 0$. Note that (4.2) depends on σ_θ^2 and σ^2 only through ρ, which is a function of $n\sigma_\theta^2/\sigma^2$.

For given k and ν, by evaluating (4.2) for different combinations of F^* and t^*, Lewis discovered that a unique pair (F^*, t^*) can be found such that (4.2) approaches the specified upper bound γ as $\sigma_\theta^2 \to 0$ and attains that upper bound as an interior maximum (as a function of $n\sigma_\theta^2/\sigma^2$). Putting $F^* = F_{k-1,\nu}^{(\alpha)}$ and $t^* = T_\nu^{(\beta)}$, Lewis tabulated the values of α and β for $\gamma = 0.025$ and for selected values of k and ν. Notice that these values are not determined so as to maximize some measure of power.

Lewis compared the performance of this modified Fisher LSD with Bohrer's (1979) Bonferroni procedure (see Section 2.3.2 of Chapter 2). The latter is equivalent to using Fisher's LSD augmented with directional decisions for $F^* = 0$ and $t^* = T_\nu^{(\gamma/k^*)}$ where $k^* = \binom{k}{2}$. He found significant power advantages for the modified LSD. For example, in one such comparison, to obtain the same expected number of correct classifications, the Bonferroni procedure required nearly twice the number of observations required by the modified LSD.

5 A Γ-MINIMAX APPROACH

In many decision problems any *a priori* information concerning the unknown parameters is likely to be incomplete and thus a fully Bayesian approach requiring a complete specification of the prior distribution may not be justified. In a Γ-minimax approach to a decision problem, it is presumed that we have partial information concerning the prior distribution, that it belongs to a specified class Γ of priors. One then finds a decision procedure that minimizes the maximum Bayes risk where the maximum is taken over Γ. Such a decision procedure is called Γ-minimax. In this generalized framework, a Bayes procedure with respect to a given prior is a Γ-minimax procedure if Γ consists of only a single prior. On the other hand, a minimax procedure is a Γ-minimax procedure if Γ consists of all priors. The present section is devoted to the description of the Γ-minimax approach adopted by Randles and Hollander (1971) to the problem of treatments versus control comparisons. See Gupta and Huang (1977, 1981) for an application of this approach to the general class of multiple comparison problems considered by Spjøtvoll (1972a).

Randles and Hollander (1971) considered the following setup: Suppose that on treatment i we observe a sufficient statistic X_i having the density function $f_i(x-\theta_i)$, $i=1,2,\ldots,k$. The $f_i(\cdot)$'s are assumed to be known, but the θ_i's are unknown. The kth treatment is a control with respect to which the first $k-1$ treatments are to be classified into two groups as follows: The ones with $\theta_i \geqq \theta_k + \Delta$ are to be classified as "better" than the control, and the ones with $\theta_i \leqq \theta_k$ are to be classified as "no better" than the control where $\Delta > 0$ is a specified constant. Randles and Hollander referred to the former treatments as "positive" and the latter treatments as "negative." This formulation is similar to those studied by Lehmann (1961) and Tong (1969).

Consider the decision rule $\boldsymbol{\delta}(\mathbf{X}) = (\delta_1(\mathbf{X}), \ldots, \delta_{k-1}(\mathbf{X}))$ where $\delta_i(\mathbf{X})$ is the conditional probability of selecting the ith treatment $(1 \leqq i \leqq k-1)$ as "positive" having observed $\mathbf{X} = (X_1, X_2, \ldots, X_k)'$. The loss function is assumed to be of the form

$$L(\boldsymbol{\theta}, \boldsymbol{\delta}(\mathbf{X})) = \sum_{i=1}^{k-1} L_i(\boldsymbol{\theta}, \boldsymbol{\delta}(\mathbf{X})) \tag{5.1}$$

where

$$L_i(\boldsymbol{\theta}, \boldsymbol{\delta}(\mathbf{X})) = \begin{cases} c_1(1-\delta_i(\mathbf{X})) & \text{if } \theta_i \geqq \theta_k + \Delta \\ c_2\delta_i(\mathbf{X}) & \text{if } \theta_i \leqq \theta_k \\ 0 & \text{otherwise}. \end{cases} \quad (1 \leqq i \leqq k-1) \tag{5.2}$$

Here $c_1 > 0$ (respectively, $c_2 > 0$) is the cost of misclassifying a "positive" (respectively, "negative") treatment. The risk function is $R(\boldsymbol{\theta}, \boldsymbol{\delta}) = c_1N_1 + c_2N_2$ where N_1 and N_2 are the expected numbers of misclassified "positive" and "negative" treatments, respectively.

Let $\pi(\boldsymbol{\theta})$ be any prior distribution on $\boldsymbol{\theta}$ and define

$$\Omega_P(i) = \{\boldsymbol{\theta} : \theta_i \geqq \theta_k + \Delta\}\ , \qquad \Omega_N(i) = \{\boldsymbol{\theta} : \theta_i \leqq \theta_k\} \qquad (1 \leqq i \leqq k-1)\ .$$

Randles and Hollander assumed the following class of priors:

$$\Gamma = \left\{ \pi(\boldsymbol{\theta}) : \int_{\Omega_P(i)} d\pi(\boldsymbol{\theta}) = \pi_i,\ \int_{\Omega_N(i)} d\pi(\boldsymbol{\theta}) = \pi'_i\ (1 \leqq i \leqq k-1) \right\}$$

where the pairs (π_i, π'_i) are specified.

The derivation of the Γ-minimax rule involves first obtaining the least favorable prior distribution in Γ and then finding the Bayes rule with respect to that prior. Because of the additive nature of the loss function (5.1), the latter problem decomposes into finding the Bayes rules for the component problems corresponding to the $k-1$ component decision rules $\delta_i(\mathbf{X})$. For the case of a known control (i.e., known θ_k), the Γ-minimax rule $\boldsymbol{\delta}^0(\mathbf{X})$ is such that $\delta_i^0(\mathbf{X}) = 1$ or 0 according as

$$c_2 \pi'_i f_i(X_i - \theta_k) - c_1 \pi_i f_i(X_i - \theta_k - \Delta) \leqq 0 \text{ or } > 0 \qquad (1 \leqq i \leqq k-1)\ . \tag{5.3}$$

Miescke (1981) generalized this result and offered an alternative proof based on the Neyman–Pearson theory of hypothesis testing. If each $f_i(\cdot)$ has a monotone nondecreasing likelihood ratio property in its parameter θ_i, then (5.3) is equivalent to checking if $X_i \geqq d_i$ or $< d_i$ for some d_i $(1 \leqq i \leqq k-1)$. Notice that this rule depends on c_1 and c_2 only through their ratio c_1/c_2 and similarly on π_i, and π'_i only through their ratio π_i/π'_i.

For the case of an unknown control, invariance considerations lead to procedures based on $X_i - X_k$ $(1 \leqq i \leqq k-1)$. Denoting by $g_i(y - (\theta_i - \theta_k))$ the density function of $Y_i = X_i - X_k$, the Γ-minimax rule $\boldsymbol{\delta}^0(\mathbf{X})$ can be shown to have the following form: $\delta_i^0(\mathbf{X}) = 1$ or 0 according as

$$c_2 \pi'_i g_i(Y_i) - c_1 \pi_i g_i(Y_i - \Delta) \leqq 0 \text{ or } > 0 \qquad (1 \leqq i \leqq k-1)\ . \tag{5.4}$$

A complete proof of this result was given by Miescke (1981). As before, if the $f_i(\cdot)$'s have the monotone likelihood ratio property, then the same is true for the $g_i(\cdot)$'s, and (5.4) is equivalent to checking if $Y_i \geqq d'_i$ or $< d'_i$ for some d'_i $(1 \leqq i \leqq k-1)$.

Example 5.1. Let $\bar{Y}_i \sim N(\theta_i, \sigma^2/n_i)$, $i = 1, 2, \ldots, k$, be the sample means of treatments in a one-way layout and suppose that θ_k, the mean of the control treatment, and the error variance σ^2 are known. The $\bar{Y}_i$'s are sufficient statistics in this problem. By applying (5.3) with $X_i = \bar{Y}_i$ we get that the ith treatment is selected as "positive" if and only if

$$\bar{Y}_i \geqq d_i = \theta_k + \frac{\Delta}{2} + \left(\frac{\sigma^2}{\Delta n_i}\right)\log_e\left(\frac{c_2 \pi'_i}{c_1 \pi_i}\right) \qquad (1 \leqq i \leqq k-1) .$$

If θ_k is unknown but σ^2 is known, then application of (5.4) results in the decision rule that selects the ith treatment as "positive" if and only if

$$\bar{Y}_i - \bar{Y}_k \geqq d'_i = \frac{\Delta}{2} + \frac{\sigma^2}{\Delta}\left(\frac{1}{n_i} + \frac{1}{n_k}\right)\log_e\left(\frac{c_2 \pi'_i}{c_1 \pi_i}\right) \qquad (1 \leqq i \leqq k-1) . \quad \square$$

Appendixes

APPENDIX 1

Some General Theory of Multiple Comparison Procedures

Chapter 2 was devoted to the theory of multiple comparison procedures (MCPs) for fixed-effects linear models with independent homoscedastic normal errors, which was the framework for Part I. Much of that theory applies with minor modifications to many of the problems considered in Part II. However, in other cases the theory of Chapter 2 needs to be supplemented and extended, which is the purpose of the present appendix. We assume that the reader is familiar with Chapter 2. Many references to that chapter are made in the sequel. As in Chapter 2, throughout this appendix we restrict to the nonsequential (fixed-sample) setting.

The following is a summary of this appendix. Section 1 discusses the theory of simultaneous test procedures in arbitrary models. This discussion is based mostly on Gabriel (1969). When a simultaneous test procedure (and more generally a single-step test procedure) addresses hypotheses concerning parametric functions, it can be inverted to obtain a simultaneous confidence procedure for those parametric functions. Conversely, from a given simultaneous confidence procedure one can obtain the associated simultaneous test procedure by applying the confidence-region test method. The relation between simultaneous confidence estimation and simultaneous testing is the topic of Section 2. Finally Section 3 discusses some theory of step-down test procedures, including the topics of error rate control, optimal choice of nominal significance levels, and directional decisions. Here no general theory for deriving the associated simultaneous confidence estimates is as yet available; some preliminary work in this direction by Kim, Stefánsson, and Hsu (1987) is discussed in Section 4.2.4 of Chapter 2.

1 SIMULTANEOUS TEST PROCEDURES

1.1 Hierarchical Families

Let $\mathbf{Y}$ denote the set of observations. The distribution function of $\mathbf{Y}$ depends on a collection of unknown entities (possibly nondenumerable) that are together denoted by $\boldsymbol{\omega}$. The space of all possible $\boldsymbol{\omega}$'s is denoted by Ω and is referred to as the *parameter space*. For example, throughout Part I we assumed the fixed-effects linear model (1.2) of Chapter 2, which states that $\mathbf{Y}$ has an N-variate normal distribution with mean vector $\mathbf{X}\boldsymbol{\beta}$ and variance-covariance matrix $\sigma^2\mathbf{I}$ where $\mathbf{X}: N \times r$ is a known matrix. Thus $\boldsymbol{\omega} = (\boldsymbol{\beta}, \sigma^2)$ and $\Omega = \mathbb{R}^r \times \mathbb{R}^1_+$ (where $\mathbb{R}^1_+$ is the positive half of the real line). For another example, in the one-way layout location model (1.8) of Chapter 9, $\boldsymbol{\omega}$ consists of the common unknown distribution function $F(\cdot)$ and the vector of location parameters $\boldsymbol{\theta} = (\theta_1, \theta_2, \ldots, \theta_k)'$.

We consider a family of distinct hypotheses $H_i: \boldsymbol{\omega} \in \Omega_i$, $i \in I$, where Ω_i is a proper subset of Ω and I is an index set, not necessarily denumerable. For convenience, instead of H_i we refer to the subset Ω_i (to which H_i restricts $\boldsymbol{\omega}$) as a hypothesis. In many problems of interest the hypothesis $\Omega_0 = \bigcap_{i \in I} \Omega_i$ belongs to the family and is referred to as the *overall hypothesis*.

A hypothesis Ω_j implies another hypothesis Ω_i ($\Omega_j \Rightarrow \Omega_i$) if $\Omega_j \subseteq \Omega_i$. In this case Ω_i is said to be a *component* of Ω_j (a *proper component* if the containment is strict) and an *implication relation* is said to hold between Ω_i and Ω_j. Those hypotheses that do not have any proper components are referred to as *minimal*; all other hypotheses are referred to as *nonminimal*. The index set of the minimal hypotheses is denoted by $I_{\min}$.

A family with at least one implication relation is referred to as a *hierarchical family*. Note that every Ω_i, $i \in I - \{0\}$, is a proper component of Ω_0; hence every family containing Ω_0 is a hierarchical family.

1.2 Testing Families

Consider a family of hypotheses $\{\Omega_i, i \in I\}$ (not necessarily hierarchical) and let $Z_i = Z_i(\mathbf{Y})$ be a real valued test statistic for Ω_i, $i \in I$. We assume that large values of Z_i are suitable for rejecting Ω_i, $i \in I$. The collection $\{(\Omega_i, Z_i), i \in I\}$ is called a *testing family* provided that for every $i \in I$, the distribution of Z_i is completely specified under Ω_i (i.e., it is the same for all $\boldsymbol{\omega} \in \Omega_i$). If the joint distribution of the Z_i's for $i \in \tilde{I}$ is completely specified under $\bigcap_{i \in \tilde{I}} \Omega_i$ for every $\tilde{I} \subseteq I$, then the testing family is called *joint*. For a hierarchical family, if the relation $Z_j(\mathbf{Y}) \geqq Z_i(\mathbf{Y})$ holds almost everywhere (a.e.) whenever $\Omega_j \Rightarrow \Omega_i$, then the testing family is called *monotone*. In a monotone testing family we have $Z_0(\mathbf{Y}) \geqq Z_i(\mathbf{Y})$ for all $i \in I - \{0\}$ a.e. where $Z_0(\mathbf{Y})$ is the test statistic for Ω_0. If the family of hypotheses is closed under intersection, then the testing family is called *closed*.

For a given testing family $\{(\Omega_i, Z_i)$, Gabriel (1969) defined a *simultaneous test procedure* as that collection of tests in which the test for Ω_i rejects if $Z_i > \xi$ $(i \in I)$ where ξ is a common critical constant for all tests. Note that in Chapter 2, Section 3.3, where we introduced simultaneous test procedures for fixed-effects linear models, we did not impose the requirement of a testing family. This allowed us to include one-sided hypotheses on scalar parameters (see Example 3.5 in Chapter 2) in the family. More generally, if this requirement is not imposed and/or the critical constant is not the same for all tests but they can be carried out in one step without reference to one another, then we refer to the corresponding procedure as a *single-step test procedure*.

1.3 Coherence and Consonance; Likelihood Ratio and Union-Intersection Methods

The concepts of coherence and consonance introduced in Section 3.2 of Chapter 2 readily extend to the general setting of this appendix. The result of Theorem 3.1 of Chapter 2 concerning the coherence of the procedure and monotonicity of the test statistics is also applicable to the general setting here.

In this book we primarily employed the likelihood ratio (LR) and union-intersection (UI) methods of test construction. As explained in Sections 3.3.1.1 and 3.3.1.2 of Chapter 2, each of these methods leads to monotone test statistics for hierarchical families, and this result is also not restricted to the fixed-effects linear model setting of Part I. There are other methods of constructing monotone test statistics. For example, instead of the LR tests of independence given in Section 1.4 of Chapter 10, one may use Pearson's chi-square statistics, which also have the monotonicity property.

Regarding consonance, we have a general result in Theorem 3.2 of Chapter 2, which states that a simultaneous test procedure is consonant if and only if it is a UI procedure. Thus UI procedures are the only ones, which guarantee that whenever a nonminimal hypothesis is rejected at least one of its components will also be rejected. It is desirable for an MCP to have this property but not essential. The lack of consonance (referred to as *dissonance*) is not as serious a drawback as the lack of coherence. This is because of the nonsymmetry in the interpretation of rejection and acceptance decisions in classical significance testing.

UI statistics are obtained by starting with suitable statistics Z_i for $i \in I_{\min}$ and then using $Z_j = \max_{i \in I_{\min}^{(j)}} Z_i$ for any nonminimal Ω_j where $I_{\min}^{(j)}$ is the index set of all minimal hypotheses implied by Ω_j, $j \in I - I_{\min}$. The Z_i's for $i \in I_{\min}$ are usually chosen to be the LR statistics. This choice is based on certain desirable properties of the LR statistics. Also often with this choice, $\{(\Omega_i, Z_i), i \in I_{\min}\}$ forms a joint testing family. The following theorem gives a useful property of the UI statistics when the latter holds.

Theorem 1.1 If $\{(\Omega_i, Z_i), i \in I_{\min}\}$ is a joint testing family, then $\{(\Omega_i, Z_i), i \in I\}$ is a joint testing family provided the Z_i's are UI statistics.

Proof. For any $I' \subseteq I$ and under any $\omega \in \bigcap_{i \in I'} \Omega_i$, the joint distribution of the Z_i's, $i \in I'$, is determined by the joint distribution of the Z_j's, $j \in I_{\min}^{(I')}$, where $I_{\min}^{(I')} = \bigcup_{i \in I'} I_{\min}^{(i)}$. But $\bigcap_{i \in I_{\min}^{(I')}} \Omega_i \supseteq \bigcap_{i \in I'} \Omega_i$, and since $\{(\Omega_i, Z_i), i \in I_{\min}\}$ is a joint testing family, it follows that the joint distribution of the Z_i's, $i \in I'$, is the same under any $\omega \in \bigcap_{i \in I'} \Omega_i$. □

1.4 Control of the Familywise Error Rate

We next turn to the question of error rate control for simultaneous test procedures. The main result is summarized in the following theorem.

Theorem 1.2 (Gabriel 1969). Consider a simultaneous test procedure based on a testing family $\{(\Omega_i, Z_i), i \in I\}$ and a critical constant ξ. Let $\Omega_0 = \bigcap_{i \in I} \Omega_i$ and $Z_0 = \max_{i \in I} Z_i$. This simultaneous test procedure controls the Type I familywise error rate (FWE) strongly (for all parameter configurations) at a designated level α provided

$$\max_{\omega \in \Omega_0} \Pr_\omega \{Z_0 > \xi\} = \alpha , \tag{1.1}$$

and the testing family is either closed or joint.

Proof. Let $\tilde{I} \subseteq I$ denote the index set of true hypotheses and let $\tilde{\Omega} = \bigcap_{i \in \tilde{I}} \Omega_i$ be the set of corresponding parameter points. For $\omega \in \tilde{\Omega}$, the FWE is given by

$$\Pr_\omega \{\Omega_i \text{ is rejected for some } i \in \tilde{I}\} = \Pr_\omega \{\max_{i \in \tilde{I}} Z_i > \xi\} .$$

When the testing family is either closed or joint, this probability is the same for all $\omega \in \tilde{\Omega}$; thus it may be denoted by $\Pr_{\tilde{\Omega}}\{\max_{i \in \tilde{I}} Z_i > \xi\}$. This equals $\Pr_{\Omega_0}\{\max_{i \in \tilde{I}} Z_i > \xi\}$ since $\Omega_0 \subseteq \tilde{\Omega}$. Now $Z_0 = \max_{i \in I} Z_i \geq$

$\max_{i \in \tilde{I}} Z_i$. Therefore

$$\Pr_{\Omega_0}\{\max_{i \in \tilde{I}} Z_i > \xi\} \leqq \Pr_{\Omega_0}\{Z_0 > \xi\} = \alpha$$

where the last step follows from (1.1) and from the fact that the testing family is either closed or joint. This proves the theorem. □

Dunn's procedure, given in Section 1.2.2 of Chapter 9, offers an example of a single-step procedure that is not based on a closed or a joint testing family. In fact, Dunn's procedure controls the FWE only under the overall null configuration.

1.5 Resolution and Power

The operating characteristics of a simultaneous test procedure depend on the family of hypotheses Ω_i, $i \in I$, and the collection of associated test statistics Z_i, $i \in I$. The choice of a family affects the operating characteristics of a simultaneous test procedure in the following way: The "wider" the family, the less power the procedure possesses for detecting nonnull hypotheses because the common critical constant corresponding to a "wider" family is larger for a fixed level α. However, a "narrow" family may not have a sufficiently rich subfamily of implied hypotheses (e.g., minimal hypotheses) and hence may often result in dissonances when a nonconsonant procedure is used. Gabriel (1969) proposed *resolution* as a criterion for comparing two simultaneous test procedures based on different testing families (i.e., either different families of hypotheses or different test statistics or both). We now discuss this criterion.

Consider two simultaneous test procedures, $\mathcal{P}$ and $\mathcal{P}'$, based on the testing families $\{(\Omega_i, Z_i),\ i \in I\}$ and $\{(\Omega_i', Z_i'),\ i \in I'\}$, and common critical constants ξ and ξ', respectively. To make a meaningful comparison between $\mathcal{P}$ and $\mathcal{P}'$, they must test the same hypotheses at least to the extent that $\Omega_0 = \bigcap_{i \in I} \Omega_i = \Omega_0' = \bigcap_{i \in I'} \Omega_i'$ and $I_{\min} \subseteq I'$, that is, both the families must have the same overall hypothesis and the minimal hypotheses of one must be included in the other. Tukey's procedure ($\mathcal{P}$) for pairwise comparisons and Scheffé's procedure ($\mathcal{P}'$) for contrasts provide an example of two such procedures.

For this setup Gabriel (1969) introduced the following definition: Procedure $\mathcal{P}$ is said to be *no less resolvent* than procedure $\mathcal{P}'$ if both have the same level, and

$$(Z_i > \xi) \supseteq (Z_i' > \xi') \text{ a.e.} \qquad \forall\, i \in I_{\min}\,. \tag{1.2}$$

$\mathcal{P}$ is *strictly more resolvent* if the containment in (1.2) is proper for at least some $i \in I_{\min}$.

The following theorem shows how to obtain a no less resolvent simultaneous test procedure from a given one.

Theorem 1.3 (Gabriel 1969). Given a simultaneous test procedure $\mathcal{P}'$ based on a monotone family $\{(\Omega_i, Z'_i), i \in I\}$ (not necessarily a testing family) with common critical constant ξ' and associated level α, a no less resolvent procedure of level α can be obtained by using UI test statistics for all the nonminimal hypotheses.

Proof. Given the test statistics Z'_i for minimal hypotheses Ω_i, the UI test statistic for a nonminimal Ω_j is given by $Z_j = \max_{i \in I^{(j)}_{\min}} Z'_i$. Because of the monotonicity of the original statistics, we have $Z_j \leqq Z'_j$ for all $j \in I - I_{\min}$. Let $Z_i = Z'_i$ for all $i \in I_{\min}$ and consider a simultaneous test procedure $\mathcal{P}$ based on $\{(\Omega_i, Z_i), i \in I\}$. To achieve the same level α as that of $\mathcal{P}'$, the new procedure $\mathcal{P}$ requires a critical constant ξ that is clearly no larger than ξ'. This proves the theorem. □

From this theorem we can also conclude that given a simultaneous test procedure, we can obtain another one that has at least the same powers for rejecting any minimal hypotheses. This is achieved by using the same test statistics for the minimal hypotheses as used by the original procedure and UI test statistics for all nonminimal hypotheses. Thus we have the following theorem.

Theorem 1.4 (Gabriel 1969). Among all simultaneous test procedures of level α that use the same set of test statistics Z_j for $j \in I_{\min}$, a UI procedure has the highest powers for all Ω_j, $j \in I_{\min}$. □

It should be noted that the superiority of a UI procedure extends only over minimal hypotheses. If powers for nonminimal hypotheses are compared, then the UI procedure is not always the winner.

2 SIMULTANEOUS CONFIDENCE PROCEDURES

2.1 Simultaneous Confidence Regions

This section is concerned with the problem of simultaneous confidence estimation of a collection of parametric functions. Usually there is some vector parametric function $\boldsymbol{\theta} = \boldsymbol{\theta}(\boldsymbol{\omega}) \in \mathbb{R}^k$ such that all parametric functions of interest can be expressed as functions of $\boldsymbol{\theta}$, say, $\gamma_i = \gamma_i(\boldsymbol{\theta})$, $i \in I$. In Chapter 2 we confined attention to linear parametric functions $\gamma_i = \mathbf{L}_i\boldsymbol{\theta}$, $i \in I$ where $\boldsymbol{\theta}$ was a subvector of $\boldsymbol{\beta}$ and hence of $\boldsymbol{\omega} = (\boldsymbol{\beta}, \sigma^2)$.

A region $\Theta(\mathbf{Y}) \subseteq \mathbb{R}^k$ is referred to as a *confidence region* for $\boldsymbol{\theta}$ of confidence level $1 - \alpha$ if

$$\inf_{\boldsymbol{\omega} \in \Omega} \Pr_{\boldsymbol{\omega}}\{\boldsymbol{\theta} \in \Theta(\mathbf{Y})\} = 1 - \alpha . \tag{2.1}$$

A collection of confidence regions $\{\Gamma_i(\mathbf{Y}), i \in I\}$ for $\{\gamma_i, i \in I\}$ is referred to as a family of *simultaneous* (or *joint*) *confidence regions* of joint confidence level $1 - \alpha$ if

$$\inf_{\boldsymbol{\omega} \in \Omega} \Pr_{\boldsymbol{\omega}}\{\gamma_i \in \Gamma_i(\mathbf{Y}) \ \forall i \in I\} = 1 - \alpha . \tag{2.2}$$

If all the γ_i's are scalar parameters, then the $\Gamma_i(\mathbf{Y})$'s, $i \in I$ are typically intervals and are referred to as *simultaneous confidence intervals*.

A procedure for constructing simultaneous confidence regions is referred to as a *simultaneous confidence procedure*. Given (2.1), a simultaneous confidence procedure can be based on the *projection method* (see Section 2.2 of Chapter 2) to yield

$$\Gamma_i(\mathbf{Y}) = \{\gamma_i(\boldsymbol{\theta}) : \boldsymbol{\theta} \in \Theta(\mathbf{Y})\} , \qquad i \in I .$$

It is clear that the joint confidence level of these simultaneous confidence regions is at least $1 - \alpha$. Alternatively one could start with (2.2) and obtain a confidence region for $\boldsymbol{\theta}$ given by $\Theta(\mathbf{Y}) = \bigcap_{i \in I} \Theta_i(\mathbf{Y})$ where

$$\Theta_i(\mathbf{Y}) = \{\boldsymbol{\theta} : \gamma_i(\boldsymbol{\theta}) \in \Gamma_i(\mathbf{Y})\} , \qquad i \in I .$$

This confidence region also has confidence level at least $1 - \alpha$.

2.2 Coherence and Consonance in Simultaneous Confidence Estimation

Consider two parametric functions, say, $\gamma_i = \gamma_i(\boldsymbol{\theta})$ and $\gamma_j = \gamma_j(\boldsymbol{\theta})$. We say that γ_j *implies* γ_i if for every value of γ_j there corresponds a unique value of γ_i, that is, if there exists a function $f_{ij}(\cdot)$ such that $\gamma_i = f_{ij}(\gamma_j)$. If there is at least one such *implication relation* in a family of parametric functions $\{\gamma_i, i \in I\}$, then that family is called a *hierarchical family*. Those γ_i's, $i \in I$, that do not imply any other γ_j's, $j \in I$, are referred to as *minimal* and their index set is denoted by $I_{\min}$.

A simultaneous confidence procedure that yields simultaneous confidence regions $\{\Gamma_i(\mathbf{Y}), i \in I\}$ for a given hierarchical family $\{\gamma_i, i \in I\}$ is said to be *coherent* if whenever γ_j implies γ_i, then a.e.

$$\{\gamma_i = f_{ij}(\gamma_j) : \gamma_j \in \Gamma_j(\mathbf{Y})\} \subseteq \Gamma_i(\mathbf{Y}) .$$

The procedure is said to be *consonant* if for every value γ_{0j} of γ_j not in $\Gamma_j(\mathbf{Y})$ there is a parametric function γ_i, $i \in I$, implied by γ_j such that its corresponding value $\gamma_{0i} = f_{ij}(\gamma_{0j})$ is not in $\Gamma_i(\mathbf{Y})$.

2.3 Deriving Simultaneous Confidence Regions from a Simultaneous Test Procedure

Let $\{\gamma_i, i \in I\}$ be a hierarchical family of parametric functions. Consider the family of hypotheses $\{\Omega_i(\gamma_{0i}), i \in I\}$ where $\Omega_i(\gamma_{0i}) = \{\omega : \gamma_i = \gamma_i(\theta(\omega)) = \gamma_{0i}\}$ for some specified value γ_{0i}, $i \in I$. We assume that the overall hypothesis $\Omega_0 = \bigcap_{i \in I} \Omega_i(\gamma_{0i})$ is nonempty.

Suppose that for each $i \in I$ we have a real valued statistic $Z_i(\gamma_{0i}) = Z_i(\mathbf{Y}, \gamma_{0i})$ for testing $\Omega_i(\gamma_{0i})$ such that large values of $Z_i(\gamma_{0i})$ are suitable for rejection of $\Omega_i(\gamma_{0i})$. Furthermore, assume that $\{(\Omega_i(\gamma_{0i}), Z_i(\gamma_{0i})), i \in I\}$ forms a joint testing family. Then the common critical point ξ for an α-level simultaneous test procedure can be found from the equation

$$\Pr_{\Omega_0}\{\max_{i \in I} Z_i(\gamma_{0i}) > \xi\} = \alpha . \tag{2.3}$$

By inverting this simultaneous test procedure we obtain the following simultaneous confidence regions:

$$\Gamma_i(\mathbf{Y}) = \{\gamma_i : Z_i(\gamma_i) = Z_i(\mathbf{Y}, \gamma_i) \leqq \xi\} , \qquad i \in I . \tag{2.4}$$

The joint confidence level of these regions is clearly $1 - \alpha$.

As in Section 3.3.2 of Chapter 2, the preceding discussion can be presented in a unified way in terms of pivotal random variables (r.v.'s). Thus suppose that for a given family of parametric functions $\{\gamma_i, i \in I\}$ we have an associated collection of pivotal r.v.'s $\{Z_i(\gamma_i), i \in I\}$ such that their joint distribution is the same under all $\omega \in \Omega$. If a critical constant ξ satisfies

$$\Pr_{\omega}\{\max_{i \in I} Z_i(\gamma_i) \leqq \xi\} = 1 - \alpha \qquad \forall\, \omega \in \Omega ,$$

then the two types of simultaneous inferences, namely, confidence regions and hypotheses tests, can be derived from this single probability statement. For given $\mathbf{Y}$, the sets (2.4) provide simultaneous confidence regions of joint confidence level $1-\alpha$. On the other hand, the sets of $\mathbf{Y}$ for which $Z_i(\mathbf{Y}, \gamma_{0i}) > \xi$ provide simultaneous tests (rejection regions) of level α for the hypothesized values γ_{0i} of γ_i, $i \in I$. More general hypotheses can also be tested by using the confidence-region tests discussed in the following section.

Theorem 2.1 **(Gabriel 1969).** The simultaneous confidence procedure (2.4) is coherent and consonant if and only if the associated simultaneous test procedure that rejects $\Omega_i(\gamma_{0i})$ if $Z_i(\gamma_{0i}) > \xi$ is coherent and consonant for all γ_{0i}, $i \in I$.

Proof. If the given simultaneous test procedure is coherent, the joint testing family $\{(\Omega_i(\gamma_{0i}), Z_i(\gamma_{0i})), i \in I\}$ must be monotone for all γ_{0i}, $i \in I$. From this it follows that if $\gamma_i = f_{ij}(\gamma_j)$, then $Z_j(\gamma_j) \geqq Z_i(\gamma_i)$ with probability 1 and hence a.e.

$$\{\gamma_i = f_{ij}(\gamma_j) : Z_j(\gamma_j) \leqq \xi\} \subseteq \{\gamma_i : Z_i(\gamma_i) \leqq \xi\} ,$$

which shows that (2.4) is coherent. The converse can also be shown similarly.

Next, the consonance of the simultaneous test procedure implies that if a nonminimal hypothesis $\Omega_j(\gamma_{0j})$ is rejected, then some component of it, say, $\Omega_i(\gamma_{0i})$, is also rejected where $\gamma_{0i} = f_{ij}(\gamma_{0j})$. The first event is equivalent to $\gamma_{0j} \not\in \Gamma_j(\mathbf{Y})$, while the second event is equivalent to $\gamma_{0i} = f_{ij}(\gamma_{0j}) \not\in \Gamma_i(\mathbf{Y})$, which implies the consonance of the simultaneous confidence procedure. Again the converse can be shown similarly. □

Clearly, a simultaneous test procedure must be based on UI statistics in order for it to be consonant. In fact, analogous to Theorem 1.4 we have the following theorem.

Theorem 2.2. Given the statistics $Z_i(\gamma_{0i})$ satisfying (2.3) and the associated $(1-\alpha)$-level simultaneous confidence regions $\Gamma_i(\mathbf{Y})$ given by (2.4), one can construct UI statistics $Z_i'(\gamma_{0i})$ and the associated $(1-\alpha)$-level simultaneous confidence regions $\Gamma_i'(\mathbf{Y})$ such that $\Gamma_i'(\mathbf{Y}) \subseteq \Gamma_i(\mathbf{Y})$ a.e. for all $i \in I$. □

2.4 Confidence-Region Tests

The basic idea of confidence-region tests (Aitchison 1964), as explained in Section 3.3.2 of Chapter 2, is to start with a $(1-\alpha)$-level confidence region in the parameter space and check whether its intersection with the set of hypothesized values for the parametric function under test is empty (in which case the given hypothesis is rejected). Any number of hypotheses can be tested in this manner and the Type I FWE for all such tests is controlled at α. In this section we discuss this idea under a more general setting than that considered in Chapter 2.

By using the projection method, from (2.1) we obtain

$$\Omega(\mathbf{Y}) = \{\boldsymbol{\omega} : \boldsymbol{\theta}(\boldsymbol{\omega}) \in \Theta(\mathbf{Y})\}$$

as a $(1-\alpha)$-level confidence region for $\boldsymbol{\omega}$. Any family of hypotheses $\{\Omega_i, i \in I\}$ can be tested using this confidence region as follows:

$$\text{Reject } \Omega_i \text{ if } \Omega_i \cap \Omega(\mathbf{Y}) = \phi , \qquad i \in I \tag{2.5}$$

where ϕ denotes an empty set. The following theorem extends Theorem 3.4 of Chapter 2.

Theorem 2.3. The simultaneous test procedure (2.5) has Type I FWE $\leqq \alpha$ for all $\boldsymbol{\omega} \in \Omega$. Furthermore, if the family $\{\Omega_i, i \in I\}$ is hierarchical, then (2.5) is coherent.

Proof. For procedure (2.5) we have, under any $\boldsymbol{\omega} \in \Omega$,

$$\begin{aligned} \text{Type I FWE} &= \Pr_{\boldsymbol{\omega}}\{\Omega_i \cap \Omega(\mathbf{Y}) = \phi \text{ for at least one true } \Omega_i\} \\ &\leqq \Pr_{\boldsymbol{\omega}}\{\boldsymbol{\omega} \not\in \Omega(\mathbf{Y})\} = \alpha , \end{aligned}$$

and hence the first part of the theorem follows. Next, if Ω_j implies Ω_i, that is, if $\Omega_j \subseteq \Omega_i$, then $\Omega_j \cap \Omega(\mathbf{Y}) \neq \phi \Rightarrow \Omega_i \cap \Omega(\mathbf{Y})) \neq \phi$, and hence the second part of the theorem follows. □

Suppose that a confidence region for $\boldsymbol{\theta}$ is constructed by inverting a simultaneous test procedure based on the LR method for a family of hypotheses on $\boldsymbol{\theta}$. Aitchison (1965) showed that the confidence-region test procedure derived from this confidence region for the given family of hypotheses is the same as the original simultaneous test procedure. A similar result can be shown to hold for procedures based on the UI method.

3 STEP-DOWN TEST PROCEDURES

In this section we discuss some theory of step-down test procedures under the general setting of the present appendix. The emphasis here is on exploring more fully and in greater generality the concepts that were only briefly dealt with in Chapter 2 (e.g., separability and error rate control); straightforward extensions of the results and methods in Chapter 2 (e.g., the closure method) are not given. We also discuss some new topics, for example, the optimal choice of nominal levels.

3.1 Control of the Familywise Error Rate

3.1.1 General Results

A step-down test procedure for an arbitrary finite hierarchical family of hypotheses, $\{\Omega_i, i \in I\}$, can be described as follows: Define a "nominal" test function $\phi_i^0(\mathbf{Y})$ for each hypothesis Ω_i where $\phi_i^0(\mathbf{Y}) = 1$ if rejection of Ω_i is indicated when $\mathbf{Y}$ is observed and $\phi_i^0(\mathbf{Y}) = 0$ otherwise; here the prefix "nominal" refers to the test of that hypothesis without reference to the tests of any others. The "true" or "actual" test function of Ω_i is given by

$$\phi_i(\mathbf{Y}) = \min_{j:\Omega_j \subseteq \Omega_i} \{\phi_j^0(\mathbf{Y})\} . \tag{3.1}$$

For a step-down test procedure with nominal tests ϕ_i^0 and true tests ϕ_i the FWE under any $\boldsymbol{\omega} \in \Omega$ is given by

$$\Pr_{\boldsymbol{\omega}}\{\max_{i \in \tilde{I}} \phi_i(\mathbf{Y}) = 1\} \tag{3.2}$$

where $\tilde{I} = \tilde{I}(\boldsymbol{\omega})$ is the set $\{i \in I: \boldsymbol{\omega} \in \Omega_i\}$, that is, the index set of true hypotheses under $\boldsymbol{\omega}$. The FWE is strongly controlled at the designated level α if (3.2) is $\leqq \alpha$ for all $\boldsymbol{\omega} \in \Omega$. The following lemma gives a simple upper bound on (3.2).

Lemma 3.1. For a step-down test procedure with nominal tests $\{\phi_i^0, i \in I\}$,

$$\Pr_{\boldsymbol{\omega}}\{\max_{i \in \tilde{I}} \phi_i(\mathbf{Y}) = 1\} \leqq \Pr_{\boldsymbol{\omega}}\{\max_{i \in \tilde{I}} \phi_i^0(\mathbf{Y}) = 1\} \qquad \forall\, \boldsymbol{\omega} \in \Omega . \tag{3.3}$$

Proof. From (3.1) we have $\phi_i(\mathbf{Y}) \leqq \phi_i^0(\mathbf{Y})$ a.e. for all $i \in I$ and hence (3.3) follows. □

For the family of subset hypotheses under the normal theory one-way layout setting, Tukey (1953, Chapter 30) noted that the task of controlling the FWE for a given step-down procedure is simplified (essentially by ignoring the step-down nature of the procedure and focusing attention only on the nominal tests) if that procedure satisfies a certain condition (see Example 3.1 below). Lehmann and Shaffer (1979) extended this condition to arbitrary one-way layouts and referred to it as the *separability* condition. We now give a definition of separability for the present general setting: Let $\tilde{I} \subseteq I$ and $\tilde{J} = I - \tilde{I}$ be the index subsets of true and false hypotheses, respectively, and let

$$\Omega(\tilde{I}) = \Big(\bigcap_{i \in \tilde{I}} \Omega_i\Big) \cap \Big(\bigcap_{j \in \tilde{J}} \Omega_j^{-1}\Big) \tag{3.4}$$

(where Ω_j^{-1} denotes the complement of Ω_j). The set of parameter configurations $\Omega(\tilde{I})$ is called *separable* by a given step-down procedure if there exists $\boldsymbol{\omega}^* \in \Omega(\tilde{I})$ such that the probability that the procedure rejects all false hypotheses is 1 under $\boldsymbol{\omega}^*$, that is,

$$\Pr_{\boldsymbol{\omega}^*}\{\phi_j(\mathbf{Y}) = 1 \ \forall\ j \in \tilde{J}\} = 1 . \tag{3.5}$$

Usually such $\boldsymbol{\omega}^*$'s are limiting parameter configurations in $\Omega(\tilde{I})$. If every $\Omega(\tilde{I})$ is separable by a given procedure, then we call that procedure *separating*.

Notice that if $j \in \tilde{J}$ and $\Omega_{j'} \subseteq \Omega_j$, then $j' \in \tilde{J}$. Therefore, from (3.1), $\phi_j(\mathbf{Y}) = 1$ for all $j \in \tilde{J}$ if and only if $\phi_j^0(\mathbf{Y}) = 1$ for all $j \in \tilde{J}$, and hence (3.5) is equivalent to

$$\Pr_{\boldsymbol{\omega}^*}\{\phi_j^0(\mathbf{Y}) = 1 \ \forall\ j \in \tilde{J}\} = 1 . \tag{3.6}$$

Example 3.1. Consider the usual normal theory one-way layout model. Let $P = \{i_1, i_2, \ldots, i_p\}$ be a subset of $K = \{1, 2, \ldots, k\}$ of cardinality p $(2 \leqq p \leqq k)$ and consider the family of subset hypotheses

$$\{\Omega_P : \theta_{i_1} = \theta_{i_2} = \cdots = \theta_{i_p},\ P \subseteq K\} . \tag{3.7}$$

An arbitrary configuration involves $r \geqq 1$ distinct groups of θ's of cardinalities $p_1, p_2, \ldots, p_r$ $(\Sigma_{i=1}^r p_i = k)$ such that

$$\theta_{i_1} = \cdots = \theta_{i_{p_1}}, \quad \theta_{i_{p_1+1}} = \cdots = \theta_{i_{p_1+p_2}}, \ldots .$$

All such configurations are separable by every step-down procedure of Chapter 4. This is because those procedures are based on the range or F-statistics, and for any subset hypothesis Ω_P that involves θ's from at least two distinct groups, these statistics tend to infinity in probability if the distance between any two unequal θ's tends to infinity. Hence Ω_P is rejected with probability 1 under such configurations. □

The configurations $\boldsymbol{\omega}^*$ that satisfy (3.6) are called *least favorable* in $\Omega(\tilde{I})$ because for separating step-down procedures, the upper bound on the FWE given in (3.4) is achieved at such $\boldsymbol{\omega}^*$'s. This is shown in the following lemma.

Lemma 3.2. For a separating step-down procedure, for any $\boldsymbol{\omega}^*$ satisfying (3.6) we have

$$\Pr_{\boldsymbol{\omega}^*}\{\max_{i\in\tilde{I}} \phi_i(\mathbf{Y}) = 1\} = \Pr_{\boldsymbol{\omega}^*}\{\max_{i\in\tilde{I}} \phi_i^0(\mathbf{Y}) = 1\} . \tag{3.8}$$

Proof. Using (3.1) we can write the left hand side of (3.8) as

$$\Pr_{\boldsymbol{\omega}^*}\{\max_{i\in\tilde{I}} \min_{j:\Omega_j\subseteq\Omega_i} \phi_j^0(\mathbf{Y}) = 1\} . \tag{3.9}$$

Let

$$\tilde{I}_i = \{j:\Omega_j \subseteq \Omega_i\} \subseteq \tilde{I} \quad \text{and} \quad \tilde{J}_i = \{j:\Omega_j \subseteq \Omega_i\} \subseteq \tilde{J} , \qquad i \in \tilde{I} .$$

Then (3.9) can be written as

$$\Pr_{\boldsymbol{\omega}^*}\{\max_{i\in\tilde{I}} \min(\min_{j\in\tilde{I}_i} \phi_j^0(\mathbf{Y}), \min_{j\in\tilde{J}_i} \phi_j^0(\mathbf{Y})) = 1\} . \tag{3.10}$$

Because of the separability condition (3.6) we have

$$\Pr_{\boldsymbol{\omega}^*}\{\min_{j\in\tilde{J}_i} \phi_j^0(\mathbf{Y}) = 1\} = 1 \qquad \forall\, i \in \tilde{I} .$$

Hence (3.10) simplifies to

$$\Pr_{\boldsymbol{\omega}^*}\{\max_{i\in\tilde{I}} \min_{j\in\tilde{I}_i} \phi_j^0(\mathbf{Y}) = 1\} ,$$

which clearly equals the right hand side of (3.8). □

The next theorem gives a sufficient condition under which controlling the FWE of a step-down procedure at all least favorable configurations $\boldsymbol{\omega}^*$ implies control of the FWE at all $\boldsymbol{\omega} \in \Omega$. This theorem makes use of the results of Lemmas 3.1 and 3.2.

Theorem 3.1. Suppose that the nominal tests ϕ_i^0 can be specified in terms of real valued test statistics Z_i, $i \in I$. Further suppose that $\{(\Omega_i, Z_i), i \in I\}$ forms a joint testing family and that the step-down procedure is separating. Under these conditions, if for all least favorable configurations $\boldsymbol{\omega}^* \in \Omega(\tilde{I})$ (where $\tilde{I} \subseteq I$ is an arbitrary subset of true hypotheses)

$$\Pr_{\boldsymbol{\omega}^*}\{\max_{i \in \tilde{I}} \phi_i^0(\mathbf{Y}) = 1\} \leqq \alpha , \tag{3.11}$$

then the FWE is strongly (for all $\boldsymbol{\omega} \in \Omega$) controlled at level α.

Proof. Because $\{(\Omega_i, Z_i), i \in I\}$ is a joint testing family, we can write

$$\Pr_{\boldsymbol{\omega}}\{\max_{i \in \tilde{I}} \phi_i^0(\mathbf{Y}) = 1\} = \Pr_{\boldsymbol{\omega}^*}\{\max_{i \in \tilde{I}} \phi_i^0(\mathbf{Y}) = 1\} \qquad \forall\, \boldsymbol{\omega}, \boldsymbol{\omega}^* \in \Omega(\tilde{I}) \tag{3.12}$$

for any $\tilde{I} \subseteq I$. Using (3.2), (3.8), and (3.11) in (3.12) we then get

$$\Pr_{\boldsymbol{\omega}}\{\max_{i \in \tilde{I}} \phi_i(\mathbf{Y}) = 1\} \leqq \alpha \qquad \forall\, \boldsymbol{\omega} \in \Omega(\tilde{I}) ,$$

which proves the theorem. □

Thus the task of controlling the FWE using a step-down test procedure reduces to guaranteeing (3.11). The probability in (3.11) depends on the nominal tests ϕ_i^0 in a simple manner because it essentially ignores the step-down testing scheme. An application of this result to the family of subset hypotheses is given in Theorem 3.2 (see also Theorem 4.4 of Chapter 2).

3.1.2 *Application to the Family of Subset Hypotheses*

We now confine attention to the case where $\mathbf{Y}$ consists of k independent samples $\mathbf{Y}_1, \mathbf{Y}_2, \ldots, \mathbf{Y}_k$. The parameter point $\boldsymbol{\omega}$ also consists of k components, $\boldsymbol{\omega}_1, \boldsymbol{\omega}_2, \ldots, \boldsymbol{\omega}_k$, such that the distribution of $\mathbf{Y}_i$ depends only on $\boldsymbol{\omega}_i$ $(1 \leqq i \leqq k)$. Let θ_i be a scalar valued function of $\boldsymbol{\omega}_i$ that is of interest $(1 \leqq i \leqq k)$. Consider the family of subset hypotheses (3.7) as in Example 3.1 but without restricting to the normal theory assumptions. Lehmann and Shaffer (1979) have given several examples of separating step-down procedures for this family in parametric as well as nonparametric problems.

Assume that $\Pr_{\boldsymbol{\omega}}\{\phi_P^0(\mathbf{Y}) = 1\}$ is the same for all $\boldsymbol{\omega} \in \Omega_P$ and denote this common value by α_P. Thus we are assuming that the nominal tests ϕ_P^0 of Ω_P are *similar* for all $P \subseteq K$. As in Section 4.3.3 of Chapter 2 we refer to the α_P's as *nominal levels*. We first state the following useful lemma. (The proofs of the results in this section are omitted as they are special cases of the general results stated above.)

Lemma 3.3 (Tukey 1953). For a separating step-down procedure, if under any subset hypothesis Ω_P we have

$$\Pr_\omega\{\phi_P^0(\mathbf{Y}) = 1\} = \alpha_P \qquad \forall\, \omega \in \Omega_P\,,$$

then

$$\sup_{\omega \in \Omega_P} \Pr_\omega\{\phi_P(\mathbf{Y}) = 1\} = \alpha_P\,.$$

This supremum is attained under any least favorable configuration such that the θ_j's, $j \not\in P$, are sufficiently different from the θ_i's, $i \in P$, to guarantee that all Ω_Q's, $Q \supseteq P$, are rejected with probability 1. □

In the following we assume that each α_P depends on P only through p, the cardinality of P, and hence we denote it by α_p. For $\Omega_{\mathbf{P}} = \bigcap_{i=1}^r \Omega_{P_i}$ where $\mathbf{P} = (P_1, P_2, \ldots, P_r)$ is any partition of $K = \{1, 2, \ldots, k\}$ consisting of disjoint subsets P_i with cardinalities $p_i \geqq 2$ $(1 \leqq i \leqq r)$ satisfying $\Sigma_{i=1}^r p_i \leqq k$, define

$$\alpha^*(\mathbf{P}) = \sup_{\omega \in \Omega_{\mathbf{P}}} \Pr_\omega\{\max_{1 \leqq i \leqq r} \phi_{P_i}(\mathbf{Y}) = 1\} \tag{3.13}$$

as in (4.15) of Chapter 2. We then have the following theorem analogous to Theorem 4.4 of Chapter 2.

Theorem 3.2 (Tukey 1953). If the given step-down procedure is separating and if the nominal tests ϕ_P^0 associated with it are statistically independent, then

$$\alpha^*(\mathbf{P}) = 1 - \prod_{i=1}^r (1 - \alpha_{p_i}) \tag{3.14}$$

where the α_{p_i}'s are nominal levels of the tests $\phi_{P_i}^0$ $(1 \leqq i \leqq r)$. Therefore

$$\alpha^* = \sup_{\mathbf{P}} \alpha^*(\mathbf{P}) = \max_{(p_1, \ldots, p_r)} \left[1 - \prod_{i=1}^r (1 - \alpha_{p_i})\right] \tag{3.15}$$

where the maximum is taken over all sets of integers $p_1, \ldots, p_r$ satisfying $p_i \geqq 2$, $\Sigma_{i=1}^r p_i \leqq k$, and $1 \leqq r \leqq k/2$. □

From (3.15) we see that to control the FWE strongly, only the choice of the nominal levels α_p matters. For given α, the nominal levels α_p

should be chosen so that

$$\max_{(p_1,\ldots,p_r)} \left[1 - \prod_{i=1}^{r} (1 - \alpha_{p_i})\right] \leqq \alpha \tag{3.16}$$

regardless of the choice of the test statistics.

3.1.3 A Shortcut Procedure

We next consider a shortcut version of the step-down test procedure for the family of subset hypotheses in the special case of the location parameter family of distributions. Thus suppose that we have k independent r.v.'s, $Y_1, Y_2, \ldots, Y_k$, where Y_i has the distribution

$$\Pr(Y_i \leqq y) = F(y - \theta_i) \qquad (1 \leqq i \leqq k)\,, \tag{3.17}$$

F being known and continuous. Consider a step-down procedure in which the nominal tests of the subset homogeneity hypotheses Ω_P are based on the range statistics $R_P = \max_{i \in P} Y_i - \min_{i \in P} Y_i$ $(P \subseteq K)$. The critical constants used in the nominal tests of Ω_P and the associated nominal levels are assumed to depend only on the cardinalities p of P; we denote them by ξ_p and α_p, respectively $(2 \leqq p \leqq k)$. As seen in Example 4.4 of Chapter 2, a shortcut step-down procedure can be employed in this setting if the critical constants ξ_p form a monotone sequence

$$\xi_2 \leqq \xi_3 \leqq \cdots \leqq \xi_k\,. \tag{3.18}$$

In this procedure whenever any Ω_P is rejected, the two θ's corresponding to $\max_{i \in P} Y_i$ and $\min_{i \in P} Y_i$ and any subset containing these two θ's are declared heterogeneous without further tests. Lehmann and Shaffer (1977) showed that (3.18) is a necessary and sufficient condition for (3.14) to hold for this shortcut procedure under the location parameter setting.

In the nonindependence case when the ranges R_P are positively dependent (e.g., when the Y_i's have independent numerators but a common Studentizing factor in the denominators), equality (3.14) is replaced by the inequality

$$\alpha^*(\mathbf{P}) \leqq 1 - \prod_{i=1}^{r} (1 - \alpha_{p_i})$$

when (3.18) is satisfied.

3.2 An Optimal Choice of the Nominal Levels

We now consider the problem of choosing the α_p's "optimally" for a given step-down procedure for the family of subset hypotheses. Here an "optimal" choice of the α_p's is one that maximizes the powers, that is, the probabilities of rejecting false hypotheses, subject to (3.16). If the statistics Z_P for the nominal tests of Ω_P are given, then fixing the α_p's fixes their critical values and hence the power functions of the procedure. Since larger α_p's are associated with higher powers, an optimal choice of the α_p's would correspond to maximizing them subject to (3.16). We now describe Lehmann and Shaffer's (1979) solution to this problem.

It can be shown that, without any restriction on $\alpha_2, \ldots, \alpha_{k-2}$, the maximal values of α_{k-1} and α_k can always be chosen as

$$\alpha^*_{k-1} = \alpha^*_k = \alpha . \tag{3.19}$$

For $k = 3$, (3.19) provides the optimal choice. For $k = 4$, (3.19) together with $\alpha^*_2 = 1 - (1 - \alpha)^{1/2}$ provides the optimal choice. For $k \geqq 5$, however, it is not possible to simultaneously maximize all α_p's.

When a uniformly best choice of the α_p's does not exist, it is convenient to define the notion of admissible α_p's. A set $(\alpha_2, \alpha_3, \ldots, \alpha_k)$ is called *inadmissible* if there exists another set $(\alpha'_2, \alpha'_3, \ldots, \alpha'_k)$ such that both satisfy (3.16) for a specified α, and $\alpha_p \leqq \alpha'_p$ for all $p = 2, \ldots, k$ with strict inequality for at least some p. If the set $(\alpha_2, \alpha_3, \ldots, \alpha_k)$ is not inadmissible, then it is said to be *admissible*.

To obtain reasonable power against any pattern of inhomogeneities, Lehmann and Shaffer proposed the criterion of maximizing $\min_{2 \leqq p \leqq k} \alpha_p$ subject to (3.16) when $k \geqq 5$. It can be shown that any admissible choice of α_p's satisfies $\alpha_2 \leqq \alpha_3 \leqq \cdots \leqq \alpha_k$. Therefore this max-min criterion requires maximization of α_2 subject to (3.16). This maximum is given by

$$\alpha^*_2 = 1 - (1 - \alpha)^{1/\lfloor k/2 \rfloor} \tag{3.20}$$

where $\lfloor x \rfloor$ denotes the integer part of x. For k odd, Lehmann and Shaffer showed that subject to (3.16) and (3.20) all α_p's are simultaneously maximized by the choice

$$\alpha^*_p = 1 - (1 - \alpha)^{\lfloor p/2 \rfloor / \lfloor k/2 \rfloor} \qquad (2 \leqq p \leqq k) . \tag{3.21}$$

Notice that this choice agrees with (3.19) for $p = k - 1$ and k.

For k even, however, it is not possible to simultaneously maximize all α_p's subject to (3.16) and (3.20) when $k \geqq 5$. Based on fairly narrow bounds on the admissible α_p's, Lehmann and Shaffer recommended the following choice for these values of k:

$$\alpha^*_p = 1 - (1 - \alpha)^{p/k} \ (2 \leqq p \leqq k - 2) , \qquad \alpha^*_{k-1} = \alpha^*_k = \alpha . \tag{3.22}$$

This choice was given in (4.22) of Chapter 2 for any k and was referred to there as the *Tukey–Welsch* (*TW*) *specification*.

3.3 Directional Decisions

In Section 4.2.3 of Chapter 2 we briefly discussed (in a restricted context) the problem that is the topic of the present section. Shaffer (1980) considered finite nonhierarchical families of hypotheses postulating specified values for scalar parametric functions against two-sided alternatives. She showed that under certain conditions it is possible to supplement step-down procedures with directional decisions in case of rejections of postulated values, and still control the Type I and Type III FWE. We now give a formal statement of this result under the more general setting of this appendix.

Consider m scalar parametric functions $\gamma_i = \gamma_i(\boldsymbol{\theta}(\boldsymbol{\omega}))$ $(1 \leqq i \leqq m)$. Suppose that we are interested in testing $H_{0i}: \gamma_i = \gamma_{0i}$ $(1 \leqq i \leqq m)$ against two-sided alternatives, the γ_{0i}'s being specified constants. Further suppose that we have independent test statistics Z_i with distributions

$$\Pr(Z_i \leqq z) = F_i(z, \gamma_i), \; F_i \text{ is nonincreasing in } \gamma_i \qquad (1 \leqq i \leqq m) . \tag{3.23}$$

Thus the Z_i's may have different distributional forms but they must be stochastically increasing in γ_i's. Let ξ'_{ij} and ξ_{ij} be the maximum and minimum values, respectively, for which

$$\Pr_{\gamma_{0i}}(Z_i < \xi'_{ij}) \leqq (1 - a_i)\alpha_j , \qquad \Pr_{\gamma_{0i}}(Z_i > \xi_{ij}) \leqq a_i \alpha_j \tag{3.24}$$

with equalities holding if the Z_i's are continuous; here the a_i's and α_j's are fixed constants satisfying $0 \leqq a_i \leqq 1$ $(1 \leqq i \leqq m)$ and

$$(1 - \alpha_j)^j = 1 - \alpha \qquad (1 \leqq j \leqq m) \tag{3.25}$$

for a designated level α. Notice that by choosing the a_i's appropriately we can have any combination of one-sided and two-sided tests. (When all tests are one-sided there are no directional decisions to be made and only the Type I FWE is to be controlled.)

The following is a generalization of Holm's (1979a) step-down procedure:

Step 1. Reject H_{0i} if $Z_i < \xi'_{im}$ or $Z_i > \xi_{im}$ $(1 \leqq i \leqq m)$. If at least one H_{0i} is rejected, then proceed to Step 2. In general, let r_{l-1} be the number of H_{0i}'s rejected at the $(l-1)$th step $(l - 1 = 1, 2, \ldots, m-1)$. If $r_{l-1} = 0$, retain all hypotheses that have not been rejected until that step and stop testing. If $r_{l-1} > 0$, proceed to Step l.

Step l. Consider all H_{0i}'s that have not been rejected up to the $(l-1)$th step. Reject any such H_{0i} if $Z_i < \xi'_{ij}$ or $Z_i > \xi_{ij}$ where $j = m - \Sigma_{k=1}^{l-1} r_k$.

Notice that, as a result of (3.23) and (3.24), if J denotes the index set of $j \geqq 1$ unrejected hypotheses at step l, then under $H_{0J} = \bigcap_{i \in J} H_{0i}$,

$$\Pr_{H_{0J}}\left\{\bigcup_{i \in J} (Z_i < \xi'_{ij} \text{ or } Z_i > \xi_{ij})\right\} \leqq \alpha$$

with equality holding if all the Z_i's are continuous.

If any $H_{0i}: \gamma_i = \gamma_{0i}$ is rejected, one would usually wish to decide whether $\gamma_i < \gamma_{0i}$ or $\gamma_i > \gamma_{0i}$. Since the Z_i's are assumed to have stochastically increasing distributions in the γ_i's, a natural way to do this is by augmenting Holm's procedure as follows: If H_{0i} is rejected at the lth step $(1 \leqq l \leqq m)$, then decide that $\gamma_i < \gamma_{0i}$ or $\gamma_i > \gamma_{0i}$ according to whether $Z_i < \xi'_{ij}$ or $Z_i > \xi_{ij}$ where $j = m - \Sigma_{k=1}^{l-1} r_k$. For the resulting procedure the following theorem gives a set of sufficient conditions for the control of the Type I and Type III FWE.

Theorem 3.3 (Shaffer 1980). The step-down procedure with directional decisions stated above controls the Type I and Type III FWE strongly (for all values of the γ_i's) at level α if the test statistics Z_i are independently distributed according to (3.23), the critical constants ξ_{ij} and ξ'_{ij} are chosen according to (3.24) and (3.25), and the F_i's satisfy the following two conditions:

(i) Let $[z^*_{i,\gamma_i}, z^{**}_{i,\gamma_i}]$ be the convex support of $F_i(z, \gamma_i)$ and let $(\gamma^*_i, \gamma^{**}_i)$ be the range of possible values of γ_i. Then for $i = 1, 2, \ldots, m$,

$$\lim_{\gamma_i \to \gamma^*_i} F_i(z, \gamma_i) = 1\,, \qquad \lim_{\gamma_i \to \gamma^{**}_i} F_i(z, \gamma_i) = 0$$
$$\forall\, z \in (z^*_{i,\gamma_{0i}}, z^{**}_{i,\gamma_{0i}})\,. \tag{3.26}$$

(ii)
$$\frac{F'_i(z', \gamma)}{F'_i(z, \gamma)} \leqq \frac{F'_i(z', \gamma')}{F'_i(z, \gamma')} \qquad \forall\, z' > z,\ \gamma' > \gamma \tag{3.27}$$

where F'_i denotes the derivative of F_i with respect to γ. □

It can be shown that condition (3.27) is satisfied for location and scale parameter families of distributions with monotone LRs, and for exponential families of distributions that satisfy (3.26).

APPENDIX 2

Some Probability Inequalities Useful in Multiple Comparisons

In any simultaneous inference problem we are concerned with making probability statements about the joint occurrences of several random events $\mathscr{E}_1, \mathscr{E}_2, \ldots, \mathscr{E}_k$ that are generally dependent. Typically event $\mathscr{E}_i$ is of the form $(X_i \leqq \xi)$ where X_i is a random variable (r.v.) and ξ is some critical constant. Thus $\bigcap_{i=1}^{k} \mathscr{E}_i$ is the event that $\max_{1 \leqq i \leqq k} X_i \leqq \xi$. Except in some special cases, it is difficult to evaluate the probability of this event exactly, even on a computer. Thus easily computable bounds (usually lower bounds) are needed (David 1981, Section 5.3). Similar problems arise in many other areas of statistics, for example, in reliability, and in ranking and selection theory. For this reason, the subject of probability inequalities has received increasing attention in recent years. Tong (1980) is an excellent comprehensive reference on this subject. Our aim here is to provide a brief review of the inequalities that are commonly employed in determining bounds on the critical constants ξ used in multiple comparison procedures (MCPs).

In Section 1 we study Bonferroni-type bounds, which are distribution-free and are thus applicable in very general settings. In Section 2 we consider the so-called multiplicative bounds, which are sharper than the first order Bonferroni bounds, but are valid under somewhat restrictive distributional assumptions. Both of these methods involve bounding the joint probability by some function of the marginal probabilities. In Section 3 we discuss some probability inequalities that can be obtained by the techniques of majorization (Marshall and Olkin 1979).

1 BONFERRONI-TYPE INEQUALITIES

Let $\mathscr{E}_1, \mathscr{E}_2, \ldots, \mathscr{E}_k$ be $k \geqq 2$ random events. By the principle of inclusion and exclusion we get the well-known Boole's (also known as Poincaré's) formula:

$$1 - \Pr\left(\bigcap_{i=1}^{k} \mathscr{E}_i\right) = \Pr\left(\bigcup_{i=1}^{k} \mathscr{E}_i^c\right)$$

$$= \sum_{i=1}^{k} \Pr(\mathscr{E}_i^c) - \sum_{i<j} \Pr(\mathscr{E}_i^c \cap \mathscr{E}_j^c) + \cdots + (-1)^{k-1} \Pr\left(\bigcap_{i=1}^{k} \mathscr{E}_i^c\right) \tag{1.1}$$

where $\mathscr{E}_i^c$ denotes the complement of event $\mathscr{E}_i$. The rth order Bonferroni approximation consists of using the first r terms of this finite series to approximate the left hand side of (1.1). Denote this approximation by S_r $(1 \leqq r \leqq k-1)$. Then $S_1, S_3, S_5, \ldots$ provide upper bounds and $S_2, S_4, S_6, \ldots$ provide lower bounds on the left hand side of (1.1). It is commonly believed that $S_1 \geqq S_3 \geqq S_5 \geqq \cdots$ and $S_2 \leqq S_4 \leqq S_6 \leqq \cdots$, and thus the bounds increase in sharpness with the order; however, this is generally not true (see Schwager 1984 for counterexamples). The popular *Bonferroni inequality* uses the first order approximation yielding

$$\Pr\left(\bigcap_{i=1}^{k} \mathscr{E}_i\right) \geqq 1 - \sum_{i=1}^{k} \Pr(\mathscr{E}_i^c). \tag{1.2}$$

Many improvements of (1.2) have been proposed that require the knowledge of joint bivariate probabilities (i.e., joint probabilities of pairs of events). One such improvement was proposed by Kounias (1968). The *Kounias inequality* gives

$$\Pr\left(\bigcap_{i=1}^{k} \mathscr{E}_i\right) \geqq 1 - \sum_{i=1}^{k} \Pr(\mathscr{E}_i^c) + \max_{j} \sum_{i \neq j} \Pr(\mathscr{E}_i^c \cap \mathscr{E}_j^c). \tag{1.3}$$

In particular, if $\Pr(\mathscr{E}_i^c) = \Pr(\mathscr{E}_1^c)$ and $\Pr(\mathscr{E}_i^c \cap \mathscr{E}_j^c) = \Pr(\mathscr{E}_1^c \cap \mathscr{E}_2^c)$ for $1 \leqq i \neq j \leqq k$, then (1.3) reduces to

$$\Pr\left(\bigcap_{i=1}^{k} \mathscr{E}_i\right) \geqq 1 - k\Pr(\mathscr{E}_1^c) + (k-1)\Pr(\mathscr{E}_1^c \cap \mathscr{E}_2^c). \tag{1.4}$$

Hunter (1976) improved upon the Kounias inequality; the same result was obtained independently by Worsley (1982). This result makes use of two concepts from graph theory that are defined here for convenience: A *tree* is a connected graph without circuits (closed paths) and a *spanning tree* is a tree in which each vertex (node) is connected to at least some other vertex (i.e., there are no isolated vertices). The following theorem summarizes the Hunter–Worsley inequality.

Theorem 1.1 (Hunter 1976, Worsley 1982). Consider a graph G with events $\mathscr{E}_1^c, \mathscr{E}_2^c, \ldots, \mathscr{E}_k^c$ as vertices with $\mathscr{E}_i^c$ and $\mathscr{E}_j^c$ joined by an edge e_{ij} if and only if $\mathscr{E}_i^c \cap \mathscr{E}_j^c \neq \phi$. Then for any spanning tree T of G,

$$\Pr\left(\bigcap_{i=1}^{k} \mathscr{E}_i\right) \geqq 1 - \sum_{i=1}^{k} \Pr(\mathscr{E}_i^c) + \sum_{(i,j):e_{ij}\in T} \Pr(\mathscr{E}_i^c \cap \mathscr{E}_j^c) . \tag{1.5}$$

□

In the class of all bounds (1.5), the sharpest bound is obtained by finding the spanning tree T^* for which the term

$$\sum_{(i,j):e_{ij}\in T} \Pr(\mathscr{E}_i^c \cap \mathscr{E}_j^c) \tag{1.6}$$

is maximum. The minimal spanning tree algorithm of Kruskal (1956) can be used to solve this problem of finding the maximum of (1.6) over the set of all spanning trees. The Kounias inequality (1.3) uses the maximum over only a subset of all spanning trees and hence is never sharper than (1.5) with $T = T^*$.

Stoline and Mitchell (1981) have given a computer algorithm for evaluating Hunter's bound when $\mathscr{E}_i$ is the event $\{T_i \leqq \xi\}$ where $T_1, T_2, \ldots, T_k$ have a joint multivariate t-distribution with a given associated correlation matrix $\{\rho_{ij}\}$ and degrees of freedom (d.f.) ν. In this case, Stoline (1983) has shown that Hunter's inequality becomes considerably sharper than the first order Bonferroni inequality when all $|\rho_{ij}| > 0.5$.

Hunter noted that if some proxies are available for the magnitudes of the probabilities $\Pr(\mathscr{E}_i^c \cap \mathscr{E}_j^c)$, then for maximizing (1.6) it is not necessary to evaluate the actual probabilities; the proxies can be used instead. An example of this occurs when the event $\mathscr{E}_i^c = (X_i > \xi)$ $(1 \leqq i \leqq k)$ where $X_1, X_2, \ldots, X_k$ are jointly distributed standard normal r.v.'s with known correlations ρ_{ij} $(1 \leqq i \neq j \leqq k)$. In this case the bivariate distribution of any pair (X_i, X_j) depends on i and j only through ρ_{ij}, and by Slepian's (1962) result (see Theorem 2.1 below) it follows that $\Pr(X_i > \xi, X_j > \xi)$ is increasing in ρ_{ij} for any ξ. Thus the ρ_{ij}'s can be used as proxies. The same conclusion holds for the events $\mathscr{E}_i^c = (|X_i| > \xi)$ $(1 \leqq i \leqq k)$

and/or if the X_i's are divided by a common independently distributed $\sqrt{\chi_\nu^2/\nu}$ r.v. (i.e., if the r.v.'s $Y_i = X_i/\sqrt{\chi_\nu^2/\nu}$ have a multivariate t-distribution with ν d.f. and associated correlation matrix $\{\rho_{ij}\}$).

Some other Bonferroni-type inequalities are discussed in Tong (1980, Section 7.1).

2 MULTIPLICATIVE INEQUALITIES

The basic aim here is to obtain multiplicative inequalities of the type

$$\Pr\left\{\bigcap_{i=1}^{k} \mathscr{E}_i\right\} \geqq \prod_{i=1}^{k} \Pr\{\mathscr{E}_i\}. \tag{2.1}$$

Kimball's (1951) inequality, which is stated in a more general form in Theorem 2.6 below, is one of the first such inequalities. Note that the bound in (2.1) is at least as sharp as the Bonferroni bound (1.2). However, while the Bonferroni bound is always valid, (2.1) is valid only under certain conditions. We study some dependence models for which multiplicative inequalities of the type (2.1) can be established.

2.1 Inequalities for Multivariate Normal and Multivariate *t*-Distributions

Let $\mathbf{X} = (X_1, X_2, \ldots, X_k)'$ have a k-variate normal distribution with zero mean vector, unit variances, and correlation matrix $\mathbf{R} = \{\rho_{ij}\}$ (denoted by $\mathbf{X} \sim N(\mathbf{0}, \mathbf{R})$). The following two probabilities are frequently encountered in the study of MCPs:

$$\Pr\{X_1 \leqq \xi_1, X_2 \leqq \xi_2, \ldots, X_k \leqq \xi_k\}, \tag{2.2}$$

and

$$\Pr\{|X_1| \leqq \xi_1, |X_2| \leqq \xi_2, \ldots, |X_k| \leqq \xi_k\} \tag{2.3}$$

where $\xi_1, \xi_2, \ldots, \xi_k$ are some constants (positive in the case of (2.3)). Often, $\xi_1 = \xi_2 = \cdots = \xi_k = \xi$ (say) where ξ is the upper α point of the distribution of $\max X_i$ (respectively, $\max|X_i|$), which is evaluated by setting (2.2) (respectively, (2.3)) equal to $1 - \alpha$.

Numerical evaluation of (2.2) and (2.3) involves a k-dimensional quadrature. Appendix 3 explains how this can be reduced to a one-dimensional quadrature when the ρ_{ij}'s have a so-called *product structure*:

$$\rho_{ij} = \lambda_i \lambda_j \qquad (1 \leqq i \neq j \leqq k) \tag{2.4}$$

where each $\lambda_i \in (-1, 1)$. However, for ρ_{ij}'s not satisfying (2.4) such a reduction is not possible and, as a result, computational costs can be prohibitively high even for moderate k. Thus easily computable bounds on (2.2) and (2.3) are required. Note that (2.4) holds when all the ρ_{ij}'s are equal.

By utilizing special properties of the multivariate normal distribution it is possible to find multiplicative lower bounds on (2.2) or (2.3) that are functions of the univariate normal distribution, and that improve upon the first order Bonferroni bound. Slepian (1962) showed the first general result along these lines, which is stated in the following theorem.

Theorem 2.1 (Slepian 1962). For $\mathbf{X} \sim N(\mathbf{0}, \mathbf{R})$, the "one-sided" probability (2.2) is a strictly increasing function of each ρ_{ij}. □

A corollary to this theorem is that a lower bound on (2.2) is obtained by replacing all the ρ_{ij}'s by $\rho_{\min} = \min \rho_{ij}$ if $\rho_{\min} > -1/(k-1)$. As noted above, since this lower bound involves an equicorrelated multivariate normal distribution, it can be evaluated by using a one-dimensional quadrature. If all the $\rho_{ij} \geqq 0$, then we obtain the well-known *Slepian inequality*:

$$\Pr\left\{\bigcap_{i=1}^{k} (X_i \leqq \xi_i)\right\} \geqq \prod_{i=1}^{k} \Pr(X_i \leqq \xi_i) . \tag{2.5}$$

It is natural to expect that monotonicity results analogous to Theorem 2.1 would hold for the "two-sided" probability (2.3). However, only partial analogs are valid here. Dunn (1958) proved the analog of (2.5), that is,

$$\Pr\left\{\bigcap_{i=1}^{k} (|X_i| \leqq \xi_i)\right\} \geqq \prod_{i=1}^{k} \Pr(|X_i| \leqq \xi_i) \tag{2.6}$$

if $k \leqq 3$ or if $k > 3$ and the correlation matrix has the product structure (2.4). (See also Halperin 1967.) Šidák (1967) later extended this result to arbitrary positive definite correlation matrices for $k > 3$. Inequality (2.6) is therefore known as the *Dunn–Šidák inequality*.

A generalization of (2.6) was given by Khatri (1967). For $\mathbf{X}$ distributed as $N(\mathbf{0}, \mathbf{R})$ where $\mathbf{R} = \{\rho_{ij}\}$ is a positive semidefinite correlation matrix, consider a partition of $\mathbf{X}$ and $\mathbf{R}$ as follows:

$$\mathbf{X} = \begin{bmatrix} \mathbf{X}^{(1)} \\ \mathbf{X}^{(2)} \end{bmatrix}, \qquad \mathbf{R} = \begin{bmatrix} \mathbf{R}_{11} & \mathbf{R}_{12} \\ \mathbf{R}'_{12} & \mathbf{R}_{22} \end{bmatrix};$$

here $\mathbf{X}^{(1)} : m \times 1$, $\mathbf{X}^{(2)} : (k-m) \times 1$, $\mathbf{R}_{11} : m \times m$, $\mathbf{R}_{12} : m \times (k-m)$, and

$\mathbf{R}_{22}:(k-m)\times(k-m)$. Then $\mathbf{X}^{(1)}\sim N(\mathbf{0},\mathbf{R}_{11})$ and $\mathbf{X}^{(2)}\sim N(\mathbf{0},\mathbf{R}_{22})$. For this setup Khatri proved the following result.

Theorem 2.2 (Khatri 1967). Let $A_1 \in \mathbb{R}^m$ and $A_2 \in \mathbb{R}^{k-m}$ be two convex regions symmetric about the origin. If $\mathbf{R}_{12}$ is of rank zero or one, then

$$\Pr\left\{\bigcap_{i=1}^{2}(\mathbf{X}^{(i)}\in A_i)\right\} \geqq \prod_{i=1}^{2}\Pr(\mathbf{X}^{(i)}\in A_i)\,. \tag{2.7}$$

□

A repeated application of (2.7) where at each stage one of the $\mathbf{X}^{(i)}$'s is a scalar X_i and the corresponding set $A_i=\{x:|x|\leqq \xi_i\}$ yields (2.6). If $\mathbf{R}$ has the product structure (2.4), then the following stronger result holds.

Theorem 2.3 (Khatri 1967). For $\mathbf{X}\sim N(\mathbf{0},\mathbf{R})$ consider a partition of $\mathbf{X}$ into r component vectors $\mathbf{X}^{(i)}$ $(1\leqq i\leqq r)$ where $\mathbf{X}^{(i)}: k_i\times 1$, $\Sigma_{i=1}^{r} k_i = k$. Let $A_i \subseteq \mathbb{R}^{k_i}$ be a convex region symmetric about the origin $(1\leqq i\leqq r)$. Then under (2.4), the inequalities

$$\Pr\left\{\bigcap_{i=1}^{r}(\mathbf{X}^{(i)}\in A_i)\right\} \geqq \Pr\left\{\bigcap_{i\in C}(\mathbf{X}^{(i)}\in A_i)\right\}\Pr\left\{\bigcap_{i\notin C}(\mathbf{X}^{(i)}\in A_i)\right\}$$
$$\geqq \prod_{i=1}^{r}\Pr(\mathbf{X}^{(i)}\in A_i) \tag{2.8}$$

and

$$\Pr\left\{\bigcap_{i=1}^{r}(\mathbf{X}^{(i)}\not\in A_i)\right\} \geqq \Pr\left\{\bigcap_{i\in C}(\mathbf{X}^{(i)}\not\in A_i)\right\}\Pr\left\{\bigcap_{i\notin C}(\mathbf{X}^{(i)}\not\in A_i)\right\}$$
$$\geqq \prod_{i=1}^{r}\Pr(\mathbf{X}^{(i)}\not\in A_i) \tag{2.9}$$

hold for every subset C of $\{1,2,\ldots,r\}$. The inequalities are strict if the A_i's are bounded sets with positive probabilities and the λ_i's in (2.4) are nonzero. □

Corollary. From (2.9) it follows that under (2.4),

$$\Pr\left\{\bigcap_{i=1}^{k}(|X_i|>\xi_i)\right\} \geqq \prod_{i=1}^{k}\Pr(|X_i|>\xi_i)\,. \tag{2.10}$$

□

Khatri (1970) attempted to extend Theorem 2.3 to cases where $\mathbf{R}$ does not possess the product structure (2.4). His proof depended on an argument of Scott (1967) that was shown to be in error by Das Gupta et al. (1972) and Šidák (1975). In fact, Šidák gave a counterexample to show that if $\mathbf{R}$ is arbitrary, then (2.9) is false while the validity of (2.8) is an open question. We refer to this open question as the *Khatri Conjecture*.

Šidák (1968) studied the monotonicity of the two-sided probability (2.3) in each ρ_{ij}. He showed by counterexamples that a complete two-sided analog of Slepian's result stated in Theorem 2.1 is false. However, he was able to obtain the following partial analog, a much simpler proof of which was later provided by Jogdeo (1970).

Theorem 2.4 (Šidák 1968). Let $\mathbf{X} \sim N(\mathbf{0}, \mathbf{R}(\lambda))$ where $\mathbf{R}(\lambda) = \{\rho_{ij}(\lambda)\}$ is a positive semidefinite correlation matrix that for a fixed correlation matrix $\mathbf{R} = \{\rho_{ij}\}$ depends on $\lambda \in [0, 1]$ in the following way: $\rho_{1j}(\lambda) = \rho_{j1}(\lambda) = \lambda\rho_{1j}$ for $j \geqq 2$ and $\rho_{ij}(\lambda) = \rho_{ij}$ for $i, j \geqq 2$. Then the two-sided probability (2.3) is a nondecreasing function of λ. If $\mathbf{R}$ is positive definite, $\rho_{1j} \neq 0$ for some $j \geqq 2$ and if all the ξ_i's are positive, then (2.3) is a strictly increasing function of λ. □

Corollary. Let $\boldsymbol{\lambda} = (\lambda_1, \lambda_2, \ldots, \lambda_k)'$. If $\mathbf{X} \sim N(\mathbf{0}, \mathbf{R}(\boldsymbol{\lambda}))$ where the correlation matrix $\mathbf{R}(\boldsymbol{\lambda}) = \{\lambda_i \lambda_j \rho_{ij}\}$ for some fixed correlation matrix $\{\rho_{ij}\}$ and $\lambda_i \in [0, 1]$ for $i = 1, 2, \ldots, k$, then the monotonicity result of the theorem holds for each λ_i. □

Šidák (1968) also showed that for the equicorrelated case, $\rho_{ij} = \rho$ for all $i \neq j$, (2.3) is locally strictly increasing in ρ. For this same case, Tong (1970) derived the following result using a moment inequality for nonnegative random variables.

Theorem 2.5 (Tong 1970). Let $\mathbf{X} \sim N(\mathbf{0}, \mathbf{R})$ where the correlation matrix $\mathbf{R}$ has all off-diagonal elements equal to ρ (say) where $\rho \geqq 0$. Define for $1 \leqq r \leqq k$,

$$P_1(r) = \Pr\left\{\bigcap_{i=1}^{r} (X_i \leqq \xi)\right\},$$

$$P_2(r) = \Pr\left\{\bigcap_{i=1}^{r} (|X_i| \leqq \xi)\right\},$$

and

$$P_3(r) = \Pr\left\{\bigcap_{i=1}^{r} (|X_i| > \xi)\right\}.$$

Then for $i = 1, 2, 3$,

$$P_i(k) \geqq \{P_i(r)\}^{k/r} \geqq \{P_i(1)\}^k \quad \text{for } 1 \leqq r \leqq k . \tag{2.11}$$

The inequalities are strict if $\rho > 0$. □

The inequalities (2.5), (2.6), (2.10), and (2.11) extend to the multivariate t-distribution by using a simple conditioning argument. Thus let $Y_i = X_i/U$ $(1 \leqq i \leqq k)$ where U is distributed as a $\sqrt{\chi_\nu^2/\nu}$ r.v. independent of the X_i's. Then $(Y_1, Y_2, \ldots, Y_k)'$ has a k-variate t-distribution with d.f. ν and associated correlation matrix $\mathbf{R}$. By conditioning on U, the abovementioned inequalities can be shown to hold with the X_i's replaced by the Y_i's. Theorem 2.5 was later generalized by Šidák (1973), and his result was further generalized by Tong (1980).

2.2 Inequalities via Association

The inequalities discussed in the preceding section are known to hold only in the case of multivariate normal and multivariate t-distributions. A question naturally arises as to whether one can establish similar multiplicative inequalities in more general settings. As noted before, a multiplicative lower bound involving marginal probabilities is of particular interest since it is sharper than the first order Bonferroni lower bound. A key idea behind the inequalities of the preceding section is that the variables with greater correlation are more likely to "hang together." The extent of dependence between two jointly distributed normal r.v.'s is completely measured by their correlation coefficient. However, for arbitrarily distributed r.v.'s this is not true and more general concepts of dependence are required. Lehmann (1966) introduced several concepts of bivariate dependence. Esary, Proschan, and Walkup (1967) introduced a very general concept of multivariate dependence called *association*. We define this concept below.

Definition 2.1. Let $\mathbf{X} = (X_1, X_2, \ldots, X_k)'$. The r.v.'s $X_1, X_2, \ldots, X_k$ are said to be *associated* if

$$\text{cov}\{g_1(\mathbf{X}), g_2(\mathbf{X})\} \geqq 0$$

for any two functions $g_1, g_2 : \mathbb{R}^k \to \mathbb{R}^1$ such that g_1 and g_2 are nondecreasing in each argument and $E\{g_i(\mathbf{X})\}$ exists for $i = 1, 2$. □

Esary, Proschan, and Walkup (1967) have given a number of results that can be used to easily verify whether a given set of r.v.'s is associated. For example, a single r.v. is associated, a set of independent r.v.'s is associated, and nondecreasing (or nonincreasing) functions of associated r.v.'s are associated.

If $X_1, X_2, \ldots, X_k$ are associated r.v.'s, then it can be shown that the multiplicative inequalities (2.5) and (2.6) hold. Another inequality that can be derived using the properties of associated r.v.'s is the following generalization of the *Kimball inequality*.

Theorem 2.6 (Kimball 1951). Let $g_1, g_2, \ldots, g_r : \mathbb{R}^k \to [0, \infty]$, and suppose that each g_i is monotone (in the same direction) in each argument. Let $\mathbf{X} = (X_1, X_2, \ldots, X_k)'$ be a vector of associated r.v.'s. Then

$$E\left\{\prod_{i=1}^{r} g_i(\mathbf{X})\right\} \geqq E\left\{\prod_{i \in C} g_i(\mathbf{X})\right\} E\left\{\prod_{i \notin C} g_i(\mathbf{X})\right\} \geqq \prod_{i=1}^{r} E\{g_i(\mathbf{X})\} \quad (2.12)$$

for every subset C of $\{1, 2, \ldots, r\}$. □

Kimball's inequality is a special case of (2.12) for $k = 1$. Inequality (2.12) is useful when dealing with probabilities associated with r.v.'s of the type $Z_i = \psi_i(Y_i, \mathbf{X})$ $(1 \leqq i \leqq r)$ where conditioned on $\mathbf{X}$, the Y_i's are independent.

3 INEQUALITIES VIA MAJORIZATION

Majorization is increasingly used as a powerful technique for obtaining probability inequalities. Marshall and Olkin (1979) is a comprehensive reference on this topic. Tong (1980) has given a nice review of the topic in his Chapter 6.

Definition 3.1. Let $\mathbf{a} = (a_1, a_2, \ldots, a_k)'$ and $\mathbf{b} = (b_1, b_2, \ldots, b_k)'$ be two real vectors and let $a_{[1]} \geqq a_{[2]} \geqq \cdots \geqq a_{[k]}$ and $b_{[1]} \geqq b_{[2]} \geqq \cdots \geqq b_{[k]}$ be the corresponding ordered values. Vector $\mathbf{a}$ is said to *majorize* vector $\mathbf{b}$ (denoted by $\mathbf{a} > \mathbf{b}$) if

$$\sum_{i=1}^{k} a_i = \sum_{i=1}^{k} b_i$$

and (3.1)

$$\sum_{i=1}^{j} a_{[i]} \geqq \sum_{i=1}^{j} b_{[i]} \qquad \text{for } 1 \leqq j \leqq k-1 .$$ □

Definition 3.2. A function $g: \mathbb{R}^k \to \mathbb{R}^1$ is called *Schur-convex* (respectively, *Schur-concave*) if $\mathbf{a} > \mathbf{b}$ implies that $g(\mathbf{a}) \geqq g(\mathbf{b})$ (respectively, $g(\mathbf{a}) \leqq g(\mathbf{b})$). Such functions are permutation-symmetric. □

Theorem 3.1 (Marshall and Olkin 1974). Suppose that $\mathbf{X} = (X_1, X_2, \ldots, X_k)'$ has a Schur-concave joint density function $f(\mathbf{x})$. Let $A \subseteq \mathbb{R}^k$ be a Lebesgue-measurable set such that

$$\mathbf{a} \in A \quad \text{and} \quad \mathbf{a} > \mathbf{b} \Rightarrow \mathbf{b} \in A . \tag{3.2}$$

Then

$$\Pr\{(\mathbf{X} - \boldsymbol{\theta}) \in A\} = \int_{A+\boldsymbol{\theta}} f(x)\,dx \tag{3.3}$$

is a Schur-concave function of $\boldsymbol{\theta}$. Here the set $A + \boldsymbol{\theta} = \{\mathbf{b} : \mathbf{b} = \mathbf{a} + \boldsymbol{\theta}, \mathbf{a} \in A\}$. □

As a consequence of this theorem we obtain that under the stated conditions

$$\boldsymbol{\theta} > \boldsymbol{\phi} \Rightarrow \Pr\{(\mathbf{X} - \boldsymbol{\phi}) \in A\} \geqq \Pr\{(\mathbf{X} - \boldsymbol{\theta}) \in A\} . \tag{3.4}$$

In particular, if $\boldsymbol{\phi} = (\bar{\theta}, \bar{\theta}, \ldots, \bar{\theta})'$ where $\bar{\theta} = \Sigma_{i=1}^k \theta_i / k$, then (3.4) holds since $\boldsymbol{\theta} > (\bar{\theta}, \bar{\theta}, \ldots, \bar{\theta})'$.

This result is useful in studying the supremums (with respect to underlying parameter values) of error probabilities of some MCPs. It can also be used to derive bounds on multivariate distribution functions. Suppose that $\mathbf{X} = (X_1, X_2, \ldots, X_k)'$ has a Schur-concave density function; then $\Pr\{\cap_{i=1}^k (X_i \leqq \xi_i)\}$ and $\Pr\{\cap_{i=1}^k (X_i > \xi_i)\}$ are Schur-concave in $\boldsymbol{\xi} = (\xi_1, \xi_2, \ldots, \xi_k)'$. Therefore,

$$\Pr\left\{\bigcap_{i=1}^k (X_i \leqq \bar{\xi})\right\} \geqq \Pr\left\{\bigcap_{i=1}^k (X_i \leqq \xi_i)\right\} \tag{3.5}$$

and a similar inequality holds for $\Pr\{\cap_{i=1}^k (X_i > \xi_i)\}$; here $\bar{\xi} = \Sigma_{i=1}^k \xi_i / k$. Marshall and Olkin (1974) attempted to apply their result to derive multiplicative inequalities of type (2.5) for r.v.'s X_i that are conditionally independent and identically distributed. However, Jogdeo (1977) pointed out a flaw in their argument and gave correct conditions under which the desired result holds.

Special cases of Theorem 3.1 were earlier proved by Anderson (1955) and Mudholkar (1966). Marshall and Olkin's (1974) result can be viewed as a generalization of Mudholkar's (1966) result in two ways: The latter assumes that $f(\mathbf{x})$ is unimodal (i.e., the level set $\{\mathbf{x}: f(\mathbf{x}) \geqq c\}$ is convex for every c), which implies Schur-concavity assumed in Theorem 3.1. Secondly, the latter assumes that set A is permutation invariant and convex, which implies condition (3.2). Mudholkar's (1966) result, in turn, generalizes the following integral inequality of Anderson (1955).

Theorem 3.2 (Anderson 1955). Let $f(\mathbf{x}): \mathbb{R}^k \rightarrow [0, \infty)$ be symmetric about the origin and unimodal. Let $A \subseteq \mathbb{R}^k$ be symmetric about the origin and convex. If $\int_A f(\mathbf{x})\, d\mathbf{x} < \infty$, then

$$\int_A f(\mathbf{x} + \lambda \mathbf{y})\, d\mathbf{x} \geqq \int_A f(\mathbf{x} + \mathbf{y})\, d\mathbf{x} \tag{3.6}$$

for all $\mathbf{y}$ and $\lambda \in [0, 1]$. □

APPENDIX 3

Some Probability Distributions and Tables Useful in Multiple Comparisons

In Section 1 of this appendix we discuss some sampling distributions (other than the elementary ones such as the normal, t, χ^2, and F) that commonly arise in multiple comparison problems. In Section 2 we discuss the tables of critical points of these distributions, which are given at the end of this appendix. In addition to these tables we have also included a table of critical points of Student's t-distribution for very small tail probabilities whose applications are indicated in the sequel, and a table of critical points for the Waller–Duncan procedure discussed Chapter 11, Section 3.4.

1 DISTRIBUTIONS

1.1 Multivariate Normal Distribution

A random vector $\mathbf{Z} = (Z_1, Z_2, \ldots, Z_k)'$ is said to have a k-variate normal distribution with mean vector $\boldsymbol{\mu}$ and covariance matrix $\boldsymbol{\Sigma}$ if its characteristic function $\psi_{\mathbf{Z}}(\mathbf{u}) = E(e^{i\mathbf{u}'\mathbf{Z}})$ is given by

$$\psi_{\mathbf{Z}}(\mathbf{u}) = \exp(i\mathbf{u}'\boldsymbol{\mu} - \tfrac{1}{2}\mathbf{u}'\boldsymbol{\Sigma}\mathbf{u})$$

where $i = \sqrt{-1}$. If $\boldsymbol{\Sigma}$ is positive definite, then the density function of $\mathbf{Z}$

exists and is given by

$$f(\mathbf{z}) = (2\pi)^{-k/2}(\det \mathbf{\Sigma})^{-1/2} \exp\{-\tfrac{1}{2}(\mathbf{z}-\boldsymbol{\mu})'\mathbf{\Sigma}^{-1}(\mathbf{z}-\boldsymbol{\mu})\} ;$$

in this case we say that $\mathbf{Z}$ has a nonsingular distribution.

If the Z_i's are standardized so that $E(Z_i) = 0$ and $\text{var}(Z_i) = 1$, then $\mathbf{\Sigma}$ is a correlation matrix; we denote it by $\mathbf{R}$ with off-diagonal elements $\rho_{ij} = \text{corr}(Z_i, Z_j)$ for $i \neq j$. Distributions of $\max Z_i$ and $\max|Z_i|$ arise frequently in multiple comparison problems. In general, the probability integrals of these distributions are of dimension k. However, if $\mathbf{R}$ satisfies the product structure condition (2.4) of Appendix 2, then using the representation

$$Z_i = \sqrt{1-\lambda_i^2}\, Y_i - \lambda_i Y_0 \qquad (1 \leqq i \leqq k)$$

where $Y_0, Y_1, \ldots, Y_k$ are independent and identically distributed (i.i.d.) $N(0, 1)$ random variables (r.v.'s), these k-variate integrals can be expressed as univariate integrals as follows:

$$\Pr\{\max Z_i \leqq z\} = \int_{-\infty}^{\infty} \prod_{i=1}^{k} \Phi\left[\frac{\lambda_i y + z}{(1-\lambda_i^2)^{1/2}}\right] d\Phi(y) \qquad (1.1\text{a})$$

and

$$\Pr\{\max|Z_i| \leqq z\} = \int_{-\infty}^{\infty} \prod_{i=1}^{k} \left\{\Phi\left[\frac{\lambda_i y + z}{(1-\lambda_i^2)^{1/2}}\right] - \Phi\left[\frac{\lambda_i y - z}{(1-\lambda_i^2)^{1/2}}\right]\right\} d\Phi(y) \qquad (1.1\text{b})$$

where $\Phi(\cdot)$ is the standard normal cumulative distribution function (c.d.f.). The case of equicorrelated Z_i's, that is, $\rho_{ij} = \rho \geqq 0$ for all $i \neq j$, is a special case of the product structure with $\lambda_i = \sqrt{\rho}$ for all i.

For arbitrary $\mathbf{R}$ we denote the upper α points of the distributions of $\max Z_i$ and $\max|Z_i|$ by $Z_{k,\mathbf{R}}^{(\alpha)}$ and $|Z|_{k,\mathbf{R}}^{(\alpha)}$, respectively; in the equicorrelated case we denote them by $Z_{k,\rho}^{(\alpha)}$ and $|Z|_{k,\rho}^{(\alpha)}$, respectively. The critical point $Z_{k,\rho}^{(\alpha)}$ (respectively, $|Z|_{k,\rho}^{(\alpha)}$) is the solution in z to the equation obtained by setting (1.1a) (respectively, (1.1b)) equal to $1-\alpha$ with $\lambda_i = \sqrt{\rho}$ for all i.

1.2 Multivariate t-Distribution

Let the Z_i's have a k-variate normal distribution with zero means, unit variances, and $\text{corr}(Z_i, Z_j) = \rho_{ij}$ for $i \neq j$. Let U be a χ_ν^2 r.v. that is distributed independently of the Z_i's and let $T_i = Z_i/\sqrt{U/\nu}$ $(1 \leq i \leq k)$. Then the joint distribution of $T_1, T_2, \ldots, T_k$ is called a k-variate t-distribution with ν degrees of freedom (d.f.) and associated correlation matrix $\mathbf{R} = \{\rho_{ij}\}$. If $\mathbf{R}$ is positive definite, then the density function of $\mathbf{T} = (T_1, T_2, \ldots, T_k)'$ exists and is given by

$$f(\mathbf{t}) = \frac{\Gamma\{(k+\nu)/2\}(\det \mathbf{R})^{-1/2}}{(\nu\pi)^{k/2}\Gamma(\nu/2)}\left[1 + \frac{1}{\nu}\mathbf{t}'\mathbf{R}^{-1}\mathbf{t}\right]^{-(k+\nu)/2}$$

This distribution was derived independently by Dunnett and Sobel (1954) and Cornish (1954). The joint distribution of $T_1^2, T_2^2, \ldots, T_k^2$ is a special case of the multivariate F-distribution (with common numerator d.f. = 1) introduced by Krishnaiah and Armitage (1970). Gupta (1963a) is a comprehensive reference on multivariate normal and multivariate t-distributions.

For arbitrary $\mathbf{R}$ we denote the upper α points of the distribution of $\max T_i$ and $\max|T_i|$ by $T_{k,\nu,\mathbf{R}}^{(\alpha)}$ and $|T|_{k,\nu,\mathbf{R}}^{(\alpha)}$, respectively; in the equicorrelated case we denote them by $T_{k,\nu,\rho}^{(\alpha)}$ and $|T|_{k,\nu,\rho}^{(\alpha)}$, respectively. By conditioning on $\sqrt{U/\nu} = x$ and using (1.1), the probability integrals of $\max T_i$ and $\max|T_i|$ in the equicorrelated case can be written as follows:

$$\Pr\{\max T_i \leq t\} = \int_0^{\infty}\left[\int_{-\infty}^{\infty}\Phi^k\left[\frac{\sqrt{\rho}y + tx}{\sqrt{1-\rho}}\right]d\Phi(y)\right]dF_\nu(x)\,, \quad (1.2a)$$

$$\Pr\{\max|T_i| \leq t\} = \int_0^{\infty}\left[\int_{-\infty}^{\infty}\left\{\Phi^k\left[\frac{\sqrt{\rho}y + tx}{\sqrt{1-\rho}}\right] - \Phi^k\left[\frac{\sqrt{\rho}y - tx}{\sqrt{1-\rho}}\right]\right\}d\Phi(y)\right]dF_\nu(x) \quad (1.2b)$$

where $F_\nu(\cdot)$ is the c.d.f. of a $\sqrt{\chi_\nu^2/\nu}$ r.v. The critical point $T_{k,\nu,\rho}^{(\alpha)}$ (respectively, $|T|_{k,\nu,\rho}^{(\alpha)}$) is the solution in t to the equation obtained by setting (1.2a) (respectively, (1.2b)) equal to $1-\alpha$.

1.3 Studentized Maximum and Maximum Modulus Distributions

The distribution of $\max_{1\leq i\leq k} T_i$ when the T_i's have a joint equicorrelated k-variate t-distribution with ν d.f. and common correlation $\rho = 0$ is referred to as the Studentized maximum distribution with parameter k and d.f. ν. The corresponding r.v. is denoted by $M_{k,\nu}$ and its upper α point by $M_{k,\nu}^{(\alpha)}$, which, of course, equals $T_{k,\nu,0}^{(\alpha)}$.

The distribution of $\max_{1\leq i\leq k}|T_i|$ in this case is known as the Studentized maximum modulus distribution with parameter k and d.f. ν. The corresponding r.v. is denoted by $|M|_{k,\nu}$ and its upper α point by $|M|_{k,\nu}^{(\alpha)}$, which, of course, equals $|T|_{k,\nu,0}^{(\alpha)}$.

1.4 Studentized Range Distribution

Let $Z_1, Z_2, \ldots, Z_k$ be i.i.d. $N(0, 1)$ r.v.'s and let U be an independently distributed χ_ν^2 r.v. Then the r.v.

$$Q_{k,\nu} = \frac{\max_{1 \leqq i < j \leqq k} |Z_i - Z_j|}{\sqrt{U/\nu}}$$

is said to have the Studentized range distribution with parameter k and d.f. ν. When $\nu = \infty$, the corresponding distribution is known as the range (of k i.i.d. standard normals) distribution. The probability integral of $Q_{k,\nu}$ can be written in a closed form as follows:

$$\begin{aligned}\Pr\{Q_{k,\nu} \leqq q\} &= \Pr\left\{|Z_i - Z_j| \leqq q\sqrt{\frac{U}{\nu}} \quad (1 \leqq i < j \leqq k)\right\} \\ &= k\int_0^\infty \left[\int_{-\infty}^{\infty} \{\Phi(y) - \Phi(y - qx)\}^{k-1}\, d\Phi(y)\right] dF_\nu(x)\,. \end{aligned} \tag{1.3}$$

The upper α point of $Q_{k,\nu}$ is denoted by $Q_{k,\nu}^{(\alpha)}$ and is given by the solution in q to the equation obtained by setting (1.3) equal to $1 - \alpha$.

In general, if $X_1, X_2, \ldots, X_k$ are independent r.v.'s with X_i having a c.d.f. $F_i(x)$ $(1 \leqq i \leqq k)$, then the distribution of $R_k = \max_{i \neq j}|X_i - X_j|$ can be written as

$$\Pr\{R_k \leqq r\} = \sum_{i=1}^{k} \int_{-\infty}^{\infty} \prod_{\substack{j=1 \\ j \neq i}}^{k} \{F_j(x) - F_j(x - r)\}\, dF_i(x)\,. \tag{1.4}$$

If the X_i's have a common c.d.f. $F(x)$, then (1.4) simplifies to the expression

$$\Pr\{R_k \leqq r\} = k\int_{-\infty}^{\infty} \{F(x) - F(x - r)\}^{k-1}\, dF(x)\,. \tag{1.5}$$

The range of k i.i.d. Student t r.v.'s arises in connection with some procedures for multiple comparison of means of normal distributions with unequal variances (see Chapter 7). For this case the upper α points of the range are given by Wilcox (1983).

1.5 Studentized Augmented Range Distribution

Consider the r.v.

$$Q'_{k,\nu} = \max(Q_{k,\nu}, |M|_{k,\nu})$$
$$= \max\left\{ \frac{\max_{1 \leq i < j \leq k} |Z_i - Z_j|}{\sqrt{U/\nu}}, \frac{\max_{1 \leq i \leq k} |Z_i|}{\sqrt{U/\nu}} \right\}$$
$$= \max_{0 \leq i < j \leq k} \frac{|Z_i - Z_j|}{\sqrt{U/\nu}}$$

where $Z_1, Z_2, \ldots, Z_k$ are i.i.d. $N(0, 1)$ r.v.'s and $Z_0 \equiv 1$. The probability integral of $Q'_{k,\nu}$ can be written as

$$\Pr\{Q'_{k,\nu} \leq q\} = \int_0^\infty [\{\Phi(qx) - \tfrac{1}{2}\}^k + k \int_0^{qx} \{\Phi(y) - \Phi(y - qx)\}^{k-1} \, d\Phi(y)] \, dF_\nu(x) . \tag{1.6}$$

The upper α point of the distribution of $Q'_{k,\nu}$ (denoted by $Q'^{(\alpha)}_{k,\nu}$) is the solution in q to the equation obtained by setting (1.6) equal to $1 - \alpha$.

A procedure based on the Studentized augmented range distribution for the family of all linear combinations of means in an unbalanced one-way layout is discussed in Section 3.1.3 of Chapter 3. However, this procedure is too conservative for the family of pairwise comparisons and hence is not recommended. Also Tukey (1953, Chapter 3) has noted that for $k \geq 3$ and $\alpha \leq 0.05$, $Q'^{(\alpha)}_{k,\nu}$ is well approximated by $Q^{(\alpha)}_{k,\nu}$. For these reasons we have not provided a table of $Q'^{(\alpha)}_{k,\nu}$-values in this book. The interested reader may refer to the tables computed by Stoline (1978).

2 TABLES

2.1 Details of the Tables

The entries in all of the tables are given to three significant places (which in most cases implies two decimal place accuracy) and are rounded to the nearest digit in the last significant place. The original sources from which these tables have been adapted give the corresponding critical points to additional significant places. (Some of the tables were computed specially for the present book as noted below.)

Table 1 lists the values of $T_\nu^{(\alpha)}$, the upper α critical point of Student's t-distribution with ν d.f., for $\alpha = 0.0001(0.0001)0.001(0.001)0.01$ and $\nu = 1(1)30(5)50, 60, 120, \infty$. Computations of these critical points were done on New York University's computer using the IMSL routine MDSTI, which gives the inverse Student's t-distribution.

The main purpose of Table 1 is to provide values of $T_\nu^{(\alpha)}$ for very small values of α, which are needed for applying many of the procedures described in this book. For example, for the Bonferroni and Dunn–Šidák procedures for pairwise comparisons of means, given in Section 3.2.3 of Chapter 3, we need the critical points $T_\nu^{(\alpha/2k^*)}$ and $T_\nu^{(\{1-(1-\alpha)^{1/k^*}\}/2)}$, respectively, where $k^* = \binom{k}{2}$. Tables of $T_\nu^{(\alpha/2k^*)}$ have been given by Bailey (1977) and of $T_\nu^{(\{1-(1-\alpha)^{1/k^*}\}/2)}$ by Games (1977). But these tables cannot be directly used in other cases where the number of comparisons is not of the form $\binom{k}{2}$ for some integer $k \geqq 2$. Moses (1978) has given charts for finding $T_\nu^{(\alpha)}$ for α ranging between 0.01 and 0.00001 that can be used in general applications, as can the present tables.

Tables 2 and 3 give the values of the critical points $Z_{k,\rho}^{(\alpha)}$ and $|Z|_{k,\rho}^{(\alpha)}$, respectively, for $\alpha = 0.01, 0.05, 0.10, 0.25$, $k = 2(1)10(2)20(4)40, 50$, and $\rho = 0.1(0.2)0.7$. Table 2 is adapted from Gupta, Nagel, and Panchapakesan (1973), while Table 3 is adapted from Odeh (1982). Previously Milton (1963) and Krishnaiah and Armitage (1965) have given tables for $Z_{k,\rho}^{(\alpha)}$ and $(|Z|_{k,\rho}^{(\alpha/2)})^2$, respectively.

Tables 4 and 5 give the values of the critical points $T_{k,\nu,\rho}^{(\alpha)}$ and $|T|_{k,\nu,\rho}^{(\alpha)}$, respectively, for $\alpha = 0.01, 0.05, 0.10, 0.20$, $k = 2(1)10(2)20$, $\nu = 2(1)10, 12(4)24, 30, 40, 60, 120, \infty$ and $\rho = 0.1(0.2)0.7$. Both of these tables were computed specially for the present book By Professor C. W. Dunnett. Using the same programs, Bechhofer and Dunnett (1986) have prepared very extensive sets of tables of $T_{k,\nu,\rho}^{(\alpha)}$ and $|T|_{k,\nu,\rho}^{(\alpha)}$. Recently Gupta, Panchapakesan, and Sohn (1985) have also published detailed tables of $T_{k,\nu,\rho}^{(\alpha)}$. Another source for tables of $T_{k,\nu,\rho}^{(\alpha)}$ is the paper by Krishnaiah and Armitage (1966). Tables of $|T|_{k,\nu,\rho}^{(\alpha)}$ can be found in Hahn and Hendrickson (1971), Dunn and Massey (1965), and Krishnaiah and Armitage (1970) (the last authors tabulated $(|T|_{k,\nu,\rho}^{(\alpha/2)})^2$).

Tables 6 and 7 give the values of the critical points $M_{k,\nu}^{(\alpha)}$ and $|M|_{k,\nu}^{(\alpha)}$, respectively, for $\alpha = 0.01, 0.05, 0.10, 0.20$, $k = 2(1)16(2)20$, and $\nu = 2(1)30(5)50, 60(20)120, 200, \infty$. Both of these tables were also computed specially for the present book by Professor C. W. Dunnett. Previously a short table of $M_{k,\nu}^{(\alpha)}$ was published by Pillai and Ramachandran (1954). Detailed tables of $|M|_{k,\nu}^{(\alpha)}$ have been published by Stoline and Ury (1979), Stoline et al. (1980), and Bechhofer and Dunnett (1982).

Table 8 gives the values of the critical points $Q_{k,\nu}^{(\alpha)}$ for $\alpha = 0.01, 0.05,$

0.10, 0.20, $k = 3(1)16(4)40$, and $\nu = 1(1)20, 24, 30, 40, 60, 120, \infty$. This table was adapted from Harter (1960, 1969). Lund and Lund (1983) have given a computer program for calculating the Studentized range integral (1.3) and the critical points $Q_{k,\nu}^{(\alpha)}$. Barnard (1978) has given a computer program for the special case $\nu = \infty$.

Finally Table 9 gives the critical points required to implement the Waller–Duncan procedure. These critical points are taken from the 1972 corrigendum to the Waller–Duncan (1969) article.

2.2 Rules for Interpolation

The following rules are recommended for interpolating in Tables 1–8.

(i) Interpolation with respect to the error d.f. ν should be done linearly in $1/\nu$.

(ii) Interpolation with respect to the upper tail probability α should be done linearly in $\log_e \alpha$.

(iii) Interpolation with respect to k in Tables 2–8 should be done linearly in $\log_e k$.

(iv) Interpolation with respect to ρ in Tables 2–5 should be done linearly in $1/(1-\rho)$.

To find the critical point $t^* = t^*(K, F, q, f)$ for F-values not listed in Table 9, Waller and Duncan (1969) recommended linear interpolation with respect to $b = [F/(F-1)]^{1/2}$ for $f \geqq 20$ and with respect to $a = 1/F^{1/2}$ for $f < 20$. For $f = 4$ the values of t^* are not given in the table but are given by the following rules: For $K = 50$, $t^* = 2.28$ for all q and F such that $F > 5.20/q$; for $K = 100$, $t^* = 2.83$ for all q and F such that $F > 8.12/q$; and for $K = 500$, $t^* = 4.52$ for all q and F such that $F > 20.43/q$. If an asterisk is shown in the table instead of a t^*-value, it means that for that combination of (K, F, q, f) all pairwise differences are not significant.

TABLE 1. Upper α Point of Student's t-Distribution with ν Degrees of Freedom ($T_\nu^{(\alpha)}$).

ν \ α	.01	.009	.008	.007	.006	.005	.004	.003	.002	.001	.0009	.0008	.0007	.0006	.0005	.0004	.0003	.0002	.0001
1	31.8	35.4	39.8	45.5	53.0	63.7	79.6	106	159	318	354	398	455	531	637	796	1062	1592	3183
2	6.96	7.35	7.81	8.36	9.05	9.92	11.1	12.9	15.8	22.3	23.5	25.0	26.7	28.8	31.6	35.3	40.8	50.0	70.7
3	4.54	4.72	4.93	5.18	5.47	5.84	6.32	6.99	8.05	10.2	10.6	11.0	11.5	12.2	12.9	13.9	15.4	17.6	22.2
4	3.75	3.87	4.01	4.17	4.37	4.60	4.91	5.32	5.95	7.17	7.38	7.61	7.88	8.21	8.61	9.13	9.83	10.9	13.0
5	3.36	3.46	3.57	3.70	3.85	4.03	4.26	4.57	5.03	5.89	6.03	6.19	6.38	6.60	6.87	7.21	7.67	8.36	9.68
6	3.14	3.23	3.32	3.43	3.55	3.71	3.90	4.15	4.52	5.21	5.32	5.44	5.59	5.75	5.96	6.22	6.56	7.07	8.02
7	3.00	3.07	3.16	3.25	3.37	3.50	3.67	3.89	4.21	4.79	4.88	4.98	5.10	5.24	5.41	5.62	5.90	6.31	7.06
8	2.90	2.97	3.04	3.13	3.23	3.36	3.51	3.70	3.99	4.50	4.58	4.67	4.77	4.90	5.04	5.22	5.46	5.81	6.44
9	2.82	2.89	2.96	3.04	3.14	3.25	3.39	3.57	3.83	4.30	4.37	4.45	4.54	4.65	4.78	4.94	5.15	5.46	6.01
10	2.76	2.83	2.89	2.97	3.06	3.17	3.30	3.47	3.72	4.14	4.21	4.28	4.37	4.47	4.59	4.73	4.93	5.20	5.69
11	2.72	2.78	2.84	2.92	3.00	3.11	3.23	3.39	3.62	4.02	4.09	4.16	4.24	4.33	4.44	4.57	4.75	5.00	5.45
12	2.68	2.74	2.80	2.87	2.96	3.05	3.17	3.33	3.55	3.93	3.99	4.05	4.13	4.21	4.32	4.44	4.61	4.85	5.26
13	2.65	2.71	2.77	2.84	2.92	3.01	3.13	3.28	3.49	3.85	3.91	3.97	4.04	4.12	4.22	4.34	4.50	4.72	5.11
14	2.62	2.68	2.74	2.81	2.88	2.98	3.09	3.23	3.44	3.79	3.84	3.90	3.97	4.05	4.14	4.26	4.40	4.62	4.99
15	2.60	2.66	2.71	2.78	2.86	2.95	3.06	3.20	3.39	3.73	3.78	3.84	3.91	3.98	4.07	4.18	4.33	4.53	4.88
16	2.58	2.64	2.69	2.76	2.83	2.92	3.03	3.17	3.36	3.69	3.74	3.79	3.86	3.93	4.01	4.12	4.26	4.45	4.79
17	2.57	2.62	2.67	2.74	2.81	2.90	3.00	3.14	3.33	3.65	3.69	3.75	3.81	3.88	3.97	4.07	4.20	4.39	4.71
18	2.55	2.60	2.66	2.72	2.79	2.88	2.98	3.11	3.30	3.61	3.66	3.71	3.77	3.84	3.92	4.02	4.15	4.33	4.65
19	2.54	2.59	2.64	2.71	2.78	2.86	2.96	3.09	3.27	3.58	3.63	3.68	3.74	3.80	3.88	3.98	4.11	4.28	4.59
20	2.53	2.58	2.63	2.69	2.76	2.85	2.95	3.07	3.25	3.55	3.60	3.65	3.71	3.77	3.85	3.95	4.07	4.24	4.54
21	2.52	2.57	2.62	2.68	2.75	2.83	2.93	3.06	3.23	3.53	3.57	3.62	3.68	3.74	3.82	3.91	4.03	4.20	4.49
22	2.51	2.56	2.61	2.67	2.74	2.82	2.92	3.04	3.21	3.50	3.55	3.60	3.65	3.72	3.79	3.88	4.00	4.17	4.45
23	2.50	2.55	2.60	2.66	2.73	2.81	2.90	3.03	3.20	3.48	3.53	3.58	3.63	3.69	3.77	3.86	3.97	4.14	4.42
24	2.49	2.54	2.59	2.65	2.72	2.80	2.89	3.01	3.18	3.47	3.51	3.56	3.61	3.67	3.75	3.83	3.95	4.11	4.38
25	2.49	2.53	2.58	2.64	2.71	2.79	2.88	3.00	3.17	3.45	3.49	3.54	3.59	3.65	3.73	3.81	3.93	4.08	4.35
26	2.48	2.53	2.58	2.63	2.70	2.78	2.87	2.99	3.16	3.43	3.48	3.52	3.58	3.64	3.71	3.79	3.90	4.06	4.32
27	2.47	2.52	2.57	2.63	2.69	2.77	2.86	2.98	3.15	3.42	3.46	3.51	3.56	3.62	3.69	3.78	3.88	4.04	4.30
28	2.47	2.51	2.56	2.62	2.69	2.76	2.86	2.97	3.14	3.41	3.45	3.49	3.55	3.60	3.67	3.76	3.87	4.02	4.28
29	2.46	2.51	2.56	2.62	2.68	2.76	2.85	2.96	3.13	3.40	3.44	3.48	3.53	3.59	3.66	3.74	3.85	4.00	4.25
30	2.46	2.50	2.55	2.61	2.67	2.75	2.84	2.96	3.12	3.39	3.43	3.47	3.52	3.58	3.65	3.73	3.83	3.98	4.23
35	2.44	2.48	2.53	2.59	2.65	2.72	2.81	2.93	3.08	3.34	3.38	3.42	3.47	3.53	3.59	3.67	3.77	3.91	4.15
40	2.42	2.47	2.52	2.57	2.63	2.70	2.79	2.90	3.05	3.31	3.34	3.39	3.43	3.49	3.55	3.63	3.73	3.86	4.09
45	2.41	2.46	2.50	2.56	2.62	2.69	2.78	2.88	3.03	3.28	3.32	3.36	3.41	3.46	3.52	3.60	3.69	3.83	4.05
50	2.40	2.45	2.49	2.55	2.61	2.68	2.76	2.87	3.02	3.26	3.30	3.34	3.38	3.44	3.50	3.57	3.66	3.79	4.01
60	2.39	2.43	2.48	2.53	2.59	2.66	2.74	2.85	2.99	3.23	3.27	3.31	3.35	3.40	3.46	3.53	3.62	3.75	3.96
120	2.36	2.40	2.44	2.49	2.55	2.62	2.70	2.80	2.93	3.16	3.19	3.23	3.27	3.32	3.37	3.44	3.53	3.64	3.84
∞	2.33	2.37	2.41	2.46	2.51	2.58	2.65	2.75	2.88	3.09	3.12	3.16	3.19	3.24	3.29	3.35	3.43	3.54	3.72

Source: Computed using the IMSL inverse Student's t-distribution program MDSTI.

TABLE 2. One-Sided Upper α Equicoordinate Point of the k-Variate Equicorrelated Standard Normal Distribution with Common Correlation Coefficient ρ ($Z_{k,\rho}^{(\alpha)}$).

	ρ = 0.1					ρ = 0.3					ρ = 0.5					ρ = 0.7			
k \ α	0.25	0.10	0.05	0.01	k \ α	0.25	0.10	0.05	0.01	k \ α	0.25	0.10	0.05	0.01	k \ α	0.25	0.10	0.05	0.01
2	1.09	1.63	1.95	2.57	2	1.06	1.61	1.94	2.57	2	1.01	1.58	1.92	2.56	2	0.95	1.53	1.88	2.53
3	1.31	1.81	2.12	2.71	3	1.28	1.78	2.10	2.70	3	1.19	1.73	2.06	2.68	3	1.09	1.66	2.00	2.64
4	1.46	1.93	2.23	2.80	4	1.39	1.90	2.20	2.80	4	1.31	1.84	2.16	2.77	4	1.18	1.75	2.08	2.72
5	1.56	2.02	2.31	2.88	5	1.49	1.98	2.28	2.86	5	1.39	1.92	2.23	2.84	5	1.25	1.81	2.14	2.78
6	1.65	2.10	2.38	2.93	6	1.57	2.05	2.35	2.92	6	1.46	1.98	2.29	2.89	6	1.30	1.86	2.19	2.82
7	1.72	2.16	2.43	2.98	7	1.63	2.11	2.40	2.97	7	1.51	2.03	2.34	2.93	7	1.35	1.90	2.23	2.86
8	1.78	2.21	2.48	3.02	8	1.69	2.16	2.45	3.01	8	1.56	2.07	2.38	2.97	8	1.39	1.94	2.27	2.89
9	1.83	2.25	2.52	3.06	9	1.73	2.20	2.49	3.04	9	1.60	2.11	2.42	3.00	9	1.42	1.97	2.30	2.92
10	1.87	2.29	2.56	3.09	10	1.77	2.24	2.52	3.07	10	1.64	2.14	2.45	3.03	10	1.45	1.99	2.32	2.94
12	1.95	2.36	2.62	3.14	12	1.84	2.30	2.58	3.12	12	1.70	2.20	2.50	3.08	12	1.49	2.04	2.37	2.98
14	2.01	2.41	2.67	3.18	14	1.90	2.35	2.63	3.17	14	1.75	2.24	2.55	3.12	14	1.53	2.08	2.40	3.02
16	2.06	2.46	2.72	3.22	16	1.95	2.40	2.67	3.20	16	1.79	2.28	2.58	3.15	16	1.56	2.11	2.43	3.05
18	2.11	2.50	2.75	3.26	18	1.99	2.43	2.71	3.24	18	1.82	2.32	2.62	3.18	18	1.59	2.13	2.46	3.07
20	2.15	2.54	2.79	3.29	20	2.03	2.47	2.74	3.27	20	1.85	2.35	2.64	3.21	20	1.62	2.16	2.48	3.09
24	2.22	2.60	2.85	3.34	24	2.09	2.53	2.79	3.32	24	1.91	2.40	2.69	3.26	24	1.66	2.20	2.52	3.13
28	2.27	2.65	2.89	3.38	28	2.14	2.57	2.84	3.36	28	1.95	2.44	2.73	3.29	28	1.70	2.23	2.56	3.16
32	2.32	2.70	2.94	3.42	32	2.18	2.61	2.88	3.39	32	1.99	2.48	2.77	3.33	32	1.72	2.26	2.58	3.19
36	2.36	2.73	2.97	3.45	36	2.22	2.65	2.91	3.42	36	2.02	2.51	2.80	3.36	36	1.75	2.29	2.61	3.21
40	2.40	2.77	3.00	3.48	40	2.26	2.68	2.94	3.45	40	2.05	2.53	2.83	3.38	40	1.77	2.31	2.63	3.23
50	2.48	2.84	3.07	3.54	50	2.33	2.75	3.01	3.51	50	2.11	2.59	2.88	3.43	50	1.82	2.35	2.67	3.28

Source: Adapted from S. S. Gupta, K. Nagel, and S. Panchapakesan (1973), "On the order statistics from equally correlated normal random variables," *Biometrika*, **60**, 403–413. Reproduced with the kind permission of Biometrika trustees.

TABLE 3. Two-Sided Upper α Equicoordinate Point of the k-Variate Equicorrelated Standard Normal Distribution with Common Correlation Coefficient ρ ($|Z|^{(\alpha)}_{k,\rho}$).

ρ = 0.1					ρ = 0.3					ρ = 0.5					ρ = 0.7				
k \ α	0.25	0.10	0.05	0.01	k \ α	0.25	0.10	0.05	0.01	k \ α	0.25	0.10	0.05	0.01	k \ α	0.25	0.10	0.05	0.01
2	1.50	1.95	2.24	2.81	2	1.48	1.94	2.23	2.80	2	1.45	1.92	2.21	2.79	2	1.40	1.88	2.18	2.77
3	1.68	2.11	2.39	2.93	3	1.66	2.10	2.38	2.93	3	1.61	2.06	2.35	2.92	3	1.54	2.00	2.30	2.88
4	1.81	2.22	2.49	3.02	4	1.78	2.20	2.48	3.02	4	1.72	2.16	2.44	3.00	4	1.62	2.08	2.38	2.95
5	1.91	2.31	2.57	3.09	5	1.87	2.28	2.55	3.08	5	1.80	2.23	2.51	3.06	5	1.69	2.14	2.44	3.01
6	1.98	2.37	2.63	3.14	6	1.94	2.35	2.61	3.13	6	1.86	2.29	2.57	3.11	6	1.74	2.19	2.48	3.05
7	2.05	2.43	2.68	3.19	7	2.00	2.40	2.66	3.18	7	1.92	2.34	2.61	3.15	7	1.78	2.23	2.52	3.09
8	2.10	2.48	2.72	3.23	8	2.05	2.45	2.70	3.22	8	1.96	2.38	2.65	3.19	8	1.82	2.27	2.55	3.12
9	2.14	2.52	2.76	3.26	9	2.09	2.49	2.74	3.25	9	2.00	2.42	2.69	3.22	9	1.85	2.30	2.58	3.15
10	2.18	2.56	2.80	3.29	10	2.13	2.52	2.77	3.28	10	2.03	2.45	2.72	3.25	10	1.87	2.32	2.61	3.17
12	2.25	2.62	2.85	3.34	12	2.20	2.58	2.83	3.33	12	2.09	2.50	2.77	3.29	12	1.92	2.37	2.65	3.21
14	2.31	2.67	2.90	3.38	14	2.25	2.63	2.88	3.37	14	2.14	2.55	2.81	3.33	14	1.96	2.40	2.69	3.24
16	2.36	2.71	2.94	3.42	16	2.29	2.67	2.92	3.41	16	2.17	2.58	2.85	3.37	16	1.99	2.43	2.72	3.27
18	2.40	2.75	2.98	3.45	18	2.33	2.71	2.95	3.44	18	2.21	2.62	2.88	3.40	18	2.02	2.46	2.74	3.30
20	2.44	2.79	3.01	3.48	20	2.37	2.74	2.98	3.47	20	2.24	2.64	2.91	3.42	20	2.04	2.48	2.76	3.32
24	2.50	2.84	3.07	3.53	24	2.43	2.79	3.03	3.51	24	2.29	2.69	2.95	3.47	24	2.08	2.52	2.80	3.35
28	2.56	2.89	3.11	3.57	28	2.48	2.84	3.08	3.55	28	2.33	2.73	2.99	3.50	28	2.12	2.56	2.84	3.39
32	2.60	2.93	3.15	3.60	32	2.52	2.88	3.11	3.59	32	2.37	2.77	3.03	3.53	32	2.15	2.58	2.86	3.41
36	2.64	2.97	3.19	3.63	36	2.55	2.91	3.15	3.62	36	2.40	2.80	3.06	3.56	36	2.17	2.61	2.89	3.44
40	2.68	3.00	3.22	3.66	40	2.59	2.94	3.18	3.64	40	2.43	2.83	3.08	3.59	40	2.19	2.63	2.91	3.46
50	2.75	3.07	3.28	3.72	50	2.65	3.01	3.24	3.70	50	2.49	2.88	3.14	3.64	50	2.24	2.67	2.95	3.50

Source: Adapted from R. E. Odeh (1982), "Tables of percentage points of the distribution of maximum absolute value of equally correlated normal random variables," *Commun. Statist., Ser. B*, **11**, 65–87. Reproduced with the kind permission of the publisher.

TABLE 4. One-Sided Upper α Equicoordinate Point of the k-Variate Equicorrelated t-Distribution with Common Correlation Coefficient ρ and Degrees of Freedom ν ($T^{(\alpha)}_{k,\nu,\rho}$)

$\rho = 0.1$

ν	α \ k	2	3	4	5	6	7	8	9	10	12	14	16	18	20
2	.01	9.36	10.9	12.1	13.0	13.0	14.4	15.0	15.4	15.9	16.6	17.2	17.7	18.2	18.6
	.05	4.03	4.75	5.28	5.70	6.04	6.32	6.57	6.79	6.98	7.31	7.58	7.82	8.02	8.20
	.10	2.71	3.23	3.61	3.91	4.15	4.36	4.53	4.69	4.82	5.06	5.25	5.42	5.56	5.69
	.20	1.71	2.10	2.38	2.59	2.77	2.92	3.04	3.15	3.25	3.42	3.55	3.67	3.77	3.86
3	.01	5.68	6.40	6.93	7.35	7.69	7.98	8.23	8.46	8.65	9.00	9.28	9.53	9.75	9.94
	.05	3.07	3.51	3.83	4.08	4.29	4.46	4.61	4.74	4.86	5.06	5.23	5.37	5.50	5.61
	.10	2.24	2.61	2.87	3.08	3.24	3.38	3.50	3.61	3.70	3.87	4.00	4.12	4.22	4.30
	.20	1.52	1.84	2.06	2.23	2.36	2.48	2.57	2.66	2.74	2.86	2.97	3.06	3.14	3.21
4	.01	4.53	5.01	5.36	5.64	5.87	6.06	6.23	6.38	6.51	6.74	6.93	7.10	7.25	7.38
	.05	2.71	3.05	3.30	3.49	3.65	3.79	3.90	4.00	4.09	4.25	4.38	4.49	4.59	4.68
	.10	2.06	2.36	2.58	2.75	2.89	3.00	3.10	3.19	3.27	3.40	3.51	3.61	3.69	3.76
	.20	1.44	1.72	1.92	2.07	2.19	2.29	2.38	2.45	2.52	2.63	2.73	2.81	2.88	2.94
5	.01	3.99	4.36	4.64	4.85	5.03	5.18	5.31	5.43	5.53	5.71	5.86	5.99	6.10	6.20
	.05	2.52	2.82	3.03	3.20	3.33	3.45	3.54	3.63	3.71	3.84	3.95	4.05	4.13	4.21
	.10	1.96	2.23	2.43	2.58	2.70	2.80	2.89	2.97	3.03	3.15	3.25	3.34	3.41	3.47
	.20	1.40	1.66	1.84	1.98	2.09	2.19	2.27	2.34	2.40	2.50	2.59	2.67	2.73	2.79
6	.01	3.68	4.00	4.23	4.41	4.56	4.68	4.79	4.89	4.98	5.13	5.25	5.36	5.46	5.54
	.05	2.41	2.68	2.87	3.02	3.14	3.24	3.33	3.41	3.47	3.59	3.69	3.78	3.85	3.92
	.10	1.89	2.15	2.33	2.47	2.58	2.68	2.76	2.83	2.89	3.00	3.09	3.17	3.24	3.30
	.20	1.37	1.62	1.80	1.93	2.03	2.12	2.20	2.26	2.32	2.42	2.50	2.58	2.64	2.69
7	.01	3.48	3.76	3.97	4.13	4.26	4.37	4.46	4.55	4.62	4.76	4.87	4.96	5.05	5.12
	.05	2.33	2.58	2.76	2.90	3.01	3.10	3.19	3.26	3.32	3.43	3.52	3.60	3.67	3.73
	.10	1.85	2.10	2.27	2.40	2.50	2.59	2.67	2.74	2.80	2.90	2.98	3.06	3.12	3.18
	.20	1.35	1.59	1.76	1.89	1.99	2.08	2.15	2.21	2.27	2.37	2.44	2.51	2.57	2.62
8	.01	3.34	3.60	3.78	3.93	4.05	4.15	4.24	4.31	4.38	4.50	4.60	4.69	4.76	4.83
	.05	2.28	2.52	2.68	2.81	2.92	3.01	3.08	3.15	3.21	3.31	3.40	3.47	3.54	3.60
	.10	1.82	2.06	2.22	2.35	2.45	2.53	2.61	2.67	2.73	2.82	2.90	2.97	3.03	3.09
	.20	1.33	1.57	1.74	1.86	1.96	2.05	2.12	2.18	2.23	2.32	2.40	2.47	2.52	2.58
9	.01	3.23	3.48	3.65	3.79	3.90	3.99	4.07	4.14	4.20	4.31	4.41	4.49	4.56	4.62
	.05	2.24	2.47	2.63	2.75	2.85	2.93	3.01	3.07	3.13	3.23	3.31	3.38	3.44	3.49
	.10	1.80	2.03	2.19	2.31	2.40	2.49	2.56	2.62	2.67	2.77	2.84	2.91	2.97	3.02
	.20	1.32	1.56	1.72	1.84	1.94	2.02	2.09	2.15	2.20	2.29	2.37	2.43	2.49	2.54
10	.01	3.16	3.39	3.55	3.68	3.78	3.87	3.94	4.01	4.07	4.17	4.26	4.33	4.40	4.46
	.05	2.20	2.43	2.58	2.70	2.80	2.88	2.95	3.01	3.06	3.16	3.24	3.30	3.36	3.42
	.10	1.78	2.00	2.16	2.28	2.37	2.45	2.52	2.58	2.63	2.72	2.80	2.86	2.92	2.97
	.20	1.31	1.55	1.71	1.83	1.92	2.00	2.07	2.13	2.18	2.27	2.34	2.40	2.46	2.51

TABLE 4. Continued.

$\rho = 0.1$

ν	α \ k	2	3	4	5	6	7	8	9	10	12	14	16	18	20
12	.01	3.04	3.26	3.40	3.52	3.61	3.69	3.76	3.82	3.88	3.97	4.05	4.12	4.18	4.23
	.05	2.16	2.37	2.52	2.63	2.72	2.80	2.86	2.92	2.97	3.06	3.13	3.20	3.25	3.30
	.10	1.75	1.97	2.12	2.23	2.32	2.40	2.46	2.52	2.57	2.66	2.73	2.79	2.85	2.89
	.20	1.30	1.53	1.69	1.80	1.89	1.97	2.04	2.09	2.15	2.23	2.30	2.36	2.42	2.46
16	.01	2.91	3.10	3.24	3.34	3.42	3.49	3.55	3.61	3.65	3.74	3.80	3.86	3.92	3.96
	.05	2.10	2.30	2.44	2.54	2.63	2.70	2.76	2.81	2.86	2.94	3.01	3.07	3.12	3.16
	.10	1.72	1.93	2.07	2.18	2.26	2.33	2.40	2.45	2.50	2.58	2.65	2.71	2.76	2.80
	.20	1.28	1.51	1.66	1.77	1.86	1.94	2.00	2.05	2.10	2.19	2.26	2.31	2.36	2.41
20	.01	2.84	3.02	3.14	3.24	3.31	3.38	3.44	3.48	3.53	3.60	3.67	3.72	3.77	3.81
	.05	2.07	2.26	2.39	2.49	2.58	2.64	2.70	2.75	2.80	2.87	2.94	2.99	3.04	3.08
	.10	1.70	1.90	2.04	2.14	2.23	2.30	2.36	2.41	2.45	2.53	2.60	2.65	2.70	2.75
	.20	1.28	1.50	1.64	1.75	1.84	1.92	1.98	2.03	2.08	2.16	2.23	2.28	2.33	2.38
24	.01	2.79	2.96	3.08	3.17	3.25	3.31	3.36	3.41	3.45	3.52	3.58	3.63	3.68	3.72
	.05	2.05	2.24	2.37	2.46	2.54	2.61	2.66	2.71	2.75	2.83	2.89	2.94	2.99	3.03
	.10	1.69	1.89	2.02	2.12	2.21	2.27	2.33	2.38	2.43	2.50	2.57	2.62	2.67	2.71
	.20	1.27	1.49	1.63	1.74	1.83	1.90	1.96	2.02	2.06	2.14	2.21	2.26	2.31	2.36
30	.01	2.75	2.91	3.02	3.11	3.18	3.24	3.29	3.33	3.37	3.44	3.50	3.54	3.59	3.63
	.05	2.03	2.21	2.34	2.43	2.51	2.57	2.62	2.67	2.71	2.79	2.85	2.90	2.94	2.98
	.10	1.67	1.87	2.00	2.10	2.18	2.25	2.31	2.36	2.40	2.47	2.54	2.59	2.64	2.68
	.20	1.26	1.48	1.62	1.73	1.82	1.89	1.95	2.00	2.05	2.13	2.19	2.25	2.29	2.34
40	.01	2.70	2.86	2.96	3.05	3.11	3.17	3.22	3.26	3.30	3.36	3.41	3.46	3.50	3.54
	.05	2.01	2.19	2.31	2.40	2.47	2.54	2.59	2.63	2.67	2.74	2.80	2.85	2.89	2.93
	.10	1.66	1.85	1.98	2.08	2.16	2.23	2.28	2.33	2.37	2.44	2.51	2.56	2.60	2.64
	.20	1.26	1.47	1.61	1.72	1.81	1.88	1.93	1.99	2.03	2.11	2.17	2.23	2.27	2.31
60	.01	2.66	2.81	2.91	2.99	3.05	3.10	3.15	3.19	3.22	3.28	3.33	3.38	3.42	3.45
	.05	1.99	2.16	2.28	2.37	2.44	2.50	2.55	2.60	2.63	2.70	2.76	2.80	2.85	2.88
	.10	1.65	1.84	1.97	2.06	2.14	2.20	2.26	2.30	2.35	2.42	2.47	2.52	2.57	2.61
	.20	1.25	1.46	1.60	1.71	1.79	1.86	1.92	1.97	2.02	2.09	2.15	2.21	2.25	2.29
120	.01	2.62	2.76	2.86	2.93	2.99	3.04	3.08	3.12	3.15	3.21	3.26	3.30	3.34	3.37
	.05	1.97	2.14	2.25	2.34	2.41	2.47	2.52	2.56	2.60	2.66	2.71	2.76	2.80	2.84
	.10	1.64	1.82	1.95	2.04	2.12	2.18	2.23	2.28	2.32	2.39	2.44	2.49	2.54	2.57
	.20	1.25	1.45	1.59	1.70	1.78	1.85	1.91	1.96	2.00	2.07	2.14	2.19	2.23	2.27
∞	.01	2.57	2.71	2.80	2.87	2.93	2.98	3.02	3.06	3.09	3.14	3.18	3.22	3.26	3.29
	.05	1.95	2.12	2.23	2.31	2.38	2.43	2.48	2.52	2.56	2.62	2.67	2.72	2.75	2.79
	.10	1.63	1.81	1.93	2.02	2.10	2.16	2.21	2.25	2.29	2.36	2.41	2.46	2.50	2.54
	.20	1.24	1.45	1.59	1.69	1.77	1.84	1.89	1.94	1.98	2.06	2.12	2.17	2.21	2.25

$\rho = 0.3$

ν	α \ k	2	3	4	5	6	7	8	9	10	12	14	16	18	20
2	.01	9.15	10.5	11.5	12.3	13.0	13.5	14.0	14.4	14.7	15.4	15.9	16.3	16.7	17.1
	.05	3.93	4.56	5.02	5.37	5.66	5.91	6.12	6.30	6.46	6.74	6.97	7.17	7.35	7.50
	.10	2.63	3.09	3.42	3.68	3.88	4.06	4.21	4.34	4.45	4.65	4.81	4.95	5.08	5.19
	.20	1.64	1.99	2.23	2.41	2.56	2.69	2.80	2.89	2.97	3.11	3.23	3.33	3.42	3.49
3	.01	5.60	6.24	6.71	7.08	7.38	7.63	7.84	8.04	8.21	8.50	8.74	8.95	9.14	9.30
	.05	3.01	3.41	3.69	3.91	4.09	4.24	4.37	4.48	4.58	4.75	4.90	5.02	5.13	5.22
	.10	2.20	2.52	2.75	2.93	3.08	3.20	3.30	3.39	3.47	3.61	3.73	3.82	3.91	3.99
	.20	1.47	1.75	1.95	2.09	2.21	2.31	2.40	2.47	2.53	2.64	2.74	2.82	2.88	2.94
4	.01	4.48	4.92	5.24	5.48	5.69	5.86	6.00	6.13	6.25	6.45	6.61	6.75	6.88	6.99
	.05	2.67	2.98	3.20	3.38	3.51	3.63	3.73	3.82	3.90	4.03	4.15	4.24	4.33	4.40
	.10	2.02	2.30	2.49	2.64	2.76	2.86	2.95	3.02	3.09	3.20	3.30	3.38	3.45	3.52
	.20	1.40	1.65	1.83	1.96	2.06	2.15	2.23	2.29	2.35	2.45	2.53	2.60	2.66	2.71
5	.01	3.95	4.30	4.55	4.75	4.91	5.04	5.15	5.26	5.35	5.50	5.63	5.75	5.84	5.93
	.05	2.49	2.76	2.95	3.10	3.22	3.32	3.41	3.49	3.55	3.67	3.77	3.85	3.92	3.98
	.10	1.92	2.17	2.35	2.48	2.59	2.68	2.76	2.83	2.89	2.99	3.07	3.15	3.21	3.27
	.20	1.36	1.60	1.76	1.88	1.98	2.06	2.13	2.19	2.25	2.34	2.41	2.48	2.54	2.59
6	.01	3.65	3.95	4.16	4.33	4.46	4.58	4.67	4.76	4.84	4.97	5.08	5.18	5.26	5.33
	.05	2.38	2.63	2.80	2.94	3.05	3.14	3.22	3.28	3.34	3.45	3.54	3.61	3.68	3.73
	.10	1.86	2.10	2.26	2.39	2.49	2.57	2.64	2.71	2.76	2.86	2.93	3.00	3.06	3.11
	.20	1.33	1.56	1.72	1.83	1.93	2.01	2.07	2.13	2.18	2.27	2.34	2.40	2.46	2.50
7	.01	3.45	3.72	3.91	4.06	4.18	4.28	4.37	4.44	4.51	4.63	4.73	4.81	4.89	4.95
	.05	2.31	2.54	2.70	2.83	2.93	3.01	3.09	3.15	3.21	3.30	3.38	3.45	3.51	3.57
	.10	1.82	2.05	2.20	2.32	2.42	2.50	2.56	2.62	2.68	2.77	2.84	2.90	2.96	3.01
	.20	1.31	1.54	1.69	1.80	1.89	1.97	2.03	2.09	2.14	2.22	2.29	2.35	2.40	2.45
8	.01	3.32	3.56	3.74	3.87	3.98	4.07	4.15	4.22	4.28	4.39	4.48	4.56	4.63	4.69
	.05	2.25	2.48	2.63	2.75	2.85	2.93	2.99	3.05	3.11	3.20	3.27	3.34	3.40	3.45
	.10	1.79	2.01	2.16	2.27	2.37	2.44	2.51	2.56	2.61	2.70	2.77	2.83	2.89	2.93
	.20	1.30	1.52	1.67	1.78	1.87	1.94	2.00	2.06	2.11	2.19	2.26	2.31	2.36	2.41
9	.01	3.22	3.45	3.61	3.74	3.84	3.92	4.00	4.06	4.12	4.22	4.30	4.37	4.44	4.49
	.05	2.21	2.43	2.58	2.69	2.78	2.86	2.93	2.98	3.03	3.12	3.19	3.26	3.31	3.36
	.10	1.77	1.98	2.13	2.24	2.33	2.40	2.46	2.52	2.57	2.65	2.72	2.78	2.83	2.88
	.20	1.29	1.50	1.65	1.76	1.85	1.92	1.98	2.03	2.08	2.16	2.23	2.28	2.33	2.38
10	.01	3.14	3.36	3.51	3.63	3.73	3.81	3.88	3.94	3.99	4.09	4.17	4.23	4.29	4.35
	.05	2.18	2.39	2.54	2.65	2.73	2.81	2.87	2.93	2.98	3.06	3.13	3.19	3.24	3.29
	.10	1.75	1.96	2.10	2.21	2.30	2.37	2.43	2.48	2.53	2.61	2.68	2.74	2.79	2.83
	.20	1.28	1.49	1.64	1.75	1.83	1.90	1.96	2.02	2.06	2.14	2.21	2.26	2.31	2.35
12	.01	3.03	3.23	3.37	3.48	3.57	3.65	3.71	3.77	3.82	3.90	3.97	4.03	4.09	4.14
	.05	2.14	2.34	2.48	2.58	2.66	2.73	2.79	2.85	2.89	2.97	3.04	3.10	3.15	3.19
	.10	1.73	1.93	2.07	2.17	2.25	2.32	2.38	2.43	2.48	2.56	2.62	2.68	2.73	2.77
	.20	1.27	1.48	1.62	1.72	1.81	1.88	1.94	1.99	2.03	2.11	2.17	2.23	2.27	2.32

TABLE 4. Continued.

$\rho = 0.3$

ν	α \ k	2	3	4	5	6	7	8	9	10	12	14	16	18	20
16	.01	2.90	3.09	3.21	3.31	3.39	3.46	3.51	3.56	3.61	3.68	3.75	3.80	3.85	3.89
	.05	2.09	2.27	2.40	2.50	2.58	2.64	2.70	2.75	2.79	2.87	2.93	2.98	3.03	3.07
	.10	1.69	1.89	2.02	2.12	2.20	2.27	2.32	2.37	2.42	2.49	2.55	2.60	2.65	2.69
	.20	1.25	1.46	1.60	1.70	1.78	1.85	1.91	1.95	2.00	2.07	2.13	2.19	2.23	2.27
20	.01	2.83	3.00	3.12	3.21	3.29	3.35	3.40	3.45	3.49	3.56	3.62	3.67	3.72	3.76
	.05	2.05	2.24	2.36	2.45	2.53	2.59	2.65	2.69	2.74	2.81	2.87	2.92	2.96	3.00
	.10	1.68	1.87	2.00	2.09	2.17	2.23	2.29	2.34	2.38	2.45	2.51	2.56	2.60	2.64
	.20	1.24	1.45	1.58	1.68	1.76	1.83	1.89	1.94	1.98	2.05	2.11	2.16	2.21	2.25
24	.01	2.78	2.95	3.06	3.15	3.22	3.28	3.33	3.37	3.41	3.48	3.54	3.59	3.63	3.67
	.05	2.03	2.21	2.33	2.43	2.50	2.56	2.61	2.66	2.70	2.77	2.82	2.87	2.92	2.95
	.10	1.66	1.85	1.98	2.07	2.15	2.21	2.27	2.31	2.35	2.42	2.48	2.53	2.57	2.61
	.20	1.24	1.44	1.57	1.67	1.75	1.82	1.87	1.92	1.96	2.04	2.10	2.15	2.19	2.23
30	.01	2.74	2.90	3.01	3.09	3.16	3.21	3.26	3.30	3.34	3.41	3.46	3.51	3.55	3.58
	.05	2.01	2.19	2.31	2.40	2.47	2.53	2.58	2.62	2.66	2.73	2.78	2.83	2.87	2.91
	.10	1.65	1.84	1.96	2.05	2.13	2.19	2.24	2.29	2.33	2.40	2.45	2.50	2.55	2.58
	.20	1.23	1.43	1.56	1.66	1.74	1.81	1.86	1.91	1.95	2.02	2.08	2.13	2.17	2.21
40	.01	2.69	2.85	2.95	3.03	3.09	3.15	3.19	3.23	3.27	3.33	3.38	3.43	3.47	3.50
	.05	1.99	2.16	2.28	2.37	2.44	2.49	2.54	2.59	2.63	2.69	2.74	2.79	2.83	2.87
	.10	1.64	1.82	1.94	2.04	2.11	2.17	2.22	2.27	2.31	2.37	2.43	2.48	2.52	2.55
	.20	1.23	1.42	1.56	1.65	1.73	1.80	1.85	1.90	1.94	2.01	2.07	2.12	2.16	2.20
60	.01	2.65	2.80	2.90	2.97	3.03	3.09	3.13	3.17	3.20	3.26	3.31	3.35	3.39	3.42
	.05	1.98	2.14	2.25	2.34	2.41	2.46	2.51	2.55	2.59	2.65	2.71	2.75	2.79	2.82
	.10	1.63	1.81	1.93	2.02	2.09	2.15	2.20	2.24	2.28	2.35	2.40	2.45	2.49	2.52
	.20	1.22	1.42	1.55	1.64	1.72	1.78	1.84	1.88	1.92	1.99	2.05	2.10	2.14	2.18
120	.01	2.61	2.75	2.85	2.92	2.98	3.03	3.07	3.10	3.14	3.19	3.24	3.28	3.31	3.34
	.05	1.96	2.12	2.23	2.31	2.38	2.43	2.48	2.52	2.56	2.62	2.67	2.71	2.75	2.78
	.10	1.62	1.79	1.91	2.00	2.07	2.13	2.18	2.22	2.26	2.32	2.38	2.42	2.46	2.50
	.20	1.22	1.41	1.54	1.63	1.71	1.77	1.83	1.87	1.91	1.98	2.04	2.08	2.13	2.16
∞	.01	2.57	2.70	2.80	2.86	2.92	2.97	3.01	3.04	3.07	3.12	3.17	3.20	3.24	3.27
	.05	1.94	2.10	2.20	2.28	2.35	2.40	2.45	2.49	2.52	2.58	2.63	2.67	2.71	2.74
	.10	1.61	1.78	1.90	1.98	2.05	2.11	2.16	2.20	2.24	2.30	2.35	2.40	2.43	2.47
	.20	1.21	1.40	1.53	1.63	1.70	1.76	1.81	1.86	1.90	1.97	2.02	2.07	2.11	2.15

ρ = 0.5

ν	α \ k	2	3	4	5	6	7	8	9	10	12	14	16	18	20
2	.01	8.88	10.0	10.9	11.5	12.0	12.5	12.9	13.2	13.5	14.0	14.4	14.7	15.1	15.3
	.05	3.80	4.34	4.71	5.00	5.24	5.43	5.60	5.75	5.88	6.11	6.29	6.45	6.59	6.72
	.10	2.54	2.92	3.20	3.40	3.57	3.71	3.83	3.94	4.03	4.19	4.32	4.44	4.54	4.62
	.20	1.57	1.85	2.05	2.21	2.33	2.43	2.52	2.59	2.66	2.77	2.87	2.95	3.02	3.08
3	.01	5.48	6.04	6.44	6.74	6.99	7.20	7.38	7.53	7.67	7.91	8.11	8.28	8.43	8.56
	.05	2.94	3.28	3.52	3.70	3.85	3.97	4.08	4.17	4.25	4.39	4.51	4.61	4.70	4.78
	.10	2.13	2.41	2.61	2.76	2.87	2.97	3.06	3.13	3.20	3.31	3.41	3.49	3.56	3.62
	.20	1.41	1.65	1.81	1.94	2.03	2.12	2.19	2.25	2.30	2.39	2.47	2.53	2.59	2.64
4	.01	4.41	4.80	5.07	5.28	5.45	5.59	5.72	5.82	5.92	6.08	6.22	6.34	6.44	6.53
	.05	2.61	2.88	3.08	3.22	3.34	3.44	3.52	3.59	3.66	3.77	3.86	3.94	4.01	4.07
	.10	1.96	2.20	2.37	2.50	2.60	2.68	2.75	2.82	2.87	2.97	3.05	3.11	3.17	3.22
	.20	1.34	1.56	1.71	1.82	1.91	1.98	2.04	2.10	2.15	2.23	2.30	2.35	2.40	2.45
5	.01	3.90	4.21	4.43	4.60	4.73	4.85	4.94	5.03	5.11	5.24	5.34	5.44	5.52	5.59
	.05	2.44	2.68	2.85	2.98	3.08	3.16	3.24	3.30	3.36	3.45	3.53	3.60	3.66	3.71
	.10	1.87	2.09	2.24	2.36	2.45	2.53	2.59	2.65	2.70	2.78	2.86	2.92	2.97	3.02
	.20	1.30	1.51	1.65	1.76	1.84	1.91	1.97	2.02	2.06	2.14	2.20	2.26	2.30	2.34
6	.01	3.61	3.88	4.06	4.21	4.32	4.42	4.51	4.58	4.64	4.76	4.85	4.93	5.00	5.06
	.05	2.34	2.56	2.71	2.83	2.92	3.00	3.06	3.12	3.17	3.26	3.33	3.40	3.45	3.50
	.10	1.82	2.02	2.17	2.27	2.36	2.43	2.49	2.54	2.59	2.67	2.74	2.79	2.84	2.89
	.20	1.28	1.48	1.61	1.71	1.79	1.86	1.92	1.97	2.01	2.08	2.14	2.19	2.24	2.28
7	.01	3.42	3.66	3.83	3.96	4.06	4.15	4.22	4.29	4.35	4.45	4.53	4.60	4.67	4.72
	.05	2.27	2.48	2.62	2.73	2.81	2.89	2.95	3.00	3.05	3.13	3.20	3.26	3.31	3.36
	.10	1.78	1.98	2.11	2.22	2.30	2.37	2.42	2.47	2.52	2.59	2.66	2.71	2.76	2.80
	.20	1.26	1.46	1.59	1.69	1.76	1.83	1.88	1.93	1.97	2.04	2.10	2.15	2.19	2.23
8	.01	3.29	3.51	3.66	3.78	3.88	3.96	4.03	4.09	4.14	4.23	4.31	4.38	4.43	4.49
	.05	2.22	2.42	2.55	2.66	2.74	2.81	2.87	2.92	2.96	3.04	3.11	3.16	3.21	3.25
	.10	1.75	1.94	2.08	2.17	2.25	2.32	2.38	2.42	2.47	2.54	2.60	2.65	2.70	2.74
	.20	1.25	1.44	1.57	1.67	1.74	1.81	1.86	1.90	1.95	2.01	2.07	2.12	2.16	2.20
9	.01	3.19	3.40	3.54	3.66	3.75	3.82	3.89	3.94	3.99	4.08	4.15	4.21	4.26	4.31
	.05	2.18	2.37	2.50	2.60	2.68	2.75	2.81	2.86	2.90	2.97	3.04	3.09	3.14	3.18
	.10	1.73	1.92	2.05	2.14	2.22	2.28	2.34	2.39	2.43	2.50	2.56	2.61	2.65	2.69
	.20	1.24	1.43	1.56	1.65	1.73	1.79	1.84	1.89	1.93	1.99	2.05	2.10	2.14	2.17
10	.01	3.11	3.31	3.45	3.56	3.64	3.72	3.78	3.83	3.88	3.96	4.03	4.08	4.14	4.18
	.05	2.15	2.34	2.47	2.56	2.64	2.70	2.76	2.81	2.85	2.92	2.98	3.03	3.08	3.12
	.10	1.71	1.90	2.02	2.12	2.19	2.26	2.31	2.35	2.40	2.46	2.52	2.57	2.61	2.65
	.20	1.23	1.42	1.55	1.64	1.71	1.77	1.83	1.87	1.91	1.98	2.03	2.08	2.12	2.15
12	.01	3.01	3.19	3.32	3.42	3.50	3.56	3.62	3.67	3.71	3.79	3.85	3.91	3.95	3.99
	.05	2.11	2.29	2.41	2.50	2.58	2.64	2.69	2.74	2.78	2.84	2.90	2.95	2.99	3.03
	.10	1.69	1.87	1.99	2.08	2.16	2.22	2.27	2.31	2.35	2.42	2.47	2.52	2.56	2.60
	.20	1.22	1.41	1.53	1.62	1.69	1.75	1.80	1.85	1.89	1.95	2.01	2.05	2.09	2.13

TABLE 4. Continued.

$\rho = 0.5$

ν	α \ k	2	3	4	5	6	7	8	9	10	12	14	16	18	20
16	.01	2.88	3.05	3.17	3.26	3.33	3.39	3.44	3.48	3.52	3.59	3.65	3.70	3.74	3.78
	.05	2.06	2.23	2.34	2.43	2.50	2.56	2.61	2.65	2.69	2.75	2.81	2.85	2.89	2.93
	.10	1.66	1.83	1.95	2.04	2.11	2.17	2.22	2.26	2.30	2.36	2.41	2.46	2.50	2.53
	.20	1.21	1.39	1.51	1.60	1.67	1.73	1.78	1.82	1.86	1.92	1.97	2.02	2.06	2.09
20	.01	2.81	2.97	3.08	3.17	3.23	3.29	3.34	3.38	3.42	3.48	3.53	3.58	3.62	3.65
	.05	2.03	2.19	2.30	2.39	2.46	2.51	2.56	2.60	2.64	2.70	2.75	2.80	2.83	2.87
	.10	1.64	1.81	1.93	2.01	2.08	2.14	2.19	2.23	2.26	2.33	2.38	2.42	2.46	2.49
	.20	1.20	1.38	1.50	1.59	1.65	1.71	1.76	1.80	1.84	1.90	1.95	2.00	2.04	2.07
24	.01	2.77	2.92	3.03	3.11	3.17	3.22	3.27	3.31	3.35	3.41	3.46	3.50	3.54	3.57
	.05	2.01	2.17	2.28	2.36	2.43	2.48	2.53	2.57	2.60	2.66	2.72	2.76	2.80	2.83
	.10	1.63	1.80	1.91	2.00	2.06	2.12	2.17	2.21	2.24	2.30	2.35	2.40	2.43	2.47
	.20	1.19	1.37	1.49	1.58	1.65	1.70	1.75	1.79	1.83	1.89	1.94	1.99	2.02	2.06
30	.01	2.72	2.87	2.97	3.05	3.11	3.16	3.21	3.25	3.28	3.34	3.39	3.43	3.46	3.50
	.05	1.99	2.15	2.25	2.34	2.40	2.45	2.50	2.54	2.57	2.63	2.68	2.72	2.76	2.79
	.10	1.62	1.79	1.90	1.98	2.05	2.10	2.15	2.19	2.22	2.28	2.33	2.37	2.41	2.44
	.20	1.19	1.36	1.48	1.57	1.64	1.69	1.74	1.78	1.82	1.88	1.93	1.97	2.01	2.04
40	.01	2.68	2.82	2.92	2.99	3.05	3.10	3.14	3.18	3.21	3.27	3.32	3.36	3.39	3.42
	.05	1.97	2.13	2.23	2.31	2.37	2.42	2.47	2.51	2.54	2.60	2.65	2.69	2.72	2.75
	.10	1.61	1.77	1.88	1.96	2.03	2.08	2.13	2.17	2.20	2.26	2.31	2.35	2.39	2.42
	.20	1.18	1.36	1.47	1.56	1.63	1.68	1.73	1.77	1.81	1.87	1.92	1.96	2.00	2.03
60	.01	2.64	2.78	2.87	2.94	3.00	3.04	3.08	3.12	3.15	3.20	3.25	3.29	3.32	3.35
	.05	1.95	2.10	2.21	2.28	2.34	2.40	2.44	2.48	2.51	2.57	2.61	2.65	2.69	2.72
	.10	1.60	1.76	1.87	1.95	2.01	2.06	2.11	2.15	2.18	2.24	2.29	2.33	2.36	2.39
	.20	1.18	1.35	1.47	1.55	1.62	1.67	1.72	1.76	1.80	1.86	1.91	1.95	1.98	2.02
120	.01	2.60	2.73	2.82	2.89	2.94	2.99	3.02	3.06	3.09	3.14	3.18	3.22	3.25	3.28
	.05	1.93	2.08	2.18	2.26	2.32	2.37	2.41	2.45	2.48	2.53	2.58	2.62	2.65	2.68
	.10	1.59	1.75	1.85	1.93	1.99	2.05	2.09	2.13	2.16	2.22	2.27	2.31	2.34	2.37
	.20	1.17	1.35	1.46	1.54	1.61	1.66	1.71	1.75	1.79	1.84	1.89	1.93	1.97	2.00
∞	.01	2.56	2.68	2.77	2.84	2.89	2.93	2.97	3.00	3.03	3.08	3.12	3.15	3.18	3.21
	.05	1.92	2.06	2.16	2.23	2.29	2.34	2.38	2.42	2.45	2.50	2.55	2.58	2.62	2.64
	.10	1.58	1.73	1.84	1.92	1.98	2.03	2.07	2.11	2.14	2.20	2.24	2.28	2.32	2.35
	.20	1.17	1.34	1.45	1.54	1.60	1.65	1.70	1.74	1.77	1.83	1.88	1.92	1.96	1.99

ρ = 0.7

ν	α \ k	2	3	4	5	6	7	8	9	10	12	14	16	18	20
2	.01	8.51	9.40	10.0	10.5	10.9	11.2	11.5	11.7	11.9	12.3	12.6	12.9	13.1	13.3
	.05	3.63	4.04	4.32	4.54	4.71	4.86	4.98	5.09	5.19	5.35	5.49	5.60	5.70	5.79
	.10	2.41	2.71	2.91	3.07	3.19	3.29	3.38	3.46	3.53	3.65	3.74	3.82	3.90	3.96
	.20	1.46	1.68	1.84	1.95	2.04	2.12	2.18	2.24	2.29	2.37	2.44	2.50	2.55	2.60
3	.01	5.32	5.76	6.07	6.30	6.49	6.65	6.78	6.90	7.00	7.18	7.32	7.45	7.56	7.66
	.05	2.83	3.10	3.29	3.43	3.54	3.63	3.71	3.78	3.84	3.95	4.03	4.11	4.17	4.23
	.10	2.04	2.26	2.41	2.52	2.61	2.69	2.75	2.81	2.86	2.94	3.01	3.07	3.12	3.17
	.20	1.33	1.51	1.64	1.73	1.81	1.87	1.92	1.97	2.01	2.08	2.13	2.18	2.22	2.26
4	.01	4.30	4.61	4.83	4.99	5.12	5.23	5.33	5.41	5.48	5.60	5.71	5.80	5.87	5.94
	.05	2.53	2.75	2.90	3.01	3.10	3.18	3.24	3.30	3.35	3.43	3.50	3.56	3.61	3.66
	.10	1.89	2.08	2.21	2.31	2.38	2.45	2.50	2.55	2.59	2.66	2.72	2.77	2.82	2.86
	.20	1.26	1.44	1.55	1.64	1.71	1.76	1.81	1.85	1.89	1.95	2.00	2.04	2.08	2.12
5	.01	3.82	4.07	4.25	4.38	4.48	4.57	4.65	4.71	4.77	4.87	4.96	5.03	5.09	5.14
	.05	2.37	2.56	2.70	2.80	2.88	2.94	3.00	3.05	3.09	3.16	3.23	3.28	3.32	3.36
	.10	1.80	1.98	2.10	2.19	2.26	2.32	2.37	2.41	2.45	2.52	2.57	2.62	2.66	2.69
	.20	1.23	1.39	1.50	1.59	1.65	1.70	1.75	1.79	1.82	1.88	1.93	1.97	2.01	2.04
6	.01	3.54	3.76	3.91	4.03	4.12	4.20	4.26	4.32	4.37	4.45	4.53	4.59	4.64	4.69
	.05	2.27	2.45	2.57	2.67	2.74	2.80	2.85	2.90	2.94	3.00	3.06	3.11	3.15	3.19
	.10	1.75	1.92	2.03	2.12	2.18	2.24	2.29	2.33	2.36	2.42	2.47	2.52	2.56	2.59
	.20	1.21	1.37	1.47	1.55	1.62	1.67	1.71	1.75	1.78	1.84	1.88	1.92	1.96	1.99
7	.01	3.36	3.56	3.70	3.80	3.88	3.95	4.01	4.06	4.11	4.19	4.25	4.31	4.35	4.40
	.05	2.21	2.38	2.49	2.58	2.65	2.71	2.75	2.80	2.83	2.90	2.95	2.99	3.03	3.07
	.10	1.72	1.88	1.99	2.07	2.13	2.18	2.23	2.27	2.30	2.36	2.41	2.45	2.49	2.52
	.20	1.19	1.35	1.45	1.53	1.59	1.64	1.68	1.72	1.75	1.81	1.85	1.89	1.92	1.95
8	.01	3.23	3.42	3.55	3.64	3.72	3.78	3.84	3.89	3.93	4.00	4.06	4.11	4.15	4.19
	.05	2.16	2.32	2.43	2.52	2.58	2.64	2.68	2.72	2.76	2.82	2.87	2.91	2.95	2.98
	.10	1.69	1.85	1.95	2.03	2.09	2.15	2.19	2.23	2.26	2.32	2.36	2.40	2.44	2.47
	.20	1.18	1.34	1.44	1.51	1.57	1.62	1.66	1.70	1.73	1.78	1.83	1.87	1.90	1.93
9	.01	3.14	3.31	3.43	3.53	3.60	3.66	3.71	3.76	3.79	3.86	3.92	3.97	4.01	4.05
	.05	2.13	2.28	2.39	2.47	2.53	2.59	2.63	2.67	2.71	2.76	2.81	2.85	2.89	2.92
	.10	1.67	1.82	1.93	2.00	2.07	2.12	2.16	2.19	2.23	2.28	2.33	2.37	2.40	2.43
	.20	1.17	1.32	1.43	1.50	1.56	1.61	1.65	1.68	1.71	1.77	1.81	1.85	1.88	1.91
10	.01	3.07	3.24	3.35	3.44	3.51	3.56	3.61	3.66	3.69	3.76	3.81	3.86	3.90	3.93
	.05	2.10	2.25	2.36	2.43	2.50	2.55	2.59	2.63	2.66	2.72	2.77	2.81	2.84	2.87
	.10	1.66	1.81	1.91	1.98	2.04	2.09	2.13	2.17	2.20	2.26	2.30	2.34	2.37	2.40
	.20	1.17	1.32	1.42	1.49	1.55	1.60	1.64	1.67	1.70	1.75	1.80	1.83	1.87	1.89
12	.01	2.97	3.12	3.23	3.31	3.37	3.43	3.47	3.51	3.55	3.61	3.66	3.70	3.74	3.77
	.05	2.06	2.21	2.31	2.38	2.44	2.49	2.53	2.57	2.60	2.65	2.70	2.74	2.77	2.80
	.10	1.63	1.78	1.88	1.95	2.01	2.06	2.10	2.13	2.16	2.22	2.26	2.30	2.33	2.36
	.20	1.16	1.30	1.40	1.48	1.53	1.58	1.62	1.65	1.68	1.73	1.78	1.81	1.84	1.87

TABLE 4. Continued.

ρ = 0.7

ν	α \ k	2	3	4	5	6	7	8	9	10	12	14	16	18	20
16	.01	2.85	2.99	3.09	3.16	3.22	3.27	3.31	3.35	3.38	3.43	3.48	3.52	3.55	3.58
	.05	2.01	2.15	2.25	2.32	2.37	2.42	2.46	2.50	2.53	2.58	2.62	2.66	2.69	2.72
	.10	1.61	1.75	1.84	1.91	1.97	2.02	2.06	2.09	2.12	2.17	2.21	2.25	2.28	2.30
	.20	1.14	1.29	1.39	1.46	1.51	1.56	1.60	1.63	1.66	1.71	1.75	1.79	1.82	1.84
20	.01	2.78	2.91	3.01	3.08	3.13	3.18	3.22	3.25	3.28	3.34	3.38	3.42	3.45	3.48
	.05	1.98	2.12	2.21	2.28	2.34	2.38	2.42	2.45	2.48	2.53	2.57	2.61	2.64	2.67
	.10	1.59	1.73	1.82	1.89	1.95	1.99	2.03	2.06	2.09	2.14	2.18	2.22	2.25	2.27
	.20	1.14	1.28	1.38	1.45	1.50	1.55	1.59	1.62	1.65	1.70	1.74	1.77	1.80	1.83
24	.01	2.73	2.87	2.96	3.02	3.08	3.12	3.16	3.19	3.22	3.27	3.31	3.35	3.38	3.41
	.05	1.96	2.10	2.19	2.26	2.31	2.36	2.39	2.43	2.45	2.50	2.54	2.58	2.61	2.63
	.10	1.58	1.72	1.81	1.88	1.93	1.98	2.01	2.05	2.08	2.12	2.16	2.20	2.23	2.25
	.20	1.13	1.27	1.37	1.44	1.49	1.54	1.58	1.61	1.64	1.69	1.73	1.76	1.79	1.82
30	.01	2.69	2.82	2.91	2.97	3.02	3.07	3.10	3.13	3.16	3.21	3.25	3.28	3.31	3.34
	.05	1.95	2.08	2.17	2.23	2.29	2.33	2.37	2.40	2.43	2.47	2.51	2.55	2.58	2.60
	.10	1.57	1.71	1.80	1.86	1.92	1.96	2.00	2.03	2.06	2.11	2.15	2.18	2.21	2.23
	.20	1.13	1.27	1.36	1.43	1.49	1.53	1.57	1.60	1.63	1.68	1.72	1.75	1.78	1.81
40	.01	2.65	2.77	2.86	2.92	2.97	3.01	3.05	3.08	3.11	3.15	3.19	3.22	3.25	3.28
	.05	1.93	2.06	2.15	2.21	2.26	2.30	2.34	2.37	2.40	2.45	2.49	2.52	2.55	2.57
	.10	1.56	1.69	1.78	1.85	1.90	1.95	1.98	2.01	2.04	2.09	2.13	2.16	2.19	2.21
	.20	1.12	1.26	1.36	1.42	1.48	1.52	1.56	1.59	1.62	1.67	1.71	1.74	1.77	1.80
60	.01	2.61	2.73	2.81	2.87	2.92	2.96	2.99	3.02	3.05	3.09	3.13	3.16	3.19	3.21
	.05	1.91	2.04	2.12	2.19	2.24	2.28	2.32	2.35	2.37	2.42	2.46	2.49	2.52	2.54
	.10	1.55	1.68	1.77	1.84	1.89	1.93	1.97	2.00	2.03	2.07	2.11	2.14	2.17	2.20
	.20	1.12	1.26	1.35	1.42	1.47	1.52	1.55	1.58	1.61	1.66	1.70	1.73	1.76	1.79
120	.01	2.57	2.69	2.76	2.82	2.87	2.91	2.94	2.97	2.99	3.04	3.07	3.10	3.13	3.15
	.05	1.89	2.02	2.10	2.17	2.22	2.26	2.29	2.32	2.35	2.39	2.43	2.46	2.49	2.51
	.10	1.54	1.67	1.76	1.82	1.87	1.92	1.95	1.98	2.01	2.05	2.09	2.12	2.15	2.18
	.20	1.11	1.25	1.34	1.41	1.46	1.51	1.54	1.58	1.60	1.65	1.69	1.72	1.75	1.78
∞	.01	2.53	2.64	2.72	2.78	2.82	2.86	2.89	2.92	2.94	2.98	3.02	3.05	3.07	3.09
	.05	1.88	2.00	2.08	2.14	2.19	2.23	2.27	2.30	2.32	2.37	2.40	2.43	2.46	2.48
	.10	1.53	1.66	1.75	1.81	1.86	1.90	1.94	1.97	1.99	2.04	2.08	2.11	2.13	2.16
	.20	1.11	1.25	1.34	1.40	1.46	1.50	1.54	1.57	1.60	1.64	1.68	1.71	1.74	1.77

Source: Computed by C. W. Dunnett.

TABLE 5. Two-Sided Upper α Equicoordinate Point of the k-Variate Equicorrelated t-Distribution with Common Correlation ρ and Degrees of Freedom ν ($|T|^{(\alpha)}_{k,\nu,\rho}$).

$\rho = 0.1$

ν	α \ k	2	3	4	5	6	7	8	9	10	12	14	16	18	20
2	.01	12.7	14.4	15.6	16.6	17.3	17.9	18.5	18.9	19.4	20.1	20.7	21.2	21.6	22.0
	.05	5.57	6.33	6.87	7.29	7.62	7.90	8.14	8.35	8.54	8.85	9.11	9.34	9.53	9.71
	.10	3.83	4.37	4.75	5.05	5.29	5.48	5.65	5.80	5.93	6.15	6.34	6.50	6.63	6.75
	.20	2.55	2.94	3.21	3.42	3.58	3.72	3.84	3.95	4.04	4.19	4.32	4.43	4.53	4.61
3	.01	7.12	7.90	8.46	8.90	9.25	9.55	9.81	10.03	10.23	10.57	10.86	11.10	11.32	11.50
	.05	3.96	4.42	4.75	5.01	5.22	5.39	5.54	5.67	5.79	5.99	6.16	6.30	6.42	6.53
	.10	2.99	3.36	3.63	3.83	4.00	4.14	4.26	4.36	4.45	4.61	4.74	4.85	4.95	5.04
	.20	2.16	2.47	2.69	2.85	2.98	3.09	3.19	3.27	3.34	3.47	3.57	3.66	3.73	3.80
4	.01	5.46	5.98	6.35	6.64	6.88	7.08	7.25	7.41	7.54	7.77	7.97	8.14	8.28	8.41
	.05	3.38	3.74	4.00	4.19	4.36	4.49	4.61	4.71	4.80	4.96	5.09	5.20	5.30	5.38
	.10	2.66	2.97	3.19	3.36	3.50	3.61	3.71	3.80	3.87	4.00	4.11	4.21	4.29	4.36
	.20	2.00	2.28	2.46	2.61	2.73	2.82	2.91	2.98	3.04	3.15	3.24	3.32	3.39	3.45
5	.01	4.70	5.18	5.39	5.62	5.80	5.96	6.09	6.21	6.32	6.50	6.65	6.78	6.90	7.00
	.05	3.09	3.40	3.61	3.78	3.92	4.03	4.13	4.22	4.30	4.43	4.54	4.64	4.72	4.80
	.10	2.49	2.76	2.96	3.11	3.23	3.33	3.42	3.50	3.56	3.68	3.78	3.86	3.93	4.00
	.20	1.92	2.17	2.34	2.48	2.58	2.67	2.75	2.82	2.88	2.98	3.06	3.14	3.20	3.26
6	.01	4.27	4.61	4.85	5.04	5.19	5.32	5.44	5.54	5.63	5.78	5.91	6.02	6.12	6.21
	.05	2.91	3.19	3.38	3.53	3.66	3.76	3.85	3.93	4.00	4.11	4.21	4.30	4.38	4.44
	.10	2.38	2.64	2.82	2.95	3.07	3.16	3.24	3.31	3.37	3.48	3.57	3.65	3.71	3.77
	.20	1.86	2.10	2.27	2.39	2.49	2.58	2.65	2.72	2.77	2.87	2.95	3.02	3.08	3.13
7	.01	4.00	4.29	4.51	4.67	4.81	4.92	5.02	5.11	5.19	5.32	5.44	5.53	5.62	5.70
	.05	2.80	3.05	3.23	3.37	3.48	3.58	3.66	3.73	3.79	3.90	4.00	4.08	4.14	4.21
	.10	2.31	2.55	2.72	2.85	2.95	3.04	3.12	3.18	3.24	3.34	3.43	3.50	3.56	3.62
	.20	1.82	2.05	2.21	2.33	2.43	2.51	2.58	2.64	2.70	2.79	2.87	2.93	2.99	3.04
8	.01	3.81	4.08	4.27	4.42	4.54	4.64	4.73	4.81	4.88	5.01	5.11	5.20	5.28	5.35
	.05	2.72	2.95	3.12	3.25	3.36	3.45	3.52	3.59	3.65	3.75	3.84	3.91	3.98	4.04
	.10	2.26	2.49	2.65	2.77	2.87	2.96	3.03	3.09	3.15	3.24	3.32	3.39	3.45	3.51
	.20	1.79	2.02	2.17	2.29	2.39	2.47	2.53	2.59	2.64	2.73	2.81	2.87	2.93	2.98
9	.01	3.67	3.92	4.10	4.23	4.35	4.44	4.53	4.60	4.66	4.78	4.87	4.95	5.02	5.09
	.05	2.66	2.88	3.04	3.17	3.27	3.35	3.42	3.49	3.54	3.64	3.72	3.79	3.85	3.91
	.10	2.22	2.44	2.60	2.72	2.81	2.89	2.96	3.02	3.08	3.17	3.25	3.31	3.37	3.42
	.20	1.77	1.99	2.14	2.26	2.35	2.43	2.49	2.55	2.60	2.69	2.76	2.83	2.88	2.93
10	.01	3.57	3.80	3.97	4.09	4.20	4.29	4.37	4.43	4.50	4.60	4.69	4.77	4.83	4.89
	.05	2.61	2.83	2.98	3.10	3.19	3.27	3.34	3.40	3.46	3.55	3.63	3.70	3.76	3.81
	.10	2.19	2.41	2.56	2.67	2.77	2.84	2.91	2.97	3.02	3.11	3.18	3.25	3.30	3.35
	.20	1.76	1.97	2.12	2.23	2.32	2.40	2.46	2.52	2.57	2.66	2.73	2.79	2.84	2.89

TABLE 5. Continued.

$\rho = 0.1$

ν	α \ k	2	3	4	5	6	7	8	9	10	12	14	16	18	20
12	.01	3.42	3.63	3.78	3.90	3.99	4.07	4.14	4.20	4.26	4.35	4.43	4.50	4.56	4.61
	.05	2.54	2.74	2.89	3.00	3.09	3.16	3.23	3.29	3.34	3.42	3.50	3.56	3.62	3.67
	.10	2.15	2.35	2.50	2.61	2.70	2.77	2.83	2.89	2.94	3.02	3.09	3.15	3.21	3.26
	.20	1.73	1.94	2.09	2.19	2.28	2.36	2.42	2.47	2.52	2.60	2.67	2.73	2.78	2.83
16	.01	3.24	3.43	3.56	3.67	3.75	3.82	3.88	3.93	3.98	4.06	4.13	4.19	4.25	4.29
	.05	2.46	2.65	2.78	2.88	2.97	3.03	3.09	3.15	3.19	3.27	3.34	3.40	3.45	3.49
	.10	2.09	2.29	2.43	2.53	2.61	2.68	2.74	2.79	2.84	2.92	2.98	3.04	3.09	3.13
	.20	1.70	1.90	2.04	2.15	2.23	2.30	2.36	2.41	2.46	2.54	2.61	2.66	2.71	2.75
20	.01	3.15	3.32	3.44	3.54	3.61	3.68	3.73	3.78	3.83	3.90	3.97	4.02	4.07	4.11
	.05	2.41	2.59	2.72	2.82	2.89	2.96	3.02	3.06	3.11	3.18	3.25	3.30	3.35	3.39
	.10	2.06	2.25	2.38	2.48	2.56	2.63	2.69	2.74	2.78	2.86	2.92	2.97	3.02	3.06
	.20	1.68	1.88	2.02	2.12	2.20	2.27	2.33	2.38	2.42	2.50	2.57	2.62	2.67	2.71
24	.01	3.09	3.25	3.37	3.46	3.53	3.59	3.64	3.69	3.73	3.80	3.86	3.91	3.96	4.00
	.05	2.38	2.56	2.68	2.77	2.85	2.91	2.96	3.01	3.05	3.13	3.19	3.24	3.28	3.32
	.10	2.04	2.23	2.36	2.45	2.53	2.59	2.65	2.70	2.74	2.81	2.88	2.93	2.97	3.02
	.20	1.67	1.87	2.00	2.10	2.18	2.25	2.31	2.36	2.40	2.48	2.54	2.59	2.64	2.68
30	.01	3.03	3.18	3.29	3.38	3.45	3.50	3.55	3.60	3.63	3.70	3.76	3.80	3.85	3.88
	.05	2.35	2.52	2.64	2.73	2.80	2.86	2.91	2.96	3.00	3.07	3.13	3.18	3.22	3.26
	.10	2.02	2.20	2.33	2.42	2.50	2.56	2.61	2.66	2.70	2.77	2.83	2.88	2.93	2.97
	.20	1.66	1.85	1.99	2.08	2.16	2.23	2.29	2.33	2.38	2.45	2.51	2.56	2.61	2.65
40	.01	2.97	3.12	3.22	3.30	3.37	3.42	3.47	3.51	3.54	3.60	3.66	3.70	3.74	3.78
	.05	2.32	2.49	2.60	2.69	2.76	2.82	2.87	2.91	2.95	3.01	3.07	3.12	3.16	3.20
	.10	2.00	2.18	2.30	2.39	2.47	2.53	2.58	2.63	2.67	2.73	2.79	2.84	2.88	2.92
	.20	1.65	1.84	1.97	2.07	2.14	2.21	2.26	2.31	2.35	2.43	2.49	2.54	2.58	2.62
60	.01	2.91	3.05	3.15	3.23	3.29	3.34	3.38	3.42	3.45	3.51	3.56	3.60	3.64	3.67
	.05	2.29	2.45	2.56	2.65	2.71	2.77	2.82	2.86	2.90	2.96	3.01	3.06	3.10	3.13
	.10	1.99	2.16	2.28	2.36	2.44	2.49	2.55	2.59	2.63	2.69	2.75	2.80	2.84	2.88
	.20	1.64	1.83	1.95	2.05	2.12	2.19	2.24	2.29	2.33	2.40	2.46	2.51	2.55	2.59
120	.01	2.86	2.99	3.09	3.16	3.21	3.26	3.30	3.34	3.37	3.42	3.47	3.51	3.54	3.57
	.05	2.26	2.42	2.53	2.61	2.67	2.72	2.77	2.81	2.85	2.91	2.96	3.00	3.04	3.07
	.10	1.97	2.13	2.25	2.34	2.40	2.46	2.51	2.55	2.59	2.66	2.71	2.76	2.80	2.83
	.20	1.63	1.81	1.94	2.03	2.11	2.17	2.22	2.27	2.31	2.38	2.43	2.48	2.52	2.56
∞	.01	2.81	2.93	3.02	3.09	3.14	3.19	3.23	3.26	3.29	3.34	3.38	3.42	3.45	3.48
	.05	2.24	2.39	2.49	2.57	2.63	2.68	2.72	2.76	2.80	2.85	2.90	2.94	2.98	3.01
	.10	1.95	2.11	2.22	2.31	2.37	2.43	2.48	2.52	2.56	2.62	2.67	2.71	2.75	2.79
	.20	1.62	1.80	1.92	2.01	2.09	2.15	2.20	2.24	2.28	2.35	2.41	2.45	2.49	2.53

ρ = 0.3

ν	α \ k	2	3	4	5	6	7	8	9	10	12	14	16	18	20
2	.01	12.6	14.2	15.4	16.2	16.9	17.5	18.0	18.4	18.8	19.3	20.0	20.5	20.9	21.2
	.05	5.52	6.24	6.75	7.14	7.45	7.71	7.93	8.12	8.29	8.58	8.82	9.03	9.20	9.36
	.10	3.79	4.31	4.67	4.94	5.16	5.35	5.50	5.64	5.76	5.96	6.13	6.27	6.40	6.51
	.20	2.52	2.89	3.15	3.34	3.50	3.63	3.74	3.83	3.91	4.06	4.18	4.28	4.36	4.44
3	.01	7.07	7.82	8.34	8.75	9.08	9.35	9.59	9.79	9.98	10.29	10.55	10.77	10.97	11.14
	.05	3.93	4.37	4.68	4.92	5.11	5.28	5.41	5.53	5.64	5.82	5.98	6.10	6.22	6.32
	.10	2.96	3.32	3.57	3.76	3.91	4.04	4.15	4.25	4.33	4.48	4.60	4.70	4.79	4.87
	.20	2.14	2.44	2.64	2.79	2.91	3.01	3.10	3.18	3.24	3.36	3.45	3.53	3.60	3.66
4	.01	5.43	5.92	6.28	6.55	6.77	6.95	7.11	7.25	7.38	7.59	7.77	7.92	8.06	8.18
	.05	3.36	3.70	3.94	4.13	4.28	4.40	4.51	4.60	4.69	4.83	4.95	5.05	5.14	5.22
	.10	2.64	2.94	3.14	3.30	3.43	3.53	3.62	3.70	3.77	3.89	3.99	4.08	4.15	4.22
	.20	1.99	2.24	2.42	2.56	2.66	2.76	2.83	2.90	2.96	3.06	3.14	3.21	3.28	3.33
5	.01	4.68	5.06	5.33	5.55	5.72	5.86	5.99	6.10	6.19	6.36	6.50	6.62	6.73	6.82
	.05	3.07	3.36	3.57	3.73	3.85	3.96	4.05	4.13	4.20	4.33	4.43	4.52	4.59	4.66
	.10	2.47	2.73	2.92	3.06	3.17	3.26	3.35	3.42	3.48	3.59	3.67	3.75	3.82	3.88
	.20	1.90	2.14	2.30	2.43	2.53	2.61	2.68	2.74	2.80	2.89	2.97	3.04	3.09	3.14
6	.01	4.25	4.57	4.80	4.98	5.13	5.25	5.35	5.45	5.53	5.67	5.79	5.89	5.98	6.06
	.05	2.90	3.16	3.34	3.49	3.60	3.70	3.78	3.85	3.91	4.02	4.12	4.19	4.26	4.32
	.10	2.37	2.61	2.78	2.91	3.01	3.10	3.17	3.24	3.30	3.39	3.48	3.55	3.61	3.66
	.20	1.84	2.07	2.23	2.35	2.44	2.52	2.59	2.65	2.70	2.79	2.86	2.92	2.98	3.03
7	.01	3.98	4.27	4.47	4.62	4.75	4.86	4.95	5.03	5.11	5.23	5.33	5.42	5.50	5.57
	.05	2.78	3.03	3.20	3.33	3.43	3.52	3.60	3.66	3.72	3.82	3.91	3.98	4.04	4.10
	.10	2.30	2.53	2.68	2.81	2.90	2.99	3.06	3.12	3.17	3.26	3.34	3.41	3.46	3.52
	.20	1.81	2.03	2.18	2.29	2.38	2.46	2.52	2.58	2.63	2.71	2.78	2.84	2.90	2.94
8	.01	3.79	4.05	4.24	4.38	4.49	4.59	4.67	4.75	4.81	4.93	5.02	5.10	5.17	5.24
	.05	2.70	2.93	3.09	3.21	3.31	3.39	3.47	3.53	3.58	3.68	3.76	3.83	3.89	3.94
	.10	2.25	2.47	2.62	2.73	2.83	2.90	2.97	3.03	3.08	3.17	3.24	3.31	3.36	3.41
	.20	1.78	1.99	2.14	2.25	2.34	2.41	2.47	2.53	2.58	2.66	2.73	2.79	2.84	2.88
9	.01	3.66	3.90	4.07	4.20	4.31	4.39	4.47	4.54	4.60	4.71	4.79	4.87	4.93	4.99
	.05	2.64	2.86	3.01	3.13	3.22	3.30	3.37	3.43	3.48	3.57	3.65	3.71	3.77	3.82
	.10	2.21	2.42	2.57	2.68	2.77	2.84	2.91	2.96	3.01	3.10	3.17	3.23	3.28	3.33
	.20	1.76	1.97	2.11	2.22	2.30	2.37	2.44	2.49	2.54	2.62	2.68	2.74	2.79	2.83
10	.01	3.56	3.78	3.94	4.06	4.16	4.25	4.32	4.38	4.44	4.54	4.62	4.69	4.75	4.81
	.05	2.60	2.80	2.95	3.06	3.15	3.23	3.29	3.35	3.40	3.49	3.56	3.62	3.68	3.72
	.10	2.18	2.38	2.53	2.63	2.72	2.79	2.86	2.91	2.96	3.04	3.11	3.17	3.22	3.27
	.20	1.74	1.95	2.09	2.19	2.28	2.35	2.41	2.46	2.51	2.58	2.65	2.71	2.75	2.80
12	.01	3.41	3.61	3.76	3.87	3.96	4.04	4.10	4.16	4.21	4.30	4.37	4.44	4.49	4.54
	.05	2.53	2.73	2.86	2.97	3.05	3.12	3.18	3.24	3.28	3.37	3.43	3.49	3.54	3.59
	.10	2.13	2.33	2.47	2.57	2.65	2.72	2.78	2.84	2.88	2.96	3.02	3.08	3.13	3.17
	.20	1.72	1.92	2.05	2.16	2.24	2.31	2.36	2.41	2.46	2.53	2.60	2.65	2.70	2.74

TABLE 5. Continued.

$\rho = 0.3$

ν	α \ k	2	3	4	5	6	7	8	9	10	12	14	16	18	20
16	.01	3.24	3.42	3.55	3.65	3.73	3.79	3.85	3.90	3.95	4.02	4.09	4.14	4.19	4.24
	.05	2.45	2.63	2.76	2.85	2.93	3.00	3.05	3.10	3.15	3.22	3.28	3.34	3.38	3.42
	.10	2.08	2.27	2.40	2.50	2.57	2.64	2.69	2.74	2.79	2.86	2.92	2.97	3.02	3.06
	.20	1.69	1.88	2.01	2.11	2.19	2.25	2.31	2.36	2.40	2.47	2.54	2.59	2.63	2.67
20	.01	3.14	3.31	3.43	3.52	3.59	3.66	3.71	3.76	3.80	3.87	3.93	3.98	4.03	4.07
	.05	2.40	2.58	2.70	2.79	2.86	2.93	2.98	3.03	3.07	3.14	3.20	3.25	3.29	3.33
	.10	2.05	2.23	2.36	2.45	2.53	2.59	2.64	2.69	2.73	2.80	2.86	2.91	2.95	2.99
	.20	1.67	1.86	1.99	2.08	2.16	2.22	2.28	2.33	2.37	2.44	2.50	2.55	2.59	2.63
24	.01	3.08	3.24	3.35	3.44	3.51	3.57	3.62	3.66	3.70	3.77	3.83	3.88	3.92	3.96
	.05	2.37	2.54	2.66	2.75	2.82	2.88	2.93	2.98	3.02	3.08	3.14	3.19	3.23	3.27
	.10	2.03	2.21	2.33	2.42	2.50	2.56	2.61	2.65	2.69	2.76	2.82	2.87	2.91	2.95
	.20	1.66	1.84	1.97	2.07	2.14	2.20	2.26	2.30	2.35	2.41	2.47	2.52	2.57	2.60
30	.01	3.02	3.18	3.28	3.36	3.43	3.49	3.53	3.57	3.61	3.68	3.73	3.77	3.82	3.85
	.05	2.34	2.51	2.62	2.71	2.78	2.83	2.88	2.93	2.96	3.03	3.08	3.13	3.17	3.21
	.10	2.01	2.19	2.30	2.39	2.47	2.52	2.58	2.62	2.66	2.73	2.78	2.83	2.87	2.91
	.20	1.65	1.83	1.96	2.05	2.12	2.18	2.24	2.28	2.32	2.39	2.45	2.50	2.54	2.58
40	.01	2.96	3.11	3.21	3.29	3.35	3.40	3.45	3.49	3.52	3.58	3.63	3.68	3.72	3.75
	.05	2.31	2.47	2.58	2.67	2.73	2.79	2.84	2.88	2.92	2.98	3.03	3.08	3.12	3.15
	.10	1.99	2.16	2.28	2.37	2.44	2.49	2.54	2.59	2.62	2.69	2.74	2.79	2.83	2.86
	.20	1.64	1.82	1.94	2.03	2.10	2.16	2.22	2.26	2.30	2.37	2.42	2.47	2.51	2.55
60	.01	2.91	3.05	3.14	3.22	3.28	3.33	3.37	3.41	3.44	3.50	3.54	3.58	3.62	3.65
	.05	2.28	2.44	2.55	2.63	2.69	2.74	2.79	2.83	2.87	2.93	2.98	3.02	3.06	3.09
	.10	1.97	2.14	2.25	2.34	2.41	2.46	2.51	2.55	2.59	2.65	2.70	2.75	2.79	2.82
	.20	1.63	1.80	1.92	2.01	2.09	2.15	2.20	2.24	2.28	2.34	2.40	2.45	2.49	2.52
120	.01	2.85	2.99	3.08	3.15	3.20	3.25	3.29	3.33	3.36	3.41	3.46	3.49	3.53	3.56
	.05	2.26	2.41	2.51	2.59	2.65	2.70	2.75	2.79	2.82	2.88	2.93	2.97	3.00	3.04
	.10	1.96	2.12	2.23	2.31	2.38	2.43	2.48	2.52	2.55	2.62	2.67	2.71	2.75	2.78
	.20	1.61	1.79	1.91	2.00	2.07	2.13	2.18	2.22	2.26	2.32	2.37	2.42	2.46	2.49
∞	.01	2.80	2.93	3.02	3.08	3.13	3.18	3.22	3.25	3.28	3.33	3.37	3.41	3.44	3.47
	.05	2.23	2.38	2.48	2.55	2.61	2.66	2.70	2.74	2.77	2.83	2.88	2.92	2.95	2.98
	.10	1.94	2.10	2.20	2.28	2.35	2.40	2.45	2.49	2.52	2.58	2.63	2.67	2.71	2.74
	.20	1.60	1.78	1.89	1.98	2.05	2.11	2.16	2.20	2.23	2.30	2.35	2.39	2.43	2.47

$\rho = 0.5$

ν	α \ k	2	3	4	5	6	7	8	9	10	12	14	16	18	20
2	.01	12.4	13.8	14.8	15.6	16.2	16.7	17.1	17.5	17.8	18.4	18.9	19.2	19.6	19.9
	.05	5.42	6.06	6.51	6.85	7.12	7.35	7.54	7.71	7.85	8.10	8.31	8.49	8.64	8.77
	.10	3.72	4.18	4.50	4.74	4.93	5.09	5.23	5.34	5.45	5.62	5.77	5.89	6.00	6.09
	.20	2.47	2.80	3.03	3.20	3.33	3.44	3.54	3.62	3.70	3.82	3.92	4.01	4.08	4.15
3	.01	6.97	7.64	8.10	8.46	8.75	8.98	9.19	9.37	9.52	9.79	10.02	10.21	10.37	10.52
	.05	3.87	4.26	4.54	4.75	4.92	5.06	5.18	5.28	5.37	5.53	5.66	5.77	5.87	5.95
	.10	2.91	3.23	3.45	3.62	3.75	3.87	3.96	4.04	4.12	4.24	4.34	4.43	4.51	4.58
	.20	2.10	2.36	2.54	2.68	2.78	2.87	2.95	3.01	3.07	3.17	3.25	3.32	3.38	3.43
4	.01	5.36	5.81	6.12	6.36	6.55	6.72	6.85	6.98	7.08	7.27	7.42	7.55	7.66	7.77
	.05	3.31	3.62	3.83	3.99	4.13	4.23	4.33	4.41	4.48	4.60	4.71	4.79	4.87	4.94
	.10	2.60	2.86	3.05	3.18	3.30	3.39	3.47	3.54	3.60	3.70	3.79	3.86	3.92	3.98
	.20	1.95	2.18	2.34	2.45	2.55	2.63	2.69	2.75	2.80	2.89	2.96	3.02	3.08	3.12
5	.01	4.63	4.97	5.22	5.41	5.56	5.68	5.79	5.89	5.97	6.12	6.24	6.34	6.43	6.51
	.05	3.03	3.29	3.48	3.62	3.73	3.82	3.90	3.97	4.03	4.14	4.23	4.30	4.37	4.42
	.10	2.43	2.67	2.83	2.96	3.05	3.14	3.21	3.27	3.32	3.41	3.49	3.56	3.61	3.66
	.20	1.86	2.08	2.22	2.33	2.42	2.49	2.56	2.61	2.66	2.74	2.80	2.86	2.91	2.95
6	.01	4.21	4.51	4.71	4.87	5.00	5.10	5.20	5.28	5.35	5.47	5.57	5.66	5.74	5.80
	.05	2.86	3.10	3.26	3.39	3.49	3.57	3.64	3.71	3.76	3.86	3.94	4.00	4.06	4.11
	.10	2.33	2.55	2.70	2.81	2.91	2.98	3.05	3.10	3.15	3.24	3.31	3.37	3.42	3.47
	.20	1.81	2.01	2.15	2.26	2.34	2.41	2.47	2.52	2.56	2.64	2.70	2.76	2.80	2.85
7	.01	3.95	4.21	4.39	4.53	4.64	4.74	4.82	4.89	4.95	5.06	5.15	5.23	5.29	5.35
	.05	2.75	2.97	3.12	3.24	3.33	3.41	3.47	3.53	3.58	3.67	3.74	3.81	3.86	3.91
	.10	2.26	2.47	2.61	2.72	2.81	2.88	2.94	2.99	3.04	3.12	3.19	3.24	3.29	3.34
	.20	1.77	1.97	2.10	2.20	2.28	2.35	2.41	2.46	2.50	2.57	2.63	2.68	2.73	2.77
8	.01	3.77	4.00	4.17	4.29	4.40	4.48	4.56	4.62	4.68	4.78	4.86	4.93	4.99	5.05
	.05	2.67	2.88	3.02	3.13	3.22	3.29	3.35	3.41	3.46	3.54	3.61	3.67	3.72	3.76
	.10	2.22	2.41	2.55	2.65	2.73	2.80	2.86	2.91	2.96	3.03	3.10	3.15	3.20	3.24
	.20	1.75	1.94	2.07	2.16	2.24	2.31	2.36	2.41	2.45	2.52	2.58	2.63	2.68	2.71
9	.01	3.63	3.85	4.01	4.12	4.22	4.30	4.37	4.43	4.48	4.57	4.65	4.71	4.77	4.82
	.05	2.61	2.81	2.95	3.05	3.14	3.20	3.26	3.32	3.36	3.44	3.51	3.56	3.61	3.65
	.10	2.18	2.37	2.50	2.60	2.68	2.74	2.80	2.85	2.89	2.97	3.03	3.08	3.13	3.17
	.20	1.73	1.91	2.04	2.14	2.21	2.27	2.33	2.37	2.41	2.48	2.54	2.59	2.63	2.67
10	.01	3.53	3.74	3.88	3.99	4.08	4.16	4.22	4.28	4.33	4.42	4.49	4.55	4.60	4.65
	.05	2.57	2.76	2.89	2.99	3.07	3.14	3.19	3.24	3.29	3.36	3.43	3.48	3.53	3.57
	.10	2.15	2.34	2.46	2.56	2.64	2.70	2.75	2.80	2.84	2.92	2.98	3.03	3.07	3.11
	.20	1.71	1.89	2.02	2.11	2.19	2.25	2.30	2.35	2.39	2.45	2.51	2.56	2.60	2.64
12	.01	3.39	3.58	3.71	3.81	3.89	3.96	4.02	4.07	4.12	4.19	4.26	4.32	4.36	4.41
	.05	2.50	2.68	2.81	2.90	2.98	3.04	3.09	3.14	3.18	3.25	3.31	3.36	3.41	3.45
	.10	2.11	2.29	2.41	2.50	2.57	2.63	2.69	2.73	2.77	2.84	2.90	2.95	2.99	3.03
	.20	1.69	1.87	1.99	2.08	2.15	2.21	2.26	2.30	2.34	2.41	2.46	2.51	2.55	2.59

TABLE 5. Continued.

$\rho = 0.5$

ν	α \ k	2	3	4	5	6	7	8	9	10	12	14	16	18	20
16	.01	3.22	3.39	3.51	3.60	3.67	3.73	3.78	3.83	3.87	3.94	4.00	4.05	4.09	4.13
	.05	2.42	2.59	2.71	2.80	2.87	2.92	2.97	3.02	3.06	3.12	3.18	3.22	3.26	3.30
	.10	2.06	2.23	2.34	2.43	2.50	2.56	2.61	2.65	2.69	2.75	2.80	2.85	2.89	2.93
	.20	1.66	1.83	1.95	2.04	2.11	2.16	2.21	2.25	2.29	2.36	2.41	2.45	2.49	2.53
20	.01	3.13	3.29	3.40	3.48	3.55	3.60	3.65	3.69	3.73	3.80	3.85	3.90	3.94	3.97
	.05	2.38	2.54	2.65	2.73	2.80	2.86	2.90	2.95	2.98	3.05	3.10	3.14	3.18	3.22
	.10	2.03	2.19	2.30	2.39	2.46	2.51	2.56	2.60	2.64	2.70	2.75	2.79	2.83	2.87
	.20	1.64	1.81	1.93	2.01	2.08	2.14	2.18	2.22	2.26	2.32	2.37	2.42	2.46	2.49
24	.01	3.07	3.22	3.32	3.40	3.47	3.52	3.57	3.61	3.64	3.70	3.76	3.80	3.84	3.87
	.05	2.35	2.51	2.61	2.70	2.76	2.81	2.86	2.90	2.94	3.00	3.05	3.09	3.13	3.16
	.10	2.01	2.17	2.28	2.36	2.43	2.48	2.53	2.57	2.60	2.66	2.71	2.76	2.79	2.83
	.20	1.63	1.80	1.91	1.99	2.06	2.12	2.16	2.20	2.24	2.30	2.35	2.39	2.43	2.47
30	.01	3.01	3.15	3.25	3.33	3.39	3.44	3.49	3.52	3.56	3.62	3.66	3.71	3.74	3.77
	.05	2.32	2.47	2.58	2.66	2.72	2.77	2.82	2.86	2.89	2.95	3.00	3.04	3.08	3.11
	.10	1.99	2.15	2.25	2.33	2.40	2.45	2.50	2.54	2.57	2.63	2.68	2.72	2.76	2.79
	.20	1.62	1.78	1.90	1.98	2.04	2.10	2.15	2.19	2.22	2.28	2.33	2.37	2.41	2.44
40	.01	2.95	3.09	3.19	3.26	3.32	3.37	3.41	3.44	3.48	3.53	3.58	3.62	3.65	3.68
	.05	2.29	2.44	2.54	2.62	2.68	2.73	2.77	2.81	2.85	2.90	2.95	2.99	3.02	3.05
	.10	1.97	2.13	2.23	2.31	2.37	2.42	2.47	2.51	2.54	2.60	2.65	2.69	2.72	2.75
	.20	1.61	1.77	1.88	1.96	2.03	2.08	2.13	2.17	2.20	2.26	2.31	2.35	2.39	2.42
60	.01	2.90	3.03	3.12	3.19	3.25	3.29	3.33	3.37	3.40	3.45	3.49	3.53	3.56	3.59
	.05	2.27	2.41	2.51	2.58	2.64	2.69	2.73	2.77	2.80	2.86	2.90	2.94	2.97	3.00
	.10	1.95	2.10	2.21	2.28	2.34	2.40	2.44	2.48	2.51	2.57	2.61	2.65	2.69	2.72
	.20	1.60	1.76	1.87	1.95	2.01	2.06	2.11	2.15	2.18	2.24	2.29	2.33	2.36	2.39
120	.01	2.85	2.97	3.06	3.12	3.18	3.22	3.26	3.29	3.32	3.37	3.41	3.45	3.48	3.50
	.05	2.24	2.38	2.47	2.55	2.60	2.65	2.69	2.73	2.76	2.81	2.86	2.89	2.93	2.95
	.10	1.93	2.08	2.18	2.26	2.32	2.37	2.41	2.45	2.48	2.53	2.58	2.62	2.65	2.68
	.20	1.59	1.75	1.85	1.93	1.99	2.05	2.09	2.13	2.16	2.22	2.27	2.30	2.34	2.37
∞	.01	2.79	2.92	3.00	3.06	3.11	3.15	3.19	3.22	3.25	3.29	3.33	3.37	3.40	3.42
	.05	2.21	2.35	2.44	2.51	2.57	2.61	2.65	2.69	2.72	2.77	2.81	2.85	2.88	2.91
	.10	1.92	2.06	2.16	2.23	2.29	2.34	2.38	2.42	2.45	2.50	2.55	2.58	2.62	2.64
	.20	1.58	1.73	1.84	1.92	1.98	2.03	2.07	2.11	2.14	2.20	2.24	2.28	2.32	2.35

ρ = 0.7

ν	α \ k	2	3	4	5	6	7	8	9	10	12	14	16	18	20
2	.01	12.0	13.2	14.0	14.6	15.0	15.4	15.8	16.1	16.3	16.8	17.1	17.4	17.7	17.9
	.05	5.24	5.76	6.12	6.39	6.61	6.78	6.94	7.07	7.18	7.38	7.54	7.67	7.79	7.90
	.10	3.59	3.97	4.22	4.41	4.57	4.69	4.80	4.89	4.97	5.11	5.22	5.32	5.40	5.48
	.20	2.38	2.64	2.83	2.96	3.07	3.16	3.23	3.30	3.36	3.45	3.53	3.60	3.66	3.71
3	.01	6.80	7.34	7.71	7.99	8.22	8.40	8.56	8.70	8.83	9.04	9.21	9.36	9.48	9.60
	.05	3.76	4.08	4.30	4.47	4.60	4.71	4.80	4.89	4.96	5.08	5.18	5.27	5.34	5.41
	.10	2.82	3.08	3.26	3.39	3.50	3.59	3.66	3.73	3.78	3.88	3.96	4.03	4.09	4.14
	.20	2.03	2.24	2.38	2.49	2.57	2.64	2.70	2.76	2.80	2.88	2.94	3.00	3.04	3.08
4	.01	5.25	5.61	5.86	6.06	6.21	6.33	6.44	6.54	6.62	6.76	6.88	6.98	7.07	7.15
	.05	3.23	3.48	3.65	3.78	3.88	3.97	4.04	4.11	4.16	4.26	4.34	4.41	4.47	4.52
	.10	2.52	2.74	2.89	3.00	3.09	3.16	3.22	3.28	3.32	3.40	3.47	3.53	3.58	3.62
	.20	1.88	2.07	2.19	2.29	2.37	2.43	2.48	2.53	2.57	2.63	2.69	2.74	2.78	2.82
5	.01	4.54	4.83	5.02	5.17	5.29	5.39	5.48	5.55	5.62	5.73	5.83	5.91	5.98	6.04
	.05	2.96	3.18	3.32	3.43	3.52	3.60	3.66	3.71	3.76	3.85	3.91	3.97	4.02	4.07
	.10	2.37	2.56	2.69	2.79	2.87	2.93	2.99	3.04	3.08	3.15	3.21	3.26	3.31	3.35
	.20	1.80	1.97	2.09	2.18	2.25	2.31	2.36	2.40	2.44	2.50	2.55	2.60	2.64	2.67
6	.01	4.14	4.38	4.55	4.68	4.78	4.87	4.94	5.00	5.06	5.15	5.23	5.30	5.36	5.41
	.05	2.80	2.99	3.13	3.23	3.31	3.38	3.43	3.48	3.52	3.60	3.66	3.71	3.76	3.80
	.10	2.27	2.45	2.57	2.66	2.74	2.80	2.85	2.89	2.93	3.00	3.05	3.10	3.14	3.18
	.20	1.75	1.92	2.03	2.11	2.18	2.23	2.28	2.32	2.35	2.41	2.46	2.51	2.54	2.58
7	.01	3.89	4.10	4.25	4.37	4.46	4.53	4.60	4.65	4.70	4.79	4.86	4.92	4.97	5.02
	.05	2.69	2.88	3.00	3.09	3.17	3.23	3.28	3.33	3.37	3.44	3.49	3.54	3.58	3.62
	.10	2.21	2.38	2.49	2.58	2.65	2.70	2.75	2.79	2.83	2.89	2.95	2.99	3.03	3.06
	.20	1.71	1.87	1.98	2.06	2.13	2.18	2.22	2.26	2.30	2.36	2.40	2.44	2.48	2.51
8	.01	3.71	3.91	4.05	4.15	4.23	4.30	4.36	4.41	4.46	4.54	4.60	4.66	4.70	4.75
	.05	2.62	2.79	2.91	3.00	3.07	3.12	3.17	3.22	3.26	3.32	3.37	3.42	3.46	3.49
	.10	2.16	2.32	2.43	2.52	2.58	2.64	2.68	2.72	2.76	2.82	2.87	2.91	2.95	2.98
	.20	1.69	1.85	1.95	2.03	2.09	2.14	2.19	2.22	2.26	2.31	2.36	2.40	2.43	2.46
9	.01	3.58	3.77	3.90	3.99	4.07	4.13	4.19	4.24	4.28	4.35	4.41	4.46	4.51	4.55
	.05	2.56	2.73	2.84	2.92	2.99	3.05	3.09	3.14	3.17	3.23	3.29	3.33	3.37	3.40
	.10	2.12	2.28	2.39	2.47	2.53	2.59	2.63	2.67	2.70	2.76	2.81	2.85	2.89	2.92
	.20	1.67	1.82	1.93	2.00	2.06	2.11	2.16	2.19	2.22	2.28	2.33	2.36	2.40	2.43
10	.01	3.48	3.66	3.78	3.87	3.95	4.01	4.06	4.10	4.14	4.21	4.27	4.32	4.36	4.40
	.05	2.52	2.68	2.79	2.87	2.93	2.99	3.03	3.07	3.11	3.17	3.22	3.26	3.30	3.33
	.10	2.10	2.25	2.36	2.43	2.50	2.55	2.59	2.63	2.66	2.72	2.77	2.81	2.84	2.87
	.20	1.65	1.81	1.91	1.98	2.04	2.09	2.13	2.17	2.20	2.25	2.30	2.34	2.37	2.40
12	.01	3.35	3.51	3.62	3.70	3.77	3.83	3.87	3.92	3.95	4.02	4.07	4.11	4.15	4.19
	.05	2.46	2.61	2.71	2.79	2.85	2.90	2.94	2.98	3.01	3.07	3.12	3.16	3.19	3.22
	.10	2.06	2.21	2.31	2.38	2.44	2.49	2.53	2.57	2.60	2.65	2.70	2.74	2.77	2.80
	.20	1.63	1.78	1.88	1.95	2.01	2.06	2.10	2.13	2.16	2.22	2.26	2.30	2.33	2.36

TABLE 5. Continued.

$\rho = 0.7$

ν	α \ k	2	3	4	5	6	7	8	9	10	12	14	16	18	20
16	.01	3.19	3.33	3.43	3.51	3.57	3.62	3.66	3.70	3.73	3.79	3.83	3.87	3.91	3.94
	.05	2.38	2.52	2.62	2.69	2.75	2.80	2.84	2.87	2.90	2.96	3.00	3.04	3.07	3.10
	.10	2.01	2.15	2.25	2.32	2.37	2.42	2.46	2.50	2.53	2.58	2.62	2.66	2.69	2.71
	.20	1.61	1.75	1.84	1.91	1.97	2.02	2.06	2.09	2.12	2.17	2.21	2.25	2.28	2.30
20	.01	3.09	3.23	3.33	3.40	3.45	3.50	3.54	3.58	3.61	3.66	3.70	3.74	3.77	3.80
	.05	2.34	2.48	2.57	2.64	2.69	2.74	2.78	2.81	2.84	2.89	2.93	2.97	3.00	3.03
	.10	1.98	2.12	2.21	2.28	2.34	2.38	2.42	2.45	2.48	2.53	2.57	2.61	2.64	2.67
	.20	1.59	1.73	1.82	1.89	1.95	1.99	2.03	2.06	2.09	2.14	2.18	2.22	2.25	2.27
24	.01	3.04	3.17	3.26	3.33	3.38	3.43	3.46	3.50	3.53	3.58	3.62	3.66	3.69	3.71
	.05	2.31	2.44	2.53	2.60	2.66	2.70	2.74	2.77	2.90	2.85	2.89	2.92	2.95	2.98
	.10	1.96	2.10	2.19	2.26	2.31	2.36	2.39	2.43	2.45	2.50	2.54	2.58	2.61	2.63
	.20	1.58	1.72	1.81	1.88	1.93	1.98	2.01	2.05	2.08	2.12	2.16	2.20	2.23	2.25
30	.01	2.98	3.11	3.19	3.26	3.31	3.35	3.39	3.42	3.45	3.50	3.54	3.57	3.60	3.63
	.05	2.28	2.41	2.50	2.57	2.62	2.66	2.70	2.73	2.76	2.81	2.85	2.88	2.91	2.93
	.10	1.95	2.08	2.17	2.23	2.29	2.33	2.37	2.40	2.43	2.47	2.51	2.55	2.58	2.60
	.20	1.57	1.71	1.80	1.86	1.92	1.96	2.00	2.03	2.06	2.11	2.15	2.18	2.21	2.23
40	.01	2.93	3.05	3.13	3.19	3.24	3.28	3.32	3.35	3.38	3.42	3.46	3.49	3.52	3.55
	.05	2.26	2.38	2.47	2.53	2.58	2.63	2.66	2.69	2.72	2.77	2.80	2.84	2.87	2.89
	.10	1.93	2.06	2.15	2.21	2.26	2.30	2.34	2.37	2.40	2.45	2.49	2.52	2.55	2.57
	.20	1.56	1.69	1.78	1.85	1.90	1.95	1.98	2.01	2.04	2.09	2.13	2.16	2.19	2.21
60	.01	2.87	2.99	3.07	3.13	3.18	3.22	3.25	3.28	3.31	3.35	3.39	3.42	3.44	3.47
	.05	2.23	2.35	2.44	2.50	2.55	2.59	2.63	2.66	2.68	2.73	2.76	2.80	2.82	2.85
	.10	1.91	2.04	2.12	2.19	2.24	2.28	2.32	2.35	2.37	2.42	2.46	2.49	2.52	2.54
	.20	1.55	1.68	1.77	1.84	1.89	1.93	1.97	2.00	2.03	2.07	2.11	2.14	2.17	2.20
120	.01	2.82	2.93	3.01	3.07	3.11	3.15	3.18	3.21	3.24	3.28	3.31	3.34	3.37	3.39
	.05	2.20	2.33	2.41	2.47	2.52	2.56	2.59	2.62	2.64	2.69	2.72	2.76	2.78	2.81
	.10	1.89	2.02	2.10	2.17	2.22	2.26	2.29	2.32	2.35	2.39	2.43	2.46	2.49	2.51
	.20	1.54	1.67	1.76	1.82	1.87	1.92	1.95	1.98	2.01	2.05	2.09	2.12	2.15	2.18
∞	.01	2.77	2.88	2.95	3.01	3.05	3.09	3.12	3.15	3.17	3.21	3.24	3.27	3.30	3.32
	.05	2.18	2.30	2.38	2.44	2.48	2.52	2.55	2.58	2.61	2.65	2.69	2.72	2.74	2.76
	.10	1.88	2.00	2.08	2.14	2.19	2.23	2.27	2.30	2.32	2.37	2.40	2.43	2.46	2.48
	.20	1.53	1.66	1.75	1.81	1.86	1.90	1.94	1.97	1.99	2.04	2.08	2.11	2.13	2.16

Source: Computed by C. W. Dunnett.

TABLE 6. Upper α Point of the Studentized Maximum Distribution with Parameter k and Degrees of Freedom ν ($M_{k,\nu}^{(\alpha)}$).

$\alpha = 0.20$

ν \ k	2	3	4	5	6	7	8	9	10	11	12	13	14	15	16	18	20
2	1.73	2.15	2.45	2.68	2.87	3.02	3.16	3.28	3.38	3.47	3.56	3.63	3.70	3.77	3.83	3.94	4.03
3	1.54	1.87	2.10	2.28	2.43	2.55	2.66	2.75	2.83	2.90	2.97	3.03	3.08	3.13	3.18	3.26	3.34
4	1.46	1.75	1.96	2.12	2.24	2.35	2.44	2.52	2.59	2.66	2.71	2.77	2.82	2.86	2.90	2.98	3.04
5	1.41	1.69	1.88	2.02	2.14	2.24	2.32	2.40	2.46	2.52	2.58	2.62	2.67	2.71	2.75	2.82	2.88
6	1.38	1.65	1.83	1.97	2.08	2.17	2.25	2.32	2.38	2.44	2.49	2.53	2.57	2.61	2.65	2.72	2.77
7	1.36	1.62	1.79	1.93	2.03	2.12	2.20	2.27	2.32	2.38	2.43	2.47	2.51	2.55	2.58	2.65	2.70
8	1.35	1.60	1.77	1.90	2.00	2.09	2.16	2.23	2.28	2.33	2.38	2.42	2.46	2.50	2.53	2.59	2.65
9	1.34	1.58	1.75	1.88	1.98	2.06	2.13	2.20	2.25	2.30	2.35	2.39	2.43	2.46	2.49	2.55	2.61
10	1.33	1.57	1.73	1.86	1.96	2.04	2.11	2.17	2.23	2.28	2.32	2.36	2.40	2.43	2.46	2.52	2.57
11	1.32	1.56	1.72	1.84	1.94	2.02	2.09	2.15	2.21	2.25	2.30	2.34	2.37	2.41	2.44	2.50	2.55
12	1.31	1.55	1.71	1.83	1.93	2.01	2.08	2.14	2.19	2.24	2.28	2.32	2.35	2.39	2.42	2.47	2.52
13	1.31	1.54	1.70	1.82	1.92	2.00	2.06	2.12	2.18	2.22	2.26	2.30	2.34	2.37	2.40	2.46	2.51
14	1.30	1.54	1.70	1.81	1.91	1.99	2.05	2.11	2.16	2.21	2.25	2.29	2.32	2.36	2.39	2.44	2.49
15	1.30	1.53	1.69	1.81	1.90	1.98	2.04	2.10	2.15	2.20	2.24	2.28	2.31	2.34	2.37	2.43	2.48
16	1.30	1.53	1.68	1.80	1.89	1.97	2.04	2.09	2.14	2.19	2.23	2.27	2.30	2.33	2.36	2.42	2.46
17	1.29	1.52	1.68	1.79	1.89	1.96	2.03	2.09	2.14	2.18	2.22	2.26	2.29	2.32	2.35	2.41	2.45
18	1.29	1.52	1.67	1.79	1.88	1.96	2.02	2.08	2.13	2.17	2.21	2.25	2.29	2.32	2.35	2.40	2.44
19	1.29	1.52	1.67	1.79	1.88	1.95	2.02	2.07	2.12	2.17	2.21	2.24	2.28	2.31	2.34	2.39	2.44
20	1.29	1.51	1.67	1.78	1.87	1.95	2.01	2.07	2.12	2.16	2.20	2.24	2.27	2.30	2.33	2.38	2.43
21	1.29	1.51	1.66	1.78	1.87	1.94	2.01	2.06	2.11	2.16	2.20	2.23	2.27	2.30	2.32	2.38	2.42
22	1.28	1.51	1.66	1.77	1.87	1.94	2.00	2.06	2.11	2.15	2.19	2.23	2.26	2.29	2.32	2.37	2.42
23	1.28	1.51	1.66	1.77	1.86	1.94	2.00	2.05	2.10	2.15	2.19	2.22	2.26	2.29	2.31	2.36	2.41
24	1.28	1.51	1.66	1.77	1.86	1.93	2.00	2.05	2.10	2.14	2.18	2.22	2.25	2.28	2.31	2.36	2.41
25	1.28	1.50	1.65	1.77	1.86	1.93	1.99	2.05	2.10	2.14	2.18	2.21	2.25	2.28	2.30	2.36	2.40
26	1.28	1.50	1.65	1.76	1.85	1.93	1.99	2.04	2.09	2.14	2.17	2.21	2.24	2.27	2.30	2.35	2.40
27	1.28	1.50	1.65	1.76	1.85	1.93	1.99	2.04	2.09	2.13	2.17	2.21	2.24	2.27	2.30	2.35	2.39
28	1.28	1.50	1.65	1.76	1.85	1.92	1.99	2.04	2.09	2.13	2.17	2.20	2.24	2.27	2.29	2.34	2.39
29	1.28	1.50	1.65	1.76	1.85	1.92	1.98	2.04	2.08	2.13	2.17	2.20	2.23	2.26	2.29	2.34	2.38
30	1.27	1.50	1.65	1.76	1.85	1.92	1.98	2.03	2.08	2.12	2.16	2.20	2.23	2.26	2.29	2.34	2.38
35	1.27	1.49	1.64	1.75	1.84	1.91	1.97	2.02	2.07	2.11	2.15	2.19	2.22	2.25	2.28	2.32	2.37
40	1.27	1.49	1.64	1.75	1.83	1.90	1.96	2.02	2.06	2.11	2.14	2.18	2.21	2.24	2.27	2.31	2.36
45	1.27	1.49	1.63	1.74	1.83	1.90	1.96	2.01	2.06	2.10	2.14	2.17	2.20	2.23	2.26	2.31	2.35
50	1.27	1.48	1.63	1.74	1.82	1.90	1.95	2.01	2.05	2.10	2.13	2.17	2.20	2.23	2.25	2.30	2.34
60	1.26	1.48	1.63	1.73	1.82	1.89	1.95	2.00	2.05	2.09	2.13	2.16	2.19	2.22	2.24	2.29	2.33
80	1.26	1.48	1.62	1.73	1.81	1.88	1.94	1.99	2.04	2.08	2.12	2.15	2.18	2.21	2.23	2.28	2.32
100	1.26	1.47	1.62	1.72	1.81	1.88	1.94	1.99	2.03	2.07	2.11	2.14	2.17	2.20	2.23	2.27	2.32
120	1.26	1.47	1.62	1.72	1.81	1.88	1.93	1.99	2.03	2.07	2.11	2.14	2.17	2.20	2.22	2.27	2.31
200	1.25	1.47	1.61	1.72	1.80	1.87	1.93	1.98	2.02	2.06	2.10	2.13	2.16	2.19	2.21	2.26	2.30
∞	1.25	1.46	1.60	1.71	1.79	1.86	1.92	1.97	2.01	2.05	2.09	2.12	2.15	2.18	2.20	2.25	2.29

TABLE 6. Continued.

$\alpha = 0.10$

ν \ k	2	3	4	5	6	7	8	9	10	11	12	13	14	15	16	18	20
2	2.74	3.29	3.70	4.02	4.28	4.50	4.69	4.85	5.00	5.13	5.25	5.36	5.46	5.55	5.64	5.79	5.93
3	2.26	2.65	2.93	3.14	3.32	3.47	3.60	3.71	3.81	3.90	3.98	4.06	4.13	4.19	4.25	4.36	4.45
4	2.07	2.39	2.62	2.80	2.94	3.06	3.17	3.26	3.34	3.42	3.49	3.55	3.60	3.66	3.71	3.79	3.87
5	1.97	2.26	2.46	2.62	2.74	2.85	2.94	3.03	3.10	3.16	3.22	3.28	3.33	3.37	3.42	3.50	3.57
6	1.90	2.17	2.36	2.50	2.62	2.72	2.81	2.88	2.95	3.01	3.06	3.11	3.16	3.20	3.24	3.31	3.38
7	1.86	2.11	2.29	2.43	2.54	2.63	2.71	2.78	2.84	2.90	2.95	3.00	3.04	3.08	3.12	3.19	3.25
8	1.83	2.07	2.24	2.37	2.48	2.57	2.64	2.71	2.77	2.82	2.87	2.92	2.96	2.99	3.03	3.09	3.15
9	1.81	2.04	2.21	2.33	2.43	2.52	2.59	2.66	2.71	2.76	2.81	2.85	2.89	2.93	2.96	3.03	3.08
10	1.79	2.02	2.18	2.30	2.40	2.48	2.55	2.61	2.67	2.72	2.76	2.81	2.84	2.88	2.91	2.97	3.02
11	1.77	2.00	2.15	2.27	2.37	2.45	2.52	2.58	2.63	2.68	2.73	2.77	2.80	2.84	2.87	2.93	2.98
12	1.76	1.98	2.14	2.25	2.35	2.43	2.49	2.55	2.60	2.65	2.69	2.73	2.77	2.80	2.84	2.89	2.94
13	1.75	1.97	2.12	2.23	2.33	2.40	2.47	2.53	2.58	2.63	2.67	2.71	2.74	2.78	2.81	2.86	2.91
14	1.74	1.96	2.11	2.22	2.31	2.39	2.45	2.51	2.56	2.61	2.65	2.68	2.72	2.75	2.78	2.84	2.88
15	1.73	1.95	2.09	2.21	2.30	2.37	2.44	2.49	2.54	2.59	2.63	2.67	2.70	2.73	2.76	2.81	2.86
16	1.73	1.94	2.08	2.20	2.28	2.36	2.42	2.48	2.53	2.57	2.61	2.65	2.68	2.71	2.74	2.80	2.84
17	1.72	1.93	2.08	2.19	2.27	2.35	2.41	2.46	2.51	2.56	2.60	2.63	2.67	2.70	2.73	2.78	2.82
18	1.71	1.92	2.07	2.18	2.26	2.34	2.40	2.45	2.50	2.54	2.58	2.62	2.65	2.68	2.71	2.76	2.81
19	1.71	1.92	2.06	2.17	2.26	2.33	2.39	2.44	2.49	2.53	2.57	2.61	2.64	2.67	2.70	2.75	2.80
20	1.71	1.91	2.06	2.16	2.25	2.32	2.38	2.43	2.48	2.52	2.56	2.60	2.63	2.66	2.69	2.74	2.78
21	1.70	1.91	2.05	2.16	2.24	2.31	2.37	2.43	2.47	2.52	2.55	2.59	2.62	2.65	2.68	2.73	2.77
22	1.70	1.90	2.04	2.15	2.24	2.31	2.37	2.42	2.47	2.51	2.54	2.58	2.61	2.64	2.67	2.72	2.76
23	1.70	1.90	2.04	2.15	2.23	2.30	2.36	2.41	2.46	2.50	2.54	2.57	2.60	2.63	2.66	2.71	2.75
24	1.69	1.90	2.04	2.14	2.22	2.29	2.35	2.41	2.45	2.49	2.53	2.56	2.60	2.63	2.65	2.70	2.75
25	1.69	1.89	2.03	2.14	2.22	2.29	2.35	2.40	2.45	2.49	2.52	2.56	2.59	2.62	2.65	2.69	2.74
26	1.69	1.89	2.03	2.13	2.22	2.28	2.34	2.39	2.44	2.48	2.52	2.55	2.58	2.61	2.64	2.69	2.73
27	1.69	1.89	2.03	2.13	2.21	2.28	2.34	2.39	2.44	2.48	2.51	2.55	2.58	2.61	2.63	2.68	2.72
28	1.68	1.89	2.02	2.13	2.21	2.28	2.33	2.39	2.43	2.47	2.51	2.54	2.57	2.60	2.63	2.68	2.72
29	1.68	1.88	2.02	2.12	2.20	2.27	2.33	2.38	2.43	2.47	2.50	2.54	2.57	2.60	2.62	2.67	2.71
30	1.68	1.88	2.02	2.12	2.20	2.27	2.33	2.38	2.42	2.46	2.50	2.53	2.56	2.59	2.62	2.66	2.71
35	1.67	1.87	2.01	2.11	2.19	2.25	2.31	2.36	2.41	2.45	2.48	2.51	2.54	2.57	2.60	2.64	2.69
40	1.67	1.86	2.00	2.10	2.18	2.24	2.30	2.35	2.39	2.43	2.47	2.50	2.53	2.56	2.58	2.63	2.67
45	1.66	1.86	1.99	2.09	2.17	2.24	2.29	2.34	2.38	2.42	2.46	2.49	2.52	2.55	2.57	2.62	2.66
50	1.66	1.86	1.99	2.09	2.16	2.23	2.28	2.33	2.38	2.41	2.45	2.48	2.51	2.54	2.56	2.61	2.65
60	1.66	1.85	1.98	2.08	2.16	2.22	2.27	2.32	2.36	2.40	2.44	2.47	2.50	2.52	2.55	2.59	2.63
80	1.65	1.84	1.97	2.07	2.14	2.21	2.26	2.31	2.35	2.39	2.42	2.45	2.48	2.51	2.53	2.57	2.61
100	1.65	1.84	1.96	2.06	2.14	2.20	2.25	2.30	2.34	2.38	2.41	2.44	2.47	2.50	2.52	2.56	2.60
120	1.64	1.83	1.96	2.06	2.13	2.20	2.25	2.30	2.34	2.37	2.41	2.44	2.46	2.49	2.51	2.56	2.60
200	1.64	1.83	1.95	2.05	2.12	2.19	2.24	2.28	2.33	2.36	2.39	2.42	2.45	2.48	2.50	2.54	2.58
∞	1.63	1.82	1.94	2.04	2.11	2.17	2.22	2.27	2.31	2.34	2.38	2.41	2.43	2.46	2.48	2.52	2.56

$\alpha = 0.05$

ν \ k	2	3	4	5	6	7	8	9	10	11	12	13	14	15	16	18	20
2	4.08	4.83	5.40	5.84	6.21	6.52	6.79	7.01	7.22	7.41	7.57	7.73	7.87	8.00	8.12	8.34	8.53
3	3.09	3.55	3.89	4.15	4.37	4.56	4.72	4.86	4.98	5.10	5.20	5.29	5.38	5.46	5.53	5.67	5.79
4	2.72	3.08	3.34	3.54	3.71	3.85	3.97	4.08	4.18	4.27	4.34	4.42	4.48	4.55	4.60	4.71	4.80
5	2.53	2.84	3.06	3.23	3.38	3.50	3.60	3.69	3.77	3.85	3.91	3.97	4.03	4.08	4.13	4.22	4.30
6	2.42	2.70	2.89	3.05	3.17	3.28	3.37	3.46	3.53	3.59	3.65	3.71	3.76	3.80	3.85	3.93	4.00
7	2.34	2.60	2.78	2.93	3.04	3.14	3.22	3.30	3.37	3.43	3.48	3.53	3.58	3.62	3.66	3.73	3.80
8	2.28	2.53	2.70	2.84	2.95	3.04	3.12	3.19	3.25	3.31	3.36	3.40	3.45	3.49	3.53	3.59	3.65
9	2.24	2.48	2.64	2.77	2.88	2.96	3.04	3.10	3.16	3.22	3.26	3.31	3.35	3.39	3.42	3.49	3.55
10	2.21	2.44	2.60	2.72	2.82	2.90	2.98	3.04	3.10	3.15	3.19	3.24	3.28	3.31	3.35	3.41	3.46
11	2.19	2.41	2.56	2.68	2.78	2.86	2.93	2.99	3.04	3.09	3.14	3.18	3.22	3.25	3.28	3.34	3.40
12	2.16	2.38	2.53	2.65	2.74	2.82	2.89	2.95	3.00	3.05	3.09	3.13	3.17	3.20	3.23	3.29	3.34
13	2.15	2.36	2.51	2.62	2.71	2.79	2.85	2.91	2.96	3.01	3.05	3.09	3.13	3.16	3.19	3.25	3.30
14	2.13	2.34	2.48	2.60	2.69	2.76	2.82	2.88	2.93	2.98	3.02	3.06	3.09	3.12	3.15	3.21	3.26
15	2.12	2.32	2.47	2.58	2.66	2.74	2.80	2.86	2.91	2.95	2.99	3.03	3.06	3.09	3.12	3.18	3.22
16	2.11	2.31	2.45	2.56	2.64	2.72	2.78	2.83	2.88	2.93	2.97	3.00	3.04	3.07	3.10	3.15	3.20
17	2.10	2.30	2.44	2.54	2.63	2.70	2.76	2.81	2.86	2.91	2.94	2.98	3.01	3.04	3.07	3.12	3.17
18	2.09	2.29	2.43	2.53	2.61	2.68	2.75	2.80	2.85	2.89	2.93	2.96	2.99	3.02	3.05	3.10	3.15
19	2.08	2.28	2.41	2.52	2.60	2.67	2.73	2.78	2.83	2.87	2.91	2.94	2.98	3.01	3.03	3.08	3.13
20	2.09	2.27	2.41	2.51	2.59	2.66	2.72	2.77	2.82	2.86	2.89	2.93	2.96	2.99	3.02	3.07	3.11
21	2.07	2.26	2.40	2.50	2.58	2.65	2.71	2.76	2.80	2.84	2.88	2.92	2.95	2.98	3.00	3.05	3.10
22	2.06	2.26	2.39	2.49	2.57	2.64	2.70	2.75	2.79	2.83	2.87	2.90	2.93	2.96	2.99	3.04	3.08
23	2.06	2.25	2.38	2.48	2.56	2.63	2.69	2.74	2.78	2.82	2.86	2.89	2.92	2.95	2.98	3.03	3.07
24	2.05	2.24	2.38	2.47	2.55	2.62	2.68	2.73	2.77	2.81	2.85	2.88	2.91	2.94	2.97	3.01	3.06
25	2.05	2.24	2.37	2.47	2.55	2.61	2.67	2.72	2.76	2.80	2.84	2.87	2.90	2.93	2.96	3.00	3.05
26	2.05	2.23	2.36	2.46	2.54	2.61	2.66	2.71	2.76	2.80	2.83	2.86	2.89	2.92	2.95	2.99	3.04
27	2.04	2.23	2.36	2.46	2.53	2.60	2.66	2.71	2.75	2.79	2.82	2.86	2.89	2.91	2.94	2.99	3.03
28	2.04	2.23	2.35	2.45	2.53	2.59	2.65	2.70	2.74	2.78	2.82	2.85	2.88	2.91	2.93	2.98	3.02
29	2.04	2.22	2.35	2.45	2.52	2.59	2.64	2.69	2.74	2.77	2.81	2.84	2.87	2.90	2.92	2.97	3.01
30	2.03	2.22	2.35	2.44	2.52	2.58	2.64	2.69	2.73	2.77	2.80	2.83	2.86	2.89	2.92	2.96	3.00
35	2.02	2.20	2.33	2.42	2.50	2.56	2.62	2.66	2.71	2.74	2.78	2.81	2.84	2.86	2.89	2.93	2.97
40	2.01	2.19	2.32	2.41	2.48	2.55	2.60	2.65	2.69	2.72	2.76	2.79	2.82	2.84	2.87	2.91	2.95
45	2.01	2.19	2.31	2.40	2.47	2.54	2.59	2.63	2.67	2.71	2.74	2.77	2.80	2.83	2.85	2.90	2.93
50	2.00	2.18	2.30	2.39	2.46	2.53	2.58	2.62	2.66	2.70	2.73	2.76	2.79	2.82	2.84	2.88	2.92
60	1.99	2.17	2.29	2.38	2.45	2.51	2.56	2.61	2.65	2.68	2.71	2.74	2.77	2.80	2.82	2.86	2.90
80	1.98	2.16	2.27	2.36	2.43	2.49	2.54	2.59	2.63	2.66	2.69	2.72	2.75	2.77	2.80	2.84	2.87
100	1.98	2.15	2.27	2.35	2.42	2.48	2.53	2.58	2.62	2.65	2.68	2.71	2.74	2.76	2.78	2.82	2.86
120	1.97	2.14	2.26	2.35	2.42	2.48	2.53	2.57	2.61	2.64	2.67	2.70	2.73	2.75	2.77	2.81	2.85
200	1.97	2.14	2.25	2.34	2.41	2.46	2.51	2.55	2.59	2.62	2.66	2.68	2.71	2.73	2.75	2.79	2.83
∞	1.95	2.12	2.23	2.32	2.39	2.44	2.49	2.53	2.57	2.60	2.63	2.66	2.68	2.71	2.73	2.77	2.80

TABLE 6. Continued.

α = 0.01

v \ k	2	3	4	5	6	7	8	9	10	11	12	13	14	15	16	18	20
2	9.45	11.11	12.35	13.34	14.15	14.83	15.43	15.94	16.40	16.82	17.19	17.54	17.85	18.14	18.41	18.90	19.33
3	5.71	6.46	7.02	7.46	7.83	8.14	8.41	8.64	8.86	9.05	9.22	9.38	9.53	9.67	9.79	10.03	10.23
4	4.54	5.04	5.41	5.70	5.94	6.15	6.33	6.48	6.62	6.75	6.87	6.98	7.07	7.17	7.25	7.41	7.55
5	4.00	4.39	4.67	4.90	5.08	5.24	5.38	5.50	5.61	5.70	5.79	5.88	5.95	6.03	6.09	6.21	6.32
6	3.68	4.01	4.25	4.44	4.59	4.73	4.84	4.94	5.03	5.12	5.19	5.26	5.32	5.38	5.44	5.54	5.63
7	3.48	3.77	3.98	4.15	4.28	4.40	4.50	4.59	4.67	4.74	4.81	4.87	4.92	4.97	5.02	5.11	5.19
8	3.34	3.61	3.80	3.95	4.07	4.17	4.27	4.35	4.42	4.48	4.54	4.60	4.65	4.69	4.74	4.82	4.89
9	3.24	3.49	3.66	3.80	3.91	4.01	4.09	4.17	4.23	4.29	4.35	4.40	4.44	4.49	4.53	4.60	4.67
10	3.16	3.39	3.56	3.69	3.80	3.89	3.96	4.03	4.09	4.15	4.20	4.25	4.29	4.33	4.37	4.44	4.50
11	3.10	3.32	3.48	3.60	3.70	3.79	3.86	3.93	3.98	4.04	4.09	4.13	4.17	4.21	4.24	4.31	4.37
12	3.05	3.26	3.41	3.53	3.63	3.71	3.78	3.84	3.90	3.95	3.99	4.03	4.07	4.11	4.14	4.21	4.26
13	3.01	3.21	3.36	3.47	3.57	3.64	3.71	3.77	3.82	3.87	3.92	3.96	3.99	4.03	4.06	4.12	4.17
14	2.97	3.17	3.31	3.42	3.51	3.59	3.65	3.71	3.76	3.81	3.85	3.89	3.93	3.96	3.99	4.05	4.10
15	2.94	3.14	3.28	3.38	3.47	3.54	3.61	3.66	3.71	3.76	3.80	3.84	3.87	3.90	3.93	3.99	4.04
16	2.92	3.11	3.24	3.35	3.43	3.50	3.56	3.62	3.67	3.71	3.75	3.79	3.82	3.85	3.88	3.94	3.98
17	2.89	3.08	3.21	3.32	3.40	3.47	3.53	3.58	3.63	3.67	3.71	3.75	3.78	3.81	3.84	3.89	3.94
18	2.87	3.06	3.19	3.29	3.37	3.44	3.50	3.55	3.60	3.64	3.68	3.71	3.74	3.77	3.80	3.85	3.90
19	2.86	3.04	3.17	3.26	3.34	3.41	3.47	3.52	3.57	3.61	3.64	3.68	3.71	3.74	3.77	3.82	3.86
20	2.84	3.02	3.15	3.24	3.32	3.39	3.44	3.49	3.54	3.58	3.62	3.65	3.68	3.71	3.74	3.79	3.83
21	2.83	3.01	3.13	3.22	3.30	3.37	3.42	3.47	3.52	3.56	3.59	3.62	3.65	3.68	3.71	3.76	3.80
22	2.82	2.99	3.11	3.21	3.28	3.35	3.40	3.45	3.49	3.53	3.57	3.60	3.63	3.66	3.69	3.73	3.78
23	2.80	2.98	3.10	3.19	3.27	3.33	3.38	3.43	3.47	3.51	3.55	3.58	3.61	3.64	3.66	3.71	3.75
24	2.79	2.97	3.09	3.18	3.25	3.31	3.37	3.42	3.46	3.50	3.53	3.56	3.59	3.62	3.64	3.69	3.73
25	2.79	2.95	3.07	3.16	3.24	3.30	3.35	3.40	3.44	3.48	3.51	3.54	3.57	3.60	3.63	3.67	3.71
26	2.78	2.94	3.06	3.15	3.23	3.29	3.34	3.39	3.43	3.46	3.50	3.53	3.56	3.58	3.61	3.65	3.69
27	2.77	2.94	3.05	3.14	3.21	3.27	3.33	3.37	3.41	3.45	3.48	3.51	3.54	3.57	3.59	3.64	3.68
28	2.76	2.93	3.04	3.13	3.20	3.26	3.31	3.36	3.40	3.44	3.47	3.50	3.53	3.55	3.58	3.62	3.66
29	2.75	2.92	3.03	3.12	3.19	3.25	3.30	3.35	3.39	3.43	3.46	3.49	3.52	3.54	3.57	3.61	3.65
30	2.75	2.91	3.03	3.11	3.18	3.24	3.29	3.34	3.38	3.41	3.45	3.48	3.50	3.53	3.55	3.60	3.64
35	2.72	2.88	2.99	3.08	3.15	3.20	3.25	3.30	3.33	3.37	3.40	3.43	3.46	3.48	3.50	3.55	3.58
40	2.70	2.86	2.97	3.05	3.12	3.17	3.22	3.26	3.30	3.34	3.37	3.39	3.42	3.44	3.47	3.51	3.54
45	2.69	2.84	2.95	3.03	3.10	3.15	3.20	3.24	3.28	3.31	3.34	3.37	3.39	3.42	3.44	3.48	3.51
50	2.68	2.83	2.93	3.01	3.08	3.13	3.18	3.22	3.26	3.29	3.32	3.35	3.37	3.40	3.42	3.46	3.49
60	2.66	2.81	2.91	2.99	3.05	3.11	3.15	3.19	3.23	3.26	3.29	3.32	3.34	3.36	3.38	3.42	3.46
80	2.64	2.78	2.88	2.96	3.02	3.08	3.12	3.16	3.19	3.22	3.25	3.28	3.30	3.32	3.34	3.38	3.41
100	2.62	2.77	2.87	2.94	3.01	3.06	3.10	3.14	3.17	3.20	3.23	3.25	3.28	3.30	3.32	3.36	3.39
120	2.62	2.76	2.86	2.93	2.99	3.04	3.09	3.12	3.16	3.19	3.21	3.24	3.26	3.28	3.30	3.34	3.37
200	2.60	2.74	2.84	2.91	2.97	3.02	3.06	3.10	3.13	3.16	3.19	3.21	3.23	3.25	3.27	3.31	3.34
∞	2.57	2.71	2.81	2.88	2.93	2.98	3.02	3.06	3.09	3.12	3.14	3.17	3.19	3.21	3.23	3.26	3.29

Source: Computed by C. W. Dunnett.

TABLE 7. Upper α Point of the Studentized Maximum Modulus Distribution with Parameter k and Degrees of Freedom ν ($|M|_{k,\nu}^{(\alpha)}$).

$\alpha = 0.20$

ν \ k	2	3	4	5	6	7	8	9	10	11	12	13	14	15	16	18	20
2	2.55	2.94	3.22	3.43	3.60	3.74	3.86	3.96	4.06	4.14	4.21	4.28	4.35	4.40	4.46	4.56	4.64
3	2.17	2.48	2.69	2.86	2.99	3.10	3.20	3.28	3.36	3.42	3.48	3.54	3.59	3.63	3.68	3.75	3.82
4	2.01	2.28	2.47	2.62	2.73	2.83	2.92	2.99	3.06	3.11	3.17	3.22	3.26	3.30	3.34	3.41	3.47
5	1.92	2.17	2.35	2.48	2.59	2.68	2.76	2.83	2.89	2.94	2.99	3.04	3.08	3.12	3.15	3.22	3.27
6	1.86	2.10	2.27	2.40	2.50	2.59	2.66	2.73	2.78	2.83	2.88	2.92	2.96	3.00	3.03	3.09	3.15
7	1.82	2.06	2.22	2.34	2.44	2.52	2.59	2.65	2.71	2.76	2.80	2.84	2.88	2.92	2.95	3.01	3.06
8	1.80	2.02	2.18	2.30	2.39	2.47	2.54	2.60	2.65	2.70	2.75	2.78	2.82	2.85	2.89	2.94	2.99
9	1.77	2.00	2.15	2.26	2.36	2.44	2.50	2.56	2.61	2.66	2.70	2.74	2.77	2.81	2.84	2.89	2.94
10	1.76	1.97	2.12	2.24	2.33	2.41	2.47	2.53	2.58	2.62	2.67	2.70	2.74	2.77	2.80	2.85	2.90
11	1.74	1.96	2.11	2.22	2.31	2.38	2.45	2.50	2.55	2.60	2.64	2.67	2.71	2.74	2.77	2.82	2.87
12	1.73	1.94	2.09	2.20	2.29	2.36	2.43	2.48	2.53	2.57	2.61	2.65	2.68	2.71	2.74	2.80	2.84
13	1.72	1.93	2.08	2.19	2.27	2.35	2.41	2.46	2.51	2.55	2.59	2.63	2.66	2.69	2.72	2.77	2.82
14	1.72	1.92	2.07	2.17	2.26	2.33	2.39	2.45	2.50	2.54	2.58	2.61	2.65	2.68	2.70	2.75	2.80
15	1.71	1.91	2.06	2.16	2.25	2.32	2.38	2.43	2.48	2.52	2.56	2.60	2.63	2.66	2.69	2.74	2.78
16	1.70	1.91	2.05	2.15	2.24	2.31	2.37	2.42	2.47	2.51	2.55	2.58	2.62	2.65	2.67	2.72	2.77
17	1.70	1.90	2.04	2.14	2.23	2.30	2.36	2.41	2.46	2.50	2.54	2.57	2.60	2.63	2.66	2.71	2.75
18	1.69	1.90	2.03	2.14	2.22	2.29	2.35	2.40	2.45	2.49	2.53	2.56	2.59	2.62	2.65	2.70	2.74
19	1.69	1.89	2.03	2.13	2.21	2.28	2.34	2.39	2.44	2.48	2.52	2.55	2.58	2.61	2.64	2.69	2.73
20	1.69	1.89	2.02	2.13	2.21	2.28	2.34	2.39	2.43	2.47	2.51	2.54	2.58	2.60	2.63	2.68	2.72
21	1.68	1.88	2.02	2.12	2.20	2.27	2.33	2.38	2.43	2.47	2.50	2.54	2.57	2.60	2.62	2.67	2.71
22	1.68	1.88	2.01	2.12	2.20	2.27	2.32	2.37	2.42	2.46	2.50	2.53	2.56	2.59	2.62	2.66	2.71
23	1.68	1.87	2.01	2.11	2.19	2.26	2.32	2.37	2.41	2.45	2.49	2.52	2.55	2.58	2.61	2.66	2.70
24	1.67	1.87	2.01	2.11	2.19	2.26	2.31	2.36	2.41	2.45	2.49	2.52	2.55	2.58	2.60	2.65	2.69
25	1.67	1.87	2.00	2.10	2.18	2.25	2.31	2.36	2.40	2.44	2.48	2.51	2.54	2.57	2.60	2.64	2.69
26	1.67	1.87	2.00	2.10	2.18	2.25	2.31	2.36	2.40	2.44	2.48	2.51	2.54	2.57	2.59	2.64	2.68
27	1.67	1.86	2.00	2.10	2.18	2.24	2.30	2.35	2.40	2.44	2.47	2.50	2.53	2.56	2.59	2.63	2.68
28	1.67	1.86	1.99	2.09	2.17	2.24	2.30	2.35	2.39	2.43	2.47	2.50	2.53	2.56	2.58	2.63	2.67
29	1.66	1.86	1.99	2.09	2.17	2.24	2.29	2.34	2.39	2.43	2.46	2.50	2.53	2.55	2.58	2.63	2.67
30	1.66	1.86	1.99	2.09	2.17	2.24	2.29	2.34	2.38	2.42	2.46	2.49	2.52	2.55	2.57	2.62	2.66
35	1.66	1.85	1.98	2.08	2.16	2.22	2.28	2.33	2.37	2.41	2.44	2.48	2.51	2.53	2.56	2.60	2.65
40	1.65	1.84	1.97	2.07	2.15	2.21	2.27	2.32	2.36	2.40	2.43	2.47	2.49	2.52	2.55	2.59	2.63
45	1.65	1.84	1.97	2.06	2.14	2.21	2.26	2.31	2.35	2.39	2.43	2.46	2.49	2.51	2.54	2.58	2.62
50	1.64	1.83	1.96	2.06	2.14	2.20	2.26	2.30	2.35	2.38	2.42	2.45	2.48	2.51	2.53	2.57	2.61
60	1.64	1.83	1.96	2.05	2.13	2.19	2.25	2.30	2.34	2.37	2.41	2.44	2.47	2.49	2.52	2.56	2.60
80	1.63	1.82	1.95	2.04	2.12	2.18	2.24	2.28	2.33	2.36	2.40	2.43	2.45	2.48	2.50	2.55	2.59
100	1.63	1.82	1.94	2.04	2.11	2.18	2.23	2.28	2.32	2.35	2.39	2.42	2.45	2.47	2.50	2.54	2.58
120	1.63	1.81	1.94	2.04	2.11	2.17	2.23	2.27	2.31	2.35	2.38	2.41	2.44	2.47	2.49	2.53	2.57
200	1.62	1.81	1.93	2.03	2.10	2.16	2.22	2.26	2.30	2.34	2.37	2.40	2.43	2.45	2.48	2.52	2.56
∞	1.62	1.80	1.92	2.02	2.09	2.15	2.20	2.25	2.29	2.32	2.36	2.39	2.41	2.44	2.46	2.50	2.54

TABLE 7. Continued.

α = 0.10

ν \ k	2	3	4	5	6	7	8	9	10	11	12	13	14	15	16	18	20
2	3.83	4.38	4.77	5.06	5.30	5.50	5.67	5.82	5.96	6.08	6.18	6.28	6.37	6.45	6.53	6.67	6.80
3	2.99	3.37	3.64	3.84	4.01	4.15	4.27	4.38	4.47	4.55	4.63	4.70	4.76	4.82	4.88	4.98	5.07
4	2.66	2.98	3.20	3.37	3.51	3.62	3.72	3.81	3.89	3.96	4.02	4.08	4.13	4.18	4.23	4.31	4.38
5	2.49	2.77	2.96	3.12	3.24	3.34	3.43	3.51	3.58	3.64	3.69	3.75	3.79	3.84	3.88	3.95	4.02
6	2.38	2.64	2.82	2.96	3.07	3.17	3.25	3.32	3.38	3.44	3.49	3.54	3.58	3.62	3.66	3.73	3.79
7	2.31	2.56	2.73	2.86	2.96	3.05	3.13	3.19	3.25	3.31	3.35	3.40	3.44	3.48	3.51	3.58	3.63
8	2.26	2.49	2.66	2.78	2.88	2.96	3.04	3.10	3.16	3.21	3.26	3.30	3.34	3.37	3.41	3.47	3.52
9	2.22	2.45	2.60	2.72	2.82	2.90	2.97	3.03	3.09	3.13	3.18	3.22	3.26	3.29	3.32	3.38	3.44
10	2.19	2.41	2.56	2.68	2.77	2.85	2.92	2.98	3.03	3.08	3.12	3.16	3.20	3.22	3.26	3.32	3.37
11	2.17	2.38	2.53	2.64	2.73	2.81	2.88	2.93	2.98	3.03	3.07	3.11	3.15	3.18	3.21	3.26	3.31
12	2.15	2.36	2.50	2.61	2.70	2.78	2.84	2.90	2.95	2.99	3.03	3.07	3.10	3.14	3.17	3.22	3.27
13	2.13	2.34	2.48	2.59	2.67	2.75	2.81	2.87	2.91	2.96	3.00	3.04	3.07	3.10	3.13	3.18	3.23
14	2.12	2.32	2.46	2.57	2.65	2.72	2.79	2.84	2.89	2.93	2.97	3.01	3.04	3.07	3.10	3.15	3.20
15	2.11	2.31	2.44	2.55	2.63	2.70	2.76	2.82	2.87	2.91	2.95	2.98	3.01	3.04	3.07	3.12	3.17
16	2.10	2.29	2.43	2.53	2.62	2.69	2.75	2.80	2.85	2.89	2.93	2.96	2.99	3.02	3.05	3.10	3.15
17	2.09	2.28	2.42	2.52	2.60	2.67	2.73	2.78	2.83	2.87	2.91	2.94	2.97	3.00	3.03	3.08	3.12
18	2.08	2.27	2.41	2.51	2.59	2.66	2.72	2.77	2.81	2.85	2.89	2.92	2.96	2.99	3.01	3.06	3.10
19	2.07	2.26	2.40	2.50	2.58	2.64	2.70	2.75	2.80	2.84	2.88	2.91	2.94	2.97	3.00	3.04	3.09
20	2.07	2.26	2.39	2.49	2.57	2.63	2.69	2.74	2.79	2.83	2.86	2.90	2.93	2.96	2.98	3.03	3.07
21	2.06	2.25	2.38	2.48	2.56	2.62	2.68	2.73	2.78	2.81	2.85	2.88	2.91	2.94	2.97	3.02	3.06
22	2.05	2.24	2.37	2.47	2.55	2.61	2.67	2.72	2.77	2.80	2.84	2.87	2.90	2.93	2.96	3.00	3.05
23	2.05	2.24	2.36	2.46	2.54	2.61	2.66	2.71	2.76	2.80	2.83	2.86	2.89	2.92	2.95	2.99	3.04
24	2.05	2.23	2.36	2.46	2.53	2.60	2.66	2.70	2.75	2.79	2.82	2.85	2.88	2.91	2.94	2.98	3.03
25	2.04	2.23	2.35	2.45	2.53	2.59	2.65	2.70	2.74	2.78	2.81	2.85	2.88	2.90	2.93	2.97	3.02
26	2.04	2.22	2.35	2.44	2.52	2.59	2.64	2.69	2.73	2.77	2.81	2.84	2.87	2.89	2.92	2.97	3.01
27	2.03	2.22	2.34	2.44	2.52	2.58	2.64	2.68	2.73	2.76	2.80	2.83	2.86	2.89	2.91	2.96	3.00
28	2.03	2.21	2.34	2.43	2.51	2.57	2.63	2.68	2.72	2.76	2.79	2.82	2.85	2.88	2.91	2.95	2.99
29	2.03	2.21	2.34	2.43	2.51	2.57	2.62	2.67	2.71	2.75	2.79	2.82	2.85	2.87	2.90	2.94	2.98
30	2.03	2.21	2.33	2.43	2.50	2.57	2.62	2.67	2.71	2.75	2.78	2.81	2.84	2.87	2.89	2.94	2.98
35	2.01	2.19	2.32	2.41	2.48	2.55	2.60	2.65	2.69	2.72	2.76	2.79	2.82	2.84	2.87	2.91	2.95
40	2.01	2.18	2.30	2.40	2.47	2.53	2.58	2.63	2.67	2.71	2.74	2.77	2.80	2.82	2.85	2.89	2.93
45	2.00	2.18	2.30	2.39	2.46	2.52	2.57	2.62	2.66	2.69	2.73	2.76	2.78	2.81	2.83	2.88	2.92
50	1.99	2.17	2.29	2.38	2.45	2.51	2.55	2.61	2.65	2.68	2.72	2.75	2.77	2.80	2.82	2.86	2.90
60	1.99	2.16	2.28	2.37	2.44	2.50	2.55	2.59	2.63	2.67	2.70	2.73	2.76	2.78	2.80	2.85	2.88
80	1.98	2.15	2.26	2.35	2.42	2.48	2.53	2.58	2.61	2.65	2.68	2.71	2.74	2.76	2.78	2.82	2.86
100	1.97	2.14	2.26	2.34	2.41	2.47	2.52	2.57	2.60	2.64	2.67	2.70	2.72	2.75	2.77	2.81	2.85
120	1.97	2.14	2.25	2.34	2.41	2.47	2.52	2.56	2.60	2.63	2.66	2.69	2.72	2.74	2.76	2.80	2.84
200	1.96	2.13	2.24	2.33	2.40	2.45	2.50	2.54	2.58	2.62	2.65	2.67	2.70	2.72	2.74	2.78	2.82
∞	1.95	2.11	2.23	2.31	2.38	2.43	2.48	2.52	2.56	2.59	2.62	2.65	2.67	2.70	2.72	2.76	2.79

α = 0.05

ν \ k	2	3	4	5	6	7	8	9	10	11	12	13	14	15	16	18	20
2	5.57	6.34	6.89	7.31	7.65	7.93	8.17	8.38	8.57	8.74	8.89	9.03	9.16	9.28	9.39	9.59	9.77
3	3.96	4.43	4.76	5.02	5.23	5.41	5.56	5.69	5.81	5.92	6.01	6.10	6.18	6.26	6.33	6.45	6.57
4	3.38	3.74	4.00	4.20	4.37	4.50	4.62	4.72	4.82	4.90	4.97	5.04	5.11	5.17	5.22	5.32	5.41
5	3.09	3.40	3.62	3.79	3.93	4.04	4.14	4.23	4.31	4.38	4.45	4.51	4.56	4.61	4.66	4.74	4.82
6	2.92	3.19	3.39	3.54	3.66	3.77	3.86	3.94	4.01	4.07	4.13	4.18	4.23	4.28	4.32	4.39	4.46
7	2.80	3.06	3.24	3.38	3.49	3.59	3.67	3.74	3.80	3.86	3.92	3.96	4.01	4.05	4.09	4.16	4.22
8	2.72	2.96	3.13	3.26	3.36	3.45	3.53	3.60	3.66	3.71	3.76	3.81	3.85	3.89	3.93	3.99	4.05
9	2.66	2.89	3.05	3.17	3.27	3.36	3.43	3.49	3.55	3.60	3.65	3.69	3.73	3.77	3.80	3.87	3.92
10	2.61	2.83	2.98	3.10	3.20	3.28	3.35	3.41	3.47	3.52	3.56	3.60	3.64	3.68	3.71	3.77	3.82
11	2.57	2.78	2.93	3.05	3.14	3.22	3.29	3.35	3.40	3.45	3.49	3.53	3.57	3.60	3.63	3.69	3.74
12	2.54	2.75	2.89	3.00	3.09	3.17	3.24	3.29	3.35	3.39	3.43	3.47	3.51	3.54	3.57	3.63	3.68
13	2.51	2.72	2.86	2.97	3.06	3.13	3.19	3.25	3.30	3.34	3.39	3.42	3.46	3.49	3.52	3.57	3.62
14	2.49	2.69	2.83	2.94	3.02	3.09	3.16	3.21	3.26	3.30	3.34	3.38	3.41	3.45	3.48	3.53	3.58
15	2.47	2.67	2.81	2.91	2.99	3.06	3.13	3.18	3.23	3.27	3.31	3.35	3.38	3.41	3.44	3.49	3.54
16	2.46	2.65	2.78	2.89	2.97	3.04	3.10	3.15	3.20	3.24	3.28	3.31	3.35	3.38	3.40	3.46	3.50
17	2.44	2.63	2.77	2.87	2.95	3.02	3.08	3.13	3.17	3.21	3.25	3.29	3.32	3.35	3.38	3.43	3.47
18	2.43	2.62	2.75	2.85	2.93	3.00	3.05	3.11	3.15	3.19	3.23	3.26	3.29	3.32	3.35	3.40	3.44
19	2.42	2.61	2.73	2.83	2.91	2.98	3.04	3.09	3.13	3.17	3.21	3.24	3.27	3.30	3.33	3.38	3.42
20	2.41	2.59	2.72	2.82	2.90	2.96	3.02	3.07	3.11	3.15	3.19	3.22	3.25	3.28	3.31	3.36	3.40
21	2.40	2.58	2.71	2.81	2.88	2.95	3.01	3.05	3.10	3.14	3.17	3.21	3.24	3.26	3.29	3.34	3.38
22	2.39	2.57	2.70	2.79	2.87	2.94	2.99	3.04	3.08	3.12	3.16	3.19	3.22	3.25	3.27	3.32	3.36
23	2.39	2.57	2.69	2.78	2.86	2.92	2.98	3.03	3.07	3.11	3.14	3.18	3.21	3.23	3.26	3.31	3.35
24	2.38	2.56	2.68	2.77	2.85	2.91	2.97	3.02	3.06	3.10	3.13	3.16	3.19	3.22	3.25	3.29	3.33
25	2.37	2.55	2.67	2.77	2.84	2.90	2.96	3.01	3.05	3.09	3.12	3.15	3.18	3.21	3.23	3.28	3.32
26	2.37	2.54	2.67	2.76	2.83	2.90	2.95	3.00	3.04	3.08	3.11	3.14	3.17	3.20	3.22	3.27	3.31
27	2.36	2.54	2.66	2.75	2.83	2.89	2.94	2.99	3.03	3.07	3.10	3.13	3.16	3.19	3.21	3.26	3.30
28	2.36	2.53	2.65	2.74	2.82	2.88	2.93	2.98	3.02	3.06	3.09	3.12	3.15	3.18	3.20	3.25	3.29
29	2.35	2.53	2.65	2.74	2.81	2.87	2.93	2.97	3.01	3.05	3.08	3.11	3.14	3.17	3.19	3.24	3.28
30	2.35	2.52	2.64	2.73	2.80	2.87	2.92	2.96	3.00	3.04	3.07	3.11	3.13	3.18	3.18	3.23	3.27
35	2.33	2.50	2.62	2.71	2.78	2.84	2.89	2.93	2.97	3.01	3.04	3.07	3.10	3.13	3.15	3.19	3.23
40	2.32	2.49	2.60	2.69	2.76	2.82	2.87	2.91	2.95	2.99	3.02	3.05	3.08	3.10	3.12	3.17	3.20
45	2.31	2.48	2.59	2.68	2.75	2.80	2.85	2.90	2.93	2.97	3.00	3.03	3.06	3.08	3.10	3.14	3.18
50	2.30	2.47	2.58	2.66	2.73	2.79	2.84	2.88	2.92	2.95	2.99	3.01	3.04	3.06	3.09	3.13	3.16
60	2.29	2.45	2.56	2.65	2.72	2.77	2.82	2.86	2.90	2.93	2.96	2.99	3.02	3.04	3.06	3.10	3.14
80	2.28	2.44	2.55	2.63	2.69	2.75	2.80	2.84	2.87	2.91	2.94	2.96	2.99	3.01	3.03	3.07	3.11
100	2.27	2.43	2.53	2.62	2.68	2.74	2.78	2.82	2.86	2.89	2.92	2.95	2.97	3.00	3.02	3.06	3.09
120	2.26	2.42	2.53	2.61	2.67	2.73	2.77	2.81	2.85	2.88	2.91	2.94	2.96	2.98	3.01	3.04	3.08
200	2.25	2.41	2.51	2.59	2.66	2.71	2.75	2.79	2.83	2.86	2.89	2.92	2.94	2.96	2.98	3.02	3.05
∞	2.24	2.39	2.49	2.57	2.63	2.68	2.73	2.77	2.80	2.83	2.86	2.88	2.91	2.93	2.95	2.98	3.02

TABLE 7. Continued.

$\alpha = 0.01$

ν \ k	2	3	4	5	6	7	8	9	10	11	12	13	14	15	16	18	20
2	12.73	14.44	15.65	16.59	17.35	17.99	18.53	19.01	19.43	19.81	20.15	20.46	20.75	21.02	21.26	21.71	22.11
3	7.13	7.91	8.48	8.92	9.28	9.58	9.84	10.06	10.27	10.45	10.61	10.76	10.90	11.03	11.15	11.37	11.56
4	5.46	5.99	6.36	6.66	6.90	7.10	7.27	7.43	7.57	7.69	7.80	7.91	8.00	8.09	8.17	8.32	8.45
5	4.70	5.11	5.40	5.63	5.81	5.97	6.11	6.23	6.33	6.43	6.52	6.60	6.67	6.74	6.81	6.93	7.03
6	4.27	4.61	4.86	5.05	5.20	5.33	5.45	5.55	5.64	5.72	5.80	5.86	5.93	5.99	6.04	6.14	6.23
7	4.00	4.30	4.51	4.68	4.81	4.93	5.03	5.12	5.20	5.27	5.33	5.39	5.45	5.50	5.55	5.64	5.72
8	3.81	4.08	4.27	4.42	4.55	4.65	4.74	4.82	4.89	4.96	5.02	5.07	5.12	5.17	5.21	5.29	5.36
9	3.67	3.92	4.10	4.24	4.35	4.45	4.53	4.61	4.67	4.73	4.79	4.84	4.88	4.92	4.96	5.04	5.10
10	3.57	3.80	3.97	4.10	4.20	4.29	4.37	4.44	4.50	4.56	4.61	4.66	4.70	4.74	4.78	4.84	4.91
11	3.48	3.71	3.87	3.99	4.09	4.17	4.25	4.31	4.37	4.42	4.47	4.51	4.55	4.59	4.63	4.69	4.75
12	3.42	3.63	3.78	3.90	4.00	4.08	4.15	4.21	4.26	4.31	4.36	4.40	4.44	4.48	4.51	4.57	4.62
13	3.36	3.57	3.71	3.83	3.92	4.00	4.05	4.12	4.18	4.22	4.27	4.31	4.34	4.38	4.41	4.47	4.52
14	3.32	3.52	3.66	3.77	3.85	3.93	3.99	4.05	4.10	4.15	4.19	4.23	4.26	4.30	4.33	4.39	4.44
15	3.28	3.47	3.61	3.71	3.80	3.87	3.93	3.99	4.04	4.08	4.12	4.16	4.20	4.23	4.26	4.31	4.36
16	3.25	3.43	3.57	3.67	3.75	3.82	3.88	3.94	3.99	4.03	4.07	4.11	4.14	4.17	4.20	4.25	4.30
17	3.22	3.40	3.53	3.63	3.71	3.78	3.84	3.89	3.94	3.98	4.02	4.06	4.09	4.12	4.15	4.20	4.25
18	3.19	3.37	3.50	3.60	3.68	3.74	3.80	3.85	3.90	3.94	3.98	4.01	4.04	4.07	4.10	4.15	4.20
19	3.17	3.35	3.47	3.57	3.65	3.71	3.77	3.82	3.86	3.90	3.94	3.97	4.01	4.03	4.06	4.11	4.15
20	3.15	3.32	3.45	3.54	3.62	3.68	3.74	3.79	3.83	3.87	3.91	3.94	3.97	4.00	4.03	4.07	4.12
21	3.13	3.30	3.42	3.52	3.59	3.66	3.71	3.76	3.80	3.84	3.88	3.91	3.94	3.97	3.99	4.04	4.08
22	3.11	3.28	3.40	3.50	3.57	3.63	3.69	3.73	3.78	3.81	3.85	3.88	3.91	3.94	3.96	4.01	4.05
23	3.10	3.27	3.39	3.48	3.55	3.61	3.66	3.71	3.75	3.79	3.83	3.86	3.89	3.91	3.94	3.98	4.03
24	3.09	3.25	3.37	3.46	3.53	3.59	3.64	3.69	3.73	3.77	3.80	3.83	3.86	3.89	3.91	3.96	4.00
25	3.07	3.24	3.35	3.44	3.51	3.57	3.63	3.67	3.71	3.75	3.78	3.81	3.84	3.87	3.89	3.94	3.98
26	3.06	3.23	3.34	3.43	3.50	3.56	3.61	3.65	3.70	3.73	3.76	3.79	3.82	3.85	3.87	3.92	3.96
27	3.05	3.21	3.33	3.41	3.48	3.54	3.59	3.64	3.68	3.71	3.75	3.78	3.81	3.83	3.86	3.90	3.94
28	3.04	3.20	3.32	3.40	3.47	3.53	3.58	3.62	3.66	3.70	3.73	3.76	3.79	3.81	3.84	3.88	3.92
29	3.04	3.19	3.30	3.39	3.46	3.52	3.57	3.61	3.65	3.68	3.72	3.75	3.77	3.80	3.82	3.87	3.90
30	3.03	3.18	3.29	3.38	3.45	3.50	3.55	3.60	3.64	3.67	3.70	3.73	3.76	3.78	3.81	3.85	3.89
35	2.99	3.15	3.25	3.33	3.40	3.46	3.50	3.55	3.58	3.62	3.65	3.68	3.70	3.73	3.75	3.79	3.83
40	2.97	3.12	3.22	3.30	3.37	3.42	3.47	3.51	3.54	3.58	3.61	3.63	3.66	3.68	3.71	3.74	3.78
45	2.95	3.10	3.20	3.28	3.34	3.39	3.44	3.48	3.51	3.55	3.58	3.60	3.63	3.65	3.67	3.71	3.74
50	2.94	3.08	3.18	3.26	3.32	3.37	3.42	3.46	3.49	3.52	3.55	3.58	3.60	3.62	3.65	3.68	3.72
60	2.91	3.06	3.15	3.23	3.29	3.34	3.38	3.42	3.46	3.49	3.51	3.54	3.56	3.59	3.61	3.64	3.68
80	2.89	3.02	3.12	3.19	3.25	3.30	3.34	3.38	3.41	3.44	3.47	3.49	3.52	3.54	3.56	3.59	3.63
100	2.87	3.01	3.10	3.17	3.23	3.28	3.32	3.36	3.39	3.42	3.44	3.47	3.49	3.51	3.53	3.56	3.60
120	2.86	2.99	3.09	3.16	3.21	3.26	3.30	3.34	3.37	3.40	3.43	3.45	3.47	3.49	3.51	3.55	3.58
200	2.84	2.97	3.06	3.13	3.19	3.23	3.27	3.31	3.34	3.37	3.39	3.41	3.44	3.46	3.47	3.51	3.54
∞	2.81	2.93	3.02	3.09	3.14	3.19	3.23	3.26	3.29	3.32	3.34	3.36	3.38	3.40	3.42	3.45	3.48

Source: Computed by C. W. Dunnett.

TABLE 8. Upper α Point of the Studentized Range Distribution with Parameter k and Degrees of Freedom ν ($Q_{k,\nu}^{(\alpha)}$).

$\alpha = 0.20$

ν \ k	3	4	5	6	7	8	9	10	11	12	13	14	15	16	20	24	28	32	36	40
1	6.62	8.08	9.14	9.97	10.6	11.2	11.7	12.1	12.5	12.8	13.1	13.4	13.7	13.9	14.7	15.4	15.9	16.3	16.7	17.1
2	3.82	4.56	5.10	5.52	5.87	6.16	6.41	6.63	6.83	7.00	7.16	7.31	7.44	7.57	7.99	8.32	8.59	8.83	9.03	9.21
3	3.25	3.83	4.26	4.60	4.87	5.10	5.31	5.48	5.64	5.78	5.91	6.02	6.13	6.23	6.57	6.84	7.06	7.24	7.41	7.55
4	3.00	3.53	3.91	4.21	4.45	4.66	4.83	4.99	5.13	5.25	5.37	5.47	5.57	5.66	5.96	6.20	6.39	6.56	6.71	6.83
5	2.87	3.36	3.71	3.99	4.21	4.41	4.57	4.72	4.84	4.96	5.07	5.16	5.25	5.33	5.61	5.84	6.02	6.18	6.31	6.43
6	2.79	3.25	3.59	3.85	4.07	4.25	4.40	4.54	4.66	4.77	4.87	4.97	5.05	5.13	5.39	5.60	5.78	5.93	6.06	6.17
7	2.73	3.18	3.50	3.76	3.96	4.14	4.29	4.42	4.54	4.64	4.74	4.83	4.91	4.98	5.24	5.44	5.61	5.75	5.88	5.99
8	2.69	3.13	3.44	3.69	3.89	4.06	4.20	4.33	4.44	4.55	4.64	4.73	4.81	4.88	5.13	5.32	5.49	5.62	5.74	5.85
9	2.66	3.09	3.39	3.63	3.83	3.99	4.14	4.26	4.37	4.87	4.56	4.65	4.72	4.80	5.04	5.23	5.39	5.52	5.64	5.75
10	2.63	3.05	3.36	3.59	3.78	3.94	4.08	4.21	4.32	4.41	4.50	4.59	4.66	4.73	4.97	5.16	5.31	5.44	5.56	5.66
11	2.61	3.03	3.33	3.56	3.75	3.91	4.04	4.16	4.27	4.37	4.45	4.53	4.61	4.68	4.91	5.10	5.25	5.38	5.49	5.59
12	2.60	3.01	3.30	3.53	3.72	3.87	4.01	4.13	4.23	4.33	4.41	4.49	4.57	4.63	4.86	5.04	5.20	5.32	5.44	5.54
13	2.58	2.99	3.28	3.51	3.69	3.84	3.98	4.10	4.20	4.29	4.38	4.46	4.53	4.60	4.82	5.00	5.15	5.28	5.39	5.49
14	2.57	2.97	3.26	3.49	3.67	3.82	3.95	4.07	4.17	4.27	4.35	4.43	4.50	4.56	4.79	4.97	5.11	5.24	5.35	5.44
15	2.56	2.96	3.25	3.47	3.65	3.80	3.93	4.05	4.15	4.24	4.32	4.40	4.47	4.54	4.76	4.93	5.08	5.20	5.31	5.41
16	2.55	2.95	3.23	3.45	3.63	3.78	3.91	4.03	4.13	4.22	4.30	4.38	4.45	4.51	4.73	4.91	5.05	5.17	5.28	5.38
17	2.54	2.94	3.22	3.44	3.62	3.77	3.90	4.01	4.11	4.20	4.28	4.36	4.43	4.49	4.71	4.88	5.03	5.15	5.25	5.35
18	2.54	2.93	3.21	3.43	3.60	3.75	3.88	3.99	4.09	4.18	4.26	4.34	4.41	4.47	4.69	4.86	5.00	5.12	5.23	5.32
19	2.53	2.92	3.20	3.42	3.59	3.74	3.87	3.98	4.08	4.17	4.25	4.32	4.39	4.45	4.67	4.84	4.98	5.10	5.21	5.30
20	2.52	2.91	3.19	3.41	3.58	3.73	3.86	3.97	4.07	4.15	4.23	4.31	4.38	4.44	4.65	4.82	4.96	5.08	5.19	5.28
24	2.51	2.89	3.17	3.38	3.55	3.69	3.82	3.93	4.02	4.11	4.19	4.26	4.33	4.39	4.60	4.77	4.90	5.02	5.12	5.21
30	2.49	2.87	3.14	3.35	3.52	3.66	3.78	3.89	3.98	4.07	4.15	4.22	4.28	4.34	4.55	4.71	4.84	4.96	5.06	5.15
40	2.47	2.85	3.11	3.32	3.48	3.62	3.74	3.85	3.94	4.03	4.10	4.17	4.23	4.29	4.49	4.65	4.78	4.90	4.99	5.08
60	2.46	2.83	3.09	3.29	3.45	3.59	3.71	3.81	3.90	3.98	4.06	4.12	4.19	4.24	4.44	4.59	4.72	4.83	4.93	5.01
120	2.44	2.81	3.06	3.26	3.42	3.55	3.67	3.77	3.86	3.94	4.01	4.08	4.14	4.19	4.38	4.54	4.66	4.77	4.86	4.94
∞	2.42	2.78	3.04	3.23	3.39	3.52	3.63	3.73	3.82	3.90	3.97	4.03	4.09	4.14	4.33	4.48	4.60	4.70	4.79	4.86

TABLE 8. Continued.

$\alpha = 0.10$

ν \ k	3	4	5	6	7	8	9	10	11	12	13	14	15	16	20	24	28	32	36	40
1	13.4	16.4	18.5	20.2	21.5	22.6	23.6	24.5	25.2	25.9	26.5	27.1	27.6	28.1	29.7	31.0	32.0	32.9	33.7	34.4
2	5.73	6.77	7.54	8.14	8.63	9.05	9.41	9.73	10.0	10.3	10.5	10.7	10.9	11.1	11.7	12.2	12.6	12.9	13.2	13.4
3	4.47	5.20	5.74	6.16	6.51	6.81	7.06	7.29	7.49	7.67	7.83	7.98	8.12	8.25	8.68	9.03	9.31	9.56	9.77	9.95
4	3.98	4.59	5.04	5.39	5.68	5.93	6.14	6.33	6.50	6.65	6.78	6.91	7.03	7.13	7.50	7.79	8.03	8.23	8.41	8.57
5	3.72	4.26	4.66	4.98	5.24	5.46	5.65	5.82	5.97	6.10	6.22	6.34	6.44	6.54	6.86	7.12	7.34	7.52	7.68	7.83
6	3.56	4.07	4.44	4.73	4.97	5.17	5.34	5.50	5.64	5.76	5.88	5.98	6.08	6.16	6.47	6.71	6.91	7.08	7.23	7.36
7	3.45	3.93	4.28	4.56	4.78	4.97	5.14	5.28	5.41	5.53	5.64	5.74	5.83	5.91	6.20	6.42	6.61	6.77	6.91	7.04
8	3.37	3.83	4.17	4.43	4.65	4.83	4.99	5.13	5.25	5.36	5.46	5.56	5.64	5.72	6.00	6.21	6.40	6.55	6.68	6.80
9	3.32	3.76	4.08	4.34	4.55	4.72	4.87	5.01	5.13	5.23	5.33	5.42	5.51	5.58	5.85	6.06	6.23	6.38	6.51	6.62
10	3.27	3.70	4.02	4.26	4.47	4.64	4.78	4.91	5.03	5.13	5.23	5.32	5.40	5.47	5.73	5.93	6.10	6.24	6.37	6.48
11	3.23	3.66	3.97	4.21	4.40	4.57	4.71	4.84	4.95	5.05	5.15	5.23	5.31	5.38	5.63	5.83	5.99	6.13	6.26	6.36
12	3.20	3.62	3.92	4.16	4.35	4.51	4.65	4.78	4.89	4.99	5.08	5.16	5.24	5.31	5.55	5.74	5.90	6.04	6.16	6.27
13	3.18	3.59	3.89	4.12	4.31	4.46	4.60	4.72	4.83	4.93	5.02	5.10	5.18	5.25	5.48	5.67	5.83	5.97	6.08	6.19
14	3.16	3.56	3.85	4.08	4.27	4.42	4.56	4.68	4.79	4.88	4.97	5.05	5.12	5.19	5.43	5.61	5.77	5.90	6.01	6.12
15	3.14	3.54	3.83	4.05	4.24	4.39	4.52	4.64	4.75	4.84	4.93	5.01	5.08	5.15	5.38	5.56	5.71	5.84	5.96	6.06
16	3.12	3.52	3.80	4.03	4.21	4.36	4.49	4.61	4.71	4.81	4.89	4.97	5.04	5.11	5.33	5.52	5.67	5.79	5.91	6.00
17	3.11	3.50	3.78	4.00	4.19	4.33	4.46	4.58	4.68	4.77	4.86	4.94	5.01	5.07	5.30	5.47	5.62	5.75	5.86	5.96
18	3.10	3.49	3.77	3.98	4.17	4.31	4.44	4.55	4.66	4.75	4.83	4.91	4.98	5.04	5.26	5.44	5.59	5.71	5.82	5.92
19	3.09	3.47	3.75	3.97	4.15	4.29	4.42	4.53	4.63	4.72	4.80	4.88	4.95	5.01	5.23	5.41	5.55	5.68	5.78	5.88
20	3.08	3.46	3.74	3.95	4.13	4.27	4.40	4.51	4.61	4.70	4.78	4.86	4.92	4.99	5.21	5.38	5.52	5.65	5.75	5.85
24	3.05	3.42	3.69	3.90	4.07	4.21	4.34	4.45	4.54	4.63	4.71	4.78	4.85	4.91	5.12	5.29	5.43	5.55	5.65	5.74
30	3.02	3.39	3.65	3.85	4.02	4.16	4.28	4.38	4.47	4.56	4.64	4.71	4.77	4.83	5.03	5.20	5.33	5.45	5.55	5.64
40	2.99	3.35	3.61	3.80	3.97	4.10	4.22	4.32	4.41	4.49	4.56	4.63	4.70	4.75	4.95	5.11	5.24	5.35	5.44	5.53
60	2.96	3.31	3.56	3.76	3.92	4.04	4.16	4.25	4.34	4.42	4.49	4.56	4.62	4.68	4.86	5.02	5.14	5.25	5.34	5.42
120	2.93	3.28	3.52	3.71	3.86	3.99	4.10	4.19	4.28	4.35	4.42	4.49	4.54	4.60	4.78	4.92	5.04	5.15	5.24	5.31
∞	2.90	3.24	3.48	3.66	3.81	3.93	4.04	4.13	4.21	4.29	4.35	4.41	4.47	4.52	4.69	4.83	4.95	5.04	5.13	5.20

$\alpha = 0.05$

ν \ k	3	4	5	6	7	8	9	10	11	12	13	14	15	16	20	24	28	32	36	40
1	27.0	32.8	37.1	40.4	43.1	45.4	47.4	49.1	50.6	52.0	53.2	54.3	55.4	56.3	59.6	62.1	64.2	66.0	67.6	68.9
2	8.33	9.80	10.9	11.7	12.4	13.0	13.5	14.0	14.4	14.8	15.1	15.4	15.7	15.9	16.8	17.5	18.0	18.5	18.9	19.3
3	5.91	6.83	7.50	8.04	8.48	8.85	9.18	9.46	9.72	9.95	10.2	10.4	10.5	10.7	11.2	11.7	12.1	12.4	12.6	12.9
4	5.04	5.76	6.29	6.71	7.05	7.35	7.60	7.83	8.03	8.21	8.37	8.53	8.66	8.79	9.23	9.58	9.88	10.1	10.3	10.5
5	4.60	5.22	5.67	6.03	6.33	6.58	6.80	7.00	7.17	7.32	7.47	7.60	7.72	7.83	8.21	8.51	8.76	8.98	9.17	9.33
6	4.34	4.90	5.31	5.63	5.90	6.12	6.32	6.49	6.65	6.79	6.92	7.03	7.14	7.24	7.59	7.86	8.09	8.28	8.45	8.60
7	4.17	4.68	5.06	5.36	5.61	5.82	6.00	6.16	6.30	6.43	6.55	6.66	6.76	6.85	7.17	7.42	7.63	7.81	7.97	8.11
8	4.04	4.53	4.89	5.17	5.40	5.60	5.77	5.92	6.05	6.18	6.29	6.39	6.48	6.57	6.87	7.11	7.31	7.48	7.63	7.76
9	3.95	4.42	4.76	5.02	5.24	5.43	5.60	5.74	5.87	5.98	6.09	6.19	6.28	6.36	6.64	6.87	7.06	7.22	7.36	7.49
10	3.88	4.33	4.65	4.91	5.12	5.31	5.46	5.60	5.72	5.83	5.94	6.03	6.11	6.19	6.47	6.69	6.87	7.02	7.16	7.28
11	3.82	4.26	4.57	4.82	5.03	5.20	5.35	5.49	5.61	5.71	5.81	5.90	5.98	6.06	6.33	6.54	6.71	6.86	6.99	7.11
12	3.77	4.20	4.51	4.75	4.95	5.12	5.27	5.40	5.51	5.62	5.71	5.80	5.88	5.95	6.21	6.41	6.59	6.73	6.86	6.97
13	3.74	4.15	4.45	4.69	4.89	5.05	5.19	5.32	5.43	5.53	5.63	5.71	5.79	5.86	6.11	6.31	6.48	6.62	6.74	6.85
14	3.70	4.11	4.41	4.64	4.83	4.99	5.13	5.25	5.36	5.46	5.55	5.64	5.71	5.79	6.03	6.22	6.39	6.53	6.65	6.75
15	3.67	4.08	4.37	4.60	4.78	4.94	5.08	5.20	5.31	5.40	5.49	5.57	5.65	5.72	5.96	6.15	6.31	6.45	6.56	6.67
16	3.65	4.05	4.33	4.56	4.74	4.90	5.03	5.15	5.26	5.35	5.44	5.52	5.59	5.66	5.90	6.08	6.24	6.37	6.49	6.59
17	3.63	4.02	4.30	4.52	4.71	4.86	4.99	5.11	5.21	5.31	5.39	5.47	5.54	5.61	5.84	6.03	6.18	6.31	6.43	6.53
18	3.61	4.00	4.28	4.50	4.67	4.82	4.96	5.07	5.17	5.27	5.35	5.43	5.50	5.57	5.79	5.98	6.13	6.26	6.37	6.47
19	3.59	3.98	4.25	4.47	4.65	4.79	4.92	5.04	5.14	5.23	5.32	5.39	5.46	5.53	5.75	5.93	6.08	6.21	6.32	6.42
20	3.58	3.96	4.23	4.45	4.62	4.77	4.90	5.01	5.11	5.20	5.28	5.36	5.43	5.49	5.71	5.89	6.04	6.17	6.28	6.37
24	3.53	3.90	4.17	4.37	4.54	4.68	4.81	4.92	5.01	5.10	5.18	5.25	5.32	5.38	5.59	5.76	5.91	6.03	6.13	6.23
30	3.49	3.85	4.10	4.30	4.46	4.60	4.72	4.82	4.92	5.00	5.08	5.15	5.21	5.27	5.48	5.64	5.77	5.89	5.99	6.08
40	3.44	3.79	4.04	4.23	4.39	4.52	4.64	4.74	4.82	4.90	4.98	5.04	5.11	5.16	5.36	5.51	5.64	5.75	5.85	5.93
60	3.40	3.74	3.98	4.16	4.31	4.44	4.55	4.65	4.73	4.81	4.88	4.94	5.00	5.06	5.24	5.39	5.51	5.62	5.71	5.79
120	3.36	3.69	3.92	4.10	4.24	4.36	4.47	4.56	4.64	4.71	4.78	4.84	4.90	4.95	5.13	5.27	5.38	5.48	5.57	5.64
∞	3.31	3.63	3.86	4.03	4.17	4.29	4.39	4.47	4.55	4.62	4.69	4.74	4.80	4.85	5.01	5.14	5.25	5.35	5.43	5.50

TABLE 8. Continued.

$\alpha = 0.01$

ν \ k	3	4	5	6	7	8	9	10	11	12	13	14	15	16	20	24	28	32	36	40
1	135	164	186	202	216	227	237	246	253	260	266	272	277	282	298	311	321	330	338	345
2	19.0	22.3	24.7	26.6	28.2	29.5	30.7	31.7	32.6	33.4	34.1	34.9	35.4	36.0	38.0	39.5	40.8	41.8	42.8	43.6
3	10.6	12.2	13.3	14.2	15.0	15.6	16.2	16.7	17.1	17.5	17.9	18.3	18.5	18.8	19.8	20.6	21.2	21.7	22.2	22.6
4	8.12	9.17	9.96	10.6	11.1	11.6	11.9	12.3	12.6	12.8	13.1	13.4	13.5	13.7	14.4	15.0	15.4	15.8	16.1	16.4
5	6.98	7.80	8.42	8.91	9.32	9.67	9.97	10.2	10.5	10.7	10.9	11.1	11.2	11.4	11.9	12.4	12.7	13.0	13.3	13.5
6	6.33	7.03	7.56	7.97	8.32	8.61	8.87	9.10	9.30	9.49	9.65	9.81	9.95	10.1	10.5	11.0	11.2	11.5	11.7	11.9
7	5.92	6.54	7.01	7.37	7.68	7.94	8.17	8.37	8.55	8.71	8.86	9.00	9.12	9.24	9.65	9.97	10.2	10.5	10.7	10.9
8	5.64	6.20	6.63	6.96	7.24	7.47	7.68	7.86	8.03	8.18	8.31	8.44	8.55	8.66	9.03	9.33	9.57	9.78	9.96	10.1
9	5.43	5.96	6.35	6.66	6.92	7.13	7.33	7.50	7.65	7.78	7.91	8.03	8.13	8.23	8.57	8.85	9.08	9.27	9.44	9.59
10	5.27	5.77	6.14	6.43	6.67	6.88	7.06	7.21	7.36	7.49	7.60	7.72	7.81	7.91	8.23	8.49	8.70	8.88	9.04	9.19
11	5.15	5.62	5.97	6.25	6.48	6.67	6.84	6.99	7.13	7.25	7.36	7.47	7.56	7.65	7.95	8.20	8.40	8.58	8.73	8.86
12	5.05	5.50	5.84	6.10	6.32	6.51	6.67	6.81	6.94	7.06	7.17	7.27	7.36	7.44	7.73	7.97	8.16	8.33	8.47	8.60
13	4.96	5.40	5.73	5.98	6.19	6.37	6.53	6.67	6.79	6.90	7.01	7.11	7.19	7.27	7.55	7.78	7.96	8.12	8.26	8.39
14	4.90	5.32	5.63	5.88	6.09	6.26	6.41	6.54	6.66	6.77	6.87	6.97	7.05	7.13	7.40	7.62	7.79	7.95	8.08	8.20
15	4.84	5.25	5.56	5.80	5.99	6.16	6.31	6.44	6.56	6.66	6.76	6.85	6.93	7.00	7.26	7.48	7.65	7.80	7.93	8.05
16	4.79	5.19	5.49	5.72	5.92	6.08	6.22	6.35	6.46	6.56	6.66	6.75	6.82	6.90	7.15	7.36	7.53	7.67	7.80	7.92
17	4.74	5.14	5.43	5.66	5.85	6.01	6.15	6.27	6.38	6.48	6.57	6.66	6.73	6.81	7.05	7.26	7.42	7.56	7.69	7.80
18	4.70	5.09	5.38	5.60	5.79	5.94	6.08	6.20	6.31	6.41	6.50	6.58	6.66	6.73	6.97	7.17	7.33	7.47	7.59	7.70
19	4.67	5.05	5.33	5.55	5.74	5.89	6.02	6.14	6.25	6.34	6.43	6.51	6.59	6.65	6.89	7.09	7.24	7.38	7.50	7.61
20	4.64	5.02	5.29	5.51	5.69	5.84	5.97	6.09	6.19	6.29	6.37	6.45	6.52	6.59	6.82	7.02	7.17	7.30	7.42	7.52
24	4.55	4.91	5.17	5.37	5.54	5.69	5.81	5.92	6.02	6.11	6.19	6.27	6.33	6.39	6.61	6.79	6.94	7.06	7.17	7.27
30	4.46	4.80	5.05	5.24	5.40	5.54	5.65	5.76	5.85	5.93	6.01	6.08	6.14	6.20	6.41	6.58	6.71	6.83	6.93	7.02
40	4.37	4.70	4.93	5.11	5.27	5.39	5.50	5.60	5.69	5.76	5.84	5.90	5.96	6.02	6.21	6.37	6.49	6.60	6.70	6.78
60	4.28	4.60	4.82	4.99	5.13	5.25	5.36	5.45	5.53	5.60	5.67	5.73	5.79	5.84	6.02	6.16	6.28	6.38	6.47	6.55
120	4.20	4.50	4.71	4.87	5.01	5.12	5.21	5.30	5.38	5.44	5.51	5.57	5.61	5.66	5.83	5.96	6.07	6.16	6.24	6.32
∞	4.12	4.40	4.60	4.76	4.88	4.99	5.08	5.16	5.23	5.29	5.35	5.40	5.45	5.49	5.65	5.77	5.87	5.95	6.03	6.09

Source: Adapted from H. L. Harter (1969), *Order Statistics and Their Use in Testing and Estimation. Vol. 1: Tests Based on Range and Studentized Range of Samples From a Normal Population*, Aerospace Research Laboratories, U.S. Air Force. Reproduced with the kind permission of the author.

TABLE 9. Minimum Average Risk *t*-Values for the Waller–Duncan Procedure $t^*(K, F, q, f)$.

q	f = 6	8	10	12	14	16	18	20	24	30	40	60	120
							$K = 50$						
							$F = 1.2$ ($a = 0.913$, $b = 2.449$)						
2–4	*	*	*	*	*	*	*	*	*	*	*	*	*
6	2.32	2.34	2.35	2.36	2.36	2.36	2.36	2.37	2.37	2.37	2.37	2.37	2.37
10	2.37	2.41	2.45	2.47	2.49	2.50	2.51	2.52	2.53	2.54	2.55	2.57	2.58
20	2.41	2.49	2.54	2.59	2.62	2.65	2.67	2.69	2.72	2.76	2.79	2.83	2.86
∞	2.45	2.58	2.68	2.77	2.84	2.90	2.95	3.00	3.09	3.19	3.32	3.48	3.68
							$F = 1.4$ ($a = 0.845$, $b = 1.871$)						
2	*	*	*	*	*	*	*	*	*	*	*	*	*
4	2.26	2.25	2.23	2.22	2.21	2.21	2.20	2.20	2.19	2.18	2.17	2.16	2.15
6	2.30	2.31	2.31	2.31	2.31	2.31	2.30	2.30	2.30	2.30	2.29	2.29	2.28
10	2.34	2.37	2.39	2.41	2.41	2.42	2.42	2.43	2.43	2.43	2.44	2.44	2.44
20	2.38	2.44	2.48	2.50	2.53	2.54	2.55	2.56	2.58	2.59	2.61	2.62	2.63
∞	2.42	2.52	2.59	2.65	2.69	2.73	2.76	2.79	2.83	2.87	2.92	2.94	2.92
							$F = 1.7$ ($a = 0.767$, $b = 1.558$)						
2	*	*	*	*	*	*	*	*	*	*	*	*	*
4	2.24	2.21	2.19	2.17	2.16	2.15	2.14	2.14	2.12	2.11	2.10	2.09	2.07
6	2.28	2.27	2.26	2.25	2.24	2.23	2.22	2.22	2.21	2.20	2.19	2.18	2.16
10	2.31	2.32	2.32	2.32	2.32	2.32	2.31	2.31	2.30	2.30	2.29	2.27	2.26
20	2.34	2.37	2.39	2.40	2.40	2.40	2.40	2.40	2.40	2.40	2.39	2.37	2.35
∞	2.35	2.44	2.47	2.50	2.52	2.53	2.53	2.54	2.54	2.53	2.51	2.47	2.40
							$F = 2.0$ ($a = 0.707$, $b = 1.414$)						
2	*	*	*	*	2.00	1.98	1.97	1.96	1.94	1.93	1.91	1.90	1.88
4	2.22	2.18	2.15	2.13	2.12	2.10	2.09	2.08	2.07	2.05	2.04	2.02	2.01
6	2.26	2.23	2.21	2.19	2.18	2.17	2.16	2.15	2.13	2.12	2.10	2.09	2.07
10	2.29	2.28	2.26	2.25	2.24	2.23	2.22	2.21	2.20	2.19	2.17	2.15	2.12
20	2.31	2.32	2.32	2.31	2.30	2.29	2.29	2.28	2.26	2.25	2.22	2.20	2.16
∞	2.38	2.37	2.38	2.38	2.38	2.37	2.37	2.36	2.34	2.31	2.27	2.22	2.16
							$F = 2.4$ ($a = 0.645$, $b = 1.309$)						
2	2.14	2.07	2.02	1.99	1.97	1.95	1.94	1.93	1.91	1.89	1.88	1.86	1.84
4	2.20	2.15	2.11	2.08	2.06	2.05	2.03	2.02	2.01	1.99	1.97	1.95	1.93
6	2.23	2.19	2.16	2.13	2.11	2.09	2.08	2.07	2.05	2.03	2.01	1.99	1.97
10	2.26	2.23	2.20	2.18	2.16	2.14	2.12	2.11	2.09	2.07	2.05	2.02	1.99
20	2.28	2.26	2.24	2.22	2.20	2.18	2.16	2.15	2.13	2.10	2.07	2.04	2.00
∞	2.30	2.30	2.28	2.26	2.24	2.22	2.21	2.19	2.16	2.12	2.08	2.03	1.99
							$F = 3.0$ ($a = 0.577$, $b = 1.225$)						
2	2.13	2.04	1.99	1.96	1.93	1.91	1.90	1.89	1.87	1.85	1.83	1.81	1.79
4	2.18	2.11	2.06	2.03	2.00	1.98	1.96	1.95	1.93	1.91	1.89	1.87	1.84
10	2.22	2.16	2.12	2.09	2.06	2.04	2.02	2.00	1.98	1.95	1.92	1.89	1.86
20	2.24	2.19	2.15	2.11	2.08	2.06	2.04	2.02	1.99	1.96	1.93	1.89	1.86
∞	2.26	2.22	2.18	2.14	2.11	2.08	2.05	2.03	2.00	1.96	1.92	1.89	1.85

TABLE 9. Continued.

q	f = 6	8	10	12	14	16	18	20	24	30	40	60	120
							$K = 50$						
							$F = 4.0$ ($a = 0.500$, $b = 1.155$)						
2	2.10	2.01	1.95	1.91	1.88	1.86	1.84	1.83	1.81	1.79	1.77	1.75	1.73
4	2.14	2.05	2.00	1.96	1.92	1.90	1.88	1.87	1.84	1.82	1.80	1.77	1.75
20	2.18	2.11	2.05	2.00	1.96	1.93	1.91	1.89	1.86	1.83	1.80	1.77	1.75
∞	2.20	2.12	2.06	2.01	1.97	1.94	1.91	1.89	1.86	1.83	1.80	1.77	1.74
							$F = 6.0$ ($a = 0.408$, $b = 1.095$)						
2	2.07	1.96	1.89	1.85	1.82	1.79	1.78	1.76	1.74	1.72	1.70	1.67	1.65
4	2.09	1.99	1.92	1.87	1.83	1.81	1.79	1.77	1.75	1.72	1.70	1.68	1.65
20	2.12	2.01	1.94	1.88	1.84	1.82	1.79	1.78	1.75	1.72	1.70	1.67	1.65
∞	2.13	2.02	1.94	1.89	1.85	1.82	1.79	1.77	1.75	1.72	1.70	1.67	1.65
							$F = 10.0$ ($a = 0.316$, $b = 1.054$)						
2	2.03	1.91	1.83	1.79	1.75	1.73	1.71	1.69	1.67	1.65	1.63	1.61	1.59
4	2.04	1.92	1.84	1.79	1.76	1.73	1.71	1.69	1.67	1.65	1.63	1.61	1.59
∞	2.06	1.93	1.85	1.79	1.76	1.73	1.71	1.69	1.67	1.65	1.63	1.61	1.59
							$F = 25.0$ ($a = 0.200$, $b = 1.021$)						
2–∞	1.98	1.84	1.76	1.72	1.68	1.66	1.64	1.63	1.61	1.59	1.57	1.55	1.53
							$F = \infty$ ($a = 0$, $b = 1$)						
2–∞	1.93	1.79	1.72	1.67	1.64	1.62	1.60	1.59	1.57	1.55	1.54	1.52	1.50
							$K = 100$						
							$F = 1.2$ ($a = 0.913$, $b = 2.449$)						
2–6	*	*	*	*	*	*	*	*	*	*	*	*	*
8	2.91	2.94	2.96	2.97	2.98	2.99	2.99	2.99	3.00	3.00	3.00	3.00	3.00
10	2.93	2.98	3.01	3.04	3.05	3.06	3.07	3.08	3.09	3.10	3.10	3.11	3.12
12	2.95	3.01	3.05	3.08	3.10	3.12	3.13	3.14	3.16	3.17	3.19	3.20	3.21
14	2.96	3.03	3.08	3.12	3.14	3.16	3.18	3.19	3.21	3.23	3.25	3.27	3.29
16	2.97	3.05	3.11	3.15	3.18	3.20	3.22	3.24	3.26	3.28	3.31	3.33	3.36
20	2.99	3.08	3.14	3.19	3.23	3.26	3.28	3.30	3.33	3.37	3.40	3.44	3.47
40	3.02	3.13	3.22	3.29	3.35	3.39	3.43	3.47	3.52	3.58	3.64	3.72	3.79
100	3.04	3.17	3.28	3.36	3.44	3.50	3.55	3.59	3.67	3.76	3.86	3.98	4.11
∞	3.05	3.20	3.32	3.42	3.50	3.58	3.64	3.70	3.80	3.91	4.06	4.24	4.45
							$F = 1.4$ ($a = 0.845$, $b = 1.871$)						
2–4	*	*	*	*	*	*	*	*	*	*	*	*	*
6	2.85	2.84	2.83	2.82	2.82	2.81	2.80	2.80	2.79	2.78	2.77	2.75	2.74
8	2.88	2.89	2.90	2.90	2.90	2.89	2.89	2.89	2.88	2.88	2.87	2.86	2.85
10	2.90	2.93	2.94	2.95	2.95	2.96	2.96	2.96	2.95	2.95	2.95	2.94	2.93
12	2.92	2.95	2.98	2.99	3.00	3.00	3.01	3.01	3.01	3.01	3.01	3.00	2.99
14	2.93	2.97	3.00	3.02	3.03	3.04	3.04	3.05	3.05	3.06	3.06	3.05	3.05
16	2.94	2.99	3.02	3.04	3.06	3.07	3.08	3.08	3.09	3.09	3.10	3.10	3.09
20	2.95	3.01	3.05	3.08	3.10	3.11	3.12	3.13	3.14	3.15	3.16	3.16	3.16
40	2.98	3.06	3.12	3.16	3.19	3.22	3.24	3.25	3.28	3.30	3.31	3.32	3.32
100	2.99	3.09	3.16	3.22	3.26	3.29	3.32	3.34	3.38	3.41	3.43	3.45	3.42
∞	3.01	3.12	3.20	3.26	3.31	3.35	3.39	3.42	3.46	3.50	3.53	3.54	3.46

TABLE 9. Continued.

q	f: 6	8	10	12	14	16	18	20	24	30	40	60	120
							K = 100						
							F = 1.7 (a = 0.767, b = 1.558)						
2	*	*	*	*	*	*	*	*	*	*	*	*	*
4	*	*	*	*	*	2.61	2.59	2.58	2.56	2.54	2.52	2.50	2.48
6	2.82	2.79	2.76	2.74	2.72	2.71	2.70	2.69	2.67	2.65	2.63	2.61	2.58
8	2.84	2.83	2.81	2.80	2.78	2.77	2.76	2.75	2.74	2.72	2.70	2.68	2.65
10	2.86	2.86	2.85	2.84	2.83	2.82	2.81	2.80	2.79	2.77	2.75	2.73	2.70
12	2.87	2.88	2.88	2.87	2.86	2.85	2.84	2.84	2.82	2.81	2.79	2.76	2.73
14	2.88	2.90	2.90	2.89	2.89	2.88	2.87	2.86	2.85	2.83	2.81	2.79	2.75
16	2.89	2.91	2.91	2.91	2.90	2.90	2.89	2.89	2.87	2.86	2.84	2.81	2.77
20	2.90	2.93	2.93	2.94	2.93	2.93	2.92	2.92	2.91	2.89	2.87	2.84	2.80
40	2.93	2.97	2.99	3.00	3.00	3.00	3.00	2.99	2.98	2.97	2.94	2.89	2.83
100	2.94	2.99	3.02	3.04	3.05	3.05	3.05	3.05	3.04	3.02	2.98	2.92	2.83
∞	2.95	3.01	3.05	3.07	3.08	3.09	3.09	3.08	3.07	3.05	3.01	2.93	2.81
							F = 2.0 (a = 0.707, b = 1.414)						
2	*	*	*	*	*	*	*	*	*	*	*	*	*
4	2.74	2.67	2.63	2.59	2.56	2.54	2.52	2.51	2.49	2.46	2.44	2.41	2.39
6	2.79	2.74	2.70	2.67	2.64	2.62	2.60	2.59	2.57	2.54	2.52	2.49	2.46
8	2.81	2.77	2.74	2.71	2.69	2.67	2.65	2.64	2.62	2.59	2.56	2.53	2.49
10	2.83	2.80	2.77	2.74	2.72	2.70	2.69	2.67	2.65	2.62	2.59	2.56	2.52
12	2.84	2.82	2.79	2.77	2.75	2.73	2.71	2.70	2.67	2.64	2.61	2.57	2.53
14	2.85	2.83	2.81	2.79	2.77	2.75	2.73	2.72	2.69	2.66	2.63	2.59	2.54
16	2.85	2.84	2.82	2.80	2.78	2.76	2.74	2.73	2.70	2.67	2.64	2.59	2.54
20	2.86	2.85	2.84	2.82	2.80	2.78	2.77	2.75	2.72	2.69	2.65	2.61	2.55
40	2.88	2.89	2.88	2.86	2.85	2.83	2.81	2.80	2.77	2.73	2.68	2.62	2.55
100	2.89	2.91	2.90	2.89	2.88	2.86	2.84	2.82	2.79	2.75	2.69	2.62	2.53
∞	2.90	2.92	2.92	2.91	2.90	2.88	2.86	2.85	2.81	2.76	2.69	2.61	2.52
							F = 2.4 (a = 0.645, b = 1.309)						
2	*	*	*	*	*	*	*	*	*	*	*	*	2.18
4	2.71	2.63	2.57	2.53	2.49	2.47	2.44	2.43	2.40	2.37	2.34	2.31	2.28
6	2.75	2.68	2.63	2.58	2.55	2.52	2.50	2.48	2.46	2.42	2.39	2.36	2.32
8	2.77	2.71	2.66	2.62	2.59	2.56	2.54	2.52	2.49	2.45	2.42	2.38	2.34
10	2.79	2.73	2.68	2.64	2.61	2.58	2.56	2.54	2.50	2.47	2.43	2.39	2.34
12	2.79	2.74	2.70	2.66	2.62	2.60	2.57	2.55	2.52	2.48	2.44	2.39	2.35
14	2.80	2.75	2.71	2.67	2.64	2.61	2.58	2.56	2.53	2.49	2.44	2.40	2.35
16	2.81	2.76	2.72	2.68	2.65	2.62	2.59	2.57	2.53	2.49	2.45	2.40	2.34
20	2.82	2.77	2.73	2.69	2.66	2.63	2.60	2.58	2.54	2.50	2.45	2.40	2.34
40	2.83	2.80	2.76	2.72	2.69	2.66	2.63	2.60	2.56	2.51	2.46	2.39	2.33
100	2.84	2.81	2.78	2.74	2.71	2.67	2.64	2.62	2.57	2.51	2.45	2.39	2.32
∞	2.85	2.83	2.79	2.76	2.72	2.68	2.65	2.62	2.57	2.51	2.45	2.38	2.31
							F = 3.0 (a = 0.577, b = 1.225)						
2	*	*	2.41	2.36	2.32	2.29	2.27	2.25	2.22	2.20	2.17	2.14	2.11
4	2.68	2.57	2.50	2.45	2.41	2.38	2.35	2.33	2.30	2.27	2.24	2.20	2.17
6	2.71	2.61	2.54	2.49	2.44	2.41	2.39	2.36	2.33	2.29	2.26	2.22	2.18
8	2.72	2.63	2.56	2.51	2.47	2.43	2.40	2.38	2.34	2.31	2.27	2.22	2.18
10	2.74	2.65	2.58	2.52	2.48	2.44	2.41	2.39	2.35	2.31	2.27	2.22	2.18
12	2.74	2.66	2.59	2.53	2.49	2.45	2.42	2.40	2.36	2.31	2.27	2.22	2.18

TABLE 9. Continued.

q	f: 6	8	10	12	14	16	18	20	24	30	40	60	120
						$K = 100$							
14	2.75	2.66	2.60	2.54	2.49	2.46	2.43	2.40	2.36	2.32	2.27	2.22	2.17
16	2.75	2.67	2.60	2.55	2.50	2.46	2.43	2.40	2.36	2.32	2.27	2.22	2.17
20	2.76	2.68	2.61	2.55	2.51	2.47	2.43	2.41	2.36	2.32	2.27	2.22	2.17
40	2.77	2.70	2.63	2.57	2.52	2.48	2.44	2.41	2.37	2.32	2.26	2.21	2.16
100	2.78	2.71	2.64	2.58	2.53	2.49	2.45	2.42	2.37	2.31	2.26	2.21	2.16
∞	2.79	2.71	2.65	2.59	2.53	2.49	2.45	2.42	2.37	2.31	2.26	2.20	2.15
						$F = 4.0$ ($a = 0.500$, $b = 1.155$)							
2	2.58	2.44	2.35	2.29	2.25	2.22	2.20	2.18	2.15	2.12	2.09	2.06	2.03
4	2.63	2.50	2.41	2.35	2.30	2.27	2.24	2.22	2.18	2.15	2.12	2.08	2.05
6	2.65	2.52	2.43	2.37	2.32	2.28	2.25	2.23	2.19	2.16	2.12	2.08	2.04
10	2.67	2.55	2.46	2.39	2.34	2.30	2.26	2.24	2.20	2.16	2.12	2.08	2.04
20	2.69	2.57	2.47	2.40	2.35	2.30	2.27	2.24	2.20	2.15	2.11	2.07	2.03
∞	2.71	2.59	2.49	2.42	2.36	2.31	2.27	2.24	2.19	2.15	2.11	2.06	2.02
						$F = 6.0$ ($a = 0.408$, $b = 1.095$)							
2	2.53	2.37	2.27	2.21	2.16	2.13	2.10	2.08	2.05	2.02	1.99	1.96	1.93
4	2.56	2.40	2.30	2.23	2.18	2.14	2.12	2.09	2.06	2.02	1.99	1.96	1.93
6	2.58	2.42	2.31	2.24	2.19	2.15	2.12	2.09	2.06	2.02	1.99	1.95	1.92
10	2.59	2.43	2.32	2.24	2.19	2.15	2.12	2.09	2.06	2.02	1.99	1.95	1.92
20	2.60	2.44	2.32	2.25	2.19	2.15	2.12	2.09	2.05	2.02	1.98	1.95	1.92
∞	2.61	2.44	2.33	2.25	2.19	2.15	2.12	2.09	2.05	2.02	1.98	1.95	1.92
						$F = 10.0$ ($a = 0.316$, $b = 1.054$)							
2	2.48	2.30	2.19	2.12	2.07	2.04	2.01	1.99	1.96	1.93	1.90	1.87	1.85
4	2.49	2.31	2.20	2.13	2.08	2.04	2.01	1.99	1.96	1.93	1.90	1.87	1.84
6	2.50	2.31	2.20	2.13	2.08	2.04	2.01	1.99	1.96	1.93	1.90	1.87	1.84
10–∞	2.51	2.32	2.20	2.13	2.08	2.04	2.01	1.99	1.96	1.93	1.90	1.87	1.84
						$F = 25.0$ ($a = 0.200$, $b = 1.021$)							
2–4	2.40	2.20	2.10	2.03	1.99	1.95	1.93	1.91	1.88	1.86	1.83	1.80	1.78
6–∞	2.41	2.21	2.10	2.03	1.99	1.95	1.93	1.91	1.88	1.86	1.83	1.80	1.78
						$F = \infty$ ($a = 0$, $b = 1$)							
2–∞	2.33	2.13	2.03	1.97	1.93	1.90	1.88	1.86	1.84	1.81	1.79	1.76	1.74
						$K = 500$							
						$F = 1.2$ ($a = 0.913$, $b = 2.449$)							
2–16	*	*	*	*	*	*	*	*	*	*	*	*	*
20	4.70	4.82	4.89	*	*	*	*	*	*	*	*	*	*
40	4.75	4.91	5.03	5.12	5.20	5.25	5.30	5.34	5.41	5.48	5.55	5.61	5.67
100	4.79	4.98	5.13	5.25	5.34	5.43	5.50	5.56	5.65	5.76	5.89	6.02	6.13
∞	4.81	5.03	5.20	5.34	5.46	5.56	5.65	5.73	5.86	6.02	6.20	6.41	6.56
						$F = 1.4$ ($a = 0.845$, $b = 1.871$)							
2–14	*	*	*	*	*	*	*	*	*	*	*	*	*
16	4.61	4.66	4.68	4.69	4.69	4.69	4.69	4.68	4.67	4.65	4.62	4.58	4.53
20	4.64	4.70	4.73	4.75	4.76	4.77	4.77	4.76	4.76	4.74	4.72	4.68	4.62
40	4.68	4.78	4.85	4.89	4.92	4.94	4.96	4.96	4.97	4.97	4.95	4.90	4.81
∞	4.74	4.88	4.99	5.06	5.12	5.17	5.20	5.23	5.26	5.28	5.26	5.16	4.82

TABLE 9. Continued.

	f												
q	6	8	10	12	14	16	18	20	24	30	40	60	120
							$K = 500$						
							$F = 1.7$ $(a = 0.767,\ b = 1.558)$						
2–8	*	*	*	*	*	*	*	*	*	*	*	*	*
10	*	*	*	*	*	*	*	*	*	4.08	4.02	3.95	3.87
12	4.50	4.46	4.42	4.38	4.34	4.30	4.27	4.24	4.19	4.14	4.07	3.99	3.90
20	4.55	4.54	4.52	4.49	4.46	4.43	4.40	4.37	4.32	4.26	4.18	4.08	3.95
40	4.59	4.61	4.61	4.60	4.57	4.55	4.52	4.49	4.44	4.36	4.26	4.12	3.93
∞	4.64	4.69	4.71	4.72	4.71	4.69	4.66	4.63	4.57	4.46	4.31	4.07	3.76
							$F = 2.0$ $(a = 0.707,\ b = 1.414)$						
2–6	*	*	*	*	*	*	*	*	*	*	*	*	*
8	*	*	*	*	*	3.98	3.93	3.89	3.83	3.76	3.69	3.60	3.51
10	4.41	4.31	4.22	4.15	4.08	4.03	3.98	3.94	3.88	3.80	3.72	3.63	3.53
20	4.48	4.41	4.34	4.27	4.21	4.16	4.10	4.06	3.98	3.89	3.78	3.65	3.51
40	4.51	4.47	4.41	4.35	4.29	4.23	4.17	4.12	4.03	3.92	3.78	3.62	3.44
∞	4.55	4.53	4.49	4.43	4.37	4.31	4.25	4.19	4.07	3.93	3.75	3.54	3.33
							$F = 2.4$ $(a = 0.645,\ b = 1.309)$						
2–4	*	*	*	*	*	*	*	*	*	*	*	*	*
6	*	*	*	*	3.77	3.71	3.65	3.61	3.54	3.47	3.39	3.30	3.22
8	4.31	4.14	4.01	3.91	3.83	3.76	3.70	3.66	3.58	3.50	3.41	3.32	3.22
10	4.33	4.18	4.05	3.95	3.87	3.79	3.73	3.68	3.60	3.51	3.42	3.31	3.21
20	4.39	4.26	4.14	4.04	3.95	3.87	3.80	3.74	3.64	3.53	3.41	3.28	3.15
∞	4.45	4.35	4.25	4.14	4.03	3.94	3.85	3.78	3.64	3.50	3.34	3.18	3.04
							$F = 3.0$ $(a = 0.577,\ b = 1.225)$						
2	*	*	*	*	*	*	*	*	*	*	*	*	*
4	*	*	*	*	*	3.43	3.38	3.33	3.26	3.19	3.12	3.04	2.97
6	4.19	3.95	3.79	3.66	3.56	3.49	3.43	3.37	3.30	3.21	3.13	3.04	2.95
10	4.24	4.02	3.85	3.72	3.62	3.53	3.46	3.40	3.31	3.21	3.12	3.02	2.92
20	4.28	4.08	3.91	3.77	3.65	3.56	3.48	3.41	3.31	3.20	3.09	2.98	2.87
∞	4.33	4.15	3.97	3.82	3.69	3.57	3.48	3.40	3.28	3.15	3.03	2.92	2.82
							$F = 4.0$ $(a = 0.500,\ b = 1.155)$						
2	*	*	*	*	*	*	*	*	*	*	*	2.81	2.75
4	*	3.74	3.54	3.40	3.30	3.22	3.16	3.11	3.04	2.96	2.89	2.81	2.74
6	4.08	3.78	3.58	3.43	3.32	3.24	3.17	3.12	3.04	2.95	2.87	2.79	2.71
10	4.12	3.83	3.62	3.46	3.34	3.25	3.17	3.11	3.03	2.94	2.85	2.77	2.69
20	4.15	3.86	3.64	3.48	3.35	3.25	3.17	3.10	3.01	2.92	2.83	2.74	2.66
∞	4.19	3.90	3.67	3.49	3.35	3.24	3.15	3.09	2.99	2.89	2.80	2.72	2.65
							$F = 6.0$ $(a = 0.408,\ b = 1.095)$						
2	*	*	3.28	3.14	3.04	2.97	2.91	2.87	2.81	2.74	2.68	2.62	2.56
4	3.90	3.54	3.32	3.17	3.06	2.98	2.92	2.87	2.80	2.73	2.66	2.60	2.53
6	3.93	3.57	3.33	3.18	3.06	2.98	2.91	2.86	2.79	2.72	2.65	2.58	2.52
10	3.95	3.59	3.34	3.18	3.06	2.97	2.91	2.85	2.78	2.71	2.64	2.57	2.51
20	3.97	3.60	3.35	3.18	3.06	2.97	2.90	2.84	2.77	2.70	2.63	2.56	2.51
∞	3.99	3.62	3.36	3.18	3.05	2.96	2.89	2.83	2.76	2.69	2.62	2.56	2.50

TABLE 9. Continued.

q	f: 6	8	10	12	14	16	18	20	24	30	40	60	120
							$K = 500$						
							$F = 10.0$ ($a = 0.316$, $b = 1.054$)						
2	3.72	3.33	3.10	2.96	2.86	2.79	2.74	2.70	2.64	2.58	2.52	2.47	2.42
4	3.75	3.35	3.11	2.96	2.86	2.79	2.73	2.69	2.63	2.57	2.51	2.46	2.41
10	3.78	3.36	3.11	2.96	2.85	2.78	2.72	2.68	2.62	2.56	2.50	2.45	2.40
20	3.79	3.36	3.11	2.96	2.85	2.78	2.72	2.68	2.62	2.56	2.50	2.45	2.40
∞	3.80	3.37	3.11	2.95	2.85	2.77	2.72	2.67	2.61	2.56	2.50	2.45	2.40
							$F = 25.0$ ($a = 0.200$, $b = 1.021$)						
2	3.55	3.14	2.92	2.79	2.70	2.64	2.59	2.56	2.51	2.46	2.41	2.36	2.32
10	3.57	3.14	2.92	2.79	2.70	2.64	2.59	2.55	2.50	2.45	2.41	2.36	2.32
∞	3.57	3.14	2.92	2.78	2.70	2.63	2.59	2.55	2.50	2.45	2.41	2.36	2.32
							$F = \infty$ ($a = 0$, $b = 1$)						
2–∞	3.39	3.00	2.80	2.69	2.61	2.55	2.51	2.48	2.44	2.39	2.35	2.31	2.27

Source: Adapted from R. A. Waller and D. B. Duncan (1972), *J. Amer. Statist. Assoc.*, **67**, 253–255, with the kind permission of the publisher.

References

Aitchison, J. (1964), "Confidence-region tests," *J. Roy. Statist. Soc., Ser. B*, **26**, 462–476.

Aitchison, J. (1965), "Likelihood-ratio and confidence region tests," *J. Roy. Statist. Soc., Ser. B*, **27**, 245–250.

Aitkin, M. (1969), "Multiple comparisons in psychological experiments," *British J. Math. Statist. Psychology*, **22**, 193–198.

Aitkin, M. (1974), "Simultaneous inference and the choice of variable subsets in multiple regression," *Technometrics*, **16**, 221–227.

Aitkin, M. (1979), "A simultaneous test procedure for contingency table models," *Appl. Statist.*, **28**, 233–242.

Alberton, Y. and Hochberg, Y. (1984), "Approximations for the distribution of a maximal pairwise *t* in some repeated measure designs," *Commun. Statist., Ser. A*, **13**, 2847–2854.

Alt, F. B. and Spruill, C. (1978), "A comparison of confidence intervals generated by the Scheffé and Bonferroni methods," *Commun. Statist., Ser. A*, **7**, 1503–1510.

Altschul, R. E. and Marcuson, R. (1979), "A generalized Scheffé method of multiple comparisons," *Commun. Statist., Ser. A*, **8**, 271–281.

Anderson, T. W. (1955), "The integral of a symmetric unimodal function over a symmetric convex set and some probability inequalities," *Proc. Amer. Math. Soc.*, **6**, 170–176.

Andrews, H. P., Snee, R. D., and Sarner, M. H. (1980), "Graphical display of means," *Amer. Statist.*, **34**, 195–199.

Anscombe, F. J. (1985), "Review of '*Simultaneous Statistical Inference*, 2nd edition' by R. G. Miller," *J. Amer. Statist. Assoc.*, **80**, 250.

Aubuchon, J. C., Gupta, S. S., and Hsu, J. C. (1986), "The RSMCB procedure," in *SUGI Supplemental Library User's Guide, Version 5 Edition*, Chapter 38, SAS Institute, Inc., Cary, NC.

Bahadur, R. R. (1952–1953), "A property of *t*-statistic," *Sankhyā*, **12**, 79–88.

Bailey, B. J. R. (1977), "Tables of the Bonferroni *t* statistic," *J. Amer. Statist. Assoc.*, **72**, 469–478.

Bailey, B. J. R. (1980), "Large sample simultaneous confidence intervals for the multinomial probabilities based on transformations of the cell frequencies," *Technometrics*, **22**, 583–589.

Balaam, L. N. (1963), "Multiple comparisons—A sampling experiment," *Austr. J. Statist.*, **5**, 62–84.

Barlow, R. E., Bartholomew, D. J., Bremner, J. M., and Brunk, H. D. (1972), *Statistical Inference Under Order Restrictions*, New York: Wiley.

Barnard, J. (1978), "Probability integral of the normal range," *Appl. Statist.*, **27**, 197–198.

Bechhofer, R. E. (1954), "A single-sample multiple decision procedure for ranking means of normal populations with known variances," *Ann. Math. Statist.*, **25**, 16–39.

Bechhofer, R. E. (1986), "Multiple comparisons with a control for multiply-classified variances of normal populations," *Technometrics*, **10**, 715–718.

Bechhofer, R. E. (1969), "Optimal allocation of observations when comparing several treatments with a control," *Multivariate Analysis*, *Vol. II*, (Ed. P. R. Krishnaiah), New York: Academic Press, pp. 465–473.

Bechhofer, R. E. and Dunnett, C. W. (1982), "Multiple comparisons for orthogonal contrasts: Examples and tables," *Technometrics*, **24**, 213–222.

Bechhofer, R. E. and Dunnett, C. W. (1986), "Tables of percentage points of multivariate Student *t* distributions," to appear in *Selected Tables in Mathematical Statistics*.

Bechhofer, R. E. and Nocturne, D. J. M. (1972), "Optimal allocation of observations when comparing several treatments with a control (II): Two-sided comparisons," *Technometrics*, **14**, 423–436.

Bechhofer, R. E. and Tamhane, A. C. (1981), "Incomplete block designs for comparing treatments with a control: General theory," *Technometrics*, **23**, 45–57.

Bechhofer, R. E. and Tamhane, A. C. (1983), "Design of experiments for comparing treatments with a control: Tables of optimal allocations of observations," *Technometrics*, **25**, 87–95.

Bechhofer, R. E. and Tamhane, A. C. (1985), "Tables of admissible and optimal BTIB designs for comparing treatments with a control," *Sel. Tables in Math. Statist.*, **8**, 41–139.

Bechhofer, R. E. and Turnbull, B. W. (1971), "Optimal allocation of observations when comparing several treatments with a control, III: Globally best one-sided intervals for unequal variances," in *Statistical Decision Theory and Related Topics*, *Vol. I* (Eds. S. S. Gupta and J. Yackel), New York: Academic Press, pp. 41–78.

Begun, J. and Gabriel, K. R. (1981), "Closure of the Newman–Keuls multiple comparisons procedure," *J. Amer. Statist. Assoc.*, **76**, 241–245.

Berger, R. L. (1982), "Multiparameter hypothesis testing and acceptance sampling," *Technometrics*, **24**, 295–300.

Bernhardson, C. S. (1975), "Type I error rates when multiple comparison procedures follow a significant *F* test of ANOVA," *Biometrics*, **31**, 229–232.

Bhapkar, V. P. and Somes, G. W. (1976), "Multiple comparisons of matched proportions," *Commun. Statist.*, *Ser. A*, **5**, 17–25.

Bhargava, R. P. and Srivastava, M. S. (1973), "On Tukey's confidence intervals for the contrasts in the means of the intraclass correlation model," *J. Roy. Statist. Soc.*, *Ser. B*, **35**, 147–152.

Bishop, T. A. (1978), "A Stein two-sample procedure for the general linear model with unequal variances," *Commun. Statist.*, *Ser. A*, **7**, 495–507.

Bishop, T. A. (1979), "Some results on simultaneous inference for analysis of variance with unequal variances," *Technometrics*, **21**, 337–340.

Bishop, T. A. and Dudewicz, E. J. (1978), "Exact analysis of variance with unequal variances: Test procedures and tables," *Technometrics*, **20**, 419–430.

Bishop, T. A., Dudewicz, E. J., Juritz, J. M., and Stephens, M. A. (1978), "Percentage points of a quadratic form in Student t variates," *Biometrika*, **65**, 435–439.

Bjørnstad, J. F. (1982), "Comparison of dependent two-way contingency tables," *Commun. Statist.*, **11**, 673–686.

Boardman, T. J. and Moffitt, D. R. (1971), "Graphical Monte Carlo Type I error rates for multiple comparison procedures," *Biometrics*, **27**, 738–744.

Bofinger, E. (1983), "Multiple comparisons and selection," *Austral. J. Statist.*, **25**, 198–207.

Bofinger, E. (1985), "Multiple comparisons and Type III errors," *J. Amer. Statist. Assoc.*, **80**, 433–437.

Bohrer, R. (1973), "An optimality property of Scheffé bounds," *Ann. Statist.*, **1**, 766–772.

Bohrer, R. (1979), "Multiple three-decision rules for parametric signs," *J. Amer. Statist. Assoc.*, **74**, 432–437.

Bohrer, R. (1982), "Optimal multiple decision problems: Some principles and procedures applicable in cancer drug screening," in *Probability Models for Cancer*, (Eds. L. LeCam and J. Neyman), Amsterdam: North Holland Publishing Co., pp. 287–301.

Bohrer, R., Chow, W., Faith, R., Joshi, V. M., and Wu, C. F. (1981), "Multiple three-decision rules for factorial effects: Bonferroni wins again!," *J. Amer. Statist. Assoc.*, **76**, 119–124.

Bohrer, R. and Schervish, M. (1980), "An optimal multiple decision rule about signs," *Proc. Nat. Acad. Sci.*, **77**, 52–56.

Boik, R. J. (1981), "A priori tests in repeated measures designs: Effects of nonsphericity," *Psychometrika*, **46**, 241–255.

Box, G. E. P. (1954), "Some theorems on quadratic forms applied in the study of analysis of variance problems II: Effects of inequality of variance and of correlation between errors in the two-way classification," *Ann. Math. Statist.*, **25**, 484–498.

Bradu, D. and Gabriel, K. R. (1974), "Simultaneous statistical inference on interactions in two-way analysis of variance," *J. Amer. Statist. Assoc.*, **69**, 428–436.

Broemeling, L. D. (1969), "Confidence regions for variance ratios of random models," *J. Amer. Statist. Assoc.*, **64**, 660–664.

Broemeling, L. D. (1978), "Simultaneous inferences for variance ratios of some mixed linear models," *Commun. Statist., Ser. A*, **7**, 297–306.

Broemeling, L. D. and Bee, D. E. (1976), "Simultaneous confidence intervals for parameters of a balanced incomplete block," *J. Amer. Statist. Assoc.*, **71**, 425–428.

Brown, L. D. (1979), "A proof that the Tukey–Kramer multiple comparison procedure for differences between treatment means is level-α for 3, 4, or 5 treatments," unpublished manuscript, Cornell Univ., Ithaca, NY.

Brown, L. D. (1984), "A note on the Tukey–Kramer procedure for pairwise comparisons of correlated means," in *Design of Experiments: Ranking and Selection (Essays in Honor of Robert E. Bechhofer)* (Eds. T. J. Santner and A. C. Tamhane), New York: Marcel-Dekker, pp. 1–6.

Brown, M. B. and Forsythe, A. B. (1974a), "The ANOVA and multiple comparisons for data with heterogeneous variances," *Biometrics*, **30**, 719–724.

Brown, M. B. and Forsythe, A. B. (1974b), "The small sample behavior of some statistics which test the equality of several means," *Technometrics*, **16**, 129–132.

Brown, R. A. (1974), "Robustness of the Studentized range statistic," *Biometrika*, **61**, 171–175.

Bryant, J. L. and Bruvold, N. T. (1980), "Multiple comparison procedures in the analysis of covariance," *J. Amer. Statist. Assoc.*, **75**, 874–880.

Bryant, J. L. and Fox, G. E. (1985), "Some comments on a class of simultaneous procedures in ANCOVA," *Commun. Statist., Ser. A*, **14**, 2511–2530.

Bryant, J. L. and Paulson, A. S. (1976), "An extension of Tukey's method of multiple comparisons to experimental designs with random concomitant variables," *Biometrika*, **63**, 631–638.

Calinski, T. (1969), "On the application of cluster analysis of experimental results," *Bull. Int. Statist. Instit.*, **42**, 101–103.

Calinski, T. and Corsten, L. C. A. (1985), "Clustering means in ANOVA by simultaneous testing," *Biometrics*, **41**, 39–48.

Campbell, G. (1980), "Nonparametric multiple comparisons," *Proc. Section Statist. Educ., Amer. Statist. Assoc.*, 24–27.

Campbell, G. and Skillings, J. H. (1985), "Nonparametric stepwise multiple comparison procedures," *J. Amer. Statist. Assoc.*, **80**, 998–1003.

Carmer, S. G. and Hsieh, W. T. (1979), "Type I error rates of divisive clustering methods for grouping means in analysis of variance," paper presented at Spring Regional Meeting of the Biometric Society, New Orleans, LA.

Carmer, S. G. and Swanson, M. R. (1973), "An evaluation of ten pairwise multiple comparison procedures by Monte Carlo techniques," *J. Amer. Statist. Assoc.*, **68**, 66–74.

Carmer, S. G. and Walker, W. M. (1982), "Baby bear's dilemma: A statistical tale," *Agronomy J.*, **74**, 122–124.

Chapman, D. G. (1950), "Some two-sample tests," *Ann. Math. Statist.*, **38**, 1466–1474.

Chase, G. R. (1974), "On testing for ordered alternatives with increased sample size for a control," *Biometrika*, **61**, 569–578.

Chernoff, H. and Savage, I. R. (1958), "Asymptotic normality and efficiency of certain nonparametric statistics," *Ann. Math. Statist.*, **29**, 972–994.

Chew, V. (1976a), "Comparing treatment means: A compendium," *Hort. Science*, **11**, 348–357.

Chew, V. (1976b), "Uses and abuses of Duncan's multiple range test," *Proc. Fla. State Hort. Soc.*, **89**, 251–253.

Cicchetti, D. V. (1972), "Extension of multiple-range tests to interaction tables in the analysis of variance: A rapid approximate solution," *Psychol. Bull.*, **77**, 405–408.

Cochran, W. G. (1964), "Approximate significance levels of the Behrens–Fisher test," *Biometrics*, **20**, 191–195.

Cochran, W. G. and Cox, G. M. (1957), *Experimental Designs*, 2nd ed., New York: Wiley.

Conforti, M. and Hochberg, Y. (1986), "Sequentially rejective pairwise testing procedures," to appear in *J. Statist. Planning and Inf.*

Cornish, E. A. (1954), "The multivariate *t*-distribution associated with a set of normal sample deviates," *Aust. J. Physics*, **7**, 531–542.

Cox, C. M., Krishnaiah, P. R., Lee, J. C., Reising, J., and Schuurmann, F. J. (1980), "A study of finite intersection tests for multiple comparisons of means," in *Multivariate Analysis, Vol. V* (Ed. P. R. Krishnaiah), Amsterdam: North Holland, 435–466.

Cox, D. R. (1958), *Planning of Experiments*, New York: Wiley.

Cox, D. R. (1965), "A remark on multiple comparison methods," *Technometrics*, **2**, 149–156.

Cox, D. R. and Spjøtvoll, E. (1982), "On partitioning means into groups," *Scand. J. Statist.*, **9**, 147–152.

Dalal, S. R. (1978), "Simultaneous confidence procedures for univariate and multivariate Behrens–Fisher type problems," *Biometrika*, **65**, 221–224.

Das Gupta, S., Eaton, M. L., Olkin, I., Perlman, M. D., Savage, L. J., and Sobel, M. (1972), "Inequalities on the probability content of convex regions for elliptically contoured distributions," in *Proc. Sixth Berkeley Symp. Math. Statist. Probability*, Vol. 2 (Eds. L. LeCam et al.), Berkeley: Univ. of California Press, pp. 241–265.

David, H. A. (1951), "Further applications of the range to the analysis of variance," *Biometrika*, **40**, 347–353.

David, H. A. (1952), "Upper 5% and 1% points of the maximum *F*-ratio," *Biometrika*, **39**, 422–424.

David, H. A. (1956), "Revised upper percentage points of the extreme Studentized deviate from the sample mean," *Biometrika*, **43**, 449–451.

David, H. A. (1981), *Order Statistics*, 2nd ed., New York: Wiley.

David, H. A., Lachenbruch, P. A., and Brandis, H. P. (1972), "The power function of range and Studentized range tests in normal samples," *Biometrika*, **59**, 161–168.

Dawkins, H. C. (1983), "Multiple comparisons misused: Why so frequently in response-curve studies?" *Biometrics*, **39**, 789–790.

Desu, M. M. (1970), "A selection problem," *Ann. Math. Statist.*, **41**, 1596–1603.

Diaconis, P. (1985), "Theories of data analysis from magical thinking through classical statistics," in *Exploring Data Tables, Trends and Shapes* (Eds. D. C. Hoaglin, F. Mosteller, and J. W. Tukey), New York: Wiley, pp. 1–36.

Dijkstra, J. B. (1983), "Robustness of multiple comparisons against variance heterogeneity," Computing Center Note 17, Eindhoven University of Technology, E. Germany.

Dijkstra, J. B. and Werter, P. S. P. J. (1981), "Testing the equality of several means when the population variances are unequal," *Commun. Statist., Ser. B*, **10**, 557–569.

Dixon, D. O. (1976), "Interval estimates derived from Bayes testing rules," *J. Amer. Statist. Assoc.*, **71**, 406–408.

Dixon, D. O. and Duncan, D. B. (1975), "Minimum Bayes risk *t*-intervals for multiple comparisons," *J. Amer. Statist. Assoc.*, **70**, 822–831.

Dudewicz, E. J. and Dalal, S. R. (1975), "Allocation of observations in ranking and selection with unequal variances, "*Sankhyā, Ser. B*, **37**, 28–78.

Dudewicz, E. J. and Dalal, S. R. (1983), "Multiple comparisons with a control when variances are unknown and unequal," *Amer. J. Math. and Management Sciences*, **4**, 275–295.

Dudewicz, E. J. and Ramberg, J. S. (1972), "Multiple comparisons with a control: Unknown variances," *Ann. Tech. Conf. Trans. Amer. Soc. Quality Control*, **26**, 483–488.

Dudewicz, E. J., Ramberg, J. S., and Chen, H. J. (1975), "New tables for multiple comparisons with a control (Unknown variances)," *Biometrische Zeitschrift*, **17**, 437–445.

Duncan, D. B. (1955), "Multiple range and multiple F tests," *Biometrics*, **11**, 1–42.

Duncan, D. B. (1957), "Multiple range tests for correlated and heteroscedastic means," *Biometrics*, **13**, 164–176.

Duncan D. B. (1961), "Bayes rules for a common multiple comparisons problem and related Student-t problems," *Ann. Math. Statist.*, **32**, 1013–1033.

Duncan, D. B. (1965), "A Bayesian approach to multiple comparisons," *Technometrics*, **7**, 171–222.

Duncan, D. B. (1975), "t tests and intervals for comparisons suggested by the data," *Biometrics*, **31**, 339–360.

Duncan, D. B. and Brant, L. J. (1983), "Adaptive t tests for multiple comparisons," *Biometrics*, **39**, 790–794.

Duncan, D. B. and Dixon, D. O. (1983), "k-ratio t tests, t intervals and point estimates for multiple comparisons," in *Encyclopedia of Statistical Sciences, Vol. 4* (Eds. S. Kotz and N. L. Johnson), New York: Wiley, pp. 403–410.

Duncan, D. B. and Godbold, J. H. (1979), "Approximate k-ratio t tests for differences between unequally replicated treatments," *Biometrics*, **35**, 749–756.

Dunn, O. J. (1958), "Estimation of the means of dependent variables," *Ann. Math. Statist.*, **29**, 1095–1111.

Dunn, O. J. (1959), "Confidence intervals for the means of dependent, normally distributed variables," *J. Amer. Statist. Assoc.*, **54**, 613–621.

Dunn, O. J. (1961), "Multiple comparisons among means," *J. Amer. Statist. Assoc.*, **56**, 52–64.

Dunn, O. J. (1964), "Multiple comparisons using rank sums," *Technometrics*, **6**, 241–252.

Dunn, O. J. and Massey, F. J., Jr. (1965), "Estimation of multiple contrasts using t-distributions," *J. Amer. Statist. Assoc.*, **60**, 573–583.

Dunnett, C. W. (1955), "A multiple comparison procedure for comparing several treatments with a control," *J. Amer. Statist. Assoc.*, **50**, 1096–1121.

Dunnett, C. W. (1964), "New tables for multiple comparisons with a control," *Biometrics*, **20**, 482–491.

Dunnett, C. W. (1970), "Multiple comparisons," in *Statistics in Endocrinology* (Eds. J. W. McArthur and T. Colton), Cambridge: MIT Press, pp. 79–103.

Dunnett, C. W. (1980a), "Pairwise multiple comparisons in the homogeneous variances, unequal sample size case," *J. Amer. Statist. Assoc.*, **75**, 789–795.

Dunnett, C. W. (1980b), "Pairwise multiple comparisons in the unequal variance case," *J. Amer. Statist. Assoc.*, **75**, 796–800.

Dunnett, C. W. (1982), "Robust multiple comparisons," *Commun. Statist.*, **11**, 2611–2629.

Dunnett, C. W. (1984), "Probability integral of multivariate t," Unpublished manuscript.

Dunnett, C. W. (1985), "Multiple comparisons between several treatments and a specified treatment," Invited talk at the Spring ENAR Meeting, Raleigh, NC.

Dunnett, C. W. and Goldsmith, C. H. (1981), "When and how to do multiple comparisons," in *Statistics in the Pharmaceutical Industry* (Eds. C. R. Buncher and J. Y. Tsay), New York: Marcel Dekker, Chapter 16, pp. 397–434.

Dunnett, C. W. and Sobel, M. (1954), "A bivariate generalization of Student's t-distribution with tables for certain special cases," *Biometrika*, **41**, 153–169.

Dutt, J. E., Mattes, K. D., Soms, A. P., and Tao, L. C. (1976), "An approximation to the trivariate t with a comparison to the exact values," *Biometrics*, **32**, 465–469.

Dwass, M. (1959), "Multiple confidence procedures," *Ann. Instit. Statist. Math.*, **10**, 277–282.

Dwass, M. (1960), "Some k-sample rank-order tests," in *Contributions to Probability and Statistics* (Eds. I. Olkin et al.), Standord: Stanford University Press, pp. 198–202.

Edwards, A. W. F. and Cavalli-Sforza, L. L. (1965), "A method for cluster analysis," *Biometrics*, **21**, 362–375.

Edwards, D. G. and Hsu, J. C. (1983), "Multiple comparisons with the best treatment," *J. Amer. Statist. Assoc.*, **78**, 965–971.

Einot. I. and Gabriel, K. R. (1975), "A study of the powers of several methods in multiple comparisons," *J. Amer. Statist. Assoc.*, **70**, 574–583.

Esary, J. D., Proschan, F., and Walkup, D. W. (1967), "Association of random variables with applications," *Ann. Math. Statist.*, **38**, 1466–1474.

Fabian, V. (1962), "On multiple decision methods for ranking population means," *Ann. Math. Statist.*, **33**, 248–254.

Fairley, D. and Pearl, D. K. (1984), "The Bahadur efficiency of paired versus joint ranking procedures for pairwise multiple comparisons," *Commun. Statist., Ser. A*, **13**, 1471–1481.

Feder, P. I. (1974), "Graphical techniques in statistical data analysis—Tools for extracting information from data," *Technometrics*, **16**, 287–299.

Feder, P. I. (1975), "Studentized range graph paper—A graphical tool for the comparison of treatment means," *Technometrics*, **17**, 181–188.

Federer, W. T. and McCulloch, C. E. (1984), "Multiple comparisons procedures for some split plot and split block designs," in *Design of Experiments: Ranking and Selection (Essays in Honor of Robert E. Bechhofer)* (Eds. T. J. Santner and A. C. Tamhane), New York: Marcel Dekker, pp. 7–22.

Felzenbaum, A., Hart, S., and Hochberg, Y. (1983), "Improving some multiple comparison procedures," *Ann. Statist.*, **11**, 121–128.

Fenech, A. P. (1979), "Tukey's method of multiple comparison in the randomized blocks model," *J. Amer. Statist. Assoc.*, **74**, 881–884.

Fisher, R. A. (1926), "The arrangement of field experiments," *J. Min. Agric. G. Br.*, **33**, 503–513.

Fisher, R. A. (1935), *The Design of Experiments*, Edinburgh and London: Oliver & Boyd.

Fligner, M. A. (1984), "A note on two-sided distribution-free treatment versus control multiple comparisons," *J. Amer. Statist. Assoc.*, **79**, 208–211.

Friedman, M. (1937), "The use of ranks to avoid the assumption of normality implicit in the analysis of variance," *J. Amer. Statist. Assoc.*, **32**, 675–701.

Gabriel, K. R. (1964), "A procedure for testing the homogeneity of all sets of means in analysis of variance," *Biometrics*, **20**, 459–477.

Gabriel, K. R. (1966), "Simultaneous test procedures for multiple comparisons on categorical data," *J. Amer. Statist. Assoc.*, **61**, 1081–1096.

Gabriel, K. R. (1969), "Simultaneous test procedures—Some theory of multiple comparisons," *Ann. Math. Statist.*, **40**, 224–250.

Gabriel, K. R. (1970), "On the relation between union-intersection and likelihood ratio tests," in *Essays in Probability and Statistics* (Eds. R. C. Bose et al.), Chapel Hill: University of North Carolina Press, pp. 251–266.

Gabriel, K. R. (1978a), "Comment on the paper by Ramsey," *J. Amer. Statist. Assoc.*, **73**, 485–487.

Gabriel, K. R. (1978b), "A simple method of multiple comparisons of means," *J. Amer. Statist. Assoc.*, **73**, 724–729.

Gabriel, K. R. and Gheva, D. (1982), "An improved graphical procedure for multiple comparisons," Tech. Rep. 82/02, Dept. of Statist., Univ. of Rochester, Rochester, NY.

Gabriel, K. R., Putter, J., and Wax, Y. (1973), "Simultaneous confidence intervals for product-type interaction contrasts," *J. Roy. Statist. Soc., Ser. B*, **35**, 234–244.

Gabriel, K. R. and Robinson, J. (1986), "A note on simultaneous inference with rerandomization tests," Tech. Rep. 86/04, Univ. of Rochester, Rochester, N.Y.

Games, P. A. (1971), "Multiple comparisons of means," *Amer. Educ. Research J.*, **8**, 531–565.

Games, P. A. (1977), "An improved table for simultaneous control on g contrasts," *J. Amer. Statist. Assoc.*, **72**, 531–534.

Games, P. A. and Howell, J. F. (1976), "Pairwise multiple comparison procedures with unequal N's and/or variances: A Monte Carlo study," *J. Educ. Statist.*, **1**, 113–125.

Genizi, A. and Hochberg, Y. (1978), "On improved extensions of the T-method of multiple comparisons for unbalanced designs," *J. Amer. Statist. Assoc.*, **73**, 879–884.

Ghosh, B. K. (1975), "A two-stage procedure for the Behrens–Fisher problem," *J. Amer. Statist. Assoc.*, **70**, 457–462.

Gibbons, J. D., Olkin, I., and Sobel, M. (1977), *Selecting and Ordering Populations*, New York: Wiley.

Gill, J. L. (1973), "Current status of multiple comparisons of means in designed experiments," *J. Dairy Sci.*, **56**, 973–977.

Gnanadesikan, R. (1959), "Equality of more than two variances and of more than two dispersion matrices against certain alternatives," *Ann. Math. Statist.*, **30**, 177–184.

Godfrey, K. (1985), "Comparing the means of several groups," *New England J. Medicine*, **311**, 1450–1456.

Gold, R. Z. (1963), "Tests auxiliary to χ^2 tests in a Markov chain," *Ann. Math. Statist.*, **34**, 56–74.

Goldman, A. S., Picard, R. R., and Shipley, J. P. (1982), "Statistical methods for nuclear materials safeguards: An overview," (with Discussion), *Technometrics*, **24**, 267–294.

Goodman, L. A. (1964a), "Simultaneous confidence intervals for contrasts among multinomial populations," *Ann. Math. Statist.*, **35**, 716–725.

Goodman, L. A. (1964b), "Simultaneous confidence limits for cross-product ratios in contingency tables," *J. Roy. Statist. Soc., Ser. B*, **26**, 86–102.

Goodman, L. A. (1965), "On simultaneous confidence intervals for multinomial proportions," *Technometrics*, **7**, 247–254.

Gupta, S. S. (1956), "On a decision rule for ranking means," Inst. Statist. Mimeo. Ser. No. 150, University of North Carolina, Chapel Hill, NC.

Gupta, S. S. (1963a), "Probability integrals of multivariate normal and multivariate t," *Ann. Math. Statist.*, **34**, 792–828.

Gupta, S. S. (1963b), "On a selection and ranking procedure for gamma populations," *Ann. Inst. Statist. Math.*, **14**, 199–216.

Gupta, S. S. (1965), "On some multiple decision (selection and ranking) rules," *Technometrics*, **7**, 225–245.

Gupta, S. S. and Hsu, J. C. (1984), "User's Guide to RS-MCB," Tech. Report, Dept. of Statistics, Ohio State Univ., Columbus, OH.

Gupta, S. S. and Huang, D. Y. (1977), "On some Γ-minimax selection and multiple comparisons procedures," in *Statistical Decision Theory and Related Topics*, Vol. II (Eds. S. S. Gupta and D. S. Moore), New York: Academic Press, pp. 139–155.

Gupta, S. S. and Huang, D. Y. (1981), *Multiple Statistical Decision Theory: Recent Developments*, New York: Springer-Verlag.

Gupta, S. S., Nagel, K., and Panchapakesan, S. (1973), "On the order statistics from equally correlated normal random variables," *Biometrika*, **60**, 403–413.

Gupta, S. S. and Panchpakesan, S. (1979), *Multiple Decision Procedures*, New York: Wiley.

Gupta, S. S., Panchapakesan, S., and Sohn, J. K. (1985), "On the distribution of the Studentized maximum of equally correlated normal random variables," *Commun. Statist., Ser. B*, **14**, 103–135.

Gupta, S. S. and Sobel, M. (1958), "On selecting a subset which contains all populations better than a standard," *Ann. Math. Statist.*, **29**, 235–244.

Gupta, S. S. and Sobel, M. (1962), "On the smallest of several correlated *F* statistics," *Biometrika*, **49**, 509–523.

Hahn, G. J. and Hendrickson, R. W. (1971), "A table of percentage points of the distribution of the largest absolute value of k Student t variates and its applications," *Biometrika*, **58**, 323–332.

Halperin, M. (1967), "An inequality on bivariate Student's t distribution," *J. Amer. Statist. Assoc.*, **62**, 603–606.

Halperin, M. and Greenhouse, S. W. (1958), "Note on multiple comparisons for adjusted means in analysis of covariance," *Biometrika*, **45**, 256–259.

Halperin, M., Greenhouse, S. W., Cornfield, J., and Zalokar, J. (1955), "Tables of percentage points for the Studentized maximum absolute deviate in normal samples," *J. Amer. Statist. Assoc.*, **50**, 185–195.

Han, C. P. (1969), "Testing homogeneity of variances in a 2-way classification," *Biometrics*, **25**, 153–158.

Hanumara, R. C. and Thompson, W. A., Jr. (1968), "Percentage points of the extreme roots of a Wishart matrix," *Biometrika*, **55**, 505–512.

Harter, H. L. (1960), "Tables of range and Studentized range," *Ann. Math. Statist.*, **31**, 1122–1147.

Harter, H. L. (1969), *Order Statistics and Their Use in Testing and Estimation, Vol.* 1: *Tests Based on Range and Studentized Range of Samples from a Normal Population*, Aerospace Reseach Laboratories, Office of Aerospace Research, U.S. Air Force.

Harter, H. L. (1970), "Multiple comparison procedures for interactions," *Amer. Statist.*, **24**, No. 5, 30–32.

Harter, H. L. (1980), "History of multiple comparisons," in *Handbook of Statistics*, Vol. 1 (Ed. P. R. Krishnaiah), Amsterdam: North Holland, pp. 617–622.

Hartley, H. O. (1950), "The maximum F-ratio as a short-cut test for heterogeneity of variance," *Biometrika*, **37**, 308–312.

Hartley, H. O. (1955), "Some recent developments in analysis of variance," *Commun. Pure and Appl. Math.*, **8**, 47–72.

Hayter, A. J. (1984), "A proof of the conjecture that the Tukey–Kramer multiple comparisons procedure is conservative," *Ann. Statist.*, **12**, 61–75.

Hayter, A. J. (1985), "A study of the Tukey multiple comparisons procedure including a proof of the Tukey conjecture for unequal sample sizes," unpublished doctoral dissertation, Cornell University, Ithaca, NY.

Hayter, A. J. (1986), "The maximum familywise error rate of Fisher's least significant difference test," *J. Amer. Statist. Assoc.*, **81**, 1000–1004.

Healy, W. C., Jr. (1956), "Two-sample procedures in simultaneous estimation," *Ann. Math. Statist.*, **27**, 687–702.

Hirotsu, C. (1978), "Multiple comparisons and grouping rows in a two-way contingency table," *Rep. Statist, Appl. Res.*, Union of Japanese Scientists and Engineers, **25/1**, 1–12.

Hirotsu, C. (1983), "Defining the pattern of association in two-way contingency tables," *Biometrika*, **70**, 579–589.

Hochberg, Y. (1974a), "Some approximate pairwise-efficient multiple-comparison procedures in general unbalanced designs," unpublished doctoral dissertation, Univ. of North Carolina, Chapel Hill, NC.

Hochberg, Y. (1974b), "Some generalizations of the *T*-method in simultaneous inference," *J. Mult. Anal.*, **4**, 224–234.

Hochberg, Y. (1974c), "The distribution of the range in general unbalanced models," *Amer. Statist.*, **28**, 137–138.

Hochberg, Y. (1975a), "An extension of the *T*-method to general unbalanced models of fixed effects," *J. Roy. Statist. Soc., Ser. B*, **37**, 426–433.

Hochberg, Y. (1975b), "Simultaneous inference under Behrens–Fisher conditions—A two sample approach," *Commun. Statist.*, **4**, 104–119.

Hochberg, Y. (1976a), "A modification of the *T*-method of multiple comparisons for a one-way layout with unequal variances," *J. Amer. Statist. Assoc.*, **71**, 200–203.

Hochberg, Y. (1976b), "On simultaneous confidence intervals for linear functions of means when the estimators have equal variances and equal covariances," *Metron*, *XXXIV*, 123–128.

Hochberg, Y. (1987), "Multiple classification rules for signs of parameters," *J. Statist. Planning Inf.*, **15**, 177–188.

Hochberg, Y. and Lachenbruch, P. A. (1976), "Two stage multiple comparison procedures based on the Studentized range," *Commun. Statist., Ser. A*, **5**, 1447–1453.

Hochberg, Y. and Marcus, R. (1978), "On partitioning successive increments in means or ratios of variances in a chain of normal populations," *Commun. Statist., Ser. A*, **7**, 1501–1513.

Hochberg, Y. and Posner, M. E. (1986), "On optimal decision rules for signs of parameters," *Ann. Statist.*, **14**, 733–742.

Hochberg, Y. and Rodriguez, G. (1977), "Intermediate simultaneous inference procedures," *J. Amer. Statist. Assoc.*, **72**, 220–225.

Hochberg, Y. and Tamhane, A. C. (1983), "Multiple comparisons in a mixed model," *Amer. Statist.*, **37**, 305–307.

Hochberg, Y. and Varon-Salomon, Y. (1984), "On simultaneous pairwise comparisons in analysis of covariance," *J. Amer. Statist. Assoc.*, **79**, 863–866.

Hochberg, Y., Weiss, G., and Hart, S. (1982), "On graphical procedures for multiple comparisons," *J. Amer. Statist. Assoc.*, **77**, 767–772.

Hocking, R. R. (1973), "A discussion of the two-way mixed model," *Amer. Statist.*, **27**, 148–152.

Hodges, J. L., Jr. and Lehmann, E. L. (1963), "Estimates of location based on rank tests," *Ann. Math. Statist.*, **34**, 598–611.

Hollander, M. (1966), "An asymptotically distribution-free multiple comparison procedure—Treatments vs. control," *Ann. Math. Statist.*, **37**, 735–738.

Hollander, M. and Wolfe, D. A. (1973), *Nonparametric Statistical Methods*, New York: Wiley.

Holm, S. A. (1977), "Sequentially rejective multiple test procedures," Statistical Research Report No. 1977-1, Institute of Mathematics and Statistics, University of Umeå.

Holm, S. (1979a), "A simple sequentially rejective multiple test procedure, *Scand. J. Statist.*, **6**, 65–70.

Holm, S. (1979b), "A stagewise directional test based on *t* statistics," unpublished report.

House, D. E. (1986), "A nonparametric version of Williams' test for a randomized block design," *Biometrics*, **42**, 187–190.

Hsu, J. C. (1981), "Simultaneous confidence intervals for all distances from the 'best'," *Ann. Statist.*, **9**, 1026–1034.

Hsu, J. C. (1982), "Simultaneous inference with respect to the best treatment in block designs," *J. Amer. Statist. Assoc.*, **77**, 461–467.

Hsu, J. C. (1984a), "Constrained simultaneous confidence intervals for multiple comparisons with the best," *Ann. Statist.*, **12**, 1136–1144.

Hsu, J. C. (1984b), "Ranking and selection and multiple comparisons with the best," in *Design of Experiments: Ranking and Selection* (*Essays in Honor of Robert E. Bechhofer*) (Eds. T. J. Santner and A. C. Tamhane), New York: Marcel Dekker, pp. 23–33.

Hsu, J. C. (1985), "A method of unconstrained multiple comparisons with the best," *Commun. Statist., Ser. A*, **14**, 2009–2028.

Hsu, J. C. (1986), "Sample size computation for designing multiple comparisons experiments," Tech. Rept., Dept. of Statistics, Ohio State University, Columbus, OH.

Hunter, D. (1976), "An upper bound for the probability of a union," *J. Appl. Prob.*, **13**, 597–603.

Huynh, H. and Feldt, L. S (1970), "Conditions under which mean square ratios in repeated measurements designs have exact *F*-distributions," *J. Amer. Statist. Assoc.*, **65**, 1582–1589.

James, G. S. (1951), "The comparison of several groups of observations when the ratios of the population variances are unknown," *Biometrika*, **38**, 324–329.

Jogdeo, K. (1970), "A simple proof of an inequality for multivariate normal probabilities of rectangles," *Ann. Math. Statist.*, **41**, 1357–1359.

Jogdeo, K. (1977), "Association and probability inequalities," *Ann. Statist.*, **5**, 495–504.

John, P. W. M. (1971), *Statistical Design and Analysis of Experiments*, New York: McGraw-Hill.

Johnson, D. E. (1976), "Some new multiple comparison procedures for the two-way AOV model with interaction," *Biometrics*, **32**, 929–934. Corrig., **33**, 766.

Johnson, D. E. and Graybill, F. A. (1972), "An analysis of a two-way model with interaction and no replication," *J. Amer. Statist. Assoc.*, **67**, 862–868.

Jolliffe, I. T. (1975), "Cluster analysis as a multiple comparison method," in *Proc. Conf. Appl. Statist.* (Ed. R. P. Gupta), Amsterdam: North Holland, pp. 159–168.

Kaiser, H. F. (1960), "Directional statistical decisions," *Psychol. Rev.*, **67**, 160–167.

Kempthorne, O. (1952), *The Design and Analysis of Experiments*, New York: Wiley.

Keselman, H. J., Games, P. A., and Rogan, J. C. (1979), "An addendum to a comparison of the modified Tukey and Scheffé methods of multiple comparisons for pairwise contrasts," *J. Amer. Statist. Assoc.*, **74**, 626–627.

Keselman, H. J., Murray, R., and Rogan, J. (1976), "Effect of very unequal group sizes on Tukey's multiple comparison test," *Educational and Psychological Measurement*, **36**, 263–270.

Keselman, H. J. and Rogan, J. C. (1978), "A comparison of the modified Tukey and Scheffé methods of multiple comparisons for pairwise contrasts," *J. Amer. Statist. Assoc.*, **73**, 47–52.

Keselman, H. J., Toothaker, L. E., and Shooter, M. (1975), "An evaluation of two unequal n_k forms of the Tukey multiple comparison statistic," *J. Amer. Statist. Assoc.*, **70**, 584–587.

Keuls, M. (1952), "The use of the 'Studentized range' in connection with an analysis of variance," *Euphytica*, **1**, 112–122.

Khatri, C. G. (1967), "On certain inequalities for normal distributions and their applications to simultaneous confidence bounds," *Ann. Math. Statist.*, **38**, 1853–1867.

Khatri, C. G. (1970), "Further contributions to some inequalities for normal distributions and their applications to simultaneous confidence bounds," *Ann. Inst. Statist. Math.*, **22**, 451–458.

Khuri, A. I. (1981), "Simultaneous confidence intervals for functions of variance components in random models," *J. Amer. Statist. Assoc.*, **76**, 878–885.

Kiefer, J. (1958), "On the nonrandomized optimality and randomized nonoptimality of symmetrical designs," *Ann. Math. Statist.*, **29**, 675–699.

Kim, W. C., Stefánsson, G., and Hsu, J. C. (1987), "On confidence sets in multiple comparisons," to appear in *Statistical Decision Theory and Related Topics*, Vol. IV (Eds. S. S. Gupta and J. O. Berger), New York: Academic Press.

Kimball, A. W. (1951), "On dependent tests of significance in the analysis of variance," *Ann. Math. Statist.*, **22**, 600–602.

Knoke, J. D. (1976), "Multiple comparisons with dichotomous data," *J. Amer. Statist. Assoc.*, **71**, 849–853.

Korhonen, M. (1979), "The robustness of some simultaneous test procedures in the one-way analysis of variance," Research Reports N:010, Computing Center, University of Helsinki.

Kounias, E. G. (1968), "Bounds for the probability of a union of events, with applications," *Ann. Math. Statist.*, **39**, 2154–2158.

Koziol, J. A. and Reid, N. (1977), "On the asymptotic equivalence of two ranking methods for k-sample linear rank statistics," *Ann. Statist.*, **5**, 1099–1106.

Kramer, C. Y. (1956), "Extension of multiple range test to group means with unequal numbers of replications," *Biometrics*, **12**, 307–310.

Kramer, C. Y. (1957), "Extension of multiple range tests to group correlated adjusted means," *Biometrics*, **13**, 13–18.

Krishnaiah, P. R. (1965), "On the simultaneous ANOVA tests," *Ann. Instit. Statist. Math.*, **17**, 167–173.

Krishnaiah, P. R. (1979), "Some developments on simultaneous test procedures," in *Developments in Statistics*, Vol. 2 (Ed. P. R. Krishnaiah), Amsterdam: North-Holland, pp. 157–201.

Krishnaiah, P. R. and Armitage, J. V. (1965), "Tables for the distribution of the maximum of correlated chi-square variates with one degree of freedom," *Trabajos Estadist.*, **16**, 91–96.

Krishnaiah, P. R. and Armitage, J. V. (1966), "Tables for multivariate *t*-distribution," *Sankhyá, Ser. B.*, **28**, 31–56.

Krishnaiah, P. R. and Armitage, J. V. (1970), "On a multivariate *F* distribution," in *Essays in Probability and Statistics* (Eds. R. C. Bose et al.), Chapel Hill: University of North Carolina Press, pp. 439–468.

Krishnaiah, P. R., Mudholkar, G. S., and Subbaiah, P. (1980), "Simultaneous test procedures for mean vectors and covariance matrices," in *Handbook of Statistics*, Vol. 1 (Ed. P. R. Krishnaiah), Amsterdam: North-Holland, pp. 631–671.

Krishnaiah, P. R. and Reising, J. M. (1985), "Multivariate multiple comparisons," in *Encyclopedia of Statistical Sciences*, Vol. 6 (Eds. S. Kotz and N. L. Johnson), New York: Wiley, pp. 88–95.

Kruskal, J. B. (1956), "On the shortest spanning subtree of a graph and the travelling salesman problem," *Proc. Amer. Math. Soc.*, **7**, 48–50.

Kruskal, W. H. and Wallis, W. A. (1952), "Use of ranks in one-criterion variance analysis," *J. Amer. Statist. Assoc.*, **47**, 583–621. Corrig. **48**, 907–911.

Kunte, S. and Rattihalli, R. N. (1984), "Rectangular regions of maximum probability content," *Ann. Statist.*, **12**, 1106–1108.

Kurtz, T. E. (1956), "An extension of a multiple comparisons procedure," unpublished doctoral dissertation, Princeton University, Princeton, NJ.

Lee, A. F. S. and Gurland, J. (1975), "Size and power of tests for equality of means of two normal populations with unequal variances," *J. Amer. Statist. Assoc.*, **70**, 933–941.

Lehmann, E. L. (1952), "Testing multiparameter hypotheses," *Ann. Math. Statist.*, **23**, 541–552.

Lehmann, E. L. (1957a,b), "A theory of some multiple decision problems," (Parts I and II), *Ann. Math. Statist.*, **28**, 1–25 and 547–572.

Lehmann, E. L. (1961), "Some model I problems of selection," *Ann. Math. Statist.*, **32**, 990–1012.

Lehmann, E. L. (1963), "Nonparametric confidence interval for a shift parameter," *Ann. Math. Statist.*, **34**, 1507–1512.

Lehmann, E. L. (1964), "Asymptotically nonparametric inference in some linear models with one observation per cell," *Ann. Math. Statist.*, **35**, 726–734.

Lehmann, E. L. (1966), "Some concepts of dependence," *Ann. Math. Statist.*, **37**, 1137–1153.

Lehmann, E. L. (1975), *Nonparametrics: Statistical Methods Based on Ranks*, San Francisco: Holden-Day.

Lehmann, E. L. (1986), *Testing Statistical Hypotheses*, 2nd ed., New York: Wiley.

Lehmann, E. L. and Shaffer, J. P. (1977), "On a fundamental theorem in multiple comparisons," *J. Amer. Statist. Assoc.*, **72**, 576–578.

Lehmann, E. L. and Shaffer, J. P. (1979), "Optimum significance levels for multistage comparison procedures," *Ann. Statist.*, **7**, 27–45.

Levy, K. J. (1975), "A multiple range procedure for correlated variances in a two-way classification," *Biometrics*, **31**, 243–246.

Levy, K. J. (1979), "Pairwise comparisons associated with the k sample median test," *Amer. Statist.*, **33**, 138–139.

Lewis, C. (1984), "Multiple comparisons: Fisher revisited," unpublished manuscript, Univ. of Groningen, Netherlands.

Lin, F. A. and Haseman, J. K. (1978), "An evaluation of some nonparametric multiple comparison procedures by Monte Carlo methods," *Commun. Statist., Ser. B*, **7**, 117–128.

Little, T. M. (1978), "If Galileo published in HortScience," *HortScience*, **13**, 504–506.

Lund, R. E. and Lund, J. R. (1983), "Probabilities and upper quantiles for the Studentized range," *Appl. Statist.*, **32**, 204–210.

Mandel, J. (1971), "A new analysis of variance model for non-additive data," *Technometrics*, **13**, 1–18.

Marcus, R. (1976), "The powers of some tests of the equality of normal means against an ordered alternative," *Biometrika*, **63**, 177–183.

Marcus, R. (1982), "Some results on simultaneous confidence intervals for monotone contrasts in one-way ANOVA model," *Commun. Statist., Ser. A*, **11**, 615–622.

Marcus, R. and Peritz, E. (1976), "Some simultaneous confidence bounds in normal models with restricted alternatives," *J. Roy. Statist. Soc., Ser. B*, **38**, 157–165.

Marcus, R., Peritz, E., and Gabriel, K. R. (1976), "On closed testing procedures with special reference to ordered analysis of variance," *Biometrika*, **63**, 655–660.

Marshall, A. W. and Olkin, I. (1974), "Majorization in multivariate distributions," *Ann. Statist.*, **2**, 1189–1200.

Marshall, A. W. and Olkin, I. (1979), *Inequalities: Theory of Majorization and Its Applications*, New York: Academic Press.

Mauchly, J. W. (1940), "Significance test for sphericity of a normal n-variate distribution," *Ann. Math. Statist.*, **11**, 204–209.

Maxwell, S. E. (1980), "Pairwise multiple comparisons in repeated measures designs." *J. Educ. Statist.*, **5**, 269–287.

McDonald, B. J. and Thompson, W. A., Jr. (1967), "Rank sum multiple comparisons in one- and two-way classifications," *Biometrika*, **54**, 487–497.

Miescke, K. J. (1981), "Γ-minimax selection procedures in simultaneous testing problems," *Ann. Statist.*, **9**, 215–220.

Miller, R. G. (1966), *Simultaneous Statistical Inference*, 1st ed., New York: McGraw-Hill.

Miller, R. G. (1977), "Developments in multiple comparisons 1966–1976," *J. Amer. Statist. Assoc.*, **72**, 779–788.

Miller, R. G. (1981), *Simultaneous Statistical Inference*, 2nd ed., New York: Springer-Verlag.

Miller, R. G. (1985), "Multiple comparisons," in *Encyclopedia of Statistical Sciences*, Vol. 5 (Eds. S. Kotz and N. L. Johnson), New York: Wiley, pp. 679–689.

Milton, R. C. (1963), "Tables of the equally correlated multivariate normal probability integral," Tech. Rep. No. 27, Dept. of Statistics, Univ. of Minnesota, Minneapolis, MN.

Mitzel, H. C. and Games, P. A. (1981), "Circularity and multiple comparisons in repeated measure designs," *British J. Math. Statist. Psychology*, **34**, 253–259.

Moses, L. E. (1978), "Charts for finding upper percentage points of Student's t in the range 0.01 to 0.00001," *Commun. Statist., Ser. B*, **7**, 479–490.

Mudholkar, G. S. (1966), "The integral of an invariant unimodal function over an invariant convex set—An inequality and applications," *Proc. Amer. Math. Soc.*, **17**, 1327–1333.

Mudholkar, G. S. and Subbaiah, P. (1976), "Unequal precision multiple comparisons for randomized block designs under nonstandard conditions," *J. Amer. Statist. Assoc.*, **71**, 429–434.

Naik, U. D. (1967), "Simultaneous confidence bounds concerning means in the case of unequal variances," *The Karnatak University Journal: Science*, **XII**, 93–104.

Naik, U. D. (1975), "Some selection rules for comparing p processes with a standard," *Commun. Statist., Ser. A*, **4**, 519–535.

Naiman, D. (1984), "Average width optimality of simultaneous confidence bounds," *Ann. Statist.*, **12**, 1199–1214.

Nair, K. R. (1948), "Distribution of the extreme deviate from the sample mean," *Biometrika*, **35**, 118–144.

Nair, K. R. (1952), "Tables of percentage points of the 'Studentized' extreme deviate from the sample mean," *Biometrika*, **39**, 189–191.

Nemenyi, P. (1963), "Distribution-free multiple comparisons," unpublished doctoral dissertation, Princeton University, Princeton, NJ.

Newman, D. (1939), "The distribution of the range in samples from a normal population, expressed in terms of an indepednent estimate of standard deviation," *Biometrika*, **31**, 20–30.

Neyman, J. (with the cooperation of K. Iwaszkiewiz and St. Kolodziesczyk) (1935a), "Statistical problems in agricultural experimentation," *J. Roy. Statist. Soc. Suppl.*, **2**, 107–180.

Neyman, J. (1935b), "Discussion of Mr. Yates' paper," *J. Roy. Statist. Soc. Suppl.*, **2**, 135–141.

Neyman, J. (1977), "Synergistic effects and the corresponding optimal version of the multiple comparison problem," in *Statistical Decision Theory and Related Topics*, Vol. II (Eds. S. S. Gupta and D. S. Moore), New York: Academic Press, pp. 297–311.

O'Brien, P. C. (1983), "The appropriateness of analysis of variance and multiple comparison procedures," *Biometrics*, **39**, 787–788.

Odeh, R. E. (1982), "Tables of percentage points of the distribution of the maximum absolute value of equally correlated normal random variables," *Commun. Statist., Ser. B*, **11**, 65–87.

Olshen, R. A. (1973), "The conditional level of the F-test," *J. Amer. Statist. Assoc.*, **68**, 692–698.

O'Neill, R. T. and Wetherill, B. G. (1971), "The present state of multiple comparisons methods (with discussion)," *J. Roy. Statist. Soc., Ser. B*, **33**, 218–241.

Oude Voshaar, J. H. (1980), "$(k-1)$-mean significance levels of nonparametric multiple comparisons procedures," *Ann. Statist.*, **8**, 75–86.

Pearce, S. C. (1960), "Supplemented balance," *Biometrika*, **47**, 263–271.

Pearson, E. S. and Harley, H. O. (1962), *Biometrika Tables for Statisticians*, 2nd ed., Vol. 1, London: Cambridge Univ. Press.

Peritz, E. (1965), "On inferring order relations in analysis of variance," *Biometrics*, **21**, 337–344.

Peritz, E. (1970), "A note on multiple comparisons," unpublished manuscript, Hebrew University, Israel.

Perry, J. N. (1986), "Multiple-comparison procedures: A dissenting view," *J. Econ. Entom.*, **79**, 1149–1155.

Petersen, R. G. (1977), "Use and misuse of multiple comparison procedures," *Agronomy J.*, **69**, 205–208.

Petrinovich, L. F. and Hardyck, C. D. (1969), "Error rates for multiple comparison methods," *Psycho. Bull.*, **71**, 43–54.

Petrondas, D. A. and Gabriel, K. R. (1983), "Multiple comparisons by rerandomization tests," *J. Amer. Statist. Assoc.*, **78**, 949–957.

Pillai, K. C. S. (1959), "Upper percentage points of the extreme Studentized deviate from the sample mean," *Biometrika*, **46**, 473–474.

Pillai, K. C. S. and Ramachandran, K. V. (1954), "On the distribution of the ratio of the ith observation in an ordered sample from a normal population to an independent estimate of the standard deviation," *Ann. Math. Statist.*, **25**, 565–572.

Pillai, K. C. S. and Tienzo, B. P. (1959), "On the distribution of the extreme Studentized deviate from the sample mean," *Biometrika*, **46**, 467–472.

Plackett, R. L. (1962), "A note of interactions in contingency tables," *J. Roy. Statist. Soc., Ser. B*, **24**, 162–166.

Puri, M. L. and Sen, P. K. (1971), *Nonparametric Methods in Multivariate Analysis*, New York: Wiley.

Putter, J. (1983), "Multiple comparisons and selective inference," in *A Festschrift for E. L. Lehmann* (Eds. P. J. Bickel, K. Doksum, and J. L. Hodges, Jr.), Belmont, CA: Wadsworth, pp. 328–347.

Pyke, R. (1965), "Spacings (with discussion)," *J. Roy. Statist. Soc., Ser. B*, **27**, 395–449.

Quesenberry, C. P. and Hurst, D. C. (1964), "Large sample simultaneous confidence intervals for multinomial proportions," *Technometrics*, **6**, 191–195.

Ramachandran, K. V. (1956a), "On the simultaneous analysis of variance test," *Ann. Math. Statist.*, **27**, 521–528.

Ramachandran, K. V. (1956b), "On the Tukey test for the equality of means and the Hartley test for the equality of variances," *Ann. Math. Statist.*, **27**, 825–831.

Ramachandran, K. V. and Khatri, C. G. (1957), "On a decision procedure based on the Tukey statistic," *Ann. Math. Statist.*, **28**, 802–806.

Ramsey, P. H. (1978), "Power differences between pairwise multiple comparisons," *J. Amer. Statist. Assoc.*, **73**, 479–485.

Ramseyer, G. C. and Tcheng, T. K. (1973), "The robustness of the Studentized range statistic to violations of the normality and homogeneity of variance assumptions," *Amer. Educ. Res. J.*, **10**, 235–240.

Randles, R. H. and Hollander, M. (1971), "γ-minimax selection procedures in treatments versus control problems," *Ann. Math. Statist.*, **42**, 330–341.

Reading, J. C. (1975), "A multiple comparison procedure for classifying all pairs out of k means as close or distant," *J. Amer. Statist. Assoc.*, **70**, 832–838.

Reiersøl, O. (1961), "Linear and nonlinear multiple comparisons in logit analysis," *Biometrika*, **48**, 359–365.

Renner, M. S. (1969), "A graphical method for making multiple comparisons of frequencies," *Technometrics*, **11**, 321–329.

Rhyne, A. L. and Steel, R. G. D. (1965), "Tables for a treatments versus control multiple comparisons sign test," *Technometrics*, **7**, 293–306.

Rhyne, A. L. and Steel, R. G. D. (1967), "A multiple comparisons sign test: All pairs of treatments," *Biometrics*, **23**, 539–549.

Richmond, J. (1982), "A general method for constructing simultaneous confidence intervals," *J. Amer. Statist. Assoc.*, **77**, 455–460.

Ringland, J. T. (1983), "Robust multiple comparisons," *J. Amer. Statist. Assoc.*, **78**, 145–151.

Rodger, R. S. (1973), "Confidence intervals for multiple comparisons and the misuse of the Bonferroni inequality," *British J. Math. Statist. Psychology*, **26**, 58–60.

Rogan, J. C., Keselman, H. J., and Mendoza, J. L. (1979), "Analysis of repeated measures," *British J. Math. Statist. Psychology*, **32**, 269–286.

Rouanet, H. and Lepine, D. (1970), "Comparison between treatments in a repeated measures design: ANOVA and multivariate methods," *British J. Math. Statist. Psychology*, **23**, 147–163.

Roy, S. N. (1953), "On a heuristic method of test construction and its use in multivariate analysis," *Ann. Math. Statist.*, **24**, 220–238.

Roy, S. N. and Bose, R. C. (1953), "Simultaneous confidence interval estimation," *Ann. Math. Statist.*, **24**, 513–536.

Ryan, T. A. (1959), "Multiple comparisons in psychological research," *Psychol. Bull.*, **56**, 26–47.

Ryan, T. A. (1960), "Significance tests for multiple comparison of proportions, variances, and other statistics," *Psychol. Bull.*, **57**, 318–328.

Ryan, T. A. and Ryan, T. A. Jr. (1980), "*k* independent sample median test," A letter to the Editor, *Amer. Statist.*, **34**, 123.

Sahai, H. (1974), "Simultaneous confidence intervals for variance components in some balanced random effects models," *Sankhyá*, *Ser. B*, **36**, 278–287.

Sahai, H. and Anderson, R L. (1973), "Confidence regions for variance ratios of random models for balanced data." *J. Amer. Statist. Assoc.*, **68**, 951–952.

Sampson, A. R. (1980), "Representations of simultaneous pairwise comparisons," *Handbook of Statistics*, Vol. 1 (Ed. P. R. Krishnaiah), Amsterdam: North-Holland, pp. 623–629.

Satterthwaite, F. E. (1946), "An approximate distribution of estimates of variance components," *Biometrics*, **2**, 110–114.

Savin, N. E. (1980), "The Bonferroni and the Scheffé multiple comparison procedures," *Review of Economic Studies*, **XLVII**, 255–273.

Schaafsma, W. (1969), "Minimax risk and unbiasedness for multiple decision problems of type I," *Ann. Math. Statist.*, **40**, 1684–1720.

Scheffé, H. (1953), "A method for judging all contrasts in the analysis of variance," *Biometrika*, **40**, 87–104.

Scheffé, H. (1956), "A 'mixed model' for the analysis of variance," *Ann. Math. Statist.*, **27**, 23–36.

Scheffé, H. (1959), *The Analysis of Variance*, New York: Wiley.

Scheffé, H. (1977), "A note on a reformulation of the *S*-method of multiple comparison" (with comment by R. A. Olshen and rejoinder by H. Scheffé), *J. Amer. Statist. Assoc.*, **72**, 143–146.

Schwager, S. J. (1984), "Bonferroni sometimes loses," *Amer. Statist.*, **38**, 192–197.

Schweder, T. and Spjøtvoll, E. (1982), "Plots of *P*-values to evaluate many tests simultaneously," *Biometrika*, **69**, 493–502.

Scott, A. J. (1967), "A note on conservative confidence regions for the mean of a multivariate normal," *Ann. Math. Statist.*, **38**, 278–280.

Scott, A. J. and Knott, M. (1974), "A cluster analysis method for grouping means in the analysis of variance," *Biometrics*, **30**, 507–512.

Sen, P. K. (1966), "On nonparametric simultaneous confidence regions and tests for the one criterion analysis of variance," *Ann. Inst. Statist. Math.*, **18**, 319–336.

Sen, P. K. (1968), "On a class of aligned rank order tests in two-way layouts," *Ann. Math. Statist.*, **39**, 1115–1124.

Sen, P. K. (1969a), "On nonparametric *T*-method of multiple comparisons in randomized blocks," *Ann. Inst. Statist. Math.*, **21**, 329–333.

Sen, P. K. (1969b), "A generalization of the *T*-method of multiple comparisons for interactions," *J. Amer. Statist. Assoc.*, **64**, 290–295.

Sen, P. K. (1980), "Nonparametric simultaneous inference for some MANOVA models," in *Handbook of Statistics*, Vol. 1 (Ed. P. R. Krishnaiah), Amsterdam: North Holland, pp. 673–702.

Shafer, G. and Olkin, I. (1983), "Adjusting *P* values to account for selection over dichotomies," *J. Amer. Statist. Assoc.*, **78**, 674–678.

Shaffer, J. P. (1977), "Multiple comparisons emphasizing selected contrasts: An extension and generalization of Dunnett's procedure," *Biometrics*, **33**, 293–304.

Shaffer, J. P. (1980), "Control of directional errors with stagewise multiple test procedures," *Ann. Statist.*, **8**, 1342–1348.

Shaffer, J. P. (1981), "Complexity: An interpretability criterion for multiple comparisons," *J. Amer. Statist. Assoc.*, **76**, 395–401.

Shaffer, J. P (1986a), "Modified sequentially rejective multiple test procedures," *J. Amer. Statist. Assoc.*, **81**, 826–831.

Shaffer, J. P. (1986b), "Simultaneous testing," to appear in *Encyclopedia of Statistical Sciences* (Eds. S. Kotz and N. L. Johnson), New York: Wiley.

Sherman, E. (1965), "A note on multiple comparisons using rank sums," *Technometrics*, **7**, 255–256.

Shirley, E. A. (1977), "A nonparametric equivalent of Williams' test for contrasting increasing dose levels of a treatment," *Biometrics*, **33**, 386–389.

Shirley, E. A. (1979), "The comparison of treatment with control group means in toxicological studies," *Appl. Statist.*, **28**, 144–151.

Shuster, J. J. and Boyett, J. M. (1979), "Nonparametric multiple comparison procedures," *J. Amer. Statist. Assoc.*, **74**, 379–382.

Šidák, Z. (1967), "Rectangular confidence regions for the means of multivariate normal distributions," *J. Amer. Statist. Assoc.*, **62**, 626–633.

Šidák, Z. (1968), "On multivariate normal probabilities of rectangles: Their dependence on correlations," *Ann. Math. Statist.*, **39**, 1425–1434.

Šidák, Z. (1973), "On probabilities in certain multivariate distributions: Their dependence on correlations," *Aplikace Matematiky*, **18**, 128–135.

Šidák, Z. (1975), "A note on C. G. Khatri and A. Scott's papers on multivariate normal distributions," *Ann. Inst. Statist. Math.*, **27**, 181–184.

Siotani, M. (1964), "Interval estimation for linear combinations of means," *J. Amer. Statist. Assoc.*, **59**, 1141–1164.

Skillings, J. H. (1983), "Nonparametric approaches to testing and multiple comparisons in a one-way ANOVA," *Commun. Statist., Ser. B*, **12**, 373–387.

Slepian, D. (1962), "The one-sided barrier problem for Guassian noise," *Bell Syst. Tech. J.*, **41**, 463–501.

Smith, R. A. (1971), "The effect of unequal group size on Tukey's HSD procedure," *Psychometrika*, **36**, 31–34.

Snedecor, G. W. and Cochran, W. G. (1976), *Statistical Methods*, 6th ed., Ames, IA: Iowa University Press.

Sobel, M. and Tong, Y. L. (1971), "Optimal allocation of observations for partitioning a set of normal populations in comparison with a control," *Biometrika*, **58**, 177–181.

Spjøtvoll, E. (1971), "On the probability of at least one false rejection for various multiple testing techniques in the one-way layout in the analysis of variance," Statistical Research Report 6, Dept. of Mathematics, University of Oslo.

Spjøtvoll, E. (1972a), "On the optimality of some multiple comparison procedures," *Ann. Math. Statist.*, **43**, 398–411.

Spjøtvoll, E. (1972b), "Joint confidence intervals for all linear functions in ANOVA with unknown variances," *Biometrika*, **59**, 684–685.

Spjøtvoll, E. (1972c), "Multiple comparisons of regression functions," *Ann. Math. Statist.*, **43**, 1076–1088.

Spjøtvoll, E. (1974), "Multiple testing in the analysis of variance," *Scand. J. Statist.*, **1**, 97–114.

Spjøtvoll, E. (1977), "Ordering ordered parameters," *Biometrika*, **64**, 327–334.

Spjøtvoll, E. and Stoline, M. R. (1973), "An extension of the *T*-method of multiple comparison to include the cases with unequal sample sizes," *J. Amer. Statist. Assoc.*, **68**, 975–978.

Spurrier, J. D. (1981), "An improved GT2 method for simultaneous confidence intervals on pairwise differences," *Technometrics*, **23**, 189–192.

Spurrier, J. D. and Isham, S. P. (1985), "Exact simultaneous confidence intervals for pairwise comparisons of three normal means," *J. Amer. Statist. Assoc.*, **80**, 438–442.

Steel, R. G. D. (1959a), "A multiple comparison sign test: Treatments vs. control," *J. Amer. Statist. Assoc.*, **54**, 767–775.

Steel, R. G. D. (1959b), "A multiple comparison rank sum test: Treatments versus control," *Biometrics*, **15**, 560–572.

Steel, R. G. D. (1960), "A rank sum test for comparing all pairs of treatments," *Technometrics*, **2**, 197–207.

Steel, R. G. D. (1961), "Some rank sum multiple comparisons tests," *Biometrics*, **17**, 326–328.

Steel, R. G. D. and Torrie, J. H. (1980), *Principles and Procedures of Statistics: A Biometrical Approach*, 2nd ed., New York: McGraw-Hill.

Stefánsson, G. and Hsu, J. C. (1985), "Exact confidence sets in multiple comparisons," Tech. Rep. No. 311, Department of Statistics, The Ohio State University, Columbus, OH.

Stein, C. (1945), "A two-sample test for a linear hypothesis whose power is indepenent of the variance," *Ann. Math. Statist.*, **16**, 243–258.

Stevenson, D. J. and Bland, R. P. (1982), "Bayesian multiple comparisons of normal populations without the assumption of equal variances," *Commun. Statist., Ser. A*, **11**, 49–57.

Stoline, M. R. (1978), "Tables of the Studentized augmented range and applications to problems of multiple comparison," *J. Amer. Statist. Assoc.*, **73**, 656–660.

Stoline, M. R. (1981), "The status of multiple comparisons: Simultaneous estimation of all pairwise comparisons in one-way ANOVA designs," *Amer. Statist.*, **35**, 134–141.

Stoline, M. R. (1983), "The Hunter method of simultaneous inference and its recommended use for applications having large correlation structures," *J. Amer. Statist. Assoc.*, **78**, 367–370.

Stoline, M. R. (1984), "Preliminary tests to determine variance-covariance structure incorporated into one-way repeated measurement design tests for means," unpublished manuscript.

Stoline, M. R. and Mitchell, B. T. (1981), "Generation of the Hunter second-order Bonferroni approximation to the multivariate *t* and its applications," Center for Statistical Services Report #1, Western Michigan University, Kalamazoo, MI.

Stoline, M. R. and Ury, H. K. (1979), "Tables of the Studentized maximum modulus distribution," *Technometrics*, **21**, 87–94.

Stoline, M. R., Ury, H. K., and Mitchell, B. T. (1980), "Further tables of the Studentized maximum modulus distribution," *Commun. Statist., Ser. B*, **9**, 167–178.

Tamhane, A. C. (1977), "Multiple comparisons in model I one-way ANOVA with unequal variances," *Commun. Statist., Ser. A*. **6**, 15–32.

Tamhane, A. C. (1979), "A comparison of procedures for multiple comparisons of means with unequal variances," *J. Amer. Statist. Assoc.*, **74**, 471–480.

Tamhane, A. C. (1987), "An optimal procedure for partitioning a set of normal populations with respect to a control," to appear in *Sankhyā, Ser. B*, **49**.

Thigpen, C. C. and Paulson, A. S. (1974), "A multiple range test for analysis of covariance," *Biometrika*, **61**, 479–484.

Thomas, D. A. H. (1973), "Multiple comparisons among means, A review," *The Statistician*, **22**, 16–42.

Thomas, D. A. H. (1974), "Error rates in multiple comparisons among means—Results of a simulation exercise," *Appl. Statist.*, **23**, 284–294.

Tobach, E., Smith, M., Rose, G., and Richter, D. (1967), "A table for making rank sum multiple paired comparisons," *Technometrics*, **9**, 561–567.

Tong, Y. L. (1969), "On partitioning a set of normal populations by their locations with respect to a control," *Ann. Math. Statist.*, **40**, 1300–1324.

Tong, Y. L. (1970), "Some probability inequalities for multivariate normal and multivariate *t*," *J. Amer. Statist. Assoc.*, **65**, 1243–1247.

Tong, Y. L. (1979), "Counterexamples to a result of Broemeling on simultaneous inferences for variance ratios of some mixed linear models," *Commun. Statist. Ser. A*, **8**, 1197–1204.

Tong, Y. L. (1980), *Probability Inequalities in Multivariate Distributions*, New York: Academic Press.

Tse, S. K. (1983), "A comparison of procedures for multiple comparisons of means with unequal sample sizes," unpublished manuscript, Dept. of Statistics, University of Wisconsin, Madison, WI.

Tukey, J. W. (1949), "Comparing individual means in the analysis of variance," *Biometrics*, **5**, 99–114.

Tukey, J. W. (1953), *The Problem of Multiple Comparisons*, Mimeographed monograph.

Tukey, J. W. (1977), "Some thoughts on clinical trials, especially problems of multiplicity," *Science*, **198**, 679–684.

Ury, H. K. (1976), "A comparison of four procedures for multiple comparisons among means," *Technometrics*, **18**, 89–97.

Ury, H. K., Stoline, M. R., and Mitchell, B. T. (1980), "Further tables of the Studentized maximum modulus distribution," *Commun. Statist., Ser. B*, **9**, 167–178.

Ury, H. K. and Wiggins, A. D. (1971), "Large sample and other comparisons among means," *British J. Math. Statist. Psychology*, **24**, 174–194.

Ury, H. K. and Wiggins, A. D. (1974), "Use of the Bonferroni inequality for multiple comparisons among means with post-hoc contrasts," *British J. Math. and Statist. Psychology*, **27**, 176–178.

Uusipaikka, E. (1985), "Exact simultaneous confidence intervals for multiple comparisons among three or four mean values," *J. Amer. Statist. Assoc.*, **80**, 196–201.

Waller, R. A. and Duncan, D. B. (1969), "A Bayes rule for the symmetric multiple comparisons," *J. Amer. Statist. Assoc.*, **64**, 1484–1503.

Waller, R. A. and Duncan, D. B. (1972), "A corrigendum to 'A Bayes rule for the symmetric multiple comparisons," *J. Amer. Statist. Assoc.*, **67**, 253–255.

Waller, R. A. and Duncan, D. B. (1974), "A Bayes rule for the symmetric multiple comparisons problem II," *Ann. Instit. Statist. Math.*, **26**, 247–264.

Waller, R. A. and Kemp, K. E. (1975), "Computations of Bayesian t-values for multiple comparisons," *J. Statist. Comp. Simul.*, **4**, 169–171.

Wang, Y. Y. (1971), "Probabilities of Type I errors of Welch tests for the Behrens–Fisher problem," *J. Amer. Statist. Assoc.*, **66**, 605–608.

Wei, L. J. (1982), "Asymptotically distribution-free simultaneous confidence region of treatment differences in a randomized block design," *J. Roy. Statist. Soc., Ser. B*, **44**, 201–208.

Welch, B. L. (1938), "The significance of the difference between two means when the population variances are unequal," *Biometrika*, **25**, 350–362.

Welch, B. L. (1947), "The generalization of Student's problem when several different population variances are involved," *Biometrika*, **34**, 28–35.

Welch, B. L. (1951), "On the comparison of several mean values: An alternative approach," *Biometrika*, **38**, 330–336.

Welsch, R. E. (1972), "A modification of the Newman–Keuls procedure for multiple comparisons," Working Paper 612–72, Sloan School of Management, M.I.T., Boston, MA.

Welsch, R. E. (1977), "Stepwise multiple comparison procedures," *J. Amer. Statist. Assoc.*, **72**, 566–575.

Wilcox, R. R. (1983), "A table of percentage points of the range of independent t variables," *Technometrics*, **25**, 201–204.

Williams, D. A. (1971), "A test for differences between treatment means when several dose levels are compared with a zero dose control." *Biometrics*, **27**, 103–117.

Williams, D. A. (1972), "The comparison of several dose levels with a zero dose control," *Biometrics*, **28**, 519–531.

Williams, D. A. (1977), "Some inference procedures for monotonically ordered normal means," *Biometrika*, **64**, 9–14.

Williams, D. A. (1986), "A note on Shirley's nonparametric test for comparing several dose levels with a zero-dose level," *Biometrics*, **42**, 183–186.

Willavize, S. A., Carmer, S. G., and Walker, W. M. (1980), "Evaluation of cluster analysis for comparing treatment means," *Agronomy J.*, **72**, 317–320.

Winer, B. J. (1971), *Statistical Principles in Experimental Design*, 2nd ed., New York: McGraw-Hill.

Working, H. and Hotelling, H. (1929), "Application of the theory of error to the interpretation of trends," *J. Amer. Statist. Assoc.*, **24**, 73–85.

Worsley, K. J. (1977), "A nonparametric extension of a cluster analysis method by Scott and Knott," *Biometrics*, **33**, 532–535.

Worsley, K. J. (1982), "An improved Bonferroni inequality and applications," *Biometrika*, **69**, 297–302.

Author Index

Subject Index

Continuous Univariate Distributions—1
Continuous Univariate Distributions—2
Continuous Multivariate Distributions

JUDGE, HILL, GRIFFITHS, LÜTKEPOHL and LEE • Introduction to the Theory and Practice of Econometrics

JUDGE, GRIFFITHS, HILL, LÜTKEPOHL and LEE • The Theory and Practice of Econometrics, *Second Edition*

KALBFLEISCH and PRENTICE • The Statistical Analysis of Failure Time Data

KISH • Statistical Design for Research

KISH • Survey Sampling

KUH, NEESE, and HOLLINGER • Structural Sensitivity in Econometric Models

KEENEY and RAIFFA • Decisions with Multiple Objectives

LAWLESS • Statistical Models and Methods for Lifetime Data

LEAMER • Specification Searches: Ad Hoc Inference with Nonexperimental Data

LEBART, MORINEAU, and WARWICK • Multivariate Descriptive Statistical Analysis: Correspondence Analysis and Related Techniques for Large Matrices

LINHART and ZUCCHINI • Model Selection

LITTLE and RUBIN • Statistical Analysis with Missing Data

McNEIL • Interactive Data Analysis

MAINDONALD • Statistical Computation

MALLOWS • Design, Data, and Analysis by Some Friends of Cuthbert Daniel

MANN, SCHAFER and SINGPURWALLA • Methods for Statistical Analysis of Reliability and Life Data

MARTZ and WALLER • Bayesian Reliability Analysis

MIKE and STANLEY • Statistics in Medical Research: Methods and Issues with Applications in Cancer Research

MILLER • Beyond ANOVA, Basics of Applied Statistics

MILLER • Survival Analysis

MILLER, EFRON, BROWN, and MOSES • Biostatistics Casebook

MONTGOMERY and PECK • Introduction to Linear Regression Analysis

NELSON • Applied Life Data Analysis

OSBORNE • Finite Algorithms in Optimization and Data Analysis

OTNES and ENOCHSON • Applied Time Series Analysis: Volume I, Basic Techniques

OTNES and ENOCHSON • Digital Time Series Analysis

PANKRATZ • Forecasting with Univariate Box-Jenkins Models: Concepts and Cases

PIELOU • Interpretation of Ecological Data: A Primer on Classification and Ordination

PLATEK, RAO, SARNDAL and SINGH • Small Area Statistics: An International Symposium

POLLOCK • The Algebra of Econometrics

RAO and MITRA • Generalized Inverse of Matrices and Its Applications

RENYI • A Diary on Information Theory

RIPLEY • Spatial Statistics

RIPLEY • Stochastic Simulation

ROSS • Introduction to Probability and Statistics for Engineers and Scientists

ROUSSEEUW and LEROY • Robust Regression and Outlier Detection

RUBIN • Multiple Imputation for Nonresponse in Surveys

RUBINSTEIN • Monte Carlo Optimization, Simulation, and Sensitivity of Queueing Networks

SCHUSS • Theory and Applications of Stochastic Differential Equations

SEARLE • Linear Models

SEARLE • Linear Models for Unbalanced Data

SEARLE • Matrix Algebra Useful for Statistics

SPRINGER • The Algebra of Random Variables

STEUER • Multiple Criteria Optimization

Applied Probability and Statistics (Continued)

Continuous Univariate Distributions—1
Continuous Univariate Distributions—2
Continuous Multivariate Distributions

JUDGE, HILL, GRIFFITHS, LÜTKEPOHL and LEE • Introduction to the Theory and Practice of Econometrics
JUDGE, GRIFFITHS, HILL, LÜTKEPOHL and LEE • The Theory and Practice of Econometrics, *Second Edition*
KALBFLEISCH and PRENTICE • The Statistical Analysis of Failure Time Data
KISH • Statistical Design for Research
KISH • Survey Sampling
KUH, NEESE, and HOLLINGER • Structural Sensitivity in Econometric Models
KEENEY and RAIFFA • Decisions with Multiple Objectives
LAWLESS • Statistical Models and Methods for Lifetime Data
LEAMER • Specification Searches: Ad Hoc Inference with Nonexperimental Data
LEBART, MORINEAU, and WARWICK • Multivariate Descriptive Statistical Analysis: Correspondence Analysis and Related Techniques for Large Matrices
LINHART and ZUCCHINI • Model Selection
LITTLE and RUBIN • Statistical Analysis with Missing Data
McNEIL • Interactive Data Analysis
MAINDONALD • Statistical Computation
MALLOWS • Design, Data, and Analysis by Some Friends of Cuthbert Daniel
MANN, SCHAFER and SINGPURWALLA • Methods for Statistical Analysis of Reliability and Life Data
MARTZ and WALLER • Bayesian Reliability Analysis
MIKÉ and STANLEY • Statistics in Medical Research: Methods and Issues with Applications in Cancer Research
MILLER • Beyond ANOVA, Basics of Applied Statistics
MILLER • Survival Analysis
MILLER, EFRON, BROWN, and MOSES • Biostatistics Casebook
MONTGOMERY and PECK • Introduction to Linear Regression Analysis
NELSON • Applied Life Data Analysis
OSBORNE • Finite Algorithms in Optimization and Data Analysis
OTNES and ENOCHSON • Applied Time Series Analysis: Volume I, Basic Techniques
OTNES and ENOCHSON • Digital Time Series Analysis
PANKRATZ • Forecasting with Univariate Box-Jenkins Models: Concepts and Cases
PIELOU • Interpretation of Ecological Data: A Primer on Classification and Ordination
PLATEK, RAO, SARNDAL and SINGH • Small Area Statistics: An International Symposium
POLLOCK • The Algebra of Econometrics
RAO and MITRA • Generalized Inverse of Matrices and Its Applications
RÉNYI • A Diary on Information Theory
RIPLEY • Spatial Statistics
RIPLEY • Stochastic Simulation
ROSS • Introduction to Probability and Statistics for Engineers and Scientists
ROUSSEEUW and LEROY • Robust Regression and Outlier Detection
RUBIN • Multiple Imputation for Nonresponse in Surveys
RUBINSTEIN • Monte Carlo Optimization, Simulation, and Sensitivity of Queueing Networks
SCHUSS • Theory and Applications of Stochastic Differential Equations
SEARLE • Linear Models
SEARLE • Linear Models for Unbalanced Data
SEARLE • Matrix Algebra Useful for Statistics
SPRINGER • The Algebra of Random Variables
STEUER • Multiple Criteria Optimization

(continued from front)

Applied Probability and Statistics (Continued)

STOYAN • Comparison Methods for Queues and Other Stochastic Models
TIJMS • Stochastic Modeling and Analysis: A Computational Approach
TITTERINGTON, SMITH, and MAKOV • Statistical Analysis of Finite Mixture Distributions
UPTON • The Analysis of Cross-Tabulated Data
UPTON and FINGLETON • Spatial Data Analysis by Example, Volume I: Point Pattern and Quantitative Data
WEISBERG • Applied Linear Regression, *Second Edition*
WHITTLE • Optimization Over Time: Dynamic Programming and Stochastic Control, Volume I and Volume II
WHITTLE • Systems in Stochastic Equilibrium
WILLIAMS • A Sampler on Sampling
WONNACOTT and WONNACOTT • Econometrics, *Second Edition*
WONNACOTT and WONNACOTT • Introductory Statistics, *Fourth Edition*
WONNACOTT and WONNACOTT • Introductory Statistics for Business and Economics, *Third Edition*
WOOLSON • Statistical Methods for The Analysis of Biomedical Data

Tracts on Probability and Statistics

AMBARTZUMIAN • Combinatorial Integral Geometry
BIBBY and TOUTENBURG • Prediction and Improved Estimation in Linear Models
BILLINGSLEY • Convergence of Probability Measures
DEVROYE and GYORFI • Nonparametric Density Estimation: The L_1 View
KELLY • Reversibility and Stochastic Networks
RAKTOE, HEDAYAT, and FEDERER • Factorial Designs
TOUTENBURG • Prior Information in Linear Models